T0215183

Galois Theory and Advanced Linear Algebra

Rajnikant Sinha

Galois Theory and Advanced Linear Algebra

 Springer

Rajnikant Sinha
Samne Ghat
Varanasi, Uttar Pradesh, India

ISBN 978-981-13-9851-3   ISBN 978-981-13-9849-0   (eBook)
https://doi.org/10.1007/978-981-13-9849-0

This Springer imprint is published by the registered company Springer Nature Singapore Pte Ltd.
The registered company address is: 152 Beach Road, #21-01/04 Gateway East, Singapore 189721, Singapore

Preface

Evariste Galois (25 October 1811–31 May 1832) was a great French mathematician. While still in his teens, he was able to determine a necessary and sufficient condition for a polynomial to be solvable by radicals, thereby solving a problem standing for 350 years. The famous ancient problem of "trisecting an angle by using solely straightedge and compass" was later solved by using the fundamental theorem of Galois theory. Many students are overwhelmed to learn this. They take keen interest in learning the theory Galois had discovered. Unfortunately, there is no literature which can lead to the fundamental theorem without going through painful learning process. Further, there is no right kind of book on linear algebra that can provide good theoretical foundation needed for later applications in Riemannian geometry, quantum mechanics, etc. These voids prompted me to write this book.

This book is meant to be an introduction to abstract algebra. The reader of this book is assumed to have some prior exposure to elementary properties of groups, rings, fields, and vector spaces. At times, we shall assume familiarity with inner product space of finite dimension, and linear transformations. For the readers who have only learned a minimum of abstract algebra will also find this book friendly. A nodding acquaintance with elementary properties of positive integers is beneficial.

Most of the material usually taught in an abstract algebra course are presented in this text. However, some results appear for the first time in a textbook form. The ordering of the topics as well as the approach we have taken sometimes deviate from the standard path, simply because of pedagogical reasons. Aside from the usual approach, we sometimes have also developed a more elementary approach that uses standard calculation techniques. Wherever required, we also have supplied abundantly "second layer of proof" (that is, proof within proof), so that comprehensibility of the proof gets enhanced. In some named theorems, rarely we need a third layer of proof.

In the first part of Chap. 1, we have developed necessary field theory. Using these theorems, we have tried to prove the fundamental theorem of Galois groups. This is the theorem for which Galois became immortal. In Chap. 2, some wonderful applications of Galois theory are presented. Solution to the ancient famous problem of "trisection of a given angle by ruler and compass" is given here with enough detailed proof. In Chap. 3, we have supplied the proofs of many celebrated theorems by using linear transformation tools as well as matrix methods. These are beautiful areas of mathematics in itself. Their applications to quantum mechanics, manifold theory, etc., are well known. In Chap. 4, we have dealt with some amazing, but forgotten, results of yesteryears. It is known that "signature" of a quadratic form is invariant, but it is difficult to find an accessible proof to it. We have tried to supply a proof which could be an effortless reading.

Finally, on a personal note, I would like to thank my lovely wife, Bina, for her patient endurance and constant encouragement.

Uttar Pradesh, India   Rajnikant Sinha

Contents

About the Author

Rajnikant Sinha is a former Professor of Mathematics at Magadh University, Bodh Gaya, India. A passionate mathematician, Prof. Sinha has published numerous interesting research findings in international journals, and has authored three textbooks with Springer Nature: *Smooth Manifolds, Real and Complex Analysis: Volume 1*, and *Real and Complex Analysis: Volume 2*; and a contributed volume on *Solutions to Weatherburn's Elementary Vector Analysis: With Applications to Geometry and Mechanics* with another publisher. His research focuses on topological vector spaces, differential geometry and manifolds.

Chapter 1
Galois Theory I

Roughly, a field is a commutative ring in which division by every nonzero element is allowed. In algebra, fields play a central role. Results about fields find important applications in the theory of numbers. The theory of fields comprises the subject matter of the theory of equations. Here, we shall deal lightly with the field of algebraic numbers. Our main emphasis will be on aspects of field theory that concern the roots of polynomials. The beautiful ideas, due to the brilliant French mathematician Évariste Galois (1811–1832), served as an inspiration for the development of abstract algebra. We shall prove the fundamental theorem of Galois theory.

1.1 Euclidean Rings

1.1.1 Definition Let R be an integral domain. Suppose that for every nonzero member a of R, $d(a)$ is a nonnegative integer. If

1. for every nonzero $a, b \in R$, $d(a) \leq d(ab)$,
2. for every nonzero $a, b \in R$, there exist $q, r \in R$ such that $a = qb + r$, and either $r = 0$ or $d(r) < d(b)$, then we say that R is a *Euclidean ring*.

1.1.2 Theorem Let R be a Euclidean ring. Let A be an ideal of R. Then there exists $a_0 \in A$ such that $a_0 R = A$.

Proof If $A = \{0\}$, then 0 serves the purpose of a_0. So we consider the case $A \neq \{0\}$.

It follows that there exists a nonzero member a of A, and hence $\{d(x) : x \in A \text{ and } x \neq 0\}$ is a nonempty set of nonnegative integers. Hence, $\min\{d(x) : x \in A \text{ and } x \neq 0\}$ exists. It follows that there exists a nonzero member a_0 of A such that

© Springer Nature Singapore Pte Ltd. 2020
R. Sinha, *Galois Theory and Advanced Linear Algebra*,
https://doi.org/10.1007/978-981-13-9849-0_1

$$d(a_0) = \min\{d(x) : x \in A \text{ and } x \neq 0\}. \quad (*)$$

It remains to show that $A = a_0 R$.

For this purpose, take an arbitrary $a \in A$. If $a = 0$, then $a = 0 = a_0 0 \in a_0 R$. Thus if $a = 0$, then $a \in a_0 R$.

Now we consider the case $a \neq 0$. Since a, a_0 are nonzero members of R, and R is a Euclidean ring, there exist $q, r \in R$ such that $a = qa_0 + r$ and either $r = 0$ or $d(r) < d(b)$. Since a, a_0 are members of the ideal A of R, we have

$$r = \underbrace{(a - qa_0)}_{} \in A,$$

and hence $r \in A$. Now, if $r \neq 0$, then from $(*)$, $d(a_0) \leq d(r)$. Further, either $r = 0$ or $d(r) < d(a_0)$. This shows that

$$a - qa_0 = \underbrace{r = 0}_{},$$

and hence $a = qa_0 = a_0 q \in a_0 R$.

Thus in all cases, $a \in A \Rightarrow a \in a_0 R$. Hence $A \subset a_0 R$. It remains to show that $a_0 R \subset A$. For this purpose, let us take an arbitrary $b \in R$. We have to show that $a_0 b \in A$. Since a_0 is a member of the ideal A of R, and $b \in R$, we have $a_0 b \in A$. ∎

Definition Let R be an integral domain. Let $a \in R$. It is clear that aR is an ideal of R. The ideal aR of R is denoted by (a).

Definition Let R be an integral domain. If

1. R has a unit element,
2. every ideal of R is of the form (a), then we say that R is a *principal ideal ring*.

1.1.3 Theorem Let R be a Euclidean ring. Then R has a unit element.

Proof Since R is an ideal of the Euclidean ring R, by 1.1.2, there exists $a_0 \in R$ such that $a_0 R = R$. Since $a_0 R = R$ and $a_0 \in R$, we have $a_0 \in a_0 R$, and hence there exists $e \in R$ such that $a_0 = a_0 e$. It suffices to show that e functions as a unit element in R. To this end, let us take an arbitrary $b \in R$. We have to show that $be = b$.

Since $b \in R = a_0 R$, there exists $c \in R$ such that $b = a_0 c$. Now,

$$\text{LHS} = be = (a_0 c)e = (ca_0)e = c(a_0 e) = ca_0 = a_0 c = b = \text{RHS},$$

where LHS and RHS are the left- and right-hand sides of the equality to be proved. ∎

1.1.4 Theorem Let R be a Euclidean ring. Then R is a principal ideal ring.

Proof By 1.1.3, R has a unit element. Next, by 1.1.2, every ideal of R is of the form (a). It follows, by the definition of principal ideal ring, that R is a principal ideal ring. ∎

1.1.5 Theorem Let R be a Euclidean ring. Let $a, b \in R$. Then the greatest common divisor (a, b) of a and b exists in R, in the sense that

1. $(a, b) \in R$,
2. $(a, b)|a$,
3. $(a, b)|b$,
4. $(c|a \text{ and } c|b) \Rightarrow c|(a, b)$.

Further, there exist $s, t \in R$ such that

$$(a, b) = as + bt.$$

Proof Since R is a Euclidean ring, by 1.1.3, R has a unit element, say e. Let

$$A \equiv \{ax + by : x, y \in R\}.$$

Clearly A is an ideal of the Euclidean ring R. Now, by 1.1.2, there exists $f \in R$ such that

$$f = fe \in \underbrace{fR = A} = \{ax + by : x, y \in R\},$$

and hence $f \in \{ax + by : x, y \in R\}$. It follows that there exist $s, t \in R$ such that

$$f = as + bt.$$

Since $fR = \{ax + by : x, y \in R\} \ni (ae + b0) = a$, we have $a \in fR$, and hence $f|a$. Similarly, $f|b$. Next suppose that $c|a$ and $c|b$. It remains to show that $c|(as + bt)$. This is clearly true. $\blacksquare$

Definition Let R be a commutative ring with unit element 1. Let $a \in R$. If there exists $b \in R$ such that $ab = 1$, then we say that *a is a unit in R*.

1.1.6 Theorem Let R be an integral domain with unit element 1. Let a, b be nonzero members of R. Suppose that $a|b$ and $b|a$. Then there exists $u \in R$ such that $au = b$, and u is a unit in R.

Proof Since $a|b$, there exists $u \in R$ such that $au = b$. Similarly, there exists $v \in R$ such that $bv = a$. It follows that

$$a(uv) = \underbrace{(au)v = a} = a1,$$

and hence $a(uv) = a1$. Now, since a is a nonzero member of the integral domain R, we have $uv = 1$, and hence u is a unit in R. $\blacksquare$

Definition Let R be a commutative ring with unit element 1. Let $a, b \in R$. If there exists a unit u in R such that $au = b$, then we say that *a and b are associates*, and we denote this relationship by $a \sim b$.

It is clear that $\sim$ is an equivalence relation over R. Hence R is partitioned by $\sim$ into equivalence classes.

1.1.7 Theorem Let R be a Euclidean ring. Let $a, b \in R$. Let c be a greatest common divisor of a and b. Let d be a greatest common divisor of a and b. Then $c \sim d$.

Proof Since c is a greatest common divisor of a and b, we have $(c|a$ and $c|b)$. Now, since d is a greatest common divisor of a and b, we have $c|d$. Similarly, $d|c$. Since $c|d$ and $d|c$, by 1.1.6 there exists a unit $u \in R$ such that $cu = d$, and hence $c \sim d$. ∎

1.1.8 Theorem Let R be a Euclidean ring with unit element 1. Let a, b be nonzero members of R. Let b be a nonunit. Then

$$d(a) < d(ab).$$

Proof Suppose to the contrary that $d(a) = d(ab)$. We seek a contradiction.

Since aR is an ideal of R, by 1.1.2 there exists $b \in R$ such that $(ab)R = aR$. Now, since $a = a1 \in aR = (ab)R$, we have $a \in (ab)R$, and hence there exists a nonzero member c of R such that $a = (ab)c$. Hence

$$a1 = \underbrace{a = a(bc)}.$$

Since $a1 = a(bc)$ and a is a nonzero member of the integral domain R, we have $1 = bc$, and hence b is a unit. This is a contradiction. ∎

Definition Let R be a Euclidean ring with unit element 1. Let p be a nonzero member of R that is not a unit. If

$$(a, b \in R \text{ and } p = ab) \Rightarrow (a \text{ is a unit or } b \text{ is a unit}),$$

then we say that p *is a prime element of* R.

1.1.9 Theorem Let R be a Euclidean ring with unit element 1. Let a be a nonzero member of R. Suppose that $d(a) = d(1)$. Then a is a unit.

Proof Suppose to the contrary that a is a nonunit. We seek a contradiction. Here $1, a$ are nonzero members of R. Also, a is not a unit. So by 1.1.8,

$$\underbrace{d(1) < d(1a)} = d(a),$$

and hence $d(1) < d(a)$. Thus $d(a) \neq d(1)$. This is a contradiction. ∎

1.1.10 Problem Let R be a Euclidean ring with unit element 1. Let a be a nonzero member of R. Let a be a unit. Then $d(a) = d(1)$.

Proof Suppose to the contrary that $d(a) \neq d(1)$. We seek a contradiction. Since a is a unit, there exists a nonzero member b of R such that $ab = 1$. Now, since R is a Euclidean ring, we have

$$\underbrace{d(a) \leq d(ab)} = d(1) \leq d(1a) = d(a),$$

and hence $d(a) = d(1)$. This is a contradiction. ∎

1.1.11 Problem Let R be a Euclidean ring with unit element 1. Then for every nonzero member a of R, either a is a unit or a can be expressed as a product of finitely many prime elements of R.

Proof (Induction on $d(a)$): Let us first consider the case that a is a nonzero member of R and $d(a) = 0$. Since

$$0 \leq d(1) \leq d(1a) = d(a) = 0,$$

we have $d(1) = d(a)$. Now by 1.1.9, a is a unit. Thus the statement "either a is a unit or a can be expressed as a product of finitely many prime elements of R" holds in this case.

Next suppose that the statement "either a is a unit or a can be expressed as a product of finitely many prime elements of R" holds for all a in R for which $d(a) \leq n$.

Next suppose that b is a nonzero member of R for which $d(b) = n + 1$. We have to show that the statement "either b is a unit or b can be expressed as a product of finitely many prime elements of R" holds.

Case I: b is a prime element of R. In this case, the statement "b can be expressed as a product of finitely many prime elements of R" holds, and hence the statement "either b is a unit or b can be expressed as a product of finitely many prime elements of R" holds.

Case II: b is not a prime element of R.

Subcase I: b is a unit. In this subcase, the statement "either b is a unit or b can be expressed as a product of finitely many prime elements of R" holds.

Subcase II: b is not a unit. Here b is not a prime element of R, so by the definition of prime element, there exist $c, e \in R$ such that

1. $b = ce$,
2. c is not a unit,
3. e is not a unit.

Since b is a nonzero member of R and $b = ce$, it follows that c, e are nonzero members of R. Since c is not a unit, by 1.1.8,

$$d(e) < d(ec) = d(ce) = d(b) = n + 1,$$

and hence $d(e) < n + 1$. Now, since $d(e)$ in an integer, we have $d(e) \leq n$. Similarly, $d(c) \leq n$. Since $d(c) \leq n$ and c is not a unit, by hypothesis, c can be expressed as a product of finitely many prime elements of R. Similarly, e can be expressed as a product of finitely many prime elements of R. It follows that ce can be expressed as a product of finitely many prime elements of R. Now, since $b = ce$, b can be expressed as a product of finitely many prime elements of R. ∎

Definition Let R be a Euclidean ring with unit element 1. Let a, b be nonzero members of R (By 1.1.5, a greatest common divisor of a and b exists in R.). If there exists a unit u in R such that u is a greatest common divisor of a and b, then we say that *a and b are relatively prime*.

1.1.12 Problem Let R be a Euclidean ring with unit element 1. Let a, b, c be any nonzero elements of R. Suppose that $a|bc$. Let a and b be relatively prime. Then $a|c$.

Proof Since a and b are relatively prime, there exists a unit u in R such that u is a greatest common divisor of a and b. Now, by 1.1.5, there exist $s, t \in R$ such that

$$u = as + bt.$$

Since u is a unit in R, there exists v in R such that $uv = 1$. It follows that

$$asv + btv = (as + bt)v = 1,$$

and hence

$$acsv + bctv = asvc + btvc = (asv + btv)c = 1c = c.$$

Thus $acsv + (bc)tv = c$. Since $a|bc$, there exists a nonzero member k of R such that $ak = bc$. It follows that

$$a(csv + ktv) = acsv + (ak)tv = c,$$

and hence $al = c$, where $l \equiv (csv + ktv) \in R$. Thus $a|c$. ∎

1.1.13 Problem Let R be a Euclidean ring with unit element 1. Let a, π be any nonzero elements of R. Suppose that π is a prime element of R. Then either $\pi|a$ or (π and a are relatively prime).

Proof By 1.1.5, there exists $e \in R$ such that $e|\pi$, $e|a$, and $((c|\pi \text{ and } c|a) \Rightarrow c|e)$, that is, e is a greatest common divisor of π and a.

Since $e|\pi$, there exists $k \in R$ such that $ek = \pi$. Now, since π is a prime element of R, either e is a unit or k is a unit.

Case I: k is a unit. It follows that there exists $l \in R$ such that $kl = 1$. Now, since $ek = \pi$, we have

$$e = e1 = e(kl) = \underbrace{(ek)l = \pi l,}$$

and hence $e = \pi l$. Now, since $e|a$, we have $(\pi l)|a$, and hence there exists $m \in R$ such that $\pi(lm) = \underbrace{(\pi l)m = a.}$ Thus $\pi n = a$, where $n \equiv lm \in R$. Hence $\pi|a$. Thus the statement "either $\pi|a$ or (π and a are relatively prime)" holds.

Case II: e is a unit. Since e is a greatest common divisor of π and a, and e is a unit, π and a are relatively prime, and hence the statement "either $\pi|a$ or (π and a are relatively prime)" holds. ∎

1.1.14 Problem Let R be a Euclidean ring with unit element 1. Let π be a nonzero element of R. Suppose that π is a prime element of R. Then

$$(a, b \in R \text{ and } \pi|(ab)) \Rightarrow (\pi|a \text{ or } \pi|b).$$

Proof Let us take arbitrary nonzero members a and b of R such that $\pi|(ab)$. It suffices to show that $\pi|a$ or $\pi|b$. Suppose to the contrary that $\pi\nmid a$ and $\pi\nmid b$. We have to arrive at a contradiction.

Since π is a prime element of R, by 1.1.13 we have $\pi|a$ or (π and a are relatively prime). Now, since $\pi\nmid a$, π and a are relatively prime. Next, since $\pi|(ab)$ and π and a are relatively prime, by 1.1.12, we have $\pi|b$. This is a contradiction. ∎

1.1.15 Theorem Let R be a Euclidean ring with unit element 1. Let a be a nonzero element of R. Suppose that a is not a unit in R (By 1.1.11, a can be expressed as a product of finitely many prime elements of R.). Let

$$a = \pi_1 \pi_2 \ldots \pi_m,$$

where each $\pi_i (i = 1, 2, \ldots, m)$ is a prime element of R. Let

$$a = \pi_1' \pi_2' \ldots \pi_n',$$

where each $\pi_j' (j = 1, 2, \ldots, n)$ is a prime element of R. Then

1. each π_i is an associate of some π_j',
2. each π_j' is an associate of some π_i,
3. $n = m$.

This theorem is known as the **unique factorization theorem**.

Proof Since $\quad \pi_1 | (\pi_1 \pi_2 \ldots \pi_m) \quad$ and $\quad \pi_1 \pi_2 \ldots \pi_m = a = \pi_1' \pi_2' \ldots \pi_n'$, we have $\pi_1 | (\pi_1' \pi_2' \ldots \pi_n')$. Now, since π_1 is a prime element of R, by 1.1.14 we have $\pi_1 | \pi_j'$ for some $j \in \{1, 2, \ldots, n\}$.

Here for some $j \in \{1, 2, \ldots, n\}$, we have $\pi_1 | \pi_j'$, and hence there exists a nonzero $k \in R$ such that $\pi_1 k = \pi_j'$. Now, since π_j' is a prime element of R, either π_1 is a unit or k is a unit. Since π_1 is a prime element of R, π_1 is not a unit. It follows that k is a unit. Now, since $\pi_1 k = \pi_j'$, we have $\pi_1 \sim \pi_j'$, where $j \in \{1, 2, \ldots, n\}$. Similarly, $\pi_2 \sim \pi_k'$, where $k \in \{1, 2, \ldots, n\}$, etc. Thus, each π_i is an associate of some π_j'. Similarly, each π_j' is an associate of some π_i. This proves (1) and (2).

For (3): Suppose to the contrary that $m < n$. We seek a contradiction.

Since each π_1 is an associate of some π_j' and

$$\pi_1 \pi_2 \ldots \pi_m = \pi_1' \pi_2' \ldots \pi_n',$$

we get an equality of the form

$$\pi_2 \pi_3 \ldots \pi_m = u \pi_1' \pi_2' \ldots \pi_{j-1}' \pi_{j+1}' \ldots \pi_n',$$

where u is a unit. Next, since $\pi_2 | (\pi_2 \pi_3 \ldots \pi_m)$, and $\pi_2 \pi_3 \ldots \pi_m = u \pi_1' \pi_2' \ldots \pi_{j-1}' \pi_{j+1}' \ldots \pi_n'$, we have $\pi_2 | \left(u \pi_1' \pi_2' \ldots \pi_{j-1}' \pi_{j+1}' \ldots \pi_n' \right)$. Now, since π_2 is a prime element of R, by 1.1.14 we have $\pi_2 | u$ or $\pi_2 | \pi_k'$ for some $k \in (\{1, 2, \ldots, n\} - \{j\})$.

We claim that $\pi_2 \nmid u$. Suppose to the contrary that $\pi_2 | u$. We seek a contradiction.

Since u is a unit, there exists a nonzero $v \in R$ such that $uv = 1$. Since $\pi_2 | u$, we have $\pi_2 | (uv)$, and hence $\pi_2 | 1$. This shows that π_2 is a unit. Since π_2 is a prime element of R, π_2 is not a unit. This is a contradiction. Thus our claim is true, that is, $\pi_2 \nmid u$.

It follows that $\pi_2 | \pi_k'$ for some $k \in (\{1, 2, \ldots, n\} - \{j\})$. Similarly, $\pi_3 | \pi_l'$ for some $l \in (\{1, 2, \ldots, n\} - \{j, k\})$, etc. On repeating this argument m times, we get an equality of the form

$$1 = w_0 \pi_{m+1}' \pi_{m+2}' \ldots \pi_n',$$

where w_0 is a unit. It follows that π_{m+1}' is a unit, and hence π_{m+1}' is not a prime element. This is a contradiction. ∎

1.1.16 Problem Let R be a Euclidean ring with unit element 1. Let π be a nonzero element of R. Suppose that π is not a unit in R. Let π be a prime element of R. Then the ideal (π) is maximal.

Proof If $(\pi) = R$, then it is clear that (π) is a maximal ideal. So we consider the case that the ideal (π) is a proper subset of R. We have to show that (π) is maximal.

Suppose to the contrary that there exists an ideal U of R such that (π) is a proper subset of U, and U is a proper subset of R. We seek a contradiction.

By 1.1.4, R is a principal ideal ring. It follows that there exists $a \in U$ such that $(a) = U$. Thus (π) is a proper subset of (a), and (a) is a proper subset of R. Since (a) is a proper subset of R, a is not a unit. Since (π) is a subset of (a) and $\pi \in (\pi)$, we have $\pi \in (a)$. Hence there exists a nonzero $u \in R$ such that $\pi = au$. Now, since π is a prime element of R, a is a unit or u is a unit. Next since a is not a unit, u is a unit. Since u is a unit and $\pi = au$, we have $(\pi) = (a)$. This contradicts the fact that (π) is a proper subset of (a).   ∎

1.1.17 Problem Let R be a Euclidean ring with unit element 1. Let π be a nonzero element of R. Suppose that π is not a unit in R. Let the ideal (π) be maximal. Then π is a prime element of R.

Proof Suppose to the contrary that there exist nonzero a, b in R such that $\pi = ab$, and neither a nor b is a unit. We seek a contradiction.

Since $\pi = ab$, we have $(\pi) \subset (a)$. Since $(\pi) \subset (a)$ and (π) is maximal, either $(\pi) = (a)$ or $(a) = R$. Since a is not a unit, we have $(a) \neq R$. It follows that $(\pi) = (a)$, and hence there exists $u \in R$ such that $a = \pi u$. Now, since $\pi = ab$, we have

$$\pi 1 = \pi = \underbrace{(\pi u)b} = \pi(ub),$$

and hence $\pi 1 = \pi(ub)$. Next, since π is a nonzero element of R, we have $1 = ub$, and hence b is a unit. This is a contradiction.   ∎

1.1.18 Notation The collection of all complex numbers $a + ib$, where a, b are integers, is denoted by $J[i]$. Its members are called *Gaussian integers*.

It is easy to see that $J[i]$ is an integral domain with unit element $1(=1 + i0 \in J[i])$. For every nonzero $a + ib \in J[i]$, by $d(a + ib)$ we shall mean the positive integer $a^2 + b^2$.

1.1.19 Note Observe that for every nonzero $(a + ib), (e + if) \in J[i]$, we have $0 < (a^2 + b^2)$, and $1 \leq (e^2 + f^2)$, and hence

$$d(a + ib) = \underbrace{(a^2 + b^2)1 \leq (a^2 + b^2)(e^2 + f^2)} = |a + ib|^2|e + if|^2$$

$$= (|a + ib||e + if|)^2 = |(a + ib)(e + if)|^2 = d((a + ib)(e + if)).$$

Thus for every nonzero $(a + ib), (e + if) \in J[i]$, we have

$$d(a + ib) \leq d((a + ib)(e + if)).$$

1.1.20 Note Let $(a + ib)$ be any nonzero member of $J[i]$, where a and b are integers. Let x be any positive integer. By the divisibility property of integers, there exist two integers q_1, r_1 such that $a = q_1 x + r_1$, where $-\frac{x}{2} \leq r_1 \leq \frac{x}{2}$.

Examples: Let $x = 7$ and $a = 42$. Since $42 = 6 \cdot 7 + 0$, we can take $q_1 = 6$ and $r_1 = 0 \in \left[-\frac{7}{2}, \frac{7}{2} \right]$.

Let $x = 7$ and $a = 47$. Since $47 = 6 \cdot 7 + 5 = (6 + 1) \cdot 7 + (-2)$, we can take $q_1 = 7$ and $r_1 = -2 \in \left[-\frac{7}{2}, \frac{7}{2} \right]$.

Let $x = 6$ and $a = 45$. Since $45 = 7 \cdot 6 + 3$, we can take $q_1 = 7$ and $r_1 = 3 \in \left[-\frac{6}{2}, \frac{6}{2} \right]$.

Similarly, there exist two integers q_2, r_2 such that $b = q_2 x + r_2$, where $-\frac{x}{2} \le r_2 \le \frac{x}{2}$.

It follows that

$$(a + ib) = (q_1 x + r_1) + i(q_2 x + r_2) = (q_1 + iq_2)x + (r_1 + ir_2),$$

and hence

$$(a + ib) = qx + r,$$

where $q \equiv (q_1 + iq_2) \in J[i]$, and $r \equiv (r_1 + ir_2) \in J[i]$.

Suppose that $r \ne 0$. Since $|r_1| \le \frac{x}{2}$ and $|r_2| \le \frac{x}{2}$, we have

$$d(r) = d(r_1 + ir_2) = (r_1)^2 + (r_2)^2 \le \frac{x^2}{2} < x^2 = d(x + i0) = d(x).$$

1.1.21 Conclusion Let $(a + ib)$ be any nonzero member of $J[i]$, where a and b are integers. Let x be any positive integer. Then there exist $q, r \in J[i]$ such that $(a + ib) = qx + r$ and (either $r = 0$ or $d(r) < d(x)$).

1.1.22 Note Let $(a + ib)$ be any nonzero member of $J[i]$, where a and b are integers. Let $(e + if)$ be any nonzero member of $J[i]$, where e and f are integers.

Since $(e + if)$ is a nonzero member of $J[i]$, $e^2 + f^2$ is a positive integer. Now, by 1.1.20, there exist $(q_1 + iq_2), (r_1 + ir_2) \in J[i]$ such that

1. $(a + ib)(e - if) = (q_1 + iq_2)(e^2 + f^2) + (r_1 + ir_2)$,
2. (either $r_1 + ir_2 = 0$ or $d(r_1 + ir_2) < d(e^2 + f^2)$).

Case I: $r_1 + ir_2 = 0$. From item 1 above,

$$(a + ib)(e - if) = (q_1 + iq_2)(e^2 + f^2).$$

Next, since $e^2 + f^2$ is a positive integer, we have $(a + ib) = (q_1 + iq_2)(e + if) + (0 + i0)$. Thus the statement "there exist $q, r \in J[i]$ such that $(a + ib) = q(e + if) + r$ and (either $r = 0$ or $d(r) < d(e + if))$" holds.

Case II: $r_1 + ir_2 \ne 0$. It follows from (2) that

$$d(r_1 + ir_2) < d(e^2 + f^2),$$

and hence from (1),

$$d\big((a+ib)(e-if) - (q_1+iq_2)(e^2+f^2)\big) < d(e^2+f^2),$$

that is,

$$\big|(a+ib)(e-if) - (q_1+iq_2)(e^2+f^2)\big|^2 < (e^2+f^2)^2,$$

that is,

$$(e^2+f^2)|(a+ib) - (q_1+iq_2)(e+if)|^2 < (e^2+f^2)^2,$$

that is,

$$|(a+ib) - (q_1+iq_2)(e+if)|^2 < (e^2+f^2). \qquad (*)$$

Let us put

$$s_1+is_2 \equiv \big((a+ib) - (q_1+iq_2)(e+if)\big) \in J[i],$$

where s_1, s_2 are integers. Here

$$(a+ib) = (q_1+iq_2)(e+if) + (s_1+is_2),$$

and from $(*)$,

$$|s_1+is_2|^2 < (e^2+f^2).$$

It follows that either $s_1+is_2 = 0$ or $d(s_1+is_2) < d(e+if)$. Thus the statement "there exist $q, r \in J[i]$ such that $(a+ib) = q(e+if)+r$, and (either $r = 0$ or $d(r) < d(e+if)$)" holds.

1.1.23 Conclusion For every nonzero $a, b \in J[i]$, there exist $q, r \in J[i]$ such that $a = qb + r$ and (either $r = 0$ or $d(r) < d(b)$). Also, we have seen that for every nonzero $a, b \in J[i]$, we have $d(a) \le d(ab)$.

Hence $J[i]$ is a Euclidean ring with unit element 1.

1.1.24 Problem It is clear that the collection $\mathbb{Z}$ of all integers is an integral domain with unit element 1. For every nonzero integer a, by $d(a)$ we shall mean the absolute value $|a|$ of a. Observe that for every nonzero $a, b \in \mathbb{Z}$,

$$d(a) = |a| \le |a||b| = |ab| = d(ab),$$

so for every nonzero $a, b \in \mathbb{Z}$, $d(a) \le d(ab)$.

Next, let us take arbitrary nonzero $a, b \in \mathbb{Z}$. By the divisibility property of integers, there exist $q, r \in \mathbb{Z}$ such that $a = qb + r$ and $0 \leq r < |b|$. Since $0 \leq r < |b|$, we have (either $r = 0$ or $0 < r < |b|$), and hence (either $r = 0$ or $|r| < |b|$). Thus,

for every nonzero $a, b \in \mathbb{Z}$, there exist $q, r \in \mathbb{Z}$ such that $a = qb + r$ and (either $r = 0$ or $d(r) < d(b)$).

This shows that $\mathbb{Z}$ is a Euclidean ring.

Further, $\mathbb{Z}$ is a subring of the Euclidean ring $J[i]$.

Proof Since for every integer a, we have $a = (a + i0) \in J[i]$, it follows that $\mathbb{Z}$ is a subset of $J[i]$. Also, $\mathbb{Z}$ is itself an integral domain. Since $\mathbb{Z}$ is a subset of $J[i]$ and $J[i]$ is a Euclidean ring, we have $d(a) \leq d(ab)$ for every nonzero $a, b \in \mathbb{Z}$.

Next, let us take arbitrary nonzero $a, b \in \mathbb{Z}$. By the divisibility property of integers, there exist $q, r \in \mathbb{Z}$ such that $a = qb + r$ and $0 \leq r < |b|$. Since $0 \leq r < |b|$, we have (either $r = 0$ or $0 < r < |b|$), and hence $\left(\text{either } r = 0 \text{ or } |r|^2 < |b|^2\right)$. Thus, for every nonzero $a, b \in \mathbb{Z}$, there exist $q, r \in \mathbb{Z}$ such that $a = qb + r$ and (either $r = 0$ or $d(r) < d(b)$).

This shows that $\mathbb{Z}$ is itself a Euclidean ring. Thus $\mathbb{Z}$ is a subring of the Euclidean ring $J[i]$.   ∎

1.1.25 Problem Let $p(\in \mathbb{Z})$ be a prime number, in the sense that $1 < p$ and $(a|p \Rightarrow (a = 1 \text{ or } -1 \text{ or } p \text{ or } -p))$. Let a, b, c be any integers satisfying

1. c is relatively prime to p,

2. $cp = (a^2 + b^2)$.

Then p is not a prime element of the Euclidean ring $J[i]$.

Proof Suppose to the contrary that p is a prime element of the Euclidean ring $J[i]$. We seek a contradiction.

From (2), $p|(a + ib)(a - ib)$. It follows, by 1.1.14, that $p|(a + ib)$ or $p|(a - ib)$.

Case I: $p|(a + ib)$. It follows that there exist integers e, f such that $p(e + if) = (a + ib)$, and hence $|a + ib|^2 = |p(e + if)|^2$. Thus $p^2(e^2 + f^2) = (a^2 + b^2)$. Now, from (2), $p^2(e^2 + f^2) = cp$, and hence $p(e^2 + f^2) = c$. This shows that $p|c$, and hence c is not relatively prime to p. This contradicts (1).

Case II: $p|(a - ib)$. This case is similar to Case I.   ∎

1.1.26 Problem Let $p(\in \mathbb{Z})$ be a prime number. Let a, b, c be any integers satisfying

1. c is relatively prime to p,

2. $cp = (a^2 + b^2)$.

Then there exist integers e and f such that $p = e^2 + f^2$.

Proof By 1.1.25, p is not a prime element of the Euclidean ring $J[i]$. Hence there exist integers e, f, g, h such that

$$p = (e + if)(g + ih),$$

and neither $(e + if)$ is a unit in $J[i]$ nor $(g + ih)$ is a unit in $J[i]$.

Since $(e + if)$ is not a unit in $J[i]$, we have $e^2 + f^2 \neq 1$.

Proof Suppose to the contrary that $e^2 + f^2 = 1$. We seek a contradiction. Since $e^2 + f^2 = 1$, we have $(e + if)(e - if) = 1$, and hence $(e + if)$ is a unit in $J[i]$. This is a contradiction. ∎

Similarly, $g^2 + h^2 \neq 1$. Since $p = (e + if)(g + ih)$, we have

$$p^2 = (e^2 + f^2)(g^2 + h^2), \qquad (*)$$

and hence $(e^2 + f^2) | p^2$. Now, since p is a prime number, we have $(e^2 + f^2) = 1$ or $(e^2 + f^2) = p$ or $(e^2 + f^2) = p^2$. And since $e^2 + f^2 \neq 1$, we have $(e^2 + f^2) = p$ or $(e^2 + f^2) = p^2$.

If $(e^2 + f^2) = p^2$, then from $(*)$, we have $p^2 = p^2(g^2 + h^2)$, and hence $g^2 + h^2 = 1$. This is a contradiction. So $(e^2 + f^2) \neq p^2$. Since $(e^2 + f^2) = p$ or $(e^2 + f^2) = p^2$, we have $(e^2 + f^2) = p$. ∎

1.2 Polynomial Rings

1.2.1 Note Let us observe that the quadratic congruence $x^2 \equiv 1 \pmod 8$ has $\{1, 3, 5, 7\}$ as a solution set. Thus the quadratic congruence $x^2 \equiv 1 \pmod 8$ has four solutions.

Definition Let n be an integer such that $n \geq 1$. By $\varphi(n)$ we mean the number of positive integers m such that $m \leq n$ and (m, n are relatively prime). Here $\varphi : \{1, 2, 3, \ldots\} \to \{1, 2, 3, \ldots\}$ is called the *Euler totient function*.

For example, $\varphi(1) = 1$, $\varphi(2) = 1$, $\varphi(3) = 2$, $\varphi(4) = 2$, $\varphi(5) = 4$, $\varphi(6) = 2$, etc.

Definition Let m be an integer such that $m \geq 1$. By a *reduced residue system modulo m* we mean a collection A of integers such that

1. the number of elements in A is $\varphi(m)$,
2. no two members of A are congruent modulo m,
3. each member of A is relatively prime to m.

Example: $\{1, 29\}$ is a reduced residue system modulo 6.

1.2.2 Problem Let m be an integer such that $m \geq 1$. Let $\{a_1, a_2, \ldots, a_{\varphi(m)}\}$ be a reduced residue system modulo m. Let k be a positive integer that is relatively prime to m. Then $\{ka_1, ka_2, \ldots, ka_{\varphi(m)}\}$ is also a reduced residue system modulo m.

Proof The proof is straightforward. ∎

1.2.3 Problem Let a, m be any integers such that $a \geq 1$, and $m \geq 1$. Suppose that a is relatively prime to m. Then

$$a^{\varphi(m)} \equiv 1 (\mathrm{mod}\, m).$$

Proof Let $\{b_1, b_2, \ldots, b_{\varphi(m)}\}$ be a reduced residue system modulo m. Since a is relatively prime to m, by 1.2.2, $\{ab_1, ab_2, \ldots, ab_{\varphi(m)}\}$ is also a reduced residue system modulo m. Since a is relatively prime to m, we have $ab_1 \equiv b_1 (\mathrm{mod}\, m)$. Similarly, $ab_2 \equiv b_2 (\mathrm{mod}\, m)$, etc. This shows that

$$((ab_1)(ab_2)\ldots(ab_{\varphi(m)})) \equiv (b_1 b_2 \ldots b_{\varphi(m)})(\mathrm{mod}\, m),$$

and hence

$$\left(\left(a^{\varphi(m)} \right) (b_1 b_2 \ldots b_{\varphi(m)}) \right) \equiv (b_1 b_2 \ldots b_{\varphi(m)})(\mathrm{mod}\, m).$$

Thus

$$\frac{\left(a^{\varphi(m)} \right) (b_1 b_2 \ldots b_{\varphi(m)}) - (b_1 b_2 \ldots b_{\varphi(m)})}{m}$$

is an integer, that is,

$$\frac{(b_1 b_2 \ldots b_{\varphi(m)}) \left(a^{\varphi(m)} - 1 \right)}{m}$$

is an integer. Since $\{b_1, b_2, \ldots, b_{\varphi(m)}\}$ is a reduced residue system modulo m, each b_i is relatively prime to m. Since each b_i is relatively prime to m, and

$$\frac{(b_1 b_2 \ldots b_{\varphi(m)}) \left(a^{\varphi(m)} - 1 \right)}{m}$$

is an integer,

$$\frac{a^{\varphi(m)} - 1}{m}$$

is an integer, and hence $a^{\varphi(m)} \equiv 1 (\mathrm{mod}\, m)$. ∎

1.2.4 Problem Let a be any integer such that $a \geq 1$. Let p be a prime. Suppose that p does not divide a. Then

$$a^{p-1} \equiv 1 (\mathrm{mod}\, p).$$

Proof Since p is a prime and p does not divide a, a is relatively prime to p. Now, by 1.2.3,

$$a^{\varphi(p)} \equiv 1(\operatorname{mod} p).$$

Since p is a prime, we have $\varphi(p) = p - 1$. Hence

$$a^{p-1} \equiv 1(\operatorname{mod} p).$$

∎

1.2.5 Theorem Let a be any integer such that $a \geq 1$. Let p be a prime. Then

$$a^p \equiv a(\operatorname{mod} p).$$

This result is known as the **little Fermat theorem**.

Proof Case I: p does not divide a. Here, by 1.2.4, $a^{p-1} \equiv 1(\operatorname{mod} p)$, and hence $a^{p-1}a \equiv 1a(\operatorname{mod} p)$. Thus $a^p \equiv a(\operatorname{mod} p)$.

Case II: p divides a. It follows that $\frac{a}{p}(a^{p-1} - 1)$ is an integer, and hence $\frac{a^p-a}{p}$ is an integer. This shows that $a^p \equiv a(\operatorname{mod} p)$.

Hence in all cases, $a^p \equiv a(\operatorname{mod} p)$.

∎

1.2.6 Problem Let a, b, m be any integers such that $a \geq 1, b \geq 1$, and $m \geq 1$. Suppose that a is relatively prime to m. Then the polynomial congruence

$$ax \equiv b(\operatorname{mod} m)$$

has a unique solution, namely, $x \equiv a^{\varphi(m)-1}b(\operatorname{mod} m)$.

Proof Existence: We must show that

$$a\left(a^{\varphi(m)-1}b\right) \equiv b(\operatorname{mod} m),$$

that is,

$$a^{\varphi(m)}b \equiv b(\operatorname{mod} m).$$

Since a is relatively prime to m, by 1.2.3, $a^{\varphi(m)} \equiv 1(\operatorname{mod} m)$, and hence $a^{\varphi(m)}b \equiv 1b(\operatorname{mod} m)$. Thus $a^{\varphi(m)}b \equiv b(\operatorname{mod} m)$.

Uniqueness: Suppose that the polynomial congruence

$$ax \equiv b(\operatorname{mod} m)$$

has two solutions x_1 and x_2, that is,

$$\left.\begin{array}{l} ax_1 \equiv b(\bmod m) \\ ax_2 \equiv b(\bmod m) \end{array}\right\}.$$

We have to show that $x_1 \equiv x_2(\bmod m)$. Since

$$\left.\begin{array}{l} ax_1 \equiv b(\bmod m) \\ ax_2 \equiv b(\bmod m) \end{array}\right\},$$

we have

$$ax_1 \equiv ax_2(\bmod m),$$

and hence

$$\frac{ax_1 - ax_2}{m}$$

is an integer, and hence $\frac{a(x_1-x_2)}{m}$ is an integer. Since a is relatively prime to m, $\frac{x_1-x_2}{m}$ is an integer, and hence

$$x_1 \equiv x_2(\bmod m).$$

$\blacksquare$

1.2.7 Theorem Let p be a prime. Let

$$f(x) \equiv c_0 + c_1x + \cdots + c_nx^n$$

be any polynomial in x with integer coefficients. Suppose that c_n is a positive integer that is not divisible by p. Then the polynomial congruence

$$f(x) \equiv 0(\bmod p)$$

has at most n solutions.

This result is due to **Lagrange**.

Proof (Induction on n): Let us take the case $n = 1$. Thus $f(x) \equiv c_0 + c_1x$. Here we have to solve the congruence

$$(c_0 + c_1x) \equiv 0(\bmod p),$$

where c_1 is not divisible by p. Since p is a prime and c_1 is not divisible by p, c_1 is relatively prime to p, and hence by 1.2.6, the polynomial congruence

$$c_1 x \equiv -c_0 (\bmod m)$$

has a unique solution. Hence the statement "$f(x) \equiv 0 (\bmod p)$ has at most n solutions" holds for $n = 1$.

Now suppose that the statement "$f(x) \equiv 0 (\bmod p)$ has at most $(n - 1)$ solutions" holds for all polynomials of degree $(n - 1)$.

Also, suppose that the polynomial congruence

$$f(x) \equiv 0 (\bmod p)$$

has $(n + 1)$ noncongruent solutions, say $x_0, x_1, \ldots, x_n$. We seek a contradiction.

Observe that

$$f(x) - f(x_0) = c_1(x - x_0) + c_1\left(x^2 - (x_0)^2\right) + \cdots + c_n(x^n - (x_0)^n)$$
$$= (x - x_0)\left(c_1 + (\cdots)x + \cdots + c_n x^{n-1}\right),$$

so

$$f(x) - f(x_0) = (x - x_0)g(x),$$

where $g(x)$ is a polynomial of degree $n - 1$, with leading coefficient c_n. Now, since c_n is a positive integer that is not divisible by p, the leading coefficient of $g(x)$ is a positive integer that is not divisible by p. It follows, by the induction hypothesis, that

$$g(x) \equiv 0 (\bmod p)$$

has at most $(n - 1)$ noncongruent solutions.

Since x_0, x_1 are solutions of $f(x) \equiv 0 (\bmod p)$, we have

$$\left.\begin{array}{l} f(x_0) \equiv 0 (\bmod p) \\ f(x_1) \equiv 0 (\bmod p) \end{array}\right\},$$

and hence

$$f(x_1) \equiv f(x_0) (\bmod p).$$

This shows that

$$(f(x_1) - f(x_0)) \equiv 0 (\bmod p),$$

that is,

$$((x_1 - x_0)g(x_1)) \equiv 0 (\bmod p).$$

Now, since, x_1, x_0 are noncongruent modulo p, we have $g(x_1) \equiv 0 \pmod{p}$, and hence x_1 is a solution of $g(x) \equiv 0 \pmod{p}$. Similarly, x_2 is a solution of $g(x) \equiv 0 \pmod{p}, \ldots, x_n$ is a solution of $g(x) \equiv 0 \pmod{p}$. Thus $g(x) \equiv 0 \pmod{p}$ has n noncongruent solutions. This is a contradiction. $\blacksquare$

1.2.8 Problem Let p be a prime. Let

$$f(x) \equiv c_0 + c_1 x + \cdots + c_n x^n$$

be any polynomial in x. Suppose that the polynomial congruence

$$f(x) \equiv 0 \pmod{p}$$

has more than n solutions. Then each c_i is divisible by p.

Proof Suppose to the contrary that there exists a largest positive integer $k \leq n$ such that c_k is not divisible by p. We seek a contradiction.

Here

$$f(x) = c_0 + c_1 x + \cdots + c_k x^k + p\big((\cdots)x^{k+1} + (\cdots)x^{k+2} + \cdots + (\cdots)x^n\big).$$

Suppose that the polynomial congruence

$$f(x) \equiv 0 \pmod{p}$$

has $(n+1)$ noncongruent solutions $x_0, x_1, \ldots, x_n$. It follows that

$$f(x_0) \equiv 0 \pmod{p},$$

that is,

$$\left(c_0 + c_1 x_0 + \cdots + c_k (x_0)^k + p\left((\cdots)(x_0)^{k+1} + (\cdots)(x_0)^{k+2} + \cdots + (\cdots)(x_0)^n \right) \right)$$
$$\equiv 0 \pmod{p}.$$

Now, since

$$p\left((\cdots)(x_0)^{k+1} + (\cdots)(x_0)^{k+2} + \cdots + (\cdots)(x_0)^n \right) \equiv 0 \pmod{p},$$

we have

$$\left(\left(c_0 + c_1 x_0 + \cdots + c_k (x_0)^k + p\left((\cdots)(x_0)^{k+1} + (\cdots)(x_0)^{k+2} + \cdots + (\cdots)(x_0)^n \right) \right) \right.$$
$$\left. - p\left((\cdots)(x_0)^{k+1} + (\cdots)(x_0)^{k+2} + \cdots + (\cdots)(x_0)^n \right) \right) \equiv (0 - 0) \pmod{p},$$

and hence

$$\left(c_0 + c_1 x_0 + \cdots + c_k (x_0)^k\right) \equiv 0 \pmod{p}.$$

This shows that x_0 is a solution of the polynomial congruence

$$\left(c_0 + c_1 x + \cdots + c_k x^k\right) \equiv 0 \pmod{p}.$$

Similarly, x_1 is a solution of the polynomial congruence

$$\left(c_0 + c_1 x + \cdots + c_k x^k\right) \equiv 0 \pmod{p},$$

$$\vdots$$

x_n is a solution of the polynomial congruence

$$\left(c_0 + c_1 x + \cdots + c_k x^k\right) \equiv 0 \pmod{p}.$$

Thus the number of solutions of the polynomial congruence

$$\left(c_0 + c_1 x + \cdots + c_k x^k\right) \equiv 0 \pmod{p}$$

is strictly greater than n. By 1.2.7, the number of solutions of the polynomial congruence

$$\left(c_0 + c_1 x + \cdots + c_k x^k\right) \equiv 0 \pmod{p}$$

is at most k. It follows that $n < k$. This contradicts $k \le n$. ■

1.2.9 Problem Let p be a prime. Let

$$f(x) \equiv \underbrace{(x-1)(x-2)\ldots(x-(p-1))}_{(p-1)\text{factors}} - \left(x^{p-1} - 1\right).$$

Then each coefficient of the $(p-2)$th-degree polynomial $f(x)$ is divisible by p.

Proof By 1.2.4, for every $x \in \{1, 2, \ldots, (p-1)\}$, we have

$$x^{p-1} - 1 \equiv 0 \pmod{p}.$$

It is clear that for every $x \in \{1, 2, \ldots, (p-1)\}$, we have

$$(x-1)(x-2)\ldots(x-(p-1)) \equiv 0 \pmod{p}.$$

Hence for every $x \in \{1, 2, \ldots, (p-1)\}$, we have

$$((x-1)(x-2)\ldots(x-(p-1)) - (x^{p-1} - 1)) \equiv (0-0)(\bmod p).$$

Thus for every $x \in \{1, 2, \ldots, (p-1)\}$, we have

$$f(x) \equiv 0(\bmod p).$$

Thus the number of solutions of the polynomial congruence $f(x) \equiv 0(\bmod p)$ is strictly greater than $(p-2)$. Now, since $f(x)$ is a polynomial of degree $(p-2)$, by 1.2.8, each coefficient of the polynomial $f(x)$ is divisible by p. ∎

1.2.10 Theorem Let p be a prime. Then $(p-1)! \equiv -1(\bmod p)$.

This result is known as **Wilson's theorem**.

Proof If $p = 2$, then $(p-1)! \equiv -1(\bmod p)$ becomes $1! \equiv -1(\bmod 2)$. This is trivially true. So we shall consider only the case in which the prime p is odd.

By   1.2.9,   each   coefficient   of   the $(p-2)$th-degree   polynomial $\underbrace{(x-1)(x-2)\cdots(x-(p-1))}_{(p-1)\,\text{factors}} - (x^{p-1} - 1)$ is divisible by p. Now, since the constant term of the polynomial

$$\underbrace{(x-1)(x-2)\cdots(x-(p-1))}_{(p-1)\,\text{factors}} - (x^{p-1} - 1)$$

is

$$(-1)^{p-1}(p-1)! + 1 \left(= (-1)^{\text{odd}-1}(p-1)! + 1 = (p-1)! + 1 \right),$$

$(p-1)! + 1$ is divisible by p, that is, $(p-1)! \equiv -1(\bmod p)$. ∎

1.2.11 Note Let p be a prime number of the form $4n+1$. It follows that $\frac{1}{2}(p-1)$ is an integer ≥ 2. Put

$$a \equiv 1 \cdot 2 \cdot 3 \cdots \cdot \left(\frac{1}{2}(p-1)\right) \left(= \left(\frac{1}{2}(p-1)\right)! \right).$$

It follows that

$$a^2 = \left(1 \cdot 2 \cdot 3 \cdots \cdot \left(\frac{1}{2}(p-1)\right)\right) \left(\left(\frac{1}{2}(p-1)\right)\left(\frac{1}{2}(p-1)-1\right) \cdots 3 \cdot 2 \cdot 1\right),$$

and hence

$$a^2 \equiv \left(1 \cdot 2 \cdot 3 \cdots \left(\frac{1}{2}(p-1)\right)\right)\left(\left(\frac{1}{2}(p-1)\right)\left(\frac{1}{2}(p-1)-1\right)\cdots 3 \cdot 2 \cdot 1\right)(\text{mod}\,p).$$

Thus

$$a^2 \equiv \left(1 \cdot 2 \cdot 3 \cdots \left(\frac{1}{2}(p-1)\right)\right)\left(\left(\frac{1}{2}(p-1)\right)\left(\frac{1}{2}(p-3)\right)\cdots 3 \cdot 2 \cdot 1\right)(\text{mod}\,p).$$

Since

$$\frac{1}{2}(p-1) \equiv \frac{-1}{2}(p+1)(\text{mod}\,p),$$
$$\frac{1}{2}(p-3) \equiv \frac{-1}{2}(p+3)(\text{mod}\,p),$$
$$\vdots$$
$$3 \equiv -(p-3)(\text{mod}\,p),$$
$$2 \equiv -(p-2)(\text{mod}\,p),$$
$$1 \equiv -(p-1)(\text{mod}\,p),$$

we have

$$\left(\left(\frac{1}{2}(p-1)\right)\left(\frac{1}{2}(p-3)\right)\cdots 3 \cdot 2 \cdot 1\right)$$
$$\equiv \left(\left(\frac{-1}{2}(p+1)\right)\left(\frac{-1}{2}(p+3)\right)\cdots(-(p-3))(-(p-2))(-(p-1))\right)(\text{mod}\,p),$$

that is,

$$\left(\left(\frac{1}{2}(p-1)\right)\left(\frac{1}{2}(p-3)\right)\cdots 3 \cdot 2 \cdot 1\right)$$
$$\equiv \left((-1)^{\frac{p-1}{2}}\left(\frac{1}{2}(p+1)\right)\left(\frac{1}{2}(p+3)\right)((p-3))((p-2))((p-1))\right)(\text{mod}\,p),$$

that is,

$$\left(\left(\frac{1}{2}(p-1)\right)\left(\frac{1}{2}(p-3)\right)\cdots 3 \cdot 2 \cdot 1\right)$$
$$\equiv \left((-1)^{\text{even}}\left(\frac{1}{2}(p+1)\right)\left(\frac{1}{2}(p+3)\right)((p-3))((p-2))((p-1))\right)(\text{mod}\,p),$$

that is,

$$\left(\left(\frac{1}{2}(p-1)\right)\left(\frac{1}{2}(p-3)\right)\cdots 3\cdot 2\cdot 1\right)$$
$$\equiv \left(\left(\frac{1}{2}(p+1)\right)\left(\frac{1}{2}(p+3)\right)((p-3))((p-2))((p-1))\right)(\bmod p).$$

Now, since

$$a^2 \equiv \left(1\cdot 2\cdot 3\cdots\left(\frac{1}{2}(p-1)\right)\right)\left(\left(\frac{1}{2}(p-1)\right)\left(\frac{1}{2}(p-3)\right)\cdots 3\cdot 2\cdot 1\right)(\bmod p),$$

we have

$$a^2 \equiv \left(1\cdot 2\cdot 3\cdots\left(\frac{1}{2}(p-1)\right)\right)$$
$$\left(\left(\frac{1}{2}(p+1)\right)\left(\frac{1}{2}(p+3)\right)((p-3))((p-2))((p-1))\right)(\bmod p),$$

that is,

$$a^2 \equiv (p-1)!(\bmod p).$$

By 1.2.10, $(p-1)! \equiv -1(\bmod p)$. Now, since $a^2 \equiv (p-1)!(\bmod p)$, we have $a^2 \equiv -1(\bmod p)$. This shows that a is a solution of the quadratic congruence $x^2 \equiv -1(\bmod p)$, that is, $(\frac{1}{2}(p-1))!$ is a solution of the quadratic congruence $x^2 \equiv -1(\bmod p)$.

1.2.12 Conclusion Let p be a prime number of the form $4n+1$. Then there exists a solution of $x^2 \equiv -1(\bmod p)$. One such solution is $(\frac{1}{2}(p-1))!$.

1.2.13 Theorem Let p be a prime number of the form $4n+1$. Then there exist integers a and b such that $p = a^2 + b^2$.
   This result is due to **Fermat**.

Proof It is clear that $5 \leq p$ and that $\frac{1}{2}(p-1)$ is an even integer. By 1.2.12, there exists an integer $x \in \{0, 1, \ldots, \frac{1}{2}(p-1), \ldots, (p-1)\}$ such that $x^2 \equiv -1(\bmod p)$. It follows that there exists an integer c such that $cp = x^2 + 1^2$.
   It follows that there exists an integer $y \in \{-\frac{1}{2}(p-1), -\frac{1}{2}(p-1) + 1, \ldots 0, \ldots, \frac{1}{2}(p-1)\}$ such that $y^2 \equiv -1(\bmod p)$.

Proof Here $x \in \{0, 1, \ldots, \frac{1}{2}(p-1), \ldots, (p-1)\}$, so

$$\text{either } x \in \left\{0, 1, \ldots, \frac{1}{2}(p-1)\right\} \text{ or } x \in \left\{\frac{1}{2}(p-1) + 1, \ldots, (p-1)\right\}.$$

Case I: $x \in \{0, 1, \ldots, \frac{1}{2}(p-1)\}$. In this case, let us take x for y. Since $x^2 \equiv -1 \pmod{p}$, we have

$$y^2 \equiv -1 \pmod{p}. \qquad \text{Since}$$

$$y = x \in \left\{0, 1, \ldots, \frac{1}{2}(p-1)\right\}$$
$$\subset \left\{-\frac{1}{2}(p-1), -\frac{1}{2}(p-1)+1, \ldots 0, \ldots, \frac{1}{2}(p-1)\right\},$$

we have $y \in \left\{-\frac{1}{2}(p-1), -\frac{1}{2}(p-1)+1, \ldots 0, \ldots, \frac{1}{2}(p-1)\right\}$.

Case II: $x \in \{\frac{1}{2}(p-1)+1, \ldots, (p-1)\}$. In this case, let us take $(p-x)$ for y. Since

$x^2 \equiv -1 \pmod{p}$, $\frac{x^2+1}{p}$ is an integer, and hence $p - 2x + \frac{x^2+1}{p}$ is an integer. Since

$$\frac{y^2+1}{p} = \frac{(p-x)^2+1}{p} = p - 2x + \frac{x^2+1}{p},$$

$\frac{y^2+1}{p}$ is an integer. Thus $y^2 \equiv -1 \pmod{p}$. It remains to show that

$$y \in \left\{-\frac{1}{2}(p-1), -\frac{1}{2}(p-1)+1, \ldots 0, \ldots, \frac{1}{2}(p-1)\right\}.$$

Since $x \in \{\frac{1}{2}(p-1)+1, \ldots, (p-1)\}$, we have

$$y = (p-x) \in \left\{p - (\tfrac{1}{2}(p-1)+1), \ldots, p - (p-1)\right\} = \left\{\tfrac{1}{2}(p-1), \ldots, 1\right\}$$
$$\subset \left\{-\tfrac{1}{2}(p-1), -\tfrac{1}{2}(p-1)+1, \ldots 0, \ldots, \tfrac{1}{2}(p-1)\right\},$$

and hence

$$y \in \left\{-\frac{1}{2}(p-1), -\frac{1}{2}(p-1)+1, \ldots 0, \ldots, \frac{1}{2}(p-1)\right\}.$$

So in all cases, there exists an integer $y \in \{-\frac{1}{2}(p-1), -\frac{1}{2}(p-1)+1, \ldots 0, \ldots, \frac{1}{2}(p-1)\}$ such that

$$y^2 \equiv -1 \pmod{p}.$$

∎

Since $y^2 \equiv -1 \pmod{p}$, there exists a positive integer e such that $ep = (y^2 + 1^2)$. In view of 1.1.26, it suffices to show that e is relatively prime to p. Since

$y \in \{-\frac{1}{2}(p-1), -\frac{1}{2}(p-1)+1, \ldots 0, \ldots, \frac{1}{2}(p-1)\}$, we have $|y| \leq \frac{1}{2}(p-1)$, and hence $y^2 \leq \frac{1}{4}(p-1)^2$. It follows that

$$(y^2 + 1^2) \leq \frac{1}{4}(p-1)^2 + 1,$$

and hence

$$e = \underbrace{\frac{1}{p}(y^2 + 1^2) \leq \frac{1}{p}\left(\frac{1}{4}(p-1)^2 + 1\right)} < \frac{1}{p-1}\left(\frac{1}{4}(p-1)^2 + 1\right)$$

$$= \frac{1}{4}(p-1) + \frac{1}{p-1} \leq \frac{1}{4}(p-1) + \frac{1}{4} = \frac{p}{4} < p.$$

Thus e is a positive integer strictly smaller than p, and since p is a prime, e must be relatively prime to p. ■

1.2.14 Definition Let F be a field. Let $F[x]$ be the collection of all polynomials $f(x)$ in the "indeterminant" x having coefficients in F. We know that $F[x]$ is an integral domain with unit element 1. For every nonzero polynomial $f(x)$, put

$$d(f(x)) \equiv \deg(f(x)).$$

It is known that

1. for every nonzero $f(x), g(x) \in F[x]$, $\deg(f(x)) \leq \deg(f(x)g(x))$,
2. for every nonzero $f(x), g(x) \in F[x]$, there exist $q(x), r(x) \in F[x]$ such that $f(x) = q(x)g(x) + r(x)$, and (either $r(x) = 0$ or $\deg(r(x)) < \deg(g(x)))$.

This shows that $F[x]$ is a Euclidean ring. Also, it is clear that
    (*) for every nonzero $f(x), g(x) \in F[x]$, $\deg(f(x)g(x)) = \deg(f(x)) + \deg(g(x))$.

1.2.15 Problem $F[x]$ is a principal ideal ring.

Proof Since $F[x]$ is a Euclidean ring, by 1.1.4, $F[x]$ is a principal ideal ring. ■

Definition Let $f(x)$ be a nonzero member of the Euclidean ring $F[x]$. If $f(x)$ is a unit or $f(x)$ is a prime element of $F[x]$, then we say that $f(x)$ *is irreducible over F.*

1.2.16 Problem Let $f(x)$ be a nonzero member of the Euclidean ring $F[x]$. Then $f(x)$ is a unit if and only if $f(x)$ is a constant.

Proof Let $f(x)$ be a unit in the Euclidean ring $F[x]$. We have to show that $f(x)$ is a constant.

Since $f(x)$ is a unit in the Euclidean ring $F[x]$, there exists $g(x) \in F[x]$ such that $f(x)g(x) = 1$. Now, since $1 \neq 0$, and $F[x]$ is an integral domain, it follows that $f(x) \neq 0$ and $g(x) \neq 0$, and then that

$$0 \leq \underbrace{\deg\left(f(x)\right) \leq \deg\left(f(x)g(x)\right)}_{} = \deg(1) = 0,$$

and hence $\deg(f(x)) = 0$. Thus $f(x)$ is a constant.

Conversely, let $f(x)$ be a constant. It follows that $\deg(f(x)) = 0$, and hence $f(x) \in F$. Now, since $f(x)$ is a nonzero member of the field F, there exists a nonzero member b of $F(\subset F[x])$ such that $f(x)b = 1$. Thus $f(x)$ is a unit. ∎

1.2.17 Problem Let $f(x)$ be a nonzero member of the Euclidean ring $F[x]$. Then $f(x)$ is irreducible if and only if

$$(g(x), h(x) \in F[x] \text{ and } f(x) = g(x)h(x)) \Rightarrow (g(x) \text{ is a unit or } h(x) \text{ is a unit}).$$

Proof Let $f(x)$ be irreducible. We have to show that

$$(g(x), h(x) \in F[x] \text{ and } f(x) = g(x)h(x)) \Rightarrow (g(x) \text{ is a unit or } h(x) \text{ is a unit}).$$

Since $f(x)$ is irreducible, $f(x)$ is a unit or $f(x)$ is a prime element of $F[x]$.

Case I: $f(x)$ is a unit. Suppose that $g(x), h(x) \in F[x]$ and $f(x) = g(x)h(x)$. We have to show that $g(x)$ is a unit or $h(x)$ is a unit. Suppose to the contrary that $g(x)$ is not a unit and $h(x)$ is not a unit. We seek a contradiction.

Since $f(x)$ is a unit, by 1.2.16, $f(x)$ is a constant, and hence $\deg(f(x)) = 0$. Since $f(x)$ is nonzero and $f(x) = g(x)h(x)$, we have $g(x) \neq 0$. Now, since $g(x)$ is not a unit, by 1.2.16, $g(x)$ is a not constant, and hence $1 \leq \deg(g(x))$. Since $f(x) = g(x)h(x)$, we have

$$1 \leq \deg(g(x)) \leq \underbrace{\deg(g(x)h(x)) = \deg(f(x))}_{} = 0.$$

This is a contradiction.

Case II: $f(x)$ is a prime element of $F[x]$. Suppose that $g(x), h(x) \in F[x]$ and $f(x) = g(x)h(x)$. We have to show that $g(x)$ is a unit or $h(x)$ is a unit.

Since $f(x)$ is a prime element of the Euclidean ring $F[x]$, $g(x), h(x) \in F[x]$, and $f(x) = g(x)h(x)$, we have that either $g(x)$ is a unit or $h(x)$ is a unit.

So in all cases,

$$(g(x), h(x) \in F[x] \text{ and } f(x) = g(x)h(x)) \Rightarrow (g(x) \text{ is a unit or } h(x) \text{ is a unit}).$$

Conversely, suppose that

$$(g(x), h(x) \in F[x] \text{ and } f(x) = g(x)h(x)) \Rightarrow (g(x) \text{ is a unit or } h(x) \text{ is a unit}). \qquad (*)$$

We have to show that $f(x)$ is irreducible, that is, $f(x)$ is a unit or $f(x)$ is a prime element of $F[x]$. Suppose to the contrary that $f(x)$ is not a unit and $f(x)$ is not a prime element of $F[x]$. We seek a contradiction.

Since $f(x)$ is not a unit and $f(x)$ is not a prime element of the Euclidean ring $F[x]$, there exist $g(x), h(x) \in F[x]$ such that $f(x) = g(x)h(x)$, $g(x)$ is not a unit and $h(x)$ is not a unit. This contradicts $(*)$. ∎

1.2.18 Problem Clearly, the polynomial $1 + x^2$ is irreducible over the field $\mathbb{R}$ of all real numbers.

Proof Suppose to the contrary that it is reducible. Then by 1.2.17, there exist $g(x), h(x) \in \mathbb{R}[x]$ such that $1 + x^2 = g(x)h(x)$, $g(x)$ is not a unit, and $h(x)$ is not a unit. We seek a contradiction.

Since $g(x)$ is not a unit, by 1.2.16, $g(x)$ is a constant, and hence $\deg(g(x)) \geq 1$. Similarly, $\deg(h(x)) \geq 1$.

Next,

$$\deg(g(x)) + \deg(h(x)) = \deg(g(x)h(x)) = \deg(1 + x^2) = 2,$$

so $\deg(g(x)) + \deg(h(x)) = 2$. Now, since $\deg(g(x)) \geq 1$ and $\deg(h(x)) \geq 1$, we have $\deg(g(x)) = 1$ and $\deg(h(x)) = 1$. So we can suppose that $g(x) \equiv x + \alpha$ and $h(x) \equiv x + \beta$, where α, β are real numbers. Thus

$$\underbrace{1 + x^2 = (x + \alpha)(x + \beta)} = \alpha\beta + (\alpha + \beta)x + x^2,$$

and hence

$$\left.\begin{array}{c} \alpha\beta = 1 \\ \alpha + \beta = 0 \end{array}\right\}.$$

This shows that $0 \leq \alpha^2 = -1$. This is a contradiction. ∎

1.2.19 Problem Clearly, the polynomial $1 + x^2$ is not irreducible over the field $\mathbb{C}$ of all complex numbers.

Proof Observe that

$$1 + x^2 = (x + i)(x - i).$$

Also, $x + i, x - i$ are members of $\mathbb{C}[x]$. Clearly, $x + i$ and $x - i$ are not units in $\mathbb{C}[x]$. Thus by 1.2.17, $1 + x^2$ is not irreducible in $\mathbb{C}[x]$. ∎

1.2.20 Problem Let $f(x)$ be a nonzero member of the Euclidean ring $F[x]$. Let $f(x)$ be a nonunit. Then $f(x)$ can be expressed as a product of finitely many irreducible polynomials of degree ≥ 1 in $F[x]$.

Proof Since $f(x)$ is not a unit, by 1.2.16, $f(x)$ is not a constant, and hence $\deg(f(x)) \geq 1$. Further, by 1.1.11, $f(x)$ can be expressed as a product of finitely many prime elements of $F[x]$. Since a prime element of $F[x]$ is not a unit, by 1.2.16, the degree of a prime element of $F[x]$ is ≥ 1. Now, by the definition of irreducibility over F, $f(x)$ can be expressed as a product of finitely many irreducible polynomials of degree ≥ 1 in $F[x]$. ∎

1.2.21 Theorem Let $f(x)$ be a nonzero member of the Euclidean ring $F[x]$. Suppose that $f(x)$ is not a unit in $F[x]$. (By 1.2.20, $f(x)$ can be expressed as a product of finitely many irreducible polynomials of degree ≥ 1 in $F[x]$.) Let

$$f(x) = \pi_1(x)\pi_2(x)\ldots\pi_m(x),$$

where each $\pi_i(x)(i = 1, 2, \ldots, m)$ is an irreducible polynomial of degree ≥ 1 in $F[x]$. Let

$$f(x) = \pi'_1(x)\pi'_2(x)\ldots\pi'_n(x),$$

where each $\pi'_j(x)(j = 1, 2, \ldots, n)$ is an irreducible polynomial of degree ≥ 1 in $F[x]$. Then

1. each $\pi_i(x)$ is an associate of some $\pi'_j(x)$,
2. each $\pi'_j(x)$ is an associate of some $\pi_i(x)$,
3. $n = m$.

This theorem is known as the **unique factorization theorem** of polynomials over F.

Proof By 1.1.15, the proof is immediate. ∎

1.2.22 Problem Let $p(x)$ be a nonzero member of the Euclidean ring $F[x]$. Let $p(x)$ be an irreducible polynomial of degree ≥ 1 in $F[x]$. Then the ideal $(p(x))$ is "maximal" in the sense that
(i) $(p(x))$ is a proper subset of $F[x]$,
(ii) if M is an ideal containing $(p(x))$ and M is a proper subset of $F[x]$, then $M = (p(x))$.

Proof By 1.2.16, $p(x)$ is not a unit. Hence by the definition of irreducible polynomial, $p(x)$ is a prime element of $F[x]$. Now by 1.1.16, the ideal $(p(x))$ is maximal. ∎

1.2.23 Problem Let $p(x)$ be a nonzero member of the Euclidean ring $F[x]$. Suppose that $\deg(p(x)) \geq 1$. Let the ideal $(p(x))$ be maximal. Then $p(x)$ is an irreducible polynomial in $F[x]$.

Proof By 1.2.16, $p(x)$ is not a unit. Hence by the definition of irreducible polynomial, it suffices to show that $p(x)$ is a prime element of $F[x]$.

Since the ideal $(p(x))$ is maximal, by 1.1.17, $p(x)$ is a prime element of $F[x]$. ∎

1.2.24 Problem Let $p(x)$ be a nonzero member of the Euclidean ring $F[x]$. Suppose that $\deg(p(x)) \geq 1$. Let $p(x)$ be irreducible over the field F. Then by 1.2.22, the ideal $(p(x))$ is maximal. Further, the quotient ring $\frac{F[x]}{(p(x))}$ is a field.

Proof Since $F[x]$ is an integral domain with unit element 1, the quotient ring $\frac{F[x]}{(p(x))}$ is a commutative ring with unit element $1 + (p(x))$. Next, let us take arbitrary nonzero elements $f(x) + (p(x))$ and $g(x) + (p(x))$ of $\frac{F[x]}{(p(x))}$, where $f(x), g(x) \in F[x]$. We have to show that $f(x)g(x) + (p(x))$ is nonzero. Suppose to the contrary that $f(x)g(x) \in (p(x))$. We seek a contradiction.

Since $f(x)g(x) \in (p(x))$, there exists $h(x) \in F[x]$ such that $f(x)g(x) = p(x)h(x)$. Now, since $p(x)$ is irreducible, by 1.2.21, $p(x)|f(x)$ or $p(x)|g(x)$. It follows that either $f(x) \in (p(x))$ or $g(x) \in (p(x))$. In other words, either $f(x) + (p(x))$ is the zero element of $\frac{F[x]}{(p(x))}$ or $g(x) + (p(x))$ is the zero element of $\frac{F[x]}{(p(x))}$. This is a contradiction.

Thus we have shown that the product of nonzero elements of $\frac{F[x]}{(p(x))}$ is nonzero.

Next, let $f(x) + (p(x))$ be a nonzero element of $\frac{F[x]}{(p(x))}$. It follows that $f(x) \notin (p(x))$, and $f(x)$ is a nonzero polynomial. Hence $p(x) \nmid f(x)$. By 1.1.5, there exists a greatest common divisor $h(x)$ of $p(x)$ and $f(x)$ in $F[x]$. Further, there exist $\lambda(x), \mu(x) \in F[x]$ such that

$$h(x) = \lambda(x)p(x) + \mu(x)f(x).$$

We claim that $h(x)$ is a unit. Suppose to the contrary that $h(x)$ is not a unit. We seek a contradiction.

Since $h(x)$ is a greatest common divisor of $p(x)$ and $f(x)$ in $F[x]$, we have $h(x)|p(x)$ and $h(x)|f(x)$. Since $h(x)|p(x)$, there exists $k(x) \in F[x]$ such that $p(x) = h(x)k(x)$. Now, since $p(x)$ is irreducible, by 1.2.17, $h(x)$ is a unit or $k(x)$ is a unit. Since $h(x)$ is not a unit, $k(x)$ is a unit. It follows from $p(x) = h(x)k(x)$ that $p(x)$ and $h(x)$ are associates. And since $p(x) \nmid f(x)$, we have $h(x) \nmid f(x)$. This is a contradiction.

Thus our claim is true, that is, $h(x)$ is a unit. It follows that there exists $l(x) \in F[x]$ such that $1 = h(x)l(x)$, and hence

$$\underbrace{1 = (\lambda(x)p(x) + \mu(x)f(x))l(x)} = (\lambda(x)l(x))p(x) + (\mu(x)l(x))f(x).$$

Thus

$$1 - g(x)f(x) = (\lambda(x)l(x))p(x) \in (p(x)),$$

where $g(x) \equiv \lambda(x)l(x) \in F[x]$. Hence $g(x) + (p(x))$ serves the purpose of the inverse element of $f(x) + (p(x))$ in $\frac{F[x]}{(p(x))}$.

Thus $\frac{F[x]}{(p(x))}$ is a field.   ∎

1.2.25 Example The field of all rational numbers is denoted by $\mathbb{Q}$. Observe that the polynomial $x^3 - 2$ is a member of the Euclidean ring $\mathbb{Q}[x]$. Also, $x^3 - 2$ is irreducible in $\mathbb{Q}[x]$.

And hence by 1.2.24, the quotient ring $\frac{\mathbb{Q}[x]}{(x^3-2)}$ is a field.

Proof Suppose to the contrary that

$$x^3 - 2 = g(x)h(x),$$

where $g(x), h(x) \in \mathbb{Q}[x]$, and neither $g(x)$ nor $h(x)$ is a unit. It follows that either $\deg(g(x)) = 1$ or $\deg(h(x)) = 1$. For definiteness, suppose that $\deg(g(x)) = 1$. Now we can suppose that

$$g(x) \equiv x - \alpha,$$

where $\alpha \in \mathbb{Q}$. It follows that

$$x^3 - 2 = (x - \alpha)h(x),$$

and hence

$$\underbrace{\alpha^3 - 2 = (\alpha - \alpha)h(\alpha)}_{} = 0h(\alpha) = 0.$$

Thus $\alpha^3 = 2$. Since $\alpha \in \mathbb{Q}$, there exist two integers r and s such that

$$r^3 = 2 \cdot s^3.$$

Observe that in the prime factorization of r^3, $2^{3(\text{integer})}$ will occur, but in the prime factorization of $(r^3 =) 2 \cdot s^3$, $2^{3(\text{integer})+1}$ will occur. But $2^{3(\text{integer})} \neq 2^{3(\text{integer})+1}$, which contradicts the uniqueness property of prime factorization of integers. ∎

1.2.26 Note Suppose that $f(x) + (x^3 - 2)$ is a member of the field $\frac{\mathbb{Q}[x]}{(x^3-2)}$, where $f(x) \in \mathbb{Q}[x]$. Let us denote the polynomial $x^3 - 2$ by $g(x)$.

It follows that there exist $q(x), r(x) \in \mathbb{Q}[x]$ such that $f(x) = q(x)g(x) + r(x)$, and (either $r(x) = 0$ or $\deg(r(x)) < \deg(g(x)) = 3$). Hence we can suppose that

$$r(x) \equiv a_0 + a_1 x + a_2 x^2,$$

where $a_0, a_1, a_2 \in \mathbb{Q}$. Thus

$$f(x) = q(x)g(x) + a_0 + a_1 x + a_2 x^2,$$

and hence

$$\underbrace{f(x) + (x^3 - 2) = q(x)g(x) + a_0 + a_1 x + a_2 x^2 + (x^3 - 2)}$$

$$= q(x)g(x) + a_0 + a_1 x + a_2 x^2 + (g(x))$$
$$= a_0 + a_1 x + a_2 x^2 + (q(x)g(x) + (g(x))).$$

Thus

$$f(x) + (x^3 - 2) = a_0 + a_1 x + a_2 x^2 + (q(x)g(x) + (g(x))).$$

Clearly, $q(x)g(x) \in (g(x))$. Now, since $(g(x))$ is an additive group, $q(x)g(x) + (g(x)) = (g(x))$. Thus

$$\underbrace{f(x) + (x^3 - 2) = a_0 + a_1 x + a_2 x^2 + (g(x))}$$

$$= (a_0 + (g(x))) + (a_1 + (g(x)))(x + (g(x))) + (a_2 + (g(x)))(x + (g(x)))^2,$$

or

$$f(x) + (x^3 - 2) = (a_0 + (g(x))) + (a_1 + (g(x)))t + (a_2 + (g(x)))t^2,$$

where $t \equiv x + (g(x))$. Next,

$$t^3 = (x + (g(x)))^3 = x^3 + (g(x)) = (g(x) + (g(x))) + (2 + (g(x)))$$
$$= (0 + (g(x))) + (2 + (g(x))) = 2 + (g(x)),$$

so

$$t^3 = 2 + (g(x)).$$

Notation Observe that the field $\frac{\mathbb{Q}[x]}{(x^3 - 2)}$ can be thought of as a vector space over the field $\mathbb{Q}$ under the obvious definition of "scalar multiplication": For every $a \in \mathbb{Q}$ and for every $f(x) \in \mathbb{Q}[x]$,

$$a(f(x) + (x^3 - 2)) \equiv (a + (x^3 - 2))(f(x) + (x^3 - 2)) (= af(x) + (x^3 - 2)).$$

That is why for every $a \in \mathbb{Q}$, it is customary to denote $a + (g(x))$ simply by a.

1.2.27 Conclusion All the elements of the field $\frac{\mathbb{Q}[x]}{(x^3 - 2)}$ can be expressed "uniquely" as

$$a_0 + a_1 t + a_2 t^2,$$

where $t \equiv x + (x^3 - 2)$, and $a_0, a_1, a_2 \in \mathbb{Q}$. Further, $t^3 - 2 = 0$.

Proof of uniqueness part To this end, suppose that

$$a_0 + a_1 t + a_2 t^2 = b_0 + b_1 t + b_2 t^2,$$

where $a_0, a_1, a_2, b_0, b_1, b_2 \in \mathbb{Q}$. We have to show that $a_i = b_i (i = 0, 1, 2)$.
Since

$$a_0 + a_1 t + a_2 t^2 = b_0 + b_1 t + b_2 t^2,$$

we have

$$(a_0 - b_0) + (a_1 - b_1)t + (a_2 - b_2)t^2 = 0,$$

and hence

$$(a_0 - b_0) + (a_1 - b_1)x + (a_2 - b_2)x^2 \in (x^3 - 2).$$

Now, since the degree of each nonzero member of $(x^3 - 2)$ is ≥ 3, $(a_0 - b_0) + (a_1 - b_1)x + (a_2 - b_2)x^2$ is the zero polynomial, and hence $a_i = b_i$ $(i = 0, 1, 2)$. ∎

1.3 The Eisenstein Criterion

1.3.1 Definition The field of all integers is denoted by $\mathbb{Z}$. Let $a_0 + a_1 x + \cdots + a_n x^n$ be a member of the ring $\mathbb{Z}[x]$, where each a_i is an integer. If 1 is a greatest common divisor of $a_0, a_1, \ldots, a_n$, then we say that $a_0 + a_1 x + \cdots + a_n x^n$ is a *primitive polynomial*.

1.3.2 Problem Let $a_0 + a_1 x + \cdots + a_n x^n$ and $b_0 + b_1 x + \cdots + b_m x^m$ be two primitive polynomials. Their product is

$$c_0 + c_1 x + \cdots + c_{n+m} x^{n+m},$$

where

$$\left. \begin{array}{c} c_0 \equiv a_0 b_0 \\ c_1 \equiv a_1 b_0 + a_0 b_1 \\ c_2 \equiv a_2 b_0 + a_1 b_1 + a_0 b_2 \\ \vdots \end{array} \right\}.$$

Then $c_0 + c_1 x + \cdots + c_{n+m} x^{n+m}$ is primitive.

Proof Suppose to the contrary that $c_0 + c_1 x + \cdots + c_{n+m} x^{n+m}$ is imprimitive. We seek a contradiction.

Since $c_0 + c_1 x + \cdots + c_{n+m} x^{n+m}$ is not primitive, 1 is not a greatest common divisor of $c_0, c_1, \cdots, c_{n+m}$, and hence there exists a prime number $p(>1)$ such that $p | c_i (i = 0, 1, \ldots, n+m)$. Since $p > 1$, and 1 is a greatest common divisor of $a_0, a_1, \ldots, a_n$, there exists $j \in \{0, 1, \ldots, n\}$ such that $p \nmid a_j$ and $p | a_l (l = 0, 1, \ldots, j-1)$. Similarly, there exists $k \in \{0, 1, \ldots, m\}$ such that $p \nmid b_k$ and $p | b_l (l = 0, 1, \ldots, k-1)$. Hence $a_0 + a_1 x + \cdots + a_n x^n$ is of the form

$$p(\cdots) + p(\cdots)x + p(\cdots)x^2 + \cdots + p(\cdots)x^{j-1} + a_j x^j + a_{j+1} x^{j+1} + \cdots,$$

and $b_0 + b_1 x + \cdots + b_m x^m$ is of the form

$$p(\cdots) + p(\cdots)x + p(\cdots)x^2 + \cdots + p(\cdots)x^{k-1} + b_k x^k + b_{k+1} x^{k+1} + \cdots.$$

Since p is a prime number, $p \nmid a_j$, and $p \nmid b_k$, we have $p \nmid a_j b_k$. Further,

$$c_{j+k} = a_{j+k} b_0 + a_{j+k-1} b_1 + \cdots + a_0 b_{j+k}$$
$$= a_{j+k} b_0 + a_{j+k-1} b_1 + \cdots + a_{j+1} b_{k-1} + a_j b_k + a_{j-1} b_{k+1} + \cdots + a_0 b_{j+k},$$

so

$$a_j b_k = c_{j+k} - \left(a_{j+k} b_0 + a_{j+k-1} b_1 + \cdots + a_{j+1} b_{k-1} \right) - \left(a_{j-1} b_{k+1} + \cdots + a_0 b_{j+k} \right)$$
$$= c_{j+k} - \left(a_{j+k} p(\cdots) + a_{j+k-1} p(\cdots) + \cdots + a_{j+1} p(\cdots) \right)$$
$$- \left(p(\cdots) b_{k+1} + \cdots + p(\cdots) b_{j+k} \right)$$
$$= c_{j+k} - p(\cdots) - p(\cdots) = c_{j+k} - p(\cdots),$$

and hence

$$a_j b_k = c_{j+k} - p(\cdots).$$

Since $p | c_{j+k}$, we have $p | a_j b_k$. This is a contradiction. ∎

1.3.3 Theorem Let $a_0 + a_1 x + \cdots + a_n x^n (\in \mathbb{Z}[x])$ be a primitive polynomial. Suppose that

$$a_0 + a_1 x + \cdots + a_n x^n = (r_0 + r_1 x + \cdots + r_m x^m)(s_0 + s_1 x + \cdots + s_{n-m} x^{n-m}),$$

where each r_i is a rational number and each s_j is a rational number. Then there exist two polynomials $\lambda(x), \mu(x) \in \mathbb{Z}[x]$ such that

$$a_0 + a_1 x + \cdots + a_n x^n = \lambda(x)\mu(x).$$

This result is known as the **Gauss's lemma**.

Proof By clearing denominators and taking out common factors, we can write

$$(r_0 + r_1 x + \cdots + r_m x^m)(s_0 + s_1 x + \cdots + s_{n-m} x^{n-m})$$

as $\frac{a}{b} \lambda(x)\mu(x)$, where a, b are positive integers and $\lambda(x), \mu(x)$ are primitive polynomials. Now, since

$$a_0 + a_1 x + \cdots + a_n x^n = (r_0 + r_1 x + \cdots + r_m x^m)(s_0 + s_1 x + \cdots + s_{n-m} x^{n-m}),$$

we have

$$a_0 + a_1 x + \cdots + a_n x^n = \frac{a}{b} \lambda(x)\mu(x),$$

or

$$b a_0 + b a_1 x + \cdots + p a_n x^n = a \lambda(x)\mu(x). \qquad (*)$$

Since $a_0 + a_1 x + \cdots + a_n x^n (\in \mathbb{Z}[x])$ is a primitive polynomial, 1 is a greatest common divisor of $a_0, a_1, \ldots, a_n$, and hence b is a greatest common divisor of $b a_0, b a_1, \ldots, b a_n$. Now, from $(*)$, b is a greatest common divisor of all the coefficients of the various powers of x in $a \lambda(x)\mu(x)$. Since $\lambda(x), \mu(x)$ are primitive, by 1.3.2, $\lambda(x)\mu(x)$ is primitive, and hence a is a greatest common divisor of all the coefficients of the various powers of x in $a \lambda(x)\mu(x)$. Since a, b are positive integers and b is a greatest common divisor of all the coefficients of the various powers of x in $a \lambda(x)\mu(x)$, we have $a = b$. Since $a = b$, by $(*)$, we have

$$a_0 + a_1 x + \cdots + a_n x^n = \lambda(x)\mu(x).$$

Also, $\lambda(x), \mu(x)$ are primitive polynomials. ∎

1.3.4 Problem Let $a_0 + a_1 x + \cdots + a_n x^n (\in \mathbb{Z}[x])$ be a primitive polynomial. Let $p(\geq 2)$ be a prime number. Suppose that $p \mid a_i (i = 0, 1, \ldots, n-1)$, $p \nmid a_n$, and $p^2 \nmid a_0$. Then $a_0 + a_1 x + \cdots + a_n x^n$ is irreducible over $\mathbb{Q}$, that is, $a_0 + a_1 x + \cdots + a_n x^n$ cannot be factored into two nontrivial polynomials with rational numbers as coefficients.

Proof Suppose to the contrary that

$$a_0 + a_1 x + \cdots + a_n x^n = (r_0 + r_1 x + \cdots + r_m x^m)(s_0 + s_1 x + \cdots + s_{n-m} x^{n-m}),$$

where each r_i is a rational number and each s_j is a rational number. We seek a contradiction.

By 1.3.3, there exist two polynomials $b_0 + b_1 x + \cdots + b_m x^m, c_0 + c_1 x + \cdots + c_{n-m} x^{n-m} \in \mathbb{Z}[x]$ such that

$$a_0 + a_1 x + \cdots + a_n x^n = (b_0 + b_1 x + \cdots + b_m x^m)(c_0 + c_1 x + \cdots + c_{n-m} x^{n-m}).$$

Since $p \mid a_i (i = 0, 1, \ldots, n - 1)$, the above equality takes the form

$$p(a_0' + a_1' x + \cdots + a_{n-1}' x^{n-1}) + a_n x^n = (b_0 + b_1 x + \cdots + b_m x^m)$$
$$(c_0 + c_1 x + \cdots + c_{n-m} x^{n-m}) \qquad (*).$$

Since $p \nmid a_n$, p does not divide any greatest common divisor of $a_0, a_1, \ldots, a_n$. Here $p a_0' = b_0 c_0$, and p is a prime, so $p \mid b_0$ or $p \mid c_0$. Since $p^2 \nmid a_0$ and $a_0 = p a_0' = b_0 c_0$, we have $p^2 \nmid b_0 c_0$, and hence $p \mid b_0$ and $p \mid c_0$ cannot be true simultaneously.
So for the sake of definiteness, suppose that $p \mid b_0$, and $p \nmid c_0$.
Now $(*)$ takes the form

$$p(a_0' + a_1' x + \cdots + a_{n-1}' x^{n-1}) + a_n x^n = (p(\cdots) + b_1 x + \cdots + b_m x^m)$$
$$(c_0 + c_1 x + \cdots + c_{n-m} x^{n-m}).$$

Since $p \nmid a_n$, from $(*)$, we find that there exists $j \in \{1, 2, \ldots, m\}$ such that

1. $p \nmid b_j$,
2. $p \mid b_k (k = 0, 1, \ldots, j - 1)$.

Now we can write

$$p(a_0' + a_1' x + \cdots + a_{n-1}' x^{n-1}) + a_n x^n (p(\cdots) + p(\cdots)x + \cdots + p(\cdots)x^{j-1}$$
$$+ b_j x^j + \cdots + b_m x^m)(c_0 + c_1 x + \cdots + c_{n-m} x^{n-m}).$$

It follows that

$$p a_j' = b_j c_0 + p(\cdots)c_1 + p(\cdots)c_2 + \cdots + p(\cdots)c_j.$$

This shows that $p \mid b_j c_0$. Now, since p is a prime number, either $p \mid b_j$ or $p \mid c_0$. This is a contradiction.   ∎

1.3.5 Theorem Let $a_0 + a_1 x + \cdots + a_n x^n (\in \mathbb{Z}[x])$ be a polynomial. Let $p(\geq 2)$ be a prime number. Suppose that $p \mid a_i (i = 0, 1, \ldots, n - 1)$, $p \nmid a_n$ and $p^2 \nmid a_0$. Then $a_0 + a_1 x + \cdots + a_n x^n$ is irreducible over $\mathbb{Q}$, that is, $a_0 + a_1 x + \cdots + a_n x^n$ cannot be factored into two nontrivial polynomials with rational numbers as coefficients.
This result is known as **Eisenstein's criterion**.

Proof Let d be the positive greatest common divisor of $a_0, a_1, \ldots, a_n$. We can write

$$a_0 + a_1 x + \cdots + a_n x^n = d(a_0' + a_1' x + \cdots + a_n' x^n),$$

where the positive greatest common divisor of $a'_0, a'_1, \ldots, a'_n$ is 1. It follows that $a'_0 + a'_1 x + \cdots + a'_n x^n$ is a primitive polynomial. Since, $p \nmid a_n$, p does not divide the positive greatest common divisor d of $a_0, a_1, \ldots, a_n$.

Since $p \mid a_0$, and $a_0 = da'_0$, we have $p \mid da'_0$. Now, since p is a prime and $p \nmid d$, we have $p \mid a'_0$.

Since $p \mid a_1$, and $a_1 = da'_1$, we have $p \mid da'_1$. Since p is a prime and $p \nmid d$, we have $p \mid a'_1$. Similarly, $p \mid a'_2$, etc. Thus $p \mid a'_i (i = 0, 1, \ldots, n - 1)$.

Since $p \nmid a_n$ and $a_n = da'_n$, we have $p \nmid da'_n$. It follows that $p \nmid a'_n$.

Since $p^2 \nmid a_0$, and $a_0 = da'_0$, we have $p^2 \nmid da'_0$. It follows that $p^2 \nmid a'_0$.

Now, by 1.3.4, $a'_0 + a'_1 x + \cdots + a'_n x^n$ is irreducible over $\mathbb{Q}$, and hence $d(a'_0 + a'_1 x + \cdots + a'_n x^n)$ is irreducible over $\mathbb{Q}$. Since

$$a_0 + a_1 x + \cdots + a_n x^n = d(a'_0 + a'_1 x + \cdots + a'_n x^n),$$

$a_0 + a_1 x + \cdots + a_n x^n$ is irreducible over $\mathbb{Q}$.   ∎

1.3.6 Definition Let R be a commutative ring with unit element 1. We know that $R[x_1]$ is a commutative ring with unit element 1. Since $R[x_1]$ is a commutative ring with unit element 1, $(R[x_1])[x_2]$ is also a commutative ring with unit element 1. Here $(R[x_1])[x_2]$ is denoted by $R[x_1, x_2]$.

Observe that the elements of $R[x_1, x_2] (= (R[x_1])[x_2])$ are of the form

$$\left(a_{00} + a_{10}x_1 + a_{20}(x_1)^2 + \cdots\right) + \left(a_{01} + a_{11}x_1 + a_{21}(x_1)^2 + \cdots\right)x_2$$
$$+ \left(a_{02} + a_{12}x_1 + a_{22}(x_1)^2 + \cdots\right)(x_2)^2 + \cdots,$$

that is,

$$a_{00} + (a_{10}x_1 + a_{01}x_2) + \left(a_{20}(x_1)^2 + a_{11}x_1x_2 + a_{02}(x_2)^2\right)$$
$$+ \left(a_{30}(x_1)^3 + a_{21}(x_1)^2 x_2 + a_{12}x_1(x_2)^2 + a_{03}(x_2)^3\right) + \cdots,$$

that is,

$$\sum_{i+j=0} \left(a_{ij}(x_1)^i(x_2)^j\right) + \sum_{i+j=1} \left(a_{ij}(x_1)^i(x_2)^j\right) + \sum_{i+j=2} \left(a_{ij}(x_1)^i(x_2)^j\right) + \cdots.$$

Thus each member of the ring $R[x_1, x_2]$ is of the form

$$\sum_{i+j=0} \left(a_{ij}(x_1)^i(x_2)^j\right) + \sum_{i+j=1} \left(a_{ij}(x_1)^i(x_2)^j\right) + \sum_{i+j=2} \left(a_{ij}(x_1)^i(x_2)^j\right) + \cdots.$$

Definition Let R be a commutative ring with unit element 1. We know that $R[x_1, x_2]$ is a commutative ring with unit element 1. Since $R[x_1, x_2]$ is a commutative ring with unit element 1, $(R[x_1, x_2])[x_3]$ is also a commutative ring with unit element 1. Here $(R[x_1, x_2])[x_3]$ is denoted by $R[x_1, x_2, x_3]$.

Observe that the elements of $R[x_1, x_2, x_3](= (R[x_1, x_2])[x_3])$ are of the form

$$
\left(a_{000} + (a_{100}x_1 + a_{010}x_2) + \left(a_{200}(x_1)^2 + a_{110}x_1x_2 + a_{020}(x_2)^2 \right) + \cdots \right)
$$
$$
+ \left(a_{001} + (a_{101}x_1 + a_{011}x_2) + \left(a_{201}(x_1)^2 + a_{111}x_1x_2 + a_{021}(x_2)^2 \right) + \cdots \right) x_3
$$
$$
+ \left(a_{002} + (a_{102}x_1 + a_{012}x_2) + \left(a_{202}(x_1)^2 + a_{112}x_1x_2 + a_{022}(x_2)^2 \right) + \cdots \right) (x_3)^2 + \cdots,
$$

that is,

$$
a_{000} + (a_{100}x_1 + a_{010}x_2 + a_{001}x_3) + \left(a_{200}(x_1)^2 + a_{110}x_1x_2 + a_{020}(x_2)^2 \right.
$$
$$
\left. + a_{101}x_1x_3 + a_{011}x_2x_3 + a_{002}(x_3)^2 \right) + \cdots,
$$

that is,

$$
\sum_{i+j+k=0} \left(a_{ijk}(x_1)^i (x_2)^j (x_3)^k \right) + \sum_{i+j+k=1} \left(a_{ijk}(x_1)^i (x_2)^j (x_3)^k \right)
$$
$$
+ \sum_{i+j+k=2} \left(a_{ijk}(x_1)^i (x_2)^j (x_3)^k \right) + \cdots.
$$

Similar definitions can be supplied for $R[x_1, x_2, x_3, x_4]$, etc. The commutative ring $R[x_1, \ldots, x_n]$ with unit element 1 is called the *ring of polynomials in n variables* $x_1, \ldots, x_n$ *over R*.

1.3.7 Problem Let R be an integral domain. Then $R[x]$ is an integral domain. And hence $R[x_1, \ldots, x_n]$ is an integral domain.

Proof Let $a_0 + a_1x + a_2x^2 + \cdots$ and $b_0 + b_1x + b_2x^2 + \cdots$ be any two nonzero members of $R[x]$. It suffices to show that their product

$$
(a_0b_0) + (a_1b_0 + a_0b_1)x + (a_2b_0 + a_1b_1 + a_0b_2)x^2 + \cdots
$$

is nonzero.

Since $a_0 + a_1x + a_2x^2 + \cdots$ is nonzero, there exists $j \in \{0, 1, 2, \ldots\}$ such that

1. a_j is nonzero,
2. $a_l = 0 (l = 0, 1, \ldots, j - 1)$.

Similarly, there exists $k \in \{0, 1, 2, \ldots\}$ such that

1′. b_k is nonzero,

2′. $b_l = 0 (l = 0, 1, \ldots, k - 1)$.

It follows that

$$a_0 + a_1 x + a_2 x^2 + \cdots = a_j x^j + a_{j+1} x^{j+1} + \cdots$$

and

$$b_0 + b_1 x + b_2 x^2 + \cdots = b_k x^k + b_{k+1} x^{k+1} + \cdots,$$

and hence the coefficient of x^{j+k} in

$$\left(a_0 + a_1 x + a_2 x^2 + \cdots \right) \left(b_0 + b_1 x + b_2 x^2 + \cdots \right)$$

is $a_j b_k$. It suffices to show that $a_j b_k$ is nonzero. Since a_j is a nonzero member of R, b_k is a nonzero member of R, and R is an integral domain, it follows that $a_j b_k$ is nonzero. ∎

Definition Let F be a field. By 1.3.7, $F[x_1, \ldots, x_n]$ is an integral domain. Now we can construct its field of quotients. This field is denoted by $F(x_1, \ldots, x_n)$ and is called the *field of rational functions in* $x_1, \ldots, x_n$ *over* F.

The field $F(x_1, \ldots, x_n)$ is important in algebraic geometry.

Definition Let R be an integral domain with unit element 1. Let p be a nonzero member of R that is not a unit. If

$$(a, b \in R \text{ and } p = ab) \Rightarrow (a \text{ is a unit or } b \text{ is a unit}),$$

then we say that p *is irreducible* (or p *is a prime element of* R).

Definition Let R be an integral domain with unit element 1. If

a. every nonzero member of R that is not a unit can be written as a product of finitely many irreducible elements of R,

b. the decomposition in part (a) is unique up to the order and associates of the irreducible elements of R, then we say that R *is a unique factorization domain.*

(∗) Since every nonzero member of a field is a unit, every field is an example of a unique factorization domain.

1.3.8 Problem Let R be a unique factorization domain with unit element 1. Let $a, b \in R$. Then clearly, a greatest common divisor of a and b exists in R.

Definition Let R be a unique factorization domain with unit element 1. Let a, b be nonzero members of R (By 1.3.8, a greatest common divisor of a and b exists in R.).

If there exists a unit u in R such that u is a greatest common divisor of a and b, then we say that *a and b are relatively prime*.

1.3.9 Problem Let R be a unique factorization domain with unit element 1. Let a, b, c be any nonzero elements of R. Suppose that $a|bc$. Let a and b be relatively prime. Then clearly, $a|c$.

1.3.10 Problem Let R be a unique factorization domain with unit element 1. Let a, b, c be any nonzero elements of R. Suppose that a is an irreducible element. Suppose that $a|bc$. Then clearly, either $a|b$ or $a|c$.

Definition Let R be a unique factorization domain with unit element 1. Let $a_0 + a_1 x + \cdots + a_n x^n$ be a member of the ring $R[x]$, where each a_i is in R. If 1 is a greatest common divisor of $a_0, a_1, \ldots, a_n$, then we say that $a_0 + a_1 x + \cdots + a_n x^n$ is *a primitive polynomial in $R[x]$*.

1.3.11 Problem Let R be a unique factorization domain with unit element 1. Let $a_0 + a_1 x + \cdots + a_n x^n$ and $b_0 + b_1 x + \cdots + b_m x^m$ be two primitive polynomials in $R[x]$. Their product is

$$c_0 + c_1 x + \cdots + c_{n+m} x^{n+m},$$

where

$$\left. \begin{array}{l} c_0 \equiv a_0 b_0 \\ c_1 \equiv a_1 b_0 + a_0 b_1 \\ c_2 \equiv a_2 b_0 + a_1 b_1 + a_0 b_2 \\ \vdots \end{array} \right\} .$$

Then $c_0 + c_1 x + \cdots + c_{n+m} x^{n+m}$ is primitive.

Proof Suppose to the contrary that $c_0 + c_1 x + \cdots + c_{n+m} x^{n+m}$ is imprimitive. We seek a contradiction.

Since $c_0 + c_1 x + \cdots + c_{n+m} x^{n+m}$ is not primitive, 1 is not a greatest common divisor of $c_0, c_1, \ldots, c_{n+m}$, and hence there exists an irreducible p such that $p|c_i (i = 0, 1, \ldots, n+m)$. Since p is irreducible, p is not a unit. Since p is not a unit and 1 is a greatest common divisor of $a_0, a_1, \ldots, a_n$, there exists $j \in \{0, 1, \ldots, n\}$ such that $p \nmid a_j$ and $p|a_l (l = 0, 1, \ldots, j-1)$. Similarly, there exists $k \in \{0, 1, \ldots, m\}$ such that $p \nmid b_k$ and $p|b_l (l = 0, 1, \ldots, k-1)$. Hence $a_0 + a_1 x + \cdots + a_n x^n$ is of the form

$$p(\cdots) + p(\cdots)x + p(\cdots)x^2 + \cdots + p(\cdots)x^{j-1} + a_j x^j + a_{j+1} x^{j+1} + \cdots,$$

and $b_0 + b_1 x + \cdots + b_m x^m$ is of the form

$$p(\cdots)+p(\cdots)x+p(\cdots)x^2 + \cdots +p(\cdots)x^{k-1} + b_k x^k + b_{k+1}x^{k+1} + \cdots.$$

Since p is irreducible, $p{\nmid}a_j$, $p{\nmid}b_k$, and R is a unique factorization domain, we have $p{\nmid}a_j b_k$. Further,

$$c_{j+k} = a_{j+k}b_0 + a_{j+k-1}b_1 + \cdots + a_0 b_{j+k}$$
$$= a_{j+k}b_0 + a_{j+k-1}b_1 + \cdots + a_{j+1}b_{k-1} + a_j b_k + a_{j-1}b_{k+1} + \cdots + a_0 b_{j+k},$$

so

$$a_j b_k = c_{j+k} - \left(a_{j+k}b_0 + a_{j+k-1}b_1 + \cdots + a_{j+1}b_{k-1}\right) - \left(a_{j-1}b_{k+1} + \cdots + a_0 b_{j+k}\right)$$
$$= c_{j+k} - \left(a_{j+k}p(\cdots) + a_{j+k-1}p(\cdots) + \cdots + a_{j+1}p(\cdots)\right)$$
$$- \left(p(\cdots)b_{k+1} + \cdots + p(\cdots)b_{j+k}\right)$$
$$= c_{j+k} - p(\cdots) - p(\cdots) = c_{j+k} - p(\cdots),$$

and hence

$$a_j b_k = c_{j+k} - p(\cdots).$$

Since $p|c_{j+k}$, we have $p|a_j b_k$. This is a contradiction. ∎

1.3.12 Note Let R be a unique factorization domain with unit element 1. It follows that R is an integral domain with unit element 1, and hence it has a field F of quotients. Since F is a field, by 1.3.7, $F[x]$ is an integral domain with unit element 1. Clearly, $R[x]$ can be considered a subring of $F[x]$. Let

$$\frac{a_0}{b_0} + \frac{a_1}{b_1}x + \frac{a_2}{b_2}x^2 + \cdots + \frac{a_n}{b_n}x^n$$

be any member of $F[x]$, where each a_i is a member of R and each b_i is a nonzero member of R. Now we can write

$$\frac{a_0}{b_0} + \frac{a_1}{b_1}x + \frac{a_2}{b_2}x^2 + \cdots + \frac{a_n}{b_n}x^n = \frac{1}{b}\left(c_0 + c_1 x + c_2 x^2 + \cdots + c_n x^n\right),$$

where $b \equiv b_0 b_1 b_2 \ldots b_n \in R$, $c_0 \equiv a_0 b_1 b_2 \ldots b_n \in R$, $c_1 \equiv b_0 a_1 b_2 \ldots b_n \in R, \ldots,$ $c_n \equiv b_0 b_1 b_2 \ldots b_{n-1}a_n \in R$. Let d be a greatest common divisor of $c_0, c_1, c_2, \ldots, c_n$. Hence

$$c_0 + c_1 x + c_2 x^2 + \cdots + c_n x^n$$

can be expressed as

$$d\left(d_0 + d_1 x + d_2 x^2 + \cdots + d_n x^n\right),$$

where each d_i is a member of R. Clearly, 1 is a greatest common divisor of $d_0, d_1, d_2, \ldots, d_n$, and hence $d_0 + d_1 x + d_2 x^2 + \cdots + d_n x^n$ is a primitive polynomial in $R[x]$.

1.3.13 Conclusion Let R be a unique factorization domain with unit element 1 and let F be its field of quotients. Then every member of $F[x]$ can be expressed as $\frac{d}{b} f(x)$, where $f(x) \in R[x]$, $b, d \in R$, and $f(x)$ is primitive in $R[x]$.

1.3.14 Problem Let R be a unique factorization domain with unit element 1, and let F be its field of quotients. Let $f(x) \in R[x]$. Suppose that

a. $f(x)$ is primitive as an element of $R[x]$,
b. $f(x)$ is irreducible as an element of $R[x]$.

Then $f(x)$ is irreducible as an element of $F[x]$.

Proof If not, then by 1.3.13, we can suppose to the contrary that

$$f(x) = \frac{d_1}{b_1} f_1(x) \cdot \frac{d_2}{b_2} f_2(x),$$

where $f_1(x), f_2(x) \in R[x]$, $b_1, b_2, d_1, d_2 \in R$, $f_1(x), f_2(x)$ are primitive as elements of $R[x]$, $\deg(f_1(x)) \geq 1$, and $\deg(f_2(x)) \geq 1$. We seek a contradiction.
Here

$$b_1 b_2 f(x) = d_1 d_2 g(x),$$

where $g(x) \equiv f_1(x) f_2(x)$. Now, since $f_1(x), f_2(x)$ are primitive as elements of $R[x]$, by 1.3.11, $g(x)$ is primitive as an element of $R[x]$. It follows that $d_1 d_2$ is a greatest common divisor of the coefficients of the various powers of x in $d_1 d_2 g(x)(=b_1 b_2 f(x))$. Thus $d_1 d_2$ is a greatest common divisor of the coefficients of the various powers of x in $b_1 b_2 f(x)$. Since $f(x)$ is primitive as an element of $R[x]$, $b_1 b_2$ is a greatest common divisor of the coefficients of the various powers of x in $b_1 b_2 f(x)$. Hence we can suppose that $d_1 d_2 = b_1 b_2 (\neq 0)$. Now, since $b_1 b_2 f(x) = d_1 d_2 f_1(x) f_2(x)$ and $R[x]$ is an integral domain, we have $f(x) = f_1(x) f_2(x)$. Next, since $\deg(f_1(x)) \geq 1$, $\deg(f_2(x)) \geq 1$, and $f_1(x), f_2(x) \in R[x]$, it follows that $f(x)$ is not irreducible as an element of $R[x]$. This is a contradiction. ∎

1.3.15 Problem Let R be a unique factorization domain with unit element 1, and let F be its field of quotients. Let $f(x) \in R[x]$. Suppose that

a. $f(x)$ is primitive as an element of $R[x]$,
b. $f(x)$ is irreducible as an element of $F[x]$.

Then $f(x)$ is irreducible as an element of $R[x]$.

Proof Suppose to the contrary that

$$f(x) = f_1(x) \cdot f_2(x),$$

where $f_1(x), f_2(x) \in R[x] (\subset F[x])$, $\deg(f_1(x)) \geq 1$, and $\deg(f_2(x)) \geq 1$. We seek a contradiction.

It follows that $f_1(x), f_2(x) \in F[x]$. Since $f(x) \in R[x] \subset F[x]$, we have $f(x) \in F[x]$. Since $f(x)$ is irreducible as an element of $F[x]$, $f(x) = f_1(x) \cdot f_2(x)$, and $f_1(x), f_2(x), f(x) \in F[x]$, we have

$(f_1(x)$ is a unit as an element of $F[x]$ or $f_2(x)$ is a unit as an element of $F[x])$,

that is, either $\deg(f_1(x)) \ngeq 1$ or $\deg(f_2(x)) \ngeq 1$. This is a contradiction. ∎

1.3.16 Problem Let R be an integral domain with unit element 1. We know, by 1.3.7, that $R[x]$ is also an integral domain with unit element 1. It is clear that
(∗) for every nonzero $f(x), g(x) \in R[x]$, $\deg(f(x)g(x)) = \deg(f(x)) + \deg(g(x))$.
Also, if $u(x)$ is a unit in $R[x]$, then $u(x)$ is also a unit in R.

Proof Let $u(x)$ be a unit in $R[x]$. We have to show that $u(x)$ is a unit in R. Since $u(x)$ is a unit in the integral domain $R[x]$, there exists a nonzero $v(x)$ in $R[x]$ such that

$$u(x)v(x) = 1,$$

and hence

$$0 \leq \deg(u(x)) + \deg(v(x)) = \underbrace{\deg(u(x)v(x)) = \deg(1)} = 0.$$

This shows that $\deg(u(x)) = 0$, and $\deg(v(x)) = 0$. So there exist nonzero α, β in R such that $u(x) = \alpha$ and $v(x) = \beta$. Now, since $u(x)v(x) = 1$, we have $\alpha\beta = 1$. It follows that $(u(x) =)\alpha$ is a unit in R, and hence $u(x)$ is a unit in R. ∎

1.3.17 Problem Let R be a unique factorization domain with unit element 1. Let $f(x)$ be a nonzero member of $R[x]$ having degree ≥ 1. Suppose that $f(x)$ is primitive as an element of $R[x]$. Then $f(x)$ can be expressed as a product of finitely many irreducible polynomials in $R[x]$.

Proof Since $R[x]$ can be considered a subring of $F[x]$, where F is the field of quotients of members of R, and $f(x) \in R[x]$, we can think of $f(x)$ as a nonzero member of $F[x]$.

Since $f(x)$ is a nonzero member of $R[x]$ having degree ≥ 1, $f(x)$ is not a unit in $F[x]$. By 1.2.20, $f(x)$ can be expressed as a product of finitely many irreducible polynomials $p_1(x), p_2(x), \ldots, p_n(x)$ of degree ≥ 1 in $F[x]$. By 1.3.13, we can suppose that

$$p_k(x) = \frac{d_k}{b_k} f_k(x) \quad (k = 1, 2, \ldots, n),$$

where $f_k(x) \in R[x]$, $b_k, d_k \in R$, and $f_k(x)$ is primitive in $R[x]$. Hence

$$b_1 b_2 \ldots b_n f(x) = d_1 d_2 \ldots d_n g(x),$$

where $g(x) \equiv f_1(x) f_2(x) \ldots f_n(x)$. Now, since $f_1(x), f_2(x), \ldots, f_n(x)$ are primitive as elements of $R[x]$, by 1.3.11, $g(x)$ is primitive as an element of $R[x]$. It follows that $d_1 d_2 \ldots d_n$ is a greatest common divisor of the coefficients of the various powers of x in $d_1 d_2 \ldots d_n g(x) (= b_1 b_2 \ldots b_n f(x))$. Thus $d_1 d_2 \ldots d_n$ is a greatest common divisor of the coefficients of the various powers of x in $b_1 b_2 \ldots b_n f(x)$. Since $f(x)$ is primitive as an element of $R[x]$, $b_1 b_2 \ldots b_n$ is a greatest common divisor of the coefficients of the various powers of x in $b_1 b_2 \ldots b_n f(x)$. Hence we can suppose that

$$d_1 d_2 \ldots d_n = b_1 b_2 \ldots b_n (\neq 0).$$

Now, since $b_1 b_2 \ldots b_n f(x) = d_1 d_2 \ldots d_n g(x)$ and $R[x]$ is an integral domain, we have $f(x) = g(x)$, and hence $f(x) = f_1(x) f_2(x) \ldots f_n(x)$. Since each $\left(\frac{d_k}{b_k} f_k(x) = \right)$ $p_k(x)$ is irreducible in $F[x]$, $f_k(x)$ is irreducible in $F[x]$. And since $f_k(x)$ is primitive in $R[x]$, by 1.3.15, $f_k(x)$ is irreducible as an element of $R[x]$.

Thus $f(x)$ is expressed as a product of finitely many irreducible polynomials in $R[x]$. ∎

1.3.18 Problem Let R be a unique factorization domain with unit element 1. Let $f(x)$ be a nonzero member of $R[x]$ having degree ≥ 1. Suppose that $f(x)$ is primitive as an element of $R[x]$. Then $f(x)$ can be expressed uniquely as a product of finitely many irreducible polynomials in $R[x]$.

Proof By 1.3.17, it suffices to prove only the uniqueness part of this theorem. To this end, let

$$f(x) = \pi_1(x) \pi_2(x) \ldots \pi_m(x),$$

where each $\pi_i(x) (i = 1, 2, \ldots, m)$ is an irreducible polynomial of degree ≥ 1 in $R[x]$. Let

$$f(x) = \pi_1'(x) \pi_2'(x) \cdots \pi_n'(x),$$

where each $\pi_j'(x) (j = 1, 2, \ldots, n)$ is an irreducible polynomial of degree ≥ 1 in $R[x]$. We have to show that

1. each $\pi_i(x)$ is an associate of some $\pi'_j(x)$ in $R[x]$,
2. each $\pi'_j(x)$ is an associate of some $\pi_i(x)$ in $R[x]$,
3. $n = m$.

Since $f(x)$ is primitive as an element of $R[x]$ and $f(x) = \pi_1(x)\pi_2(x)\ldots\pi_m(x)$, each $\pi_i(x)$ is primitive as an element of $R[x]$. Similarly, each $\pi'_j(x)$ is primitive as an element of $R[x]$. Now, since $\pi_1(x)$ is an irreducible polynomial in $R[x]$, by 1.3.14, $\pi_1(x)$ is irreducible as an element of $F[x]$, where F is the field of quotients of members of R. Similarly, each $\pi_i(x)$ is irreducible as an element of $F[x]$, and each $\pi'_j(x)$ is irreducible as an element of $F[x]$.

Since $f(x)$ is a nonzero member of $R[x] (\subset F[x])$ having degree ≥ 1, by 1.2.16, $f(x)$ is not a unit in $F[x]$. By 1.2.21,

1. each $\pi_i(x)$ is an associate of some $\pi'_j(x)$ in $F[x]$,
2. each $\pi'_j(x)$ is an associate of some $\pi_i(x)$ in $F[x]$,
3. $n = m$.

Suppose that $\pi_i(x)$ is an associate of some $\pi'_j(x)$ in $F[x]$. It follows that there exist $a, b \in R$ such that

$$\pi_i(x) = \frac{a}{b}\pi'_j(x),$$

and hence

$$b\pi_i(x) = a\pi'_j(x).$$

Since $\pi_i(x)$ is primitive as an element of $R[x]$, b is a greatest common divisor of coefficients of the various powers of x in $b\pi_i(x)\left(= a\pi'_j(x)\right)$. Thus b is a greatest common divisor of the coefficients of the various powers of x in $a\pi'_j(x)$. Since $\pi'_j(x)$ is primitive as an element of $R[x]$, a is a greatest common divisor of the coefficients of the various powers of x in $a\pi'_j(x)$. Hence we can suppose that

$$a = b(\neq 0).$$

Now, since $b\pi_i(x) = a\pi'_j(x)$ and $R[x]$ is an integral domain, $\pi_i(x) = \pi'_j(x)$, and hence $\pi_i(x)$ is an associate of some $\pi'_j(x)$ in $R[x]$. Thus each $\pi_i(x)$ is an associate of some $\pi'_j(x)$ in $R[x]$. Similarly, each $\pi'_j(x)$ is associate of some $\pi_i(x)$ in $R[x]$. ∎

1.3.19 Problem Let R be a unique factorization domain with unit element 1. Then $R[x]$ is also a unique factorization domain with unit element 1.

Proof Since R is a unique factorization domain with unit element 1, R is an integral domain with unit element 1, and hence by 1.3.7, $R[x]$ is an integral domain with unit element 1. It remains to show that

a. every nonzero member of $R[x]$ that is not a unit can be written as a product of finitely many irreducible elements of $R[x]$,

b. the decomposition in part (a) is unique up to the order and associates of the irreducible elements of $R[x]$.

To this end, let us take a nonzero member $f(x)$ of $R[x]$ that is not a unit in $R[x]$. We have to show that $f(x)$ can be expressed uniquely as a product of finitely many irreducible polynomials in $R[x]$.

Let d be a greatest common divisor of the coefficients of the various powers of x in $f(x)$. It follows that $f(x)$ takes the form $dg(x)$, where $g(x)$ is primitive as an element of $R[x]$.

Case I: $\deg(f(x)) \geq 1$. Since $f(x) = dg(x)$ and $d \in R$, by 1.3.16, we have $\deg(g(x)) \geq 1$. It follows, by 1.3.18, that $g(x)$ can be expressed uniquely as a product of finitely many irreducible polynomials in $R[x]$. Since $d \in R$, d is irreducible in $R[x]$. Now, since $f(x) = dg(x)$, $f(x)$ can be expressed uniquely as a product of finitely many irreducible polynomials in $R[x]$.

Case II: $\deg(f(x)) = 0$, that is, $f(x)$ is a nonzero member of R. It follows that $f(x)$ is irreducible in $R[x]$.

So in all cases, $f(x)$ can be expressed uniquely as a product of finitely many irreducible polynomials in $R[x]$.   ∎

1.3.20 Problem Let R be a unique factorization domain with unit element 1. Then $R[x_1, x_2, \cdots, x_n]$ is also a unique factorization domain with unit element 1.

Proof Since R is a unique factorization domain with unit element 1, by 1.3.19, $R[x_1]$ is also a unique factorization domain with unit element 1. Now, again by 1.3.19, $(R[x_1, x_2] =)(R[x_1])[x_2]$ is a unique factorization domain with unit element 1. and hence $R[x_1, x_2]$ is a unique factorization domain with unit element 1. Similarly, $R[x_1, x_2, x_3]$ is a unique factorization domain with unit element 1. Finally, $R[x_1, x_2, \ldots, x_n]$ is a unique factorization domain with unit element 1.   ∎

1.3.21 Problem Let F be a field. Then $F[x_1, x_2, \cdots, x_n]$ is a unique factorization domain.

Proof From 1.3.7, F is a unique factorization domain, and hence by 1.3.20, $F[x_1, x_2, \ldots, x_n]$ is a unique factorization domain.   ∎

1.4 Roots of Polynomials

1.4.1 Definition Let F be a field. Let K be a field such that $F \subset K$. If F is a subfield of K, then we say that K *is an extension of* F.

Examples

1. The field $\mathbb{R}$ of all real numbers is an extension of the field $\mathbb{Q}$ of all rational numbers.
2. The field $\mathbb{C}$ of all complex numbers is an extension of the field $\mathbb{R}$ of all real numbers.
3. The field $\{a + \sqrt{2}b : a, b \in \mathbb{Q}\}$ is an extension of $\mathbb{Q}$.

1.4.2 Problem Let F and K be any fields such that K is an extension of F. Let us treat every member of K as a vector and every member of F as a scalar. We define the operation of scalar multiplication as follows:

for every $f \in F(\subset K)$ and every $v \in K$, we say that the product fv in the field K is the result of scalar multiplication of the scalar f and vector v.

Then K is a vector space over the field F.

Proof It suffices to show that

1. for every $f_1, f_2 \in F$ and for every $v \in K$, $(f_1 + f_2)v = f_1v + f_2v$ and $(f_1f_2)v = f_1(f_2v)$,
2. for every $f \in F$ and for every $v, w \in K$, $f(v + w) = fv + fw$,
3. for every $v \in K$, $1v = v$.

For 1: Let us take arbitrary $f_1, f_2 \in F$ and $v \in K$. We have to show that $(f_1 + f_2)v = f_1v + f_2v$ and $(f_1f_2)v = f_1(f_2v)$.

We see that $(f_1 + f_2)v = f_1v + f_2v$ is trivially true, in view of the facts that $F \subset K$ and the right distributive law holds in the field K. Similarly, $(f_1f_2)v = f_1(f_2v)$ is trivially true, in view of the facts that $F \subset K$ and the associative law of multiplication holds in the field K.

For 2: Let us take arbitrary $f \in F$ and $v, w \in K$. We have to show that $f(v + w) = fv + fw$.

This is trivially true, in view of the facts that $F \subset K$ and the left distributive law holds in the field K.

For 3: Let us take an arbitrary $v \in K$. We have to show that $1v = v$.

Since F is a subfield of K, the unit element 1 of F is also the unit element of K. Now, $1v = v$ is trivially true, in view of the fact that $F \subset K$ and the existence of the unit element 1 in the field K. ∎

Definition Let F and K be any fields such that K is an extension of F. By 1.4.2, K is a vector space over the field F. If the dimension of this vector space is finite, then we say that K *is a finite extension of* F. In this case, the dimension of the vector space K over F is denoted by $[K : F]$ and is called the *degree of* K *over* F.

Example: We have seen that the field $\mathbb{C}$ of all complex numbers is an extension of the field $\mathbb{R}$ of all real numbers. Hence $\mathbb{C}$ is a vector space over the field $\mathbb{R}$. Here

$\{1, \sqrt{-1}\} \subset \mathbb{C}$, and every member of $\mathbb{C}$ can be expressed as a linear combination of vectors $1, \sqrt{-1}$. Further $\{1, \sqrt{-1}\}$ is a linearly independent set of vectors, in the sense that

$$\left(a1 + b\sqrt{-1} = 0, \text{and } a, b \in \mathbb{R} \right) \Rightarrow (a = 0 \text{ and } b = 0).$$

Thus $\{1, \sqrt{-1}\}$ is a basis of $\mathbb{C}$. Since the number of elements in the basis $\{1, \sqrt{-1}\}$ is 2, the dimension of the vector space $\mathbb{C}$ over $\mathbb{R}$ is 2, which is of course finite. Hence $\mathbb{C}$ is a finite extension of $\mathbb{R}$. Also, $[\mathbb{C} : \mathbb{R}] = 2$.

1.4.3 Problem Let F, K, and L be any fields such that $F \subset K \subset L$. Suppose that K is a finite extension of F and L is a finite extension of K. Then

a. L is a finite extension of F,
b. $[L : F] = [L : K][K : F]$.

Proof Let $[L : K] = m$ and $[K : F] = n$. It suffices to construct a basis of the vector space L over F that has mn elements. For the sake of simplicity, let us take $m = 2$, and $n = 3$.

Since $[L : K] = 2$, there exists a basis $\{v_1, v_2\}(\subset L)$ of the vector space L over K. Similarly, there exists a basis $\{w_1, w_2, w_3\}(\subset K \subset L)$ of the vector space K over F. It follows that $\{w_1, w_2, w_3\} \subset L$ and $\{v_1, v_2\} \subset L$. Since L is a field, $\{v_1 w_1, v_1 w_2, v_1 w_3, v_2 w_1, v_2 w_2, v_2 w_3\} \subset L$. It suffices to show that $\{v_1 w_1, v_1 w_2, v_1 w_3, v_2 w_1, v_2 w_2, v_2 w_3\}$ is a basis of the vector space L over F. To this end, we must prove the following:

1. every element of L can be expressed as a linear combination of $v_1 w_1, v_1 w_2, v_1 w_3, v_2 w_1, v_2 w_2, v_2 w_3$ with coefficients in F,
2. $\{v_1 w_1, v_1 w_2, v_1 w_3, v_2 w_1, v_2 w_2, v_2 w_3\}$ is a linearly independent set of vectors in the vector space L over F.

For 1: Let us take an arbitrary $v \in L$. Now, since $\{v_1, v_2\}$ is a basis of the vector space L over K, there exist $k_1, k_2 \in K$ such that $v = k_1 v_1 + k_2 v_2$. Since $k_1 \in K$, and $\{w_1, w_2, w_3\}$ is a basis of the vector space K over F, there exist $f_{11}, f_{12}, f_{13} \in F$ such that $k_1 = f_{11} w_1 + f_{12} w_2 + f_{13} w_3$. Similarly, there exist $f_{21}, f_{22}, f_{23} \in F$ such that $k_2 = f_{21} w_1 + f_{22} w_2 + f_{23} w_3$. Hence

$$v = (f_{11} w_1 + f_{12} w_2 + f_{13} w_3)v_1 + (f_{21} w_1 + f_{22} w_2 + f_{23} w_3)v_2,$$

that is,

$$v = (f_{11} w_1 v_1 + f_{12} w_2 v_1 + f_{13} w_3 v_1) + (f_{21} w_1 v_2 + f_{22} w_2 v_2 + f_{23} w_3 v_2),$$

that is,

$$v = f_{11}(v_1 w_1) + f_{12}(v_1 w_2) + f_{13}(v_1 w_3) + f_{21}(v_2 w_1) + f_{22}(v_2 w_2) + f_{23}(v_2 w_3).$$

Thus v is expressed as a linear combination of $v_1 w_1, v_1 w_2, v_1 w_3, v_2 w_1, v_2 w_2, v_2 w_3$ having coefficients in F.

For 2: Suppose that

$$f_{11}(v_1 w_1) + f_{12}(v_1 w_2) + f_{13}(v_1 w_3) + f_{21}(v_2 w_1) + f_{22}(v_2 w_2) + f_{23}(v_2 w_3) = 0,$$

where each $f_{ij} \in F$. We have to show that each f_{ij} is zero. We have

$$(f_{11}(v_1 w_1) + f_{12}(v_1 w_2) + f_{13}(v_1 w_3)) + (f_{21}(v_2 w_1) + f_{22}(v_2 w_2) + f_{23}(v_2 w_3)) = 0,$$

that is,

$$(f_{11} w_1 + f_{12} w_2 + f_{13} w_3) v_1 + (f_{21} w_1 + f_{22} w_2 + f_{23} w_3) v_2 = 0 \, (*).$$

Since each $f_{ij} \in F(\subset K)$, each $w_k \in K$, and K is a field, it follows that $(f_{11} w_1 + f_{12} w_2 + f_{13} w_3) \in K$. Similarly, $(f_{21} w_1 + f_{22} w_2 + f_{23} w_3) \in K$. Since $\{v_1, v_2\}$ is a basis of the vector space L over K, $\{v_1, v_2\}$ is linearly independent. Now from $(*)$,

$$\left. \begin{array}{l} f_{11} w_1 + f_{12} w_2 + f_{13} w_3 = 0 \\ f_{21} w_1 + f_{22} w_2 + f_{23} w_3 = 0 \end{array} \right\}.$$

Since $\{w_1, w_2, w_3\}$ is a basis of the vector space K over F, $\{w_1, w_2, w_3\}$ is linearly independent. Since

$$f_{11} w_1 + f_{12} w_2 + f_{13} w_3 = 0,$$

we have $f_{11} = f_{12} = f_{13} = 0$. Similarly, $f_{21} = f_{22} = f_{23} = 0$. ∎

1.4.4 Problem Let F, K, and L be any fields such that L is a finite extension of F, K is an extension of F, and L is an extension of K. Then

a. K is a finite extension of F, and L is a finite extension of K,

b. $[K : F] | [L : F]$.

Proof Since L is a finite extension of F, the dimension of the vector space L over F is finite. So let $\{v_1, v_2, \ldots, v_n\} (\subset L)$ be a basis of the vector space L over K. Since K is an extension of F, it follows by 1.4.2 that K is a vector space over the field F. Since $K \subset L$, K is a vector space over the field F, and L is a vector space over the field F, we have that K is a linear subspace of L. Since $\{v_1, v_2, \ldots, v_n\} (\subset L)$ is a basis of the vector space L over F, the dimension of the vector space K over F is $\leq n$, and hence K is a finite extension of F.

Since $\{v_1, v_2, \ldots, v_n\} (\subset L)$ is a basis of the vector space L over F, each element of L is a linear combination of $v_1, v_2, \ldots, v_n$ with coefficients in $F(\subset K)$, and hence each element of L is a linear combination of $v_1, v_2, \ldots, v_n$ with coefficients in K. This shows that the dimension of the vector space L over K is $\leq n$, and hence L is a finite extension of K. This proves (a). Now, since K is a finite extension of F, by 1.4.3, $[L : F] = [L : K][K : F]$, and hence $[K : F] \,|\, [L : F]$. This proves (b). ∎

Definition Let F and K be any fields such that K is an extension of F. Let $\alpha \in K$. If there exists a nonzero polynomial $q(x) \in F[x]$ such that $(K \ni)q(\alpha) = 0$, then we say that α *is algebraic over* F (Caution: Here the polynomial $q(x)$ is a symbol, while $q(\alpha)$ is a member of the field K.).

1.4.5 Problem Let F and K be any fields such that K is an extension of F. Then every element of F is algebraic over F.

Proof Let us take an arbitrary $\alpha \in F$. We have to show that α is algebraic over F. Let us take $\alpha + (-1)x$ for $q(x)(\in F[x])$. Clearly, $q(\alpha) = 0$. Thus, α is algebraic over F. ∎

1.4.6 Problem Let F and K be any fields such that K is an extension of F. Let $\alpha \in K$. Let $\mathcal{M}$ be the collection of all fields L satisfying

a. $F \cup \{\alpha\} \subset L \subset K$,
b. L is a subfield of K.

Clearly, $K \in \mathcal{M}$, and hence $\mathcal{M}$ is a nonempty collection. Also, $\cap \mathcal{M}$ is a member of $\mathcal{M}$.

Thus, $\cap \mathcal{M}$ is the smallest member of $\mathcal{M}$.

Proof Since each member of $\mathcal{M}$ is a field, $\cap \mathcal{M}$ is also a field. Since each member of $\mathcal{M}$ contains $F \cup \{\alpha\}$, $\cap \mathcal{M}$ also contains $F \cup \{\alpha\}$. Since each member of $\mathcal{M}$ is contained in K, $\cap \mathcal{M}$ is also contained in the field K. Now, since $\cap \mathcal{M}$ is a field, $\cap \mathcal{M}$ is a subfield of K. Thus by the definition of $\mathcal{M}$, $\cap \mathcal{M}$ is a member of $\mathcal{M}$. ∎

1.4.7 Problem Let F and K be any fields such that K is an extension of F. Let $\alpha \in K$. Let N be the set of all elements of K of the form $(g(\alpha))^{-1}f(\alpha)$, where $f(x), g(x)$ are members of $F[x]$ and $g(\alpha)$ is a nonzero member of K. Then N is a field.

Proof Let $(g(\alpha))^{-1}f(\alpha) \in N$, where $f(x), g(x)$ are members of $F[x]$ and $g(\alpha)$ is a nonzero member of K. Let $(g_1(\alpha))^{-1}f_1(\alpha) \in N$, where $f_1(x), g_1(x)$ are members of $F[x]$ and $g_1(\alpha)$ is a nonzero member of K. It suffices to show the following:

1. $\left((g(\alpha))^{-1}f(\alpha) + (g_1(\alpha))^{-1}f_1(\alpha) \right) \in N$,

2. $\left((g(\alpha))^{-1}f(\alpha) \right)\left((g_1(\alpha))^{-1}f_1(\alpha) \right) \in N$,

3. if $(g(\alpha))^{-1}f(\alpha), (g_1(\alpha))^{-1}f_1(\alpha)$ are nonzero elements of N, then their product is nonzero,

4. $1 \in N$,

5. if $(g(\alpha))^{-1}f(\alpha)$ is a nonzero element of N, then there exists β in N such that $\left((g(\alpha))^{-1}f(\alpha) \right)\beta = 1$.

For 1: Observe that

$$(g(\alpha))^{-1}f(\alpha) + (g_1(\alpha))^{-1}f_1(\alpha) = (g(\alpha))^{-1}(g_1(\alpha))^{-1}(f(\alpha)g_1(\alpha) + g(\alpha)f_1(\alpha))$$
$$= (g(\alpha)g_1(\alpha))^{-1}(f(\alpha)g_1(\alpha) + g(\alpha)f_1(\alpha)) = (g(\alpha)g_1(\alpha))^{-1}(k_1(\alpha) + k_2(\alpha)),$$

where $k_1(x) \equiv f(x)g_1(x) (\in F[x])$ and $k_2(x) \equiv g(x)f_1(x) (\in F[x])$, and hence

$$(g(\alpha))^{-1}f(\alpha) + (g_1(\alpha))^{-1}f_1(\alpha) = (g(\alpha)g_1(\alpha))^{-1}k(\alpha),$$

where $k(x) \equiv (k_1(x) + k_2(x)) (\in F[x])$. Since $g(x), g_1(x)$ are members of $F[x]$, $h(x)$ is a member of $F[x]$, where $h(x) \equiv g(x)g_1(x)$. It follows that $h(\alpha) = g(\alpha)g_1(\alpha)$. Since $g(\alpha), g_1(\alpha)$ are nonzero members of K and K. is a field, $(h(\alpha) =)g(\alpha)g_1(\alpha)$ is a nonzero member of K, and hence $h(\alpha)$ is a nonzero member of K. Thus

$$(g(\alpha))^{-1}f(\alpha) + (g_1(\alpha))^{-1}f_1(\alpha) = (h(\alpha))^{-1}k(\alpha),$$

where $h(x), k(x)$ are members of $F[x]$ and $h(\alpha)$ is a nonzero member of K. It follows that

$$\left((g(\alpha))^{-1}f(\alpha) + (g_1(\alpha))^{-1}f_1(\alpha) =\right)(h(\alpha))^{-1}k(\alpha) \in N,$$

and hence

$$\left((g(\alpha))^{-1}f(\alpha) + (g_1(\alpha))^{-1}f_1(\alpha)\right) \in N.$$

For 2: Observe that

$$\left((g(\alpha))^{-1}f(\alpha)\right)\left((g_1(\alpha))^{-1}f_1(\alpha)\right) = (g(\alpha))^{-1}(g_1(\alpha))^{-1}(f(\alpha)f_1(\alpha))$$
$$= (g(\alpha)g_1(\alpha))^{-1}(f(\alpha)f_1(\alpha)) = (g(\alpha)g_1(\alpha))^{-1}(k(\alpha)),$$

where $k(x) \equiv f(x)f_1(x) (\in F[x])$. Since $g(x), g_1(x)$ are members of $F[x]$, $h(x)$ is a member of $F[x]$, where $h(x) \equiv g(x)g_1(x)$. It follows that $h(\alpha) = g(\alpha)g_1(\alpha)$. Since $g(\alpha), g_1(\alpha)$ are nonzero members of K, and K is a field, $(h(\alpha) =)g(\alpha)g_1(\alpha)$ is a nonzero member of K, and hence $h(\alpha)$ is a nonzero member of K. Thus

$$\left((g(\alpha))^{-1}f(\alpha)\right)\left((g_1(\alpha))^{-1}f_1(\alpha)\right) = (h(\alpha))^{-1}k(\alpha),$$

where $h(x), k(x)$ are members of $F[x]$, and $h(\alpha)$ is a nonzero member of K. It follows that

$$\left(\left((g(\alpha))^{-1}f(\alpha)\right)\left((g_1(\alpha))^{-1}f_1(\alpha)\right) =\right)(h(\alpha))^{-1}k(\alpha) \in N,$$

and hence

$$\left(\left((g(\alpha))^{-1}f(\alpha)\right)\left((g_1(\alpha))^{-1}f_1(\alpha)\right)\right) \in N.$$

For 3: Let $(g(\alpha))^{-1}f(\alpha), (g_1(\alpha))^{-1}f_1(\alpha)$ be nonzero elements of N. We have to show that

$$\left(\left((g(\alpha))^{-1}f(\alpha)\right)\left((g_1(\alpha))^{-1}f_1(\alpha)\right)\right)$$

is a nonzero element of N. Suppose to the contrary that

$$\left(\left((g(\alpha))^{-1}f(\alpha)\right)\left((g_1(\alpha))^{-1}f_1(\alpha)\right)\right) = 0. (*)$$

We seek a contradiction. We have seen above that

$$\left(\left((g(\alpha))^{-1}f(\alpha)\right)\left((g_1(\alpha))^{-1}f_1(\alpha)\right)\right) = (h(\alpha))^{-1}k(\alpha),$$

where $h(x) \equiv g(x)g_1(x) (\in F[x])$, $k(x) \equiv f(x)f_1(x) (\in F[x])$, and $h(\alpha)$ is a nonzero member of K. Now from $(*)$, $(h(\alpha))^{-1}k(\alpha) = 0$. Since $h(\alpha)$ is a nonzero member of the field K, we have $(f(\alpha)f_1(\alpha) =)k(\alpha) = 0$, and hence either $f(\alpha) = 0$ or $f_1(\alpha) = 0$. It follows that either $(g(\alpha))^{-1}f(\alpha) = 0$ or $(g_1(\alpha))^{-1}f_1(\alpha) = 0$. This is a contradiction.

For 4: Let us take the constant polynomial 1 for $f(x)$, and again the constant polynomial 1 for $g(x)$. Clearly, $(1 = 1^{-1}1 =)(g(\alpha))^{-1}f(\alpha) \in N$, and hence $1 \in N$.

For 5: Let us take an arbitrary nonzero element $(g(\alpha))^{-1}f(\alpha)$ of $N(\subset K)$, where $f(x), g(x)$ are members of $F[x]$, and $g(\alpha)$ is a nonzero member of K. Since $g(\alpha)$ is a nonzero member of the field K, $(g(\alpha))^{-1}$ is a nonzero member of K. Next, since $(g(\alpha))^{-1}f(\alpha)$ is a nonzero member of the field K, $f(\alpha)$ is a nonzero member of the field K. This shows that $(f(\alpha))^{-1}g(\alpha) \in N$. Further, it is clear that $\left((g(\alpha))^{-1}f(\alpha)\right)$ $\left((f(\alpha))^{-1}g(\alpha)\right) = 1$. Hence $(f(\alpha))^{-1}g(\alpha)$ serves the purpose of β. ∎

1.4.8 Problem Let F and K be any fields such that K is an extension of F. Let $\alpha \in K$. Let N be the symbol as described in 1.4.7, and $\mathcal{M}$ the symbol as described in 1.4.6. Then $N = \cap \mathcal{M}$.

Proof We must prove:

1. $N \subset \cap \mathcal{M}$,
2. $\cap \mathcal{M} \subset N$.

For 1: By 1.4.6, $\cap \mathcal{M}$ is a member of $\mathcal{M}$, so it suffices to show that every member of $\mathcal{M}$ contains N. To this end, let us take an arbitrary $L \in \mathcal{M}$. We have to show that $N \subset L$.

Next let us take an arbitrary $(g(\alpha))^{-1}f(\alpha)$, where $f(x), g(x)$ are members of $F[x]$, and $g(\alpha)$ is a nonzero member of K. We have to show that $(g(\alpha))^{-1}f(\alpha) \in L$.

Since $L \in \mathcal{M}$, by the definition of $\mathcal{M}$, L is a field satisfying

a. $F \cup \{\alpha\} \subset L \subset K$,
b. L is a subfield of K.

Since $f(x)$ is a member of $F[x]$ and L is a field containing $F \cup \{\alpha\}$, we have $f(\alpha) \in L$. Similarly, $g(\alpha) \in L$. Now, since $g(\alpha)$ is nonzero, $(g(\alpha))^{-1} \in L$. Next, since $f(\alpha) \in L$ and L is a field, we have $(g(\alpha))^{-1}f(\alpha) \in L$.

For 2: By 1.4.6, $\cap \mathcal{M}$ is the smallest member of $\mathcal{M}$, so it suffices to show that N is a member of $\mathcal{M}$. By the definition of $\mathcal{M}$, we must prove:

a. N is a field,
b. $F \cup \{\alpha\} \subset N \subset K$,
c. N is a subfield of K.

For a: By 1.4.7, N is a field.
For b: Let us take an arbitrary $a \in F$. We want to show that $a \in N$. To this end, let us take the constant polynomial a as $f(x) (\in F[x])$, and the constant polynomial 1 as $g(x) (\in F[x])$. It is clear that $\left(a = (1)^{-1}a = \right)(g(\alpha))^{-1}f(\alpha) \in N$, and hence $a \in N$. Thus we have shown that $F \subset N$.

Now we want to show that $\alpha \in N$. To this end, let us take the polynomial $0 + 1x$ as $f(x) (\in F[x])$, and the constant polynomial 1 as $g(x) (\in F[x])$. It is clear that $\left(\alpha = (1)^{-1}(0 + 1\alpha) = \right)(g(\alpha))^{-1}f(\alpha) \in N$, and hence $\alpha \in N$. Thus we have shown that $F \cup \{\alpha\} \subset N$.

By the definition of N, $N \subset K$.
For c: Since K, N are fields and $N \subset K$, it follows that N is a subfield of K. $\blacksquare$

Definition Let F and K be any fields such that $F \subset K$. Suppose that K is an extension of F. Let $\alpha \in K$. The smallest subfield of K that contains both F and α is denoted by $F(\alpha)$, and we say that $F(\alpha)$ *is the subfield obtained by adjoining α to F.*

Thus $F \subset F(\alpha)$, $\alpha \in F(\alpha)$, and F is a field. It follows that $\{f(\alpha) : f(x) \in F[x]\} \subset F(\alpha)$.

(∗) By 1.4.8, $F(\alpha)$ is equal to the set of all elements of K of the form $(g(\alpha))^{-1}f(\alpha)$, where $f(x), g(x)$ are members of $F[x]$, and $g(\alpha)$ is a nonzero member of K Further, since $F(\alpha)$ is a field containing the field F as a subfield, $F(\alpha)$ is an extension of F. It follows, by 1.4.2, that $F(\alpha)$ is a vector space over the field F.

1.4.9 Problem Let F and K be any fields such that K is an extension of F. Let $\alpha \in K$. Let $F(\alpha)$ be a finite extension of F. Then α is algebraic over F.

Proof Case I: $\alpha \in F$. Clearly $f(\alpha) = 0$, where $f(x)$ is the polynomial $\alpha + (-1)x (\in F[x])$, and hence α is algebraic over F.

Case II: $\alpha \notin F$. We have to show that α is algebraic over F. Suppose to the contrary that α is not algebraic over F. We seek a contradiction.

Since α be not algebraic over F, $\{1, \alpha, \alpha^2, \ldots\}$ is a collection of distinct elements of $F(\alpha)$. Thus $\{1, \alpha, \alpha^2, \ldots\}$ is an infinite subset of $F(\alpha)$. Since $F(\alpha)$ is a finite extension of F, the dimension of the vector space $F(\alpha)$ over F is finite. Now, since $\{1, \alpha, \alpha^2, \ldots\}$ is an infinite subset of $F(\alpha)$, $\{1, \alpha, \alpha^2, \ldots\}$ is linearly dependent over F. It follows that there exists a positive integer n such that $\{1, \alpha, \alpha^2, \ldots, \alpha^n\}$ is linearly dependent over F. Hence, there exist $a_0, a_1, a_2, \ldots, a_n \in F$ such that not all $a_i (i = 0, 1, \ldots, n)$ are 0, and

$$a_0 1 + a_1 \alpha + a_2 \alpha^2 + \cdots + a_n \alpha^n = 0.$$

Thus $f(\alpha) = 0$, where $f(x) \equiv a_0 + a_1 x + a_2 x^2 + \cdots + a_n x^n (\in F[x])$ is such that not all $a_i (i = 0, 1, \cdots, n)$ are 0. Hence $f(x)$ is nonzero. Thus α is algebraic over F. ∎

1.4.10 Problem Let F be a field. Let $g(x) \in F[x]$ and $g(x) \neq 0$. Let n be the degree of the polynomial $g(x)$. Let us denote the ideal $(g(x)) (= \{f(x)g(x) : f(x) \in F[x]\})$ by V. Then the quotient ring $\frac{F[x]}{V}$ is a vector space over the field F under the usual vector addition and scalar multiplication. Further, $\{1 + V, x + V, x^2 + V, \cdots, x^{n-1} + V\}$ is a basis of $\frac{F[x]}{V}$. And hence n is the dimension of the vector space $\frac{F[x]}{V}$.

Proof It suffices to show that

1. for every $a, b \in F$, and for every $v(x) \in F[x]$, we have $(a + b)(v(x) + V) = a(v(x) + V) + b(v(x) + V)$ and $(ab)(v(x) + V) = a(b(v(x) + V))$,
2. for every $a \in F$ and for every $v(x), w(x) \in F[x]$, we have $a((v(x) + V) + (w(x) + V)) = a(v(x) + V) + a(w(x) + V)$,
3. for every $v(x) \in F[x]$, $1(v(x) + V) = (v(x) + V)$.

For 1: Let us take arbitrary $a, b \in F$ and $v(x) \in F[x]$. We have to show that $(a + b)(v(x) + V) = a(v(x) + V) + b(v(x) + V)$ and $(ab)(v(x) + V) = a(b(v(x) + V))$. Here

$$\text{LHS} = (a + b)(v(x) + V) = (a + b)v(x) + V = (av(x) + bv(x)) + V$$
$$= (av(x) + V) + (bv(x) + V) = a(v(x) + V) + b(v(x) + V) = \text{RHS}$$

and

$$\text{LHS} = (ab)(v(x) + V) = (ab)v(x) + V = a(bv(x)) + V$$
$$= a(bv(x) + V) = a(b(v(x) + V)) = \text{RHS}.$$

For 2: Let us take arbitrary $a \in F$ and $v(x), w(x) \in F[x]$. We have to show that $a((v(x) + V) + (w(x) + V)) = a(v(x) + V) + a(w(x) + V)$. Here

$$\text{LHS} = a((v(x) + V) + (w(x) + V)) = a((v(x) + w(x)) + V)$$
$$= a(v(x) + w(x)) + V = (av(x) + aw(x)) + V$$
$$= (av(x) + V) + (aw(x) + V) = a(v(x) + V) + a(w(x) + V) = \text{RHS}.$$

For 3: Let us take an arbitrary $v(x) \in F[x]$. We have to show that $1(v(x) + V) = v(x) + V$. Here

$$\text{LHS} = 1(v(x) + V) = 1v(x) + V = v(x) + V = \text{RHS}.$$

Thus we have shown that $\frac{F[x]}{V}$ is a vector space over the field $F (\subset F[x])$. It is clear that $\{1 + V, x + V, x^2 + V, \cdots, x^{n-1} + V\}$ is a subset of $\frac{F[x]}{V}$. We shall try to show that $\{1 + V, x + V, x^2 + V, \cdots, x^{n-1} + V\}$ is a basis of $\frac{F[x]}{V}$. To this end, we must show that

1. $\{1 + V, x + V, x^2 + V, \cdots, x^{n-1} + V\}$ is linearly independent,
2. $\{1 + V, x + V, x^2 + V, \cdots, x^{n-1} + V\}$ generates every element of $\frac{F[x]}{V}$.

For 1: Suppose that

$$a_0(1 + V) + a_1(x + V) + \cdots a_{n-1}(x^{n-1} + V) = 0 + V.$$

We have to show that each a_k is 0. Since

$$(a_0 + a_1 x + \cdots + a_{n-1}x^{n-1}) + V = (a_0 1 + V) + (a_1 x + V) + \cdots + (a_{n-1}x^{n-1} + V)$$
$$= \underbrace{a_0(1 + V) + a_1(x + V) + \cdots a_{n-1}(x^{n-1} + V) = 0 + V,}$$

we have
$$a_0 + a_1 x + \cdots + a_{n-1}x^{n-1} = \underbrace{(a_0 + a_1 x + \cdots + a_{n-1}x^{n-1}) - 0 \in V}$$

$$= \{f(x)g(x) : f(x) \in F[x]\},$$

and hence $a_0 + a_1 x + \cdots + a_{n-1} x^{n-1}$ is a member of $\{f(x)g(x) : f(x) \in F[x]\}$. By 1.2.14, every nonzero member of $\{f(x)g(x) : f(x) \in F[x]\}$ is of degree $\geq \deg(g(x)) = n$. It follows that

$$\text{either } a_0 + a_1 x + \cdots + a_{n-1} x^{n-1} = 0 \text{ or } \deg\left(a_0 + a_1 x + \cdots + a_{n-1} x^{n-1}\right)$$
$$\geq n(> (n-1)).$$

It follows that $a_0 + a_1 x + \cdots + a_{n-1} x^{n-1} = 0$, that is, each a_k is 0.

For 2: Let us take an arbitrary nonzero member $h(x)$ of $F[x]$, where $h(x) \equiv b_0 + b_1 x + b_2 x^2 + \cdots$. We have to show that $h(x) + V$ can be expressed as a linear combination of $1 + V, x + V, x^2 + V, \cdots, x^{n-1} + V$.

By 1.2.14, there exist $q(x), r(x) \in F[x]$ such that $h(x) = q(x)g(x) + r(x)$ and (either $r(x) = 0$ or $\deg(r(x)) < \deg(g(x))$).

Case I: $r(x) = 0$. It follows that $h(x) = q(x)g(x)(\in \{f(x)g(x) : f(x) \in F[x]\} = V)$, and hence $h(x) \in V$. We have

$$h(x) + V = 0(1 + V) + 0(x + V) + 0\left(x^2 + V\right) + \cdots + 0\left(x^{n-1} + V\right).$$

Case II: $\deg(r(x)) < \deg(g(x))(= n)$. We suppose that $r(x) \equiv c_0 + c_1 x + \cdots + c_{n-1} x^{n-1}$, where not all $c_i(i = 0, 1, \ldots, n-1)$ are 0. It suffices to show that

$$(q(x)g(x) = h(x) - r(x) =)h(x) - \left(c_0 + c_1 x + \cdots + c_{n-1} x^{n-1}\right)$$

is a member of $V(= \{f(x)g(x) : f(x) \in F[x]\})$, that is, $q(x)g(x)$ is a member of $\{f(x)g(x) : f(x) \in F[x]\}$. This is clearly true.

Thus we have shown that $\{1 + V, x + V, x^2 + V, \cdots, x^{n-1} + V\}$ is a basis of $\frac{F[x]}{V}$.

Since $\{1 + V, x + V, x^2 + V, \cdots, x^{n-1} + V\}$ is linearly independent, $\{1 + V, x + V, x^2 + V, \cdots, x^{n-1} + V\}$ is a set of distinct elements, and hence the number of elements in the basis $\{1 + V, x + V, x^2 + V, \cdots, x^{n-1} + V\}$ of $\frac{F[x]}{V}$ is n. Thus n is the dimension of the vector space $\frac{F[x]}{V}$.   ∎

Definition Let F and K be any fields such that K is an extension of F. Let α be a member of K. Let α be algebraic over F. It follows that there exists a nonzero polynomial $q(x) \in F[x](\supset F)$ such that

1. $(K \ni)q(\alpha) = 0$,
2. $\deg(q(x)) \geq 1$,
3. the leading coefficient of $q(x)$ is 1.

If $n(\geq 1)$ is the smallest degree of all such $q(x)$, then we say that α *is algebraic of degree n over* F.

Clearly, every member of F is algebraic of degree 1 over F.

1.4.11 Problem Let F and K be any fields such that K is an extension of F. Let α be a member of K. Let α be algebraic of degree n over F. Then there exists a unique polynomial $q(x) \in F[x](\supset F)$ such that

1. $(K \ni)q(\alpha) = 0$,
2. $n = \deg(q(x)) \geq 1$,
3. the leading coefficient of $q(x)$ is 1.

The unique polynomial $q(x)$ is called the *minimal polynomial of α over F.*

Proof Existence of $q(x)$ is clear from the definition of "algebraic of degree n over F."

Uniqueness: Suppose that there exist $q_1(x), q_2(x) \in F[x]$ such that

1. $q_1(\alpha) = 0, q_2(\alpha) = 0$,
2. $n = \deg(q_1(x)) = \deg(q_2(x)) \geq 1$,
3. the leading coefficient of $q_1(x)$ is 1, and the leading coefficient of $q_2(x)$ is 1.

We have to show that $q_1(x) = q_2(x)$. Suppose to the contrary that $q_1(x) \neq q_2(x)$, that is, $q_1(x) - q_2(x) \neq 0$. We seek a contradiction.

Let us put $h(x) \equiv q_1(x) - q_2(x)$. Clearly, $h(x) \neq 0$. Since $q_1(x), q_2(x) \in F[x]$ and $F[x]$ is a ring, we have $q_1(x) - q_2(x) \in F[x]$, and hence $h(x) \in F[x]$. Since $n = \deg(q_1(x)) = \deg(q_2(x))$, the leading coefficient of $q_1(x)$ is 1, and the leading coefficient of $q_2(x)$ is 1, we have $\deg(q_1(x) - q_2(x)) < n$, and hence $\deg(h(x)) < n$. Since $q_1(\alpha) = 0, q_2(\alpha) = 0$, and $h(x) = q_1(x) - q_2(x)$, we have $\underbrace{h(\alpha) = q_1(\alpha) - q_2(\alpha)}_{} = 0 - 0 = 0$, and hence $h(\alpha) = 0$.

Since α is algebraic of degree n over F, $h(x) \in F[x], h(x) \neq 0$, $h(\alpha) = 0$, $\deg(h(x)) < n$, we find that either $\deg(h(x)) < 1$ or the leading coefficient of $h(x)$ is different from 1.

Case I: $\deg(h(x)) < 1$. It follows that $h(x)$ is a constant polynomial, and since $h(\alpha) = 0$, we have $h(x) = 0$. This is a contradiction.

Case II: the leading coefficient of $h(x)$ is different from 1. Here, we can suppose that

$$h(x) \equiv b_0 + b_1 x + \cdots + b_k x^k,$$

where k is a positive integer strictly smaller than n, and b_k is a nonzero member of F. Put $q(x) \equiv \frac{1}{b_k} h(x)$. Clearly, $q(x) \in F[x], q(x) \neq 0$, $q(\alpha) = 0$, the leading coefficient of $q(x)$ is 1, and $1 \leq k = \deg(q(x))$. Now, since α is algebraic of degree n over F, we have $n \leq k$. This is a contradiction. ∎

1.4.12 Problem Let F and K be any fields such that K is an extension of F. Let α be a member of K. Let α be algebraic of degree n over F. Let $q(x) \in F[x](\supset F)$. Let $q(x)$ be the minimal polynomial of α over F. Then $q(x)$ is irreducible over F.

Proof Suppose to the contrary that $q(x)$ is not irreducible over F. We seek a contradiction.

Since $q(x)$ is not irreducible over F, there exist $r(x), s(x) \in F[x]$ such that

1. $q(x) = r(x)s(x)$,
2. $1 \le \deg(r(x)) < \deg(q(x))(= n)$, and $1 \le \deg(s(x)) < \deg(q(x))(= n)$.

Since α is algebraic of degree n over F, $q(x) \in F[x] (\supset F)$, and $q(x)$ is the minimal polynomial of α over F, we have $q(\alpha) = 0$, $n = \deg(q(x)) \ge 1$, and the leading coefficient of $q(x)$ is 1. Since $q(x) = r(x)s(x)$, we have $0 = q(\alpha) = r(\alpha)s(\alpha)$, and hence $r(\alpha)s(\alpha) = 0$. Now, since $r(\alpha), s(\alpha) \in K$ and K is a field, either $r(\alpha) = 0$ or $s(\alpha) = 0$.

Case I: $r(\alpha) = 0$. Here $1 \le \deg(r(x))$, so $\deg(r_1(x)) = \deg(r(x))(< n)$, where $r_1(x) \equiv \frac{1}{\text{leading coefficient of } r(x)} r(x)$. Thus $r_1(x) \in F[x]$, $r_1(\alpha) = 0$, and the leading coefficient of $r_1(x)$ is 1. Further, $1 \le \deg(r(x)) = \deg(r_1(x))$, so $1 \le \deg(r_1(x))$. Now, since α is algebraic of degree n over F, we have $n \le \deg(r_1(x))$. This is a contradiction.
Case II: $s(\alpha) = 0$. This case is similar to Case I.

Thus in all cases, we get a contradiction.   ∎

1.4.13 Problem Let F and K be any fields such that K is an extension of F. Let α be a member of K. Let α be algebraic of degree n over F. Let $p(x) \in F[x] (\supset F)$. Let $p(x)$ be the minimal polynomial of α over F. By 1.4.12, $p(x)$ is irreducible over F, and hence by 1.2.22, the ideal $(p(x))(\equiv \{p(x)f(x) : f(x) \in F[x]\})$ is a maximal ideal of the ring $F[x]$. Further, by 1.2.24, the quotient ring $\frac{F[x]}{(p(x))}$ is a field. Also, $\{f(\alpha) : f(x) \in F[x]\} \subset F(\alpha)$.

Let $\psi : f(x) \mapsto f(\alpha)$ be a mapping from the ring $F[x]$ to the field $F(\alpha)$. Then

1. $\psi : F[x] \to F(\alpha)$ is a ring homomorphism,
2. $\ker(\psi) = (p(x))$, where $\ker(\psi)(\equiv \{f(x) : f(x) \in F[x] \text{ and } \psi(f(x)) = 0\} = \{f(x) : f(x) \in F[x] \text{ and } f(\alpha) = 0\})$ denotes the kernel of the homomorphism ψ, and $(p(x))(= \{p(x)f(x) : f(x) \in F[x]\})$ is the ideal of the ring $F[x]$ generated by $p(x)$.

Proof 1. Let us take arbitrary $f(x), g(x) \in F[x]$, where

$$f(x) \equiv a_0 + a_1 x + a_2 x^2 + \cdots, g(x) \equiv b_0 + b_1 x + b_2 x^2 + \cdots,$$

each $a_i \in F$, and each $b_i \in F$. We have to show that

a. $\psi(f(x) + g(x)) = \psi(f(x)) + \psi(g(x))$,
b. $\psi(f(x)g(x)) = \psi(f(x))\psi(g(x))$.

For (a): Here

$$\text{LHS} = \psi(f(x) + g(x)) = \psi\left((a_0 + a_1 x + a_2 x^2 + \cdots) + (b_0 + b_1 x + b_2 x^2 + \cdots)\right)$$
$$= \psi\left((a_0 + b_0) + (a_1 + b_1)x + (a_2 + b_2)x^2 + \cdots\right)$$
$$= (a_0 + b_0) + (a_1 + b_1)\alpha + (a_2 + b_2)\alpha^2 + \cdots$$
$$= (a_0 + b_0) + (a_1 \alpha + b_1 \alpha) + (a_2 \alpha^2 + b_2 \alpha^2) + \cdots$$
$$= \left(a_0 + a_1 \alpha + a_2 \alpha^2 + \cdots\right) + \left(b_0 + b_1 \alpha + b_2 \alpha^2 + \cdots\right)$$
$$= f(\alpha) + g(\alpha) = \psi(f(x)) + \psi(g(x)) = \text{RHS}.$$

For (b): Here

$$\text{LHS} = \psi(f(x)g(x)) = \psi\left((a_0 + a_1 x + a_2 x^2 + \cdots)(b_0 + b_1 x + b_2 x^2 + \cdots)\right)$$
$$= \psi\left(a_0 b_0 + (a_0 b_1 + a_1 b_0)x + (a_0 b_2 + a_1 b_1 + a_2 b_0)x^2 + \cdots\right)$$
$$= a_0 b_0 + (a_0 b_1 + a_1 b_0)\alpha + (a_0 b_2 + a_1 b_1 + a_2 b_0)\alpha^2 + \cdots$$
$$= a_0 b_0 + (a_0 b_1 \alpha + a_1 b_0 \alpha) + (a_0 b_2 \alpha^2 + a_1 b_1 \alpha^2 + a_2 b_0 \alpha^2) + \cdots$$
$$= (a_0 b_0 + a_0 b_1 \alpha + a_0 b_2 \alpha^2 + \cdots) + (a_1 b_0 \alpha + a_1 b_1 \alpha^2 + a_1 b_2 \alpha^3 + \cdots)$$
$$+ (a_2 b_0 \alpha^2 + a_2 b_1 \alpha^3 + a_2 b_2 \alpha^4 + \cdots) + \cdots = a_0 (b_0 + b_1 \alpha + b_2 \alpha^2 + \cdots)$$
$$+ a_1 \alpha (b_0 + b_1 \alpha + b_2 \alpha^2 + \cdots) + a_2 \alpha^2 (b_0 + b_1 \alpha + b_2 \alpha^2 + \cdots) + \cdots$$
$$= (a_0 + a_1 \alpha + a_2 \alpha^2 + \cdots)(b_0 + b_1 \alpha + b_2 \alpha^2 + \cdots) = f(\alpha)g(\alpha)$$
$$= \psi(f(x))\psi(g(x)) = \text{RHS}.$$

2. We have to show that

a. $\{p(x)g(x) : g(x) \in F[x]\} \subset \{f(x) : f(x) \in F[x] \text{ and } f(\alpha) = 0\}$,
b. $\{f(x) : f(x) \in F[x] \text{ and } f(\alpha) = 0\} \subset \{p(x)g(x) : g(x) \in F[x]\}$.

For (a): Let us take an arbitrary $g(x) \in F[x]$. We have to show that $p(x)g(x) \in \{f(x) : f(x) \in F[x] \text{ and } f(\alpha) = 0\}$, that is,

$$p(\alpha)g(\alpha) = \psi(p(x))\psi(g(x)) = \underbrace{\psi(p(x)g(x)) = 0},$$

that is, $p(\alpha)g(\alpha) = 0$. It suffices to show that $p(\alpha) = 0$.

Since α is algebraic of degree n over F and $p(x)$ is the minimal polynomial of α over F, we have $p(\alpha) = 0$.

For (b): Let us take an arbitrary nonzero $f(x) \in F[x]$ such that $\psi(f(x)) = 0$, that is, $f(\alpha) = 0$. We have to show that $f(x) \in \{p(x)g(x) : g(x) \in F[x]\}$.

Since $p(x)$ is the minimal polynomial of α over F, we have $p(\alpha) = 0$, $n = \deg(p(x)) \geq 1$, and the leading coefficient of $p(x)$ is 1. Since $\deg(p(x)) \geq 1$,

$p(x)$ is a nonzero member of $F[x]$. Now by 1.2.14, there exist $q(x), r(x) \in F[x]$ such that $f(x) = q(x)p(x) + r(x)$, and (either $r(x) = 0$ or $\deg(r(x)) < \deg(p(x))$). Since $f(x) = q(x)p(x) + r(x)$, we have

$$0 = f(\alpha) = \underbrace{\psi(f(x)) = \psi(q(x)p(x) + r(x))}$$

$$= \psi(q(x)p(x)) + \psi(r(x)) = \psi(q(x))\psi(p(x)) + \psi(r(x))$$
$$= q(\alpha)p(\alpha) + r(\alpha) = q(\alpha)0 + r(\alpha) = r(\alpha),$$

and hence $r(\alpha) = 0$.

We claim that $r(x) = 0$. Suppose to the contrary that $r(x) \neq 0$. We seek a contradiction.

Since $r(x) \neq 0$ and $r(\alpha) = 0$, $r(x)$ is not a constant polynomial, and hence $\deg(r(x)) \geq 1$. Since $r(x) \neq 0$ and (either $r(x) = 0$ or $\deg(r(x)) < \deg(p(x))$), we have $1 \leq \underbrace{\deg(r(x)) < \deg(p(x))} = n$, and hence $1 \leq \deg(r(x)) < n$.

Put

$$r_1(x) \equiv \frac{1}{\text{leading coefficient of } r(x)} r(x).$$

Thus $r_1(x) \in F[x]$, $r_1(\alpha) = 0$, and the leading coefficient of $r_1(x)$ is 1. Further, $1 \leq \deg(r(x)) = \deg(r_1(x))$, so $1 \leq \deg(r_1(x))$. Now, since α is algebraic of degree n over F, we have $n \leq \deg(r_1(x))$, and hence $n \leq \deg(r(x))$. This is a contradiction.

Thus our claim is true, that is, $(f(x) - q(x)p(x) =)r(x) = 0$. Hence $f(x) = q(x)p(x) \in \{p(x)g(x) : g(x) \in F[x]\}$. ∎

1.4.14 Problem Let F and K be any fields such that K is an extension of F. Let α be a member of K. Let α be algebraic of degree n over F. Let $p(x) \in F[x](\supset F)$. Let $p(x)$ be the minimal polynomial of α over F. Let $\psi : f(x) \mapsto f(\alpha)$ be a mapping from the ring $F[x]$ to the field $F(\alpha)$. By 1.4.13, $\psi : F[x] \to F(\alpha)$ is a ring homomorphism, and hence by the fundamental theorem of ring homomorphisms, the mapping $\eta : (f(x) + \ker(\psi)) \mapsto \psi(f(x))(= f(\alpha))$ is a ring isomorphism from the quotient ring $\frac{F[x]}{\ker(\psi)}$ to $F(\alpha)$. Thus η maps $\frac{F[x]}{\ker(\psi)}$ onto $F(\alpha)$.

In short, the field $\frac{F[x]}{\ker(\psi)}$ is isomorphic to the field $F(\alpha)$.

Proof Recall that $F(\alpha)$ is equal to the set of all elements of K of the form $(g(\alpha))^{-1}f(\alpha)$, where $f(x), g(x)$ are members of $F[x]$, and $g(\alpha)$ is a nonzero member of K. Next, let us take an arbitrary $(g(\alpha))^{-1}f(\alpha) \in F(\alpha)$, where $f(x), g(x)$ are members of $F[x]$, and $g(\alpha)$ is a nonzero member of K. From 1.4.13, $\frac{F[x]}{(p(x))}$ is a field, and $\ker(\psi) = (p(x))$, so $\frac{F[x]}{\ker(\psi)}$ is a field. Since $f(x), g(x)$ are members of $F[x]$, $(f(x) + \ker(\psi)), (g(x) + \ker(\psi)) \in \frac{F[x]}{\ker(\psi)}$. Since $(\psi(g(x)) =)g(\alpha)$ is nonzero,

$g(x) \notin \ker{(\psi)}$, and hence $g(x) + \ker{(\psi)}$ is a nonzero member of the field $\frac{F[x]}{\ker{(\psi)}}$. It follows that

$$(f(x)h(x) + \ker{(\psi)}) = (f(x) + \ker{(\psi)})(h(x) + \ker{(\psi)})$$
$$= \underbrace{(f(x) + \ker{(\psi)})(g(x) + \ker{(\psi)})^{-1}} \in \frac{F[x]}{\ker{(\psi)}},$$

where $(h(x) + \ker{(\psi)})(g(x) + \ker{(\psi)}) = (1 + \ker{(\psi)})$. Thus $(f(x)h(x) + \ker{(\psi)})$ $\in \frac{F[x]}{\ker{(\psi)}}$, and $(h(x)g(x) + \ker{(\psi)}) = (1 + \ker{(\psi)})$. It suffices to show that $\eta(f(x)h(x) + \ker{(\psi)}) = (g(\alpha))^{-1}f(\alpha)$. Since

$$\eta(f(x)h(x) + \ker{(\psi)}) = \psi(f(x)h(x)) = \psi(f(x))\psi(h(x)) = f(\alpha)h(\alpha),$$

it suffices to show that $f(\alpha)h(\alpha) = (g(\alpha))^{-1}f(\alpha)$, that is, $g(\alpha)f(\alpha)h(\alpha) = f(\alpha)$, that is, $f(\alpha)g(\alpha)h(\alpha) = f(\alpha)$. Again, it suffices to show that $h(\alpha)g(\alpha) = 1$.
Since $(h(x)g(x) + \ker{(\psi)}) = (1 + \ker{(\psi)})$, we have

$$h(\alpha)g(\alpha) = \psi(h(x))\psi(g(x)) = \psi(h(x)g(x))$$
$$= \underbrace{\eta(h(x)g(x) + \ker{(\psi)}) = \eta(1 + \ker{(\psi)})} = \psi(1) = 1,$$

and hence $h(\alpha)g(\alpha) = 1$.   ∎

1.4.15 Note Let F and K be any fields such that $F \subset K$. Suppose that K is an extension of F. Let α be a member of K. Let α be algebraic of degree n over F. Let $p(x) \in F[x](\supset F)$. Let $p(x)$ be the minimal polynomial of α over F. It follows that $n = \deg(p(x)) \geq 1$. Now by 1.4.10, $\dim\left(\frac{F[x]}{(p(x))}\right) = n$.

Let $\psi : f(x) \mapsto f(\alpha)$ be a mapping from ring $F[x]$ to the field $F(\alpha)$. By 1.4.14, the mapping $\eta : (f(x) + \ker{(\psi)}) \mapsto \psi(f(x))(= f(\alpha))$ is a ring isomorphism from the quotient ring $\frac{F[x]}{\ker{(\psi)}}$ onto $F(\alpha)$. Further, by 1.4.13, $\ker{(\psi)} = (p(x))$. Thus $\dim\left(\frac{F[x]}{\ker{(\psi)}}\right) = n$.

We can think of $\frac{F[x]}{\ker{(\psi)}}$ as a vector space over the field F under the usual operations of vector addition and scalar multiplication:
For every $f(x), g(x) \in F[x]$ and for every $a \in F(\subset F[x])$,

$$(f(x) + \ker{(\psi)}) + (g(x) + \ker{(\psi)}) \equiv (f(x) + g(x)) + \ker{(\psi)}$$

and

$$a(f(x) + \ker(\psi)) \equiv (a + \ker(\psi))(f(x) + \ker(\psi))(= af(x) + \ker(\psi)).$$

It suffices to show the following:

1. For every $f(x) \in F[x]$ and for every $a, b \in F$,

$$(a + b)(f(x) + \ker(\psi)) = a(f(x) + \ker(\psi)) + b(f(x) + \ker(\psi))$$

and

$$(ab)(f(x) + \ker(\psi)) = a(b(f(x) + \ker(\psi))).$$

2. For every $f(x), g(x) \in F[x]$ and for every $a \in F$,

$$a((f(x) + \ker(\psi)) + (g(x) + \ker(\psi))) = a(f(x) + \ker(\psi)) + a(g(x) + \ker(\psi)).$$

3. For every $f(x) \in F[x]$,

$$1(f(x) + \ker(\psi)) = (f(x) + \ker(\psi)).$$

For 1:

$$\begin{aligned}
\text{LHS} &= (a + b)(f(x) + \ker(\psi)) = (a + b)f(x) + \ker(\psi) \\
&= (af(x) + bf(x)) + \ker(\psi) = (af(x) + \ker(\psi)) + (bf(x) + \ker(\psi)) \\
&= a(f(x) + \ker(\psi)) + b(f(x) + \ker(\psi)) = \text{RHS}.
\end{aligned}$$

Next,

$$\begin{aligned}
\text{LHS} &= (ab)(f(x) + \ker(\psi)) = (ab)f(x) + \ker(\psi) = a(bf(x)) + \ker(\psi) \\
&= a(bf(x) + \ker(\psi)) = a(b(f(x) + \ker(\psi))) = \text{RHS}.
\end{aligned}$$

For 2:

$$\begin{aligned}
\text{LHS} &= a((f(x) + \ker(\psi)) + (g(x) + \ker(\psi))) = a((f(x) + g(x)) + \ker(\psi)) \\
&= a(f(x) + g(x)) + \ker(\psi) = (af(x) + ag(x)) + \ker(\psi) \\
&= (af(x) + \ker(\psi)) + (ag(x) + \ker(\psi)) = a(f(x) + \ker(\psi)) + a(g(x) + \ker(\psi)) \\
&= \text{RHS}.
\end{aligned}$$

For 3:

$$\text{LHS} = 1(f(x) + \ker(\psi)) = 1f(x) + \ker(\psi) = f(x) + \ker(\psi) = \text{RHS}.$$

Thus $\frac{F[x]}{\ker(\psi)}$ is a vector space over the field F.

Since $F \subset F(\alpha)$ and $F, F(\alpha)$ are fields, by 1.4.2, $F(\alpha)$ can be thought of as a vector space over the field F.

We shall show that the mapping $\eta : (f(x) + \ker(\psi)) \mapsto \psi(f(x))(= f(\alpha))$ is an isomorphism from the vector space $\frac{F[x]}{\ker(\psi)}$ onto the vector space $F(\alpha)$.

Since η is an isomorphism from $\frac{F[x]}{\ker(\psi)}$ to $F(\alpha)$, the map η from $\frac{F[x]}{\ker(\psi)}$ to $F(\alpha)$ is one-to-one and onto. Hence it suffices to show that for every $f(x), g(x) \in F[x]$ and for every $a, b \in F$,

$$\eta(a(f(x) + \ker(\psi)) + b(g(x) + \ker(\psi))) = a\eta(f(x) + \ker(\psi)) + b\eta(g(x) + \ker(\psi)),$$

$$\begin{aligned}
\text{LHS} &= \eta(a(f(x) + \ker(\psi)) + b(g(x) + \ker(\psi))) = \eta((af(x) + bg(x)) + \ker(\psi)) \\
&= \psi(af(x) + bg(x)) = \psi(af(x)) + \psi(bg(x)) = \psi(a)\psi(f(x)) + \psi(b)\psi(g(x)) \\
&= a\psi(f(x)) + b\psi(g(x)) = a\eta(f(x) + \ker(\psi)) + b\eta(g(x) + \ker(\psi)) = \text{RHS}.
\end{aligned}$$

Thus the vector space $\frac{F[x]}{\ker(\psi)}$ over F is isomorphic to the vector space $F(\alpha)$ over F. It follows that

$$n = \dim\left(\underbrace{\frac{F[x]}{\ker(\psi)}}\right) = \dim(F(\alpha)) = [F(\alpha) : F],$$

and hence $[F(\alpha) : F] = n$.

1.4.16 Conclusion Let F and K be any fields such that K is an extension of F Let α be a member of K. Let α be algebraic of degree n over F. Then $[F(\alpha) : F] = n$. In short, α is algebraic of degree $[F(\alpha) : F]$ over F.

1.4.17 Problem Let F and K be any fields such that K is an extension of F. Let α be a member of K. Let α be algebraic over F. Then $F(\alpha)$ is a finite extension of F.

Proof Since α is algebraic over F, there exists a nonzero polynomial $q(x) \in F[x]$ such that $(K \ni) q(\alpha) = 0$. It follows that $\deg(q(x)) \geq 1$. Let n be the smallest degree of all such polynomials $q(x)$. Hence α is algebraic of degree n over F. Now by 1.4.16, $[F(\alpha) : F] = n < \infty$. Hence $F(\alpha)$ is a finite extension of F.   ∎

1.4.18 Note Let F and K be any fields such that K is an extension of F. Let A be the collection of all elements of K that are algebraic over F. By 1.4.5, $F \subset A$. Thus $F \subset A \subset K$. We shall show that A is a field.

To this end, let us take arbitrary $a, b \in A$. It suffices to show the following:

1. $(a - b) \in A$,
2. $ab \in A$,
3. $ab^{-1} \in A$, provided a, b are nonzero.

Since $a \in A$, a is algebraic over F, and hence there exists a positive integer m such that a is algebraic of degree m over F. Similarly, there exists a positive integer n such that b is algebraic of degree n over F. Since b is algebraic of degree n over F, by 1.4.11, there exists a unique polynomial $q(x) \in F[x] (\supset F)$ such that

1. $(K \ni) q(b) = 0$,
2. $n = \deg(q(x)) \geq 1$,
3. the leading coefficient of $q(x)$ is 1.

Suppose that b is algebraic of degree k over the field $F(a) (\subset K)$.

Since $F \subset F(a)$, and $F(a)$ is a field, we have $F[x] \subset (F(a))[x]$. Now, since $q(x) \in F[x]$, we have $q(x) \in (F(a))[x]$. Since $q(x) \in (F(a))[x]$, $q(b) = 0$, the leading coefficient of $q(x)$ is 1, and b is algebraic of degree k over the field $F(a)$, we have

$$\underbrace{k \leq \deg(q(x))}_{} = n.$$

Thus $k \leq n$. Since b is algebraic of degree k over the field $F(a) (\subset K)$, by 1.4.16, $[(F(a))(b) : F(a)] = k$. Since a is algebraic of degree m over the field $F(\subset K)$, by 1.4.16, $[F(a) : F] = m$. Since $F \subset F(a) \subset (F(a))(b) \subset K(b) = K$, by 1.4.3, we have

$$\underbrace{[(F(a))(b) : F] = [(F(a))(b) : F(a)][F(a) : F]}_{} = k[F(a) : F] \leq n[F(a) : F] = nm.$$

Thus $[(F(a))(b) : F] \leq nm (< \infty)$. Since $(F(a))(b)$ is a field containing b and all the elements of $F(a) (\ni a)$, $(F(a))(b)$ is a field containing a, b.

1. Since $(F(a))(b)$ is a field containing a, b, $(F(a))(b)$ is a field containing $a - b$, and hence $F(a - b) \subset (F(a))(b)$. Thus $F(a - b)$ is a linear subspace of the vector space $(F(a))(b)$. It follows that

$$[F(a - b) : F] = \underbrace{\dim (F(a - b)) \leq \dim ((F(a))(b))}_{}$$

$$= [(F(a))(b) : F] (\leq nm < \infty),$$

and hence $[F(a - b) : F] \leq nm$. Now, by 1.4.9, $a - b$ is algebraic over F, and hence $(a - b) \in A$.

Next, by 1.4.16, $a - b$ is algebraic of degree $[F(a - b) : F] (\leq nm)$ over F, and hence $a - b$ is algebraic of degree $\leq nm$ over F.

2. Since $(F(a))(b)$ is a field containing a, b, $(F(a))(b)$ is a field containing ab. It follows, as above, that $[F(ab) : F] \leq nm$. Now by 1.4.9, ab is algebraic over F, and hence $ab \in A$.

Next, by 1.4.16, ab is algebraic of degree $[F(ab) : F] (\leq nm)$ over F, and hence ab is algebraic of degree $\leq nm$ over F.

3. Suppose that a, b are nonzero. Since $(F(a))(b)$ is a field containing a, b, $(F(a))(b)$ is a field containing ab^{-1}. It follows, as above, that $[F(ab^{-1}) : F] \leq nm$. Now by 1.4.9, ab^{-1} is algebraic over F, and hence $ab^{-1} \in A$.

Next, by 1.4.16, ab^{-1} is algebraic of degree $[F(ab^{-1}) : F] (\leq nm)$ over F, and hence ab^{-1} is algebraic of degree $\leq nm$ over F.

1.4.19 Conclusion Let F and K be any fields such that K is an extension of F. Let A be the collection of all elements of K that are algebraic over F. Then

1. $F \subset A \subset K$,
2. A is a subfield of K, and F is a subfield of A,
3. if a is algebraic of degree m over F, and b is algebraic of degree n over F, then all of $a \pm b, ab, ab^{-1}$ (provided b is nonzero) are algebraic of degree $\leq mn$ over F.

Definition Let F and K be any fields such that $F \subset K$. Suppose that K is an extension of F. Let $a, b \in K$. Since $F \cup \{a\} \subset F(a) \subset (F(a))(b)$ and $b \in (F(a))(b)$, we have $F \cup \{a, b\} \subset (F(a))(b)$. So $(F(a))(b)$ is a field containing $F \cup \{a, b\}$. The smallest field containing $F \cup \{a, b\}$ is denoted by $F(a, b)$.

We have seen that $F(a, b) \subset (F(a))(b)$. It is clear that $F(a) \subset F(a, b)$, and hence $(F(a)) \cup \{b\} \subset F(a, b)$. It follows that $(F(a))(b) \subset F(a, b)$. Thus we have shown that $(F(a))(b) = F(a, b)$. Similarly, $(F(b))(a) = F(a, b)$. Thus

$$(F(a))(b) = (F(b))(a) = F(a, b).$$

A similar definition can be supplied for $F(a, b, c)$, etc.

Definition Let F and K be any fields such that K is an extension of F. If every element of K is algebraic over F, then we say that K is an algebraic extension of F.

1.4.20 Problem Let F, K, and L be any fields such that $F \subset K \subset L$. Suppose that K is an algebraic extension of F, and L is an algebraic extension of K. Then L is an algebraic extension of F.

Proof Let us take an arbitrary $l \in L$. We have to show that l is algebraic over F.

Since L is an algebraic extension of K, and $l \in L$, there exists a nonzero polynomial

$$k_0 + k_1 x + k_2 x^2 + \cdots + k_n x^n$$

such that each $k_i \in K$, n is a positive integer, and

$$k_0 + k_1 l + k_2 l^2 + \cdots + k_n l^n = 0.$$

Since $k_0 \in K$ and K is an algebraic extension of F, k_0 is algebraic over F, and hence by 1.4.17, $F(k_0)(\subset K)$ is a finite extension of F. Since $k_1 \in K$ and K is an algebraic extension of F, k_1 is algebraic over F. Now, since $F(k_0)$ is an extension of F, by 1.4.17, $(F(k_0, k_1) =)(F(k_0))(k_1)(\subset K)$ is a finite extension of F. Thus $F(k_0, k_1)$ is an extension of F. Since $k_2 \in K$ and K is an algebraic extension of F, k_2 is algebraic over F. Now, since $F(k_0, k_1)$ is an extension of F, by 1.4.17, $(F(k_0, k_1, k_2) =)(F(k_0, k_1))(k_2)(\subset K)$ is a finite extension of F. Thus $F(k_0, k_1, k_2)$ is a finite extension of F, etc.

It follows that $F(k_0, k_1, \cdots, k_n)(\subset K)$ is a finite extension of F. Since each $k_i \in F(k_0, k_1, \ldots, k_n)$, the nonzero polynomial

$$k_0 + k_1 x + k_2 x^2 + \cdots + k_n x^n$$

is a member of $(F(k_0, k_1, \ldots, k_n))[x]$. Next,

$$k_0 + k_1 l + k_2 l^2 + \ldots + k_n l^n = 0,$$

so l is algebraic over $F(k_0, k_1, \ldots, k_n)(\subset K)$. It follows, by 1.4.17, that $(F(k_0, k_1, \ldots, k_n))(l)$ is a finite extension of $F(k_0, k_1, \ldots, k_n)$. Further, $F(k_0, k_1, \ldots, k_n)$ is a finite extension of F, so by 1.4.4, $(F(k_0, k_1, \cdots, k_n))(l)$ is a finite extension of F. Now by 1.4.9, l is algebraic over F.   ∎

Definition Let $\alpha \in \mathbb{C}$. Recall that $\mathbb{Q} \subset \mathbb{C}$, and the field $\mathbb{C}$ is an extension of the field $\mathbb{Q}$. If α is algebraic over $\mathbb{Q}$, then we say that α *is an algebraic number*. A complex number that is not an algebraic number is called a *transcendental number*.

(∗) By 1.4.5, every rational number is an algebraic number. By 1.4.19, the collection of all algebraic numbers is a subfield of $\mathbb{C}$. Thus if a is an algebraic number and b is an algebraic number, then all of $a \pm b, ab$, ab^{-1} (provided b is nonzero) are algebraic numbers.

1.4.21 Problem Recall that $\mathbb{Q} \subset \mathbb{C}$, and the field $\mathbb{C}$ is an extension of the field $\mathbb{Q}$. Let $\mathbb{A}$ be the collection of all algebraic numbers. We know from 1.4.19 that $\mathbb{Q} \subset \mathbb{A} \subset \mathbb{C}$, $\mathbb{C}$ is an extension of $\mathbb{A}$, and $\mathbb{A}$ is an algebraic extension of $\mathbb{Q}$. Let $f(x)$ be a nonzero member of $\mathbb{A}[x]$, and let $\alpha(\in \mathbb{C})$ be a root of the polynomial $f(x)$, in the sense that $f(\alpha) = 0$. Then $\alpha \in \mathbb{A}$.

In short, the roots of a polynomial whose coefficients are algebraic numbers.

Proof Suppose to the contrary that $\alpha \notin \mathbb{A}$. We seek a contradiction.

Since $\mathbb{C}$ is an extension of $\mathbb{A}$, $\alpha \in \mathbb{C}$, $f(x)$ is a nonzero member of $\mathbb{A}[x]$, and $f(\alpha) = 0$, α is algebraic over $\mathbb{A}$, and hence each member of $\mathbb{A} \cup \{\alpha\}$ is algebraic over $\mathbb{A}$. Observe that the field $\mathbb{A}(\alpha)$ is an extension of $\mathbb{A}$.

Let B be the collection of all elements of $\mathbb{A}(\alpha)$ that are algebraic over $\mathbb{A}$. By 1.4.19, $\mathbb{A} \subset B \subset \mathbb{A}(\alpha)$, B is a subfield of $\mathbb{A}(\alpha)$, and $\mathbb{A}$ is a subfield of B. Since each member of $\mathbb{A} \cup \{\alpha\}$ is algebraic over $\mathbb{A}$, we have $\mathbb{A} \cup \{\alpha\} \subset B$, and hence $\mathbb{A}(\alpha) \subset B$. Now, since $B \subset \mathbb{A}(\alpha)$, we have $B = \mathbb{A}(\alpha)$.

Thus every element of $\mathbb{A}(\alpha)$ is algebraic over $\mathbb{A}$, and hence $\mathbb{A}(\alpha)$ is an algebraic extension of $\mathbb{A}$. Now, since $\mathbb{A}$ is an algebraic extension of $\mathbb{Q}$, by 1.4.20, $\mathbb{A}(\alpha)$ is an algebraic extension of $\mathbb{Q}$. And since $\alpha \in \mathbb{A}(\alpha)$, α is algebraic over $\mathbb{Q}$, and hence α is an algebraic number. Thus $\alpha \in \mathbb{A}$. This is a contradiction. ∎

1.5   Splitting Fields

1.5.1   Theorem The number $e \left(\equiv 1 + \frac{1}{1!} + \frac{1}{2!} + \frac{1}{3!} + \cdots \right)$ is a transcendental number.

Proof (due to **Hermite**) Suppose to the contrary that e is not a transcendental number. We seek a contradiction.

Since e is not a transcendental number, e is an algebraic number, and hence there exists a nonzero polynomial

$$c_0 + c_1 x + c_2 x^2 + \cdots + c_n x^n$$

such that

1. each c_i is an integer,
2. c_0 is a positive integer,
3. n is a positive integer,
4. $c_0 + c_1 e + c_2 e^2 + \cdots + c_n e^n = 0$,
5. c_n is a nonzero integer.

Let us take a polynomial $f(x) \in \mathbb{R}[x]$, and let $\deg (f(x)) = r > 1$.

It follows that the $(r+1)$th derivative $f^{(r+1)}(x)$ of $f(x)$ is the zero polynomial. Similarly, $f^{(r+2)}(x)$ is the zero polynomial, etc.

Put

$$F(x) \equiv f(x) + f'(x) + f''(x) + \cdots + f^{(r)}(x).$$

By the mean value theorem, there exists a real number $\theta_1 \in (0, 1)$ such that

$$e^{-1}F(1) - e^{-0}F(0) = (1-0)\frac{d(e^{-x}F(x))}{dx}\Big|_{x=\theta_1},$$

that is,

$$e^{-1}F(1) - F(0) = (1-0)(-e^{-x}F(x) + e^{-x}F'(x))\big|_{x=\theta_1},$$

that is,

$$e^{-1}F(1) - F(0) = (1-0)e^{-\theta_1}(F'(\theta_1) - F(\theta_1)),$$

that is,

$$e^{-1}F(1) - F(0) = (1-0)e^{-\theta_1}\left(\left(f'(\theta_1) + f''(\theta_1) + \cdots + f^{(r)}(\theta_1) + f^{(r+1)}(\theta_1)\right)\right.$$
$$\left. - \left(f(\theta_1) + f'(\theta_1) + f''(\theta_1) + \cdots + f^{(r)}(\theta_1)\right)\right),$$

that is,

$$e^{-1}F(1) - F(0) = (1-0)e^{-\theta_1}\left(f^{(r+1)}(\theta_1) - f(\theta_1)\right),$$

that is,

$$e^{-1}F(1) - F(0) = (1-0)e^{-\theta_1}(0 - f(\theta_1)),$$

that is,

$$e^{-1}F(1) - F(0) = -(1-0)e^{-1\theta_1}f(1\theta_1).$$

Similarly, there exists a real number $\theta_2 \in (0,1)$ such that

$$e^{-2}F(2) - F(0) = -(2-0)e^{-2\theta_2}f(2\theta_2).$$

Also, there exists a real number $\theta_3 \in (0,1)$ such that

$$e^{-3}F(3) - F(0) = -(3-0)e^{-3\theta_3}f(3\theta_3).$$

$$\vdots$$

There exists a real number $\theta_n \in (0,1)$ such that

$$e^{-n}F(n) - F(0) = -(n-0)e^{-n\theta_n}f(n\theta_n).$$

Thus

$$F(1) - eF(0) = -e^{1-\theta_1}f(1\theta_1),$$
$$F(2) - e^2F(0) = -2e^{2-2\theta_2}f(2\theta_2),$$
$$F(3) - e^3F(0) = -3e^{3-3\theta_3}f(3\theta_3),$$
$$\vdots$$
$$F(n) - e^nF(0) = -ne^{n-n\theta_n}f(n\theta_n).$$

It follows that

$$c_0F(0) + c_1F(1) + c_2F(2) + \cdots + c_nF(n)$$
$$= c_0F(0) + c_1\left(eF(0) - e^{1-\theta_1}f(1\theta_1)\right) + c_2\left(e^2F(0) - 2e^{2-2\theta_2}f(2\theta_2)\right)$$
$$+ \cdots + c_n\left(e^nF(0) - ne^{n-n\theta_n}f(n\theta_n)\right) = \left(c_0 + c_1e + c_2e^2 + \cdots + c_ne^n\right)F(0)$$
$$- \left(c_1e^{1-\theta_1}1f(\theta_1) + c_22e^{2-2\theta_2}f(2\theta_2) + \cdots + c_nne^{n-n\theta_n}f(n\theta_n)\right)$$
$$= 0 \cdot F(0) - \left(c_11e^{1(1-\theta_1)}f(1\theta_1) + c_22e^{2(1-\theta_2)}f(2\theta_2) + \cdots + c_nne^{n(1-\theta_n)}f(n\theta_n)\right),$$

and hence

$$c_0F(0) + c_1F(1) + c_2F(2) + \cdots + c_nF(n)$$
$$= -\left(c_11e^{1(1-\theta_1)}f(1\theta_1) + c_22e^{2(1-\theta_2)}f(2\theta_2) + \cdots + c_nne^{n(1-\theta_n)}f(n\theta_n)\right). \qquad (*)$$

Let us take an arbitrary prime p such that $1 \leq n < p$, and $1 \leq c_0 < p$. It follows that p divides neither the integer c_0 nor $n!$, and hence p does not divide the integer $c_0(n!)^p$.

Next let us take

$$f(x) \equiv \frac{1}{(p-1)!}x^{p-1}(1-x)^p(2-x)^p \cdots (n-x)^p (\in \mathbb{R}[x]).$$

Here

$$r = \deg(f(x)) = (p-1) + \underbrace{p + p + \cdots + p}_{n \text{ terms}} = (n+1)p - 1.$$

Thus $r = (n+1)p - 1$. Observe that

$$f(x) = \frac{1}{(p-1)!} x^{p-1} (1-x)^p (2-x)^p \cdots (n-x)^p$$

$$= \frac{1}{(p-1)!} x^{p-1} \left(1^p - \binom{p}{1} 1^{p-1} x + \binom{p}{2} 1^{p-2} x^2 - \cdots \right)$$

$$\times \left(2^p - \binom{p}{1} 2^{p-1} x + \binom{p}{2} 2^{p-2} x^2 - \cdots \right)$$

$$\times \left(3^p - \binom{p}{1} 3^{p-1} x + \binom{p}{2} 3^{p-2} x^2 - \cdots \right)$$

$$\vdots$$

$$\times \left(n^p - \binom{p}{1} n^{p-1} x + \binom{p}{2} n^{p-2} x^2 - \cdots \right)$$

$$= \frac{1}{(p-1)!} x^{p-1} (1^p \cdot 2^p \cdots n^p + (\text{integer}) x + (\text{integer}) x^2 + \cdots)$$

$$= \frac{(n!)^p}{(p-1)!} x^{p-1} + \frac{\text{integer}}{(p-1)!} x^p + \frac{\text{integer}}{(p-1)!} x^{p+1} + \cdots + \frac{\text{integer}}{(p-1)!} x^{(n+1)p-1},$$

so

$$f(x) = \frac{(n!)^p}{(p-1)!} x^{p-1} + \frac{a_0}{(p-1)!} x^p + \frac{a_1}{(p-1)!} x^{p+1} + \cdots + \frac{a_{np-1}}{(p-1)!} x^{(n+1)p-1},$$

where each a_i is an integer. Now,

$$f^{(p)}(x) = \frac{(n!)^p}{(p-1)!} D^p \left(x^{p-1} \right) + \frac{a_0}{(p-1)!} D^p (x^p) + \frac{a_1}{(p-1)!} D^p \left(x^{p+1} \right) + \cdots$$

$$= \frac{(n!)^p}{(p-1)!} \cdot 0 + \frac{a_0}{(p-1)!} \cdot p! + \frac{a_1}{(p-1)!} \frac{(p+1)!}{((p+1)-p)!} x^{(p+1)-p}$$

$$+ \frac{a_2}{(p-1)!} \frac{(p+2)!}{((p+2)-p)!} x^{(p+2)-p} + \cdots$$

$$= \frac{a_0}{(p-1)!} \cdot p! + \frac{a_1}{(p-1)!} \frac{(p+1)!}{1!} x + \frac{a_2}{(p-1)!} \frac{(p+2)!}{2!} x^2 + \cdots$$

$$= a_0 p + a_1 (p+1) p x + a_2 \frac{(p+2)(p+1)p}{2!} x^2 + \cdots$$

$$= a_0 p + a_1 \binom{p+1}{1} p x + a_2 \binom{p+2}{2} p x^2 + \cdots,$$

so

$$f^{(p)}(x) = a_0 p + a_1 \binom{p+1}{1} p x + a_2 \binom{p+2}{2} p x^2 + \cdots.$$

Here we observe that each coefficient of $f^{(p)}(x)$ is an integer that is divisible by p.

Further,

$$f^{(p+1)}(x) = a_1 \binom{p+1}{1} p + a_2 \binom{p+2}{2} p2x + a_3 \binom{p+3}{3} p3x^2 + \cdots.$$

Again, we observe that each coefficient of $f^{(p+1)}(x)$ is an integer that is divisible by p.

Thus for every integer j, and for every integer $i \geq p$, $f^{(i)}(j)$ is an integer that is divisible by p.

[Before going ahead, let us recall the Leibniz rule of differentiation:

$$(uv)' = u'v + uv',$$

$$(uvw)' = u'vw + uv'w + uvw',$$

$$(uvw)'' = (u''vw + u'v'w + u'vw') + (u'v'w + uv''w + uv'w') + (u'vw' + uv'w' + uvw''),$$

$$(uvw)'' = u''vw + + uv''w + uvw'' + 2uv'w' + 2u'vw' + 2u'v'w$$
$$= u'' \cdot vw + u'(2vw' + 2v'w) + u(v''w + vw'' + 2v'w').$$

Similarly,

$$(uvw)^{(n)} = \sum_{\substack{(i,j,k) \\ i,j,k \text{ are nonnegative integers} \\ i+j+k=n}} (\text{positive integer}) u^{(i)} v^{(j)} w^{(k)}$$

$$= u^{(n)}(\cdots) + u^{(n-1)}(\cdots) + u^{(n-2)}(\cdots) + \cdots$$
$$= v^{(n)}(\cdots) + v^{(n-1)}(\cdots) + v^{(n-2)} + \cdots,$$

etc. Also, for every integer $i \in \{0, 1, 2, \ldots, p-2\}$, we have $D^i(x^{p-1})\big|_{x=0} = 0$. Also $D^{p-1}(x^{p-1}) = (p-1)!$, and $D^p(x^{p-1}) = 0$.

Next, for every integer $i \in \{0, 1, 2, \ldots, p-1\}$, $D^i((1-x)^p)\big|_{x=1} = 0$. Also $D^p((1-x)^p) = (-1)^p(p!)$.

Similarly, for every integer $i \in \{0, 1, 2, \cdots, p-1\}$, $D^i((2-x)^p)\big|_{x=2} = 0$. Also, $D^p((2-x)^p) = (-1)^p(p!)$, etc.]

Now, since

$$f(x) = \frac{1}{(p-1)!} x^{p-1}(1-x)^p(2-x)^p \cdots (n-x)^p,$$

we have

$$f^{(p-1)}(1) = \frac{1}{(p-1)!}(0+0+\cdots) = 0..$$

Similarly, $f^{(p-1)}(2) = 0$, $f^{(p-1)}(3) = 0$, etc. Also, $f^{(p-2)}(1) = 0$, $f^{(p-2)}(2) = 0$, etc. In short, for every $i \in \{0,1,2,\ldots,p-1\}$, and for every $j \in \{1,2,\cdots,n\}$, $f^i(j) = 0$.

Since

$$f(x) = \frac{1}{(p-1)!}x^{p-1}(1-x)^p(2-x)^p\cdots(n-x)^p,$$

we have

$$f^{(p-1)}(0) = \frac{1}{(p-1)!}((p-1)!(1-0)^p(2-0)^p\cdots(n-0)^p + 0+0+\cdots) = (n!)^p.$$

Similarly, $f^{(p-2)}(0) = 0$, $f^{(p-3)}(0) = 0$, etc. In short, for every $i \in \{0,1,2,\ldots, p-2\}$, we have $f^{(i)}(0) = 0$, and $f^{(p-1)}(0) = (n!)^p$.

Since

$$F(x) = f(x) + f'(x) + f''(x) + \cdots + f^{(r)}(x),$$

we have, for every $j \in \{1,2,\ldots,n\}$,

$$\begin{aligned}
F(j) &= f(j) + f'(j) + f''(j) + \cdots + f^{(r)}(j) \\
&= f(j) + f'(j) + f''(j) + \cdots + f^{((n+1)p-1)}(j) \\
&= f(j) + f'(j) + f''(j) + \cdots + f^{(p-1)}(j) + \left(f^p(j) + \cdots + f^{((n+1)p-1)}(j)\right) \\
&= f(j) + f'(j) + f''(j) + \cdots + f^{(p-1)}(j) + p \cdot (\text{integer}) \\
&= 0 + 0 + 0 + \cdots + 0 + p \cdot (\text{integer}) = p \cdot (\text{integer}),
\end{aligned}$$

and hence for every $j \in \{1,2,\ldots,n\}$, $F(j)$ is an integer that is a multiple of p.

Since

$$F(x) = f(x) + f'(x) + f''(x) + \cdots + f^{(r)}(x),$$

we have

$$
\begin{aligned}
F(0) &= f(0) + f'(0) + f''(0) + \cdots + f^{(r)}(0) \\
&= f(0) + f'(0) + f''(0) + \cdots + f^{((n+1)p-1)}(0) \\
&= f(0) + f'(0) + f''(0) + \cdots + f^{(p-2)}(0) \\
&\quad + f^{p-1}(0) + \left(f^p(0) + \cdots + f^{((n+1)p-1)}(0) \right) \\
&= f(0) + f'(0) + f''(0) + \cdots + f^{(p-2)}(0) + f^{p-1}(0) + p \cdot (\text{integer}) \\
&= 0 + 0 + 0 + \cdots + 0 + (n!)^p + p \cdot (\text{integer}) = (n!)^p + p \cdot (\text{integer}),
\end{aligned}
$$

and hence $F(0)$ is an integer of the form $(n!)^p + p \cdot (\text{integer})$.

Since for every $j \in \{1, 2, \cdots, n\}$, $F(j)$ is a multiple of p, $F(0)$ is of the form $(n!)^p + p \cdot (\text{integer})$, and each c_i is an integer, it follows that

$$
c_0 F(0) + c_1 F(1) + c_2 F(2) + \cdots + c_n F(n)
$$
$$
\left(= -\left(c_1 1 e^{1(1-\theta_1)} f(1\theta_1) + c_2 2 e^{2(1-\theta_2)} f(2\theta_2) + \cdots + c_n n e^{n(1-\theta_n)} f(n\theta_n) \right) \right)
$$

is an integer of the form $c_0 (n!)^p + p \cdot (\text{integer})$. Thus

$$
-\left(c_1 1 e^{1(1-\theta_1)} f(1\theta_1) + c_2 2 e^{2(1-\theta_2)} f(2\theta_2) + \cdots + c_n n e^{n(1-\theta_n)} f(n\theta_n) \right)
$$

is an integer of the form $c_0 (n!)^p + p \cdot (\text{integer})$.

Observe that

$$
1 e^{1(1-\theta_1)} f(1\theta_1) = 1 e^{1(1-\theta_1)} \frac{1}{(p-1)!} (1\theta_1)^{p-1} (1 - 1\theta_1)^p (2 - 1\theta_1)^p \cdots (n - 1\theta_1)^p,
$$

so

$$
\begin{aligned}
\left| 1 e^{1(1-\theta_1)} f(1\theta_1) \right| &= \left| 1 e^{1(1-\theta_1)} \tfrac{1}{(p-1)!} (1\theta_1)^{p-1} (1 - 1\theta_1)^p (2 - 1\theta_1)^p \cdots (n - 1\theta_1)^p \right| \\
&= 1 e^{1(1-\theta_1)} \tfrac{1}{(p-1)!} (1\theta_1)^{p-1} (|1 - 1\theta_1| |2 - 1\theta_1| \cdots |n - 1\theta_1|)^p.
\end{aligned}
$$

Now, since $\theta_1 \in (0, 1)$, we have

$$
|1 - 1\theta_1| |2 - 1\theta_1| \cdots |n - 1\theta_1| \le 1 \cdot 2 \cdots \cdots n = n!,
$$

and hence

$$\left|1e^{1(1-\theta_1)}f(1\theta_1)\right| \leq 1e^{1(1-\theta_1)}\frac{1}{(p-1)!}(1\theta_1)^{p-1}\cdot(n!)^p \leq 1e^{1(1-\theta_1)}\frac{1}{(p-1)!}n^{p-1}\cdot(n!)^p$$
$$\leq 1e^{1(1-\theta_1)}\frac{1}{(p-1)!}n^p\cdot(n!)^p = 1e^{1(1-\theta_1)}(n(n!))\frac{(n(n!))^{p-1}}{(p-1)!} \to 1e^{1(1-\theta_1)}(n(n!))\cdot 0$$

as $p \to \infty$. Thus $1e^{1(1-\theta_1)}f(1\theta_1) \to 0$ as $p \to \infty$.

Since

$$2e^{2(1-\theta_2)}f(2\theta_2) = 2e^{2(1-\theta_2)}\frac{1}{(p-1)!}(2\theta_2)^{p-1}(1-2\theta_2)^p(2-2\theta_2)^p\cdots(n-2\theta_2)^p,$$

we have

$$\left|2e^{2(1-\theta_2)}f(2\theta_2)\right| = \left|2e^{2(1-\theta_2)}\frac{1}{(p-1)!}(2\theta_2)^{p-1}(1-2\theta_2)^p(2-2\theta_2)^p\cdots(n-2\theta_2)^p\right|$$
$$= 2e^{2(1-\theta_2)}\frac{1}{(p-1)!}(2\theta_2)^{p-1}(|1-2\theta_2||2-2\theta_2|\cdots|n-2\theta_2|)^p.$$

Now, since $\theta_2 \in (0,1)$, we have

$$|1-2\theta_2||2-2\theta_2|\cdots|n-2\theta_2| \leq 1\cdot 2\cdots\cdots n = n!,$$

and hence

$$\left|2e^{2(1-\theta_2)}f(2\theta_2)\right| \leq 2e^{2(1-\theta_2)}\frac{1}{(p-1)!}(2\theta_2)^{p-1}\cdot(n!)^p \leq 2e^{2(1-\theta_2)}\frac{1}{(p-1)!}n^{p-1}\cdot(n!)^p$$
$$\leq 2e^{2(1-\theta_2)}\frac{1}{(p-1)!}n^p\cdot(n!)^p = 2e^{2(1-\theta_2)}(n(n!))\frac{(n(n!))^{p-1}}{(p-1)!} \to 2e^{2(1-\theta_2)}(n(n!))\cdot 0$$

as $p \to \infty$. Thus $2e^{2(1-\theta_2)}f(2\theta_2) \to 0$ as $p \to \infty$. Similarly, $3e^{3(1-\theta_3)}f(3\theta_3) \to 0$ as $p \to \infty$, etc. It follows that

$$-\left(c_1 1e^{1(1-\theta_1)}f(1\theta_1) + c_2 2e^{2(1-\theta_2)}f(2\theta_2) + \cdots + c_n ne^{n(1-\theta_n)}f(n\theta_n)\right) \to 0 \text{ as } p$$
$$\to \infty.$$

Since

$$-\left(c_1 1e^{1(1-\theta_1)}f(1\theta_1) + c_2 2e^{2(1-\theta_2)}f(2\theta_2) + \cdots + c_n ne^{n(1-\theta_n)}f(n\theta_n)\right)$$

is an integer of the form $c_0(n!)^p + p\cdot$ (integer) and p does not divide the integer $c_0(n!)^p$, it follows that

$$-\left(c_1 1 e^{1(1-\theta_1)} f(1\theta_1) + c_2 2 e^{2(1-\theta_2)} f(2\theta_2) + \cdots + c_n n e^{n(1-\theta_n)} f(n\theta_n)\right)$$

is a nonzero integer, and hence

$$-\left(c_1 1 e^{1(1-\theta_1)} f(1\theta_1) + c_2 2 e^{2(1-\theta_2)} f(2\theta_2) + \cdots + c_n n e^{n(1-\theta_n)} f(n\theta_n)\right) \not\to 0 \text{ as } p$$
$$\to \infty.$$

This is a contradiction. ∎

Definition Let F and K be any fields such that K is an extension of F. Let $f(x)$ be a nonzero member of $F[x]$ with $\deg(f(x)) \geq 1$. Let $\alpha \in K$. If $(K \ni) f(\alpha) = 0$, then we say that α *is a root of* $f(x)$.

1.5.2 Theorem Let F and K be any fields such that K is an extension of F. Let $f(x)$ be a nonzero member of $F[x]$ with $\deg(f(x)) \geq 1$. Let $\alpha \in K$. Then there exists a nonzero $q(x) \in K[x]$ such that

1. $f(x) = (x - \alpha)q(x) + f(\alpha)$,
2. $\deg(q(x)) = \deg(f(x)) - 1$.

This theorem is known as the **remainder theorem**.

Proof Since $F \subset K$, we have $(f(x) \in)F[x] \subset K[x]$, and hence $f(x) \in K[x]$. It is given that $f(x)$ is nonzero. Since $1, -\alpha \in K$, the polynomial $x - \alpha$ is a nonzero member of $K[x]$. Now, by 1.2.14, there exist $q(x), r(x) \in K[x]$ such that

$$f(x) = q(x)(x - \alpha) + r(x),$$

and (either $r(x) = 0$ or $\deg(r(x)) < \deg(x - \alpha)(= 1)$). It follows that either $r(x) = 0$ or $\deg(r(x)) = 0$. Since $r(x) \in K[x]$, $r(x)$ is a member of K, and hence $r(x) = r(\alpha) \in K$. Since

$$f(x) = q(x)(x - \alpha) + r(x),$$

we have

$$\underbrace{f(\alpha) = q(\alpha)(\alpha - \alpha) + r(\alpha)} = q(\alpha) \cdot 0 + r(\alpha) = r(\alpha) = r(x),$$

and hence $f(\alpha) = r(x)$. Thus $f(x) = q(x)(x - \alpha) + f(\alpha)$. This proves (1). Since $f(\alpha) \in K$, we have

$$\deg(f(x)) = \underbrace{\deg(q(x)(x - \alpha) + f(\alpha)) = \deg(q(x)(x - \alpha))}$$
$$= \deg(q(x)) + \deg(x - \alpha) = \deg(q(x)) + 1,$$

and hence $\deg(f(x)) = \deg(q(x)) + 1$. ∎

1.5.3 Theorem Let F and K be any fields such that K is an extension of F. Let $f(x)$ be a nonzero member of $F[x]$ with $\deg(f(x)) \geq 1$. Let $\alpha \in K$. Let α be a root of $f(x)$. Then $(x - \alpha)|f(x)$ in $K[x]$.

Proof By 1.5.2, there exists a nonzero $q(x) \in K[x]$ such that

1. $f(x) = (x - \alpha)q(x) + f(\alpha)$,
2. $\deg(q(x)) = \deg(f(x)) - 1$.

Since α is a root of $f(x)$, we have

$$f(x) - (x - \alpha)q(x) = \underbrace{f(\alpha) = 0,}$$

and hence $f(x) - (x - \alpha)q(x) = 0$, that is, $f(x) = (x - \alpha)q(x)$. Since $F \subset K$, we have $(f(x) \in)F[x] \subset K[x]$, and hence $f(x) \in K[x]$. Since $\alpha \in K$, we have $(x - \alpha) \in K[x]$. Also, $q(x) \in K[x]$. Next, since $f(x) = (x - \alpha)q(x)$, it follows that $(x - \alpha)|f(x)$ in $K[x]$. ∎

Definition Let F and K be any fields such that K is an extension of F. Let $f(x)$ be a nonzero member of $F[x]$ with $\deg(f(x)) \geq 1$. Let $\alpha \in K$. Let m be a positive integer.

If $(x - \alpha)^m|f(x)$ in $K[x]$, then clearly, $f(\alpha) = 0$, and hence α is a root of $f(x)$.

If $(x - \alpha)^m|f(x)$ in $K[x]$ and $(x - \alpha)^{m+1} \nmid f(x)$ in $K[x]$, then we say that α *is a root of $f(x)$ of multiplicity m*.

Caution We count α as m roots.

1.5.4 Theorem Let F and K be any fields such that $F \subset K$. Suppose that K is an extension of F. Let $f(x)$ be a nonzero member of $F[x]$ with $\deg(f(x)) \geq 1$. Suppose that $\deg(f(x)) = n$. Then the number of roots of $f(x)$ in K is $\leq n$.

Proof (Induction on n) If $f(x)$ has no root in K, then the number of roots of $f(x)$ in K is 0, and hence the result is trivially true. So we consider the case that there exists a root of $f(x)$ in K.

Suppose that $\deg(f(x)) = 1$. We can suppose that $f(x) \equiv a + bx$, where $a, b \in F$ and $b \neq 0$. Next, let $\alpha, \beta \in K$ such that

$$\left.\begin{array}{c} a + b\alpha = 0 \\ a + b\beta = 0 \end{array}\right\}.$$

We shall show that $\alpha = \beta$. Since $a + b\alpha = 0$ and $b \neq 0$, we have $\alpha = -b^{-1}a$. Similarly, $\beta = -b^{-1}a$. It follows that $\alpha = \beta$. Thus the result is true for $n = 1$.

Now let us suppose that the result is true for all positive integer values $<n$. It suffices to show that the result is true for n.

Let α be a root of $f(x)$ in K, and let $m(\geq 1)$ be its multiplicity. Hence $(x - \alpha)^m|f(x)$ in $K[x]$, and $(x - \alpha)^{m+1} \nmid f(x)$ in $K[x]$. It follows that there exists

$g(x) \in K[x]$ such   that $f(x) = (x - \alpha)^m g(x)$ and $(x - \alpha) \nmid g(x)$.   Since $f(x) = (x - \alpha)^m g(x)$, we have

$$n = \underbrace{\deg(f(x)) = \deg((x - \alpha)^m g(x))} = \deg((x - \alpha)^m) + \deg(g(x))$$

$$= m + \deg(g(x)) \geq 1 + \deg(g(x)) > \deg(g(x)),$$

and hence $\deg(g(x)) < n$. By the induction hypothesis, the number of roots of $g(x)$ in K is $\leq \deg(g(x))$. Since

$$f(x) = (x - \alpha)^m g(x),$$

the number of roots of $f(x)$ in K is equal to

$$m + (\text{the number of roots of } g(x) \text{ in } K)(\leq m + \deg(g(x)) = n),$$

and hence the number of roots of $f(x)$ in K is $\leq n$.   ∎

1.5.5 Note Let F and K be any fields such that K is an extension of F. Let $p(x)$ be a nonzero member of $F[x]$ with $\deg(p(x)) \geq 1$. Suppose that $\deg(p(x)) = n$. Let $p(x)$ be irreducible over F.

By 1.2.24, the quotient ring $\frac{F[x]}{V}$ is a field, where V denotes the ideal $(p(x))(= \{f(x)p(x) : f(x) \in F[x]\})$.

Let $\psi : a \mapsto (a + V)$ be a mapping from the field F to the field $\frac{F[x]}{V}$. It is clear that ψ is a ring isomorphism:

1. $\psi : F \to \frac{F[x]}{V}$ is one-to-one: To prove this, let $\psi(a) = \psi(b)$. We have to show that $a = b$.

Since $a + V = \underbrace{\psi(a) = \psi(b)} = b + V$,   we   have $a + V = b + V$,   and   hence $(a - b) \in V$.   Since   each   nonzero   element   of $((a - b) \in V =)$ $\{f(x)p(x) : f(x) \in F[x]\}$ is   of   degree $\geq \deg(p(x))(\geq 1)$,   we have $a - b = 0$ or $\deg(a - b) \geq 1$. Since $(a - b) \in V$, either $a - b = 0$ or $\deg(a - b) = 0$. It follows that $a - b = 0$, that is, $a = b$.

2. It is clear that ψ is a ring homomorphism.

Thus we have shown that $\psi : F \to \frac{F[x]}{V}$ is a ring isomorphism from the field F to the field $\frac{F[x]}{V}$. It follows that we can identify each element a of F with $\psi(a) (= (a + V))$ of the field $\frac{F[x]}{V}$. It is in this sense that we write $F \subset \frac{F[x]}{V}$ and treat $\frac{F[x]}{V}$ as an extension of F.

Since $\deg(p(x)) = n$, by 1.4.10, n is the dimension of the vector space $\frac{F[x]}{V}$, and hence $\left[\frac{F[x]}{V} : F\right] = n$. Also, by 1.4.10, $\{1 + V, x + V, x^2 + V, \cdots, x^{n-1} + V\}$ is a basis of $\frac{F[x]}{V}$. From the definition of addition and scalar multiplication over the quotient ring $\frac{F[x]}{V}$, it is clear that

$$\underbrace{p(x+V) = p(x) + V} = p(x) + (p(x)) = (p(x)) = V = 0 + V \in \frac{F[x]}{V},$$

and hence $p(x+V) = 0 + V$. Here $p(x+V) = 0 + V$, $x + V$ is a member of $\frac{F[x]}{V}$, and $0 + V$ is the zero element of the field $\frac{F[x]}{V}$, so $x + V$ is a root of the given polynomial. Thus $p(x)$ has a root in $\frac{F[x]}{V}$.

1.5.6 Conclusion Let F and K be any fields such that K is an extension of F. Let $p(x)$ be a nonzero member of $F[x]$ with $\deg(p(x)) \geq 1$. Suppose that $\deg(p(x)) = n$. Let $p(x)$ be irreducible over F. Then there exists a field E such that

1. E is an extension of F,
2. $[E : F] = n$,
3. $p(x)$ has a root in E.

1.5.7 Problem Let F and K be any fields such that K is an extension of F. Let $f(x)$ be a nonzero member of $F[x]$ with $\deg(f(x)) \geq 1$. Then there exists a field E such that

1. E is a finite extension of F,
2. $[E : F] \leq \deg(f(x))$,
3. $f(x)$ has a root in E.

Proof Since $\deg(f(x)) \geq 1$, $f(x)$ is not a unit in $F[x]$, and hence by 1.2.20, there exists an irreducible $p(x) \in F[x]$ such that $1 \leq \deg(p(x)) \leq \deg(f(x))$ and $p(x)|f(x)$. It follows, by 1.5.6, that there exists a field E such that

1. E is an extension of F,
2. $[E : F] = \deg(p(x))(\leq \deg(f(x)) < \infty)$,
3. $p(x)$ has a root in E.

Since $[E : F] < \infty$, E is a finite extension of F. Also $[E : F] \leq \deg(f(x))$. Since $p(x)$ has a root, say α, in E, and $p(x)|f(x)$, α is also a root of $f(x)$. ∎

1.5.8 Note Let F and K be any fields such that K is an extension of F. Let $f(x)$ be a nonzero member of $F[x]$ and $\deg(f(x)) \geq 1$. Let $\deg(f(x)) = n$.

By 1.5.8, there exists a field E_1 such that

1. E_1 is a finite extension of F,
2. $[E_1 : F] \leq n$,
3. $f(x)$ has a root, say α_1, in E_1.

It follows, by 1.5.3, that $(x - \alpha_1)|f(x)$ in $E_1[x]$, and hence there exists $f_1(x) \in E_1[x]$ such that $f(x) = (x - \alpha_1)f_1(x)$ and $\deg(f_1(x)) = n - 1$.

By 1.5.9, there exists a field E_1 such that

1. E_2 is a finite extension of E_1,
2. $[E_2 : E_1] \leq n - 1$,
3. $f_1(x)$ has a root, say α_2, in E_2.

It follows, by 1.5.3, that $(x - \alpha_2)|f_1(x)$ in $E_2[x]$, and hence there exists $f_2(x) \in E_2[x]$ such that $f_1(x) = (x - \alpha_2)f_2(x)$ and $\deg(f_2(x)) = (n - 1) - 1(= n - 2)$. It follows, by 1.4.3, that E_2 is a finite extension of F, and

$$\underbrace{[E_2 : F] = [E_2 : E_1][E_1 : F]}_{} \leq [E_2 : E_1]n \leq (n - 1)n = n(n - 1).$$

Thus $[E_2 : F] \leq n(n - 1)$. Also,

$$f(x) = (x - \alpha_1)f_1(x) = f(x) = (x - \alpha_1)(x - \alpha_2)f_2(x),$$

so the field E_2 contains two roots, α_1, α_2 of $f(x)$.

Similarly, the field E_3 contains three roots of $f(x)$, $[E_3 : F] \leq n(n - 1)(n - 2)$, and E_3 is a finite extension of F.

Finally, there exists a field E such that

1. E is a finite extension of F,
2. E contains all the roots of $f(x)$ in K,
3. $[E : F] \leq n(n - 1)(n - 2) \cdots 2 \cdot 1(= n!)$.

1.5.9 Conclusion Let F and K be any fields such that K is an extension of F. Let $f(x)$ be a nonzero member of $F[x]$ with $\deg(f(x)) \geq 1$. Let $\deg(f(x)) = n$. Suppose that K contains n roots of $f(x)$. Then there exists a field E such that

1. E is a finite extension of F,
2. E contains all the roots of $f(x)$ in K,
3. if G is a proper subfield of E, then G does not contain all the roots of $f(x)$ in K,
4. $[E : F] \leq n!$.

Definition Let F and K be any fields such that $F \subset K$. Suppose that K is an extension of F. Let $f(x)$ be a nonzero member of $F[x]$ with $\deg(f(x)) \geq 1$. Let $\deg(f(x)) = n$. Suppose that K contains n roots of $f(x)$. Let E be a field such that

1. E is a finite extension of F,
2. E contains all the roots of $f(x)$ in K,
3. if G is a proper subfield of E that contains F, then G does not contain all the roots of $f(x)$ in K.

Then we say that E is a *splitting field over F for $f(x)$*.

Thus a field E is a splitting field over F for $f(x)$ if and only if E is a minimal finite extension of F in which $f(x)$ can be factored as a product of linear factors in $E[x]$. From 1.5.9,

$$[(\text{splitting field over } F \text{ for } f(x)) : F] \le (\deg(f(x)))!.$$

1.5.10 Problem Let F, F' be any fields. Let $\tau : \alpha \mapsto \alpha'$ be a ring isomorphism from F onto F'. Then the map

$$\tau^* : \left(\alpha_0 + \alpha_1 x + \alpha_2 x^2 + \cdots + \alpha_n x^n\right) \mapsto \left(\alpha_0' + \alpha_1' t + \alpha_2' t^2 + \cdots + \alpha_n' t^n\right)$$

from the polynomial ring $F[x]$ to the polynomial ring $F'[t]$ is a ring isomorphism from $F[x]$ onto $F'[t]$ such that for every $\alpha \in F$, we have $\tau^*(\alpha) = \alpha'$.

Proof $\tau^* : F[x] \to F'[t]$ is one-to-one: To show this, suppose that

$$\alpha_0' + \alpha_1' t + \alpha_2' t^2 + \cdots + \alpha_n' t^n = \beta_0' + \beta_1' t + \beta_2' t^2 + \cdots + \beta_n' t^n.$$

We have to show that $\alpha_i = \beta_i (i = 0, 1, \ldots, n)$. Since

$$\alpha_0' + \alpha_1' t + \alpha_2' t^2 + \cdots + \alpha_n' t^n = \beta_0' + \beta_1' t + \beta_2' t^2 + \cdots + \beta_n' t^n,$$

we have $\alpha_i' = \beta_i'(i = 0, 1, \ldots, n)$, and hence $\tau(\alpha_i) = \tau(\beta_i)(i = 0, 1, \ldots, n)$. Since $\tau : F \to F'$ is a ring isomorphism, $\tau : F \to F'$ is one-to-one, and since $\tau(\alpha_i) = \tau(\beta_i)(i = 0, 1, \ldots, n)$, we have $\alpha_i = \beta_i(i = 0, 1, \ldots, n)$.

$\tau^* : F[x] \to F'[t]$ is onto: This is clear.

$\tau^* : F[x] \to F'[t]$ is a ring homomorphism: This is clear.

Thus, τ^* is a ring isomorphism from $F[x]$ onto $F'[t]$. Also, it is clear that for every $\alpha \in F$, $\tau^*(\alpha) = \alpha'$. ∎

1.5.11 Problem Let F, F' be any fields. Let $\tau : \alpha \mapsto \alpha'$ be a ring isomorphism from F onto F'. For every $f(x) \in F[x]$, we shall denote $\tau^*(f(x))$ by $f'(t)$, where τ^* is the same as discussed in 1.5.10. Thus $\tau^* : f(x) \mapsto f'(t)$ from the ring $F[x]$ onto the ring $F'[t]$ is an isomorphism. Let $p(x) \in F[x]$. It follows that $p'(t) \in F'[t]$. Put $V \equiv (p(x))$, where $(p(x))$ denotes the ideal generated by $p(x)$ in $F[x]$. Put $V' \equiv (p'(t))$, where $(p'(t))$ denotes the ideal generated by $p'(t)$ in $F'[t]$. Let

$$\tau^{**} : f(x) + V \mapsto f'(t) + V'$$

be the mapping from the quotient ring $\frac{F[x]}{V}$ to the quotient ring $\frac{F'[t]}{V'}$. Then τ^{**} is an isomorphism from $\frac{F[x]}{V}$ onto $\frac{F'[t]}{V'}$. Also, for every $\alpha \in F$, we have $\tau^{**}(\alpha + V) = \alpha'$ and

$$\tau^{**}(x + V) = t + V'.$$

Proof $\tau^{**} : \frac{F[x]}{V} \to \frac{F'[t]}{V'}$ is well defined. To show this, let $f(x), g(x) \in F[x]$ be such that $f(x) - g(x) \in V(= (p(x)))$. We have to show that $f'(t) - g'(t) \in V'$. Since

$f(x) - g(x) \in (p(x))$, there exists $h(x) \in F[x]$ such that $f(x) - g(x) = p(x)h(x)$, and hence

$$f'(t) - g'(t) = \tau^*(f(x)) - \tau^*(g(x)) = \underline{\tau^*(f(x) - g(x)) = \tau^*(p(x)h(x))} = \tau^*(p(x))\tau^*(h(x))$$

$$= p'(t)\tau^*(h(x)) = p'(t)h'(t) \in (p'(t)) = V'.$$

Thus $f'(t) - g'(t) \in V'$.

$\tau^{**} : \frac{F[x]}{V} \to \frac{F'[t]}{V'}$ is one-to-one. To show this, let $f(x), g(x) \in F[x]$ be such that $f'(t) - g'(t) \in V'(= (p'(t)))$. We have to show that $f(x) - g(x) \in V$. Since $f'(t) - g'(t) \in (p'(t))$, there exists $h(x) \in F[x]$ such that

$$\tau^*(f(x) - g(x)) = \tau^*(f(x)) - \tau^*(g(x))$$
$$= \underline{f'(t) - g'(t) = p'(t)h'(t)} = \tau^*(p(x))\tau^*(h(x)) = \tau^*(p(x)h(x)),$$

and hence

$$\tau^*(f(x) - g(x)) = \tau^*(p(x)h(x)).$$

Since τ^* is one-to-one, we have $f(x) - g(x) = p(x)h(x) \in (p(x)) = V$, and hence $f(x) - g(x) \in V$.

$\tau^{**} : \frac{F[x]}{V} \to \frac{F'[t]}{V'}$ is onto. This is clear.

$\tau^{**} : \frac{F[x]}{V} \to \frac{F'[t]}{V'}$ is a ring homomorphism. This is clear.

Thus τ^{**} is an isomorphism from $\frac{F[x]}{V}$ onto $\frac{F'[t]}{V'}$.

By the definition of τ^{**}, for every $\alpha \in F$, $\underline{\tau^{**}(\alpha + V) = \tau^*(\alpha) + V' = \alpha' + V'}$, so

for every $\alpha \in F$, $\tau^{**}(\alpha + V) = \alpha' + V'$. Since for every $\alpha \in F$, we have $\alpha' \in F'$, and we identify $(\tau^{**}(\alpha + V) =)\alpha' + V'$ with α', we can write $\tau^{**}(\alpha + V) = \alpha'$.

Since the polynomial x is a member of $F[x]$, we have

$$\tau^{**}(x + V) = t + V'. \qquad \blacksquare$$

1.5.12 Problem Let F and K be any fields such that $F \subset K$. Suppose that K is an extension of F. Let α be a member of K. Let $p(x) \in F[x](\supset F)$. Let $p(x)$ be irreducible over F. Let n be a positive integer. Let n be the degree of $p(x)$. Let α be a root of $p(x)$ in K. Then

1. α is algebraic of degree n over F.
2. $\frac{1}{\text{leading coefficient of } p(x)} p(x)$ is the minimal polynomial of α over F.

Proof Put

$$p_1(x) \equiv \frac{1}{\text{leading coefficient of } p(x)} p(x).$$

Clearly, $(K \ni) p_1(\alpha) = 0$, $1 \leq n = \deg(p_1(x))$, and the leading coefficient of $p_1(x)$ is 1.

Let α be algebraic of degree m over F. It follows that $m \leq n$. We have to show that $m = n$. Suppose to the contrary that $m < n$. We seek a contradiction.

Since α is algebraic of degree m over F, there exists $f(x) \in F[x]$ such that $(K \ni) f(\alpha) = 0$,

$$\underbrace{1 \leq \deg(f(x)) = m}_{} < n = \deg(p_1(x)),$$

and the leading coefficient of $f(x)$ is 1. It follows that there exist $q(x), r(x) \in F[x]$ such that

$$p_1(x) = f(x)q(x) + r(x)$$

and

$$(r(x) = 0 \text{ or } \deg(r(x)) < \deg(f(x))).$$

Since $\deg(f(x)) < \deg(p_1(x))$, $p_1(x) = f(x)q(x) + r(x)$, and $p_1(x)$ is irreducible over F, we have $r(x) \neq 0$. Also

$$0 = \underbrace{p_1(\alpha) = f(\alpha)q(\alpha) + r(\alpha)}_{} = 0q(\alpha) + r(\alpha) = r(\alpha),$$

so $r(\alpha) = 0$. Since α is algebraic of degree m over F, $r(x) \neq 0$, and $r(\alpha) = 0$, we have $(\deg(f(x)) =)m \leq \deg(r(x))$. Since $(r(x) = 0 \text{ or } \deg(r(x)) < \deg(f(x)))$, we have $r(x) = 0$. This is a contradiction.

Thus α is algebraic of degree n over F. Also

$$\frac{1}{\text{leading coefficient of } p(x)} p(x)$$

is the minimal polynomial of α over F.   ∎

1.5.13 Note Let F and K be any fields such that K is an extension of F. Let α be a member of K. Let $p(x) \in F[x](\supset F)$. Let $p(x)$ be irreducible over F. Let n be a positive integer. Let n be the degree of $p(x)$. Let α be a root of $p(x)$ in K. By 1.5.12,

1. α is algebraic of degree n over F.

2. $\frac{1}{\text{leading coefficient of } p(x)} p(x)$ is the minimal polynomial of α over F.

Let $\psi : f(x) \mapsto f(\alpha)$ be the mapping from the ring $F[x]$ to the field $F(\alpha)$. By 1.4.13, $\psi : F[x] \to F(\alpha)$ is a ring homomorphism, and hence by the fundamental theorem of ring homomorphism, the mapping $\psi^* : (f(x) + \ker(\psi)) \mapsto \psi(f(x))(= f(\alpha))$ is a ring isomorphism from the quotient ring $\frac{F[x]}{\ker(\psi)}$ to $F(\alpha)$. Also, ψ^* maps $\frac{F[x]}{\ker(\psi)}$ onto $F(\alpha)$. Put

$$p_1(x) \equiv \frac{1}{\text{leading coefficient of } p(x)} p(x).$$

By 1.4.12, $p_1(x)$ is irreducible over F, and hence by 1.2.22, the ideal

$$(p_1(x)) \left(\begin{aligned} &\equiv \{p_1(x)f(x) : f(x) \in F[x]\} \\ &= \left\{ \frac{1}{\text{leading coefficient of } p(x)} p(x)f(x) : f(x) \in F[x] \right\} \end{aligned} \right.$$

$$= \left\{ p(x) \cdot \frac{1}{\text{leading coefficient of } p(x)} f(x) : f(x) \in F[x] \right\}$$

$$= \{p(x) \cdot f(x) : f(x) \in F[x]\} = (p(x)))$$

is a maximal ideal of the ring $F[x]$, and hence $(p(x))$ is a maximal ideal of the ring $F[x]$.

Further, by Note 1.2.24, the quotient ring $\frac{F[x]}{(p_1(x))} \left(= \frac{F[x]}{(p(x))} \right)$ is a field. Also, $\{f(\alpha) : f(x) \in F[x]\} \subset F(\alpha)$. By 1.4.13,

$$\underbrace{\ker(\psi) = (p_1(x))} = (p(x)).$$

Now, since $\ker(\psi) = (p(x))$, it follows that $\psi^* : (f(x) + \ker(\psi)) \mapsto f(\alpha)$ is a ring isomorphism from the field $\frac{F[x]}{(p(x))}$ onto the field $F(\alpha)$.

Further, for every $\beta \in F$,

$$\psi^*(\beta + (p(x))) = \underbrace{\psi^*(\beta + \ker(\psi)) = \psi(\beta)} = \beta$$

and

$$\psi^*(x + (p(x))) = \psi(x) = \alpha.$$

1.5.14 Conclusion Let F and K be any fields such that K is an extension of F. Let α be a member of K. Let $p(x) \in F[x](\supset F)$. Let $p(x)$ be irreducible over F. Let α be a root of $p(x)$ in K. Let $\psi : f(x) \mapsto f(\alpha)$ be the mapping from the ring $F[x]$ to the field $F(\alpha)$. Then

1. $\psi^* : (f(x) + \ker(\psi)) \mapsto f(\alpha)$ is a ring isomorphism from the field $\frac{F[x]}{(p(x))}$ onto the field $F(\alpha)$,
2. for every $\beta \in F$, $\psi^*(\beta + (p(x))) = \beta$,
3. $\psi^*(x + (p(x))) = \alpha$.

In short, ψ^* is an isomorphism from $\frac{F[x]}{(p(x))}$ onto $F(\alpha)$ such that every element of F is fixed, and the ψ^*-image of $x + (p(x))$ is α.

1.5.15 Note Let F and K be any fields such that K is an extension of F. Let α be a member of K. Let $p(x) \in F[x](\supset F)$. Let $p(x)$ be irreducible over F. Let α be a root of $p(x)$ in K. Let F' be any field. Let $\tau : a \mapsto a'$ be a ring isomorphism from F onto F'. For every $f(x) \in F[x]$, we shall denote $\tau^*(f(x))$ by $f'(t)$, where τ^* is the same as discussed in 1.5.10. We know that $\tau^* : f(x) \mapsto f'(t)$ is an isomorphism from the ring $F[x]$ onto the ring $F'[t]$.

Since $p(x) \in F[x]$, we have $p'(t) \in F'[t]$. Let β be a root of $p'(t)$ in some extension K' of F'.

Suppose that $(p(x))$ denotes the ideal generated by $p(x)$ in $F[x]$. Suppose that $(p'(t))$ denotes the ideal generated by $p'(t)$ in $F'[t]$. Let

$$\tau^{**} : f(x) + (p(x)) \mapsto f'(t) + (p'(t))$$

be a mapping from the quotient ring $\frac{F[x]}{(p(x))}$ to the quotient ring $\frac{F'[t]}{(p'(t))}$. By 1.5.11, τ^{**} is an isomorphism from $\frac{F[x]}{(p(x))}$ onto $\frac{F'[t]}{(p'(t))}$. Also, for every $\alpha \in F$, we have $\tau^{**}(\alpha + (p(x))) = \alpha' + (p'(t))$ and

$$\tau^{**}(x + (p(x))) = t + (p'(t)).$$

Let $\psi : f(x) \mapsto f(\alpha)$ be a mapping from the ring $F[x]$ to the field $F(\alpha)$. Then by 1.5.14,

1. $\psi^* : (f(x) + (p(x))) \mapsto f(\alpha)$ is a ring isomorphism from the field $\frac{F[x]}{(p(x))}$ onto the field $F(\alpha)$,
2. for every $a \in F$, $\psi^*(a + (p(x))) = a$,
3. $\psi^*(x + (p(x))) = \alpha$.

Let $\theta : f'(t) \mapsto f'(\beta)$ be a mapping from the ring $F'[t]$ to the field $F'(\beta)$. Then by 1.5.14,

1. $\theta^* : (f'(t) + (p'(t))) \mapsto f'(\beta)$ is a ring isomorphism from the field $\frac{F'[t]}{(p'(t))}$ onto the field $F'(\beta)$,

2. for every $b \in F'$, $\theta^*(b + (p'(t))) = b$,
3. $\theta^*(t + (p'(t))) = \beta$.

Since ψ^* is a ring isomorphism from $\frac{F[x]}{(p(x))}$ onto $F(\alpha)$, $(\psi^*)^{-1}$ is a ring isomor-phism from $F(\alpha)$ onto $\frac{F[x]}{(p(x))}$. Now, since τ^{**} is an isomorphism from $\frac{F[x]}{(p(x))}$ onto $\frac{F'[t]}{(p'(t))}$, and θ^* is a ring isomorphism from $\frac{F'[t]}{(p'(t))}$ onto $F'(\beta)$, the composite $\theta^* \deg \tau^{**} \deg\left((\psi^*)^{-1}\right)$ is an isomorphism from $F(\alpha)$ onto $F'(\beta)$.

For every $a \in F$,

$$\left(\theta^* \circ \tau^{**} \circ \left((\psi^*)^{-1}\right)\right)(a) = \theta^*\left(\tau^{**}\left((\psi^*)^{-1}(a)\right)\right)$$
$$= \theta^*(\tau^{**}(a + (p(x)))) = \theta^*(a' + (p'(t))) = a',$$

so for every $a \in F$, we have $\sigma(a) = a'$, where $\sigma \equiv \theta^* \circ \tau^{**} \circ \left((\psi^*)^{-1}\right)$.

Next,

$$\sigma(\alpha) = \left(\theta^* \circ \tau^{**} \circ \left((\psi^*)^{-1}\right)\right)(\alpha) = \theta^*\left(\tau^{**}\left((\psi^*)^{-1}(\alpha)\right)\right) = \theta^*(\tau^{**}(x + (p(x))))$$
$$= \theta^*(t + (p'(t))) = \beta,$$

so $\sigma(\alpha) = \beta$.

1.5.16 Conclusion Let F and K be any fields such that K is an extension of F. Let α be a member of K. Let $p(x) \in F[x](\supset F)$. Let $p(x)$ be irreducible over F. Let α be a root of $p(x)$ in K. Let F' be any field. Let $\tau : a \mapsto a'$ be a ring isomorphism from F onto F'. For every $f(x) \in F[x]$, we shall denote $\tau^*(f(x))$ by $f'(t)$, where τ^* is the same as discussed in 1.5.10. Let β be a root of $p'(t)$ in some extension K' of F'. Then there exists an isomorphism σ from the field $F(\alpha)$ onto the field $F'(\beta)$ such that

1. $\sigma(\alpha) = \beta$,
2. for every $a \in F$, $\sigma(a) = a'$.

1.5.17 Note In 1.5.16, let us take F for F' and the identity map $i : F \rightarrow F$ for τ. Thus for every $a \in F$, a' means a, and for every $f(x) \in F[x]$, $f'(t)$ means $f(t)$. Also, $p'(t)$ means $p(t)$. Thus α, β are the roots of the same polynomial $p(x)$. By 1.5.16, there exists an isomorphism σ from the field $F(\alpha)$ onto the field $F(\beta)$ such that

1. $\sigma(\alpha) = \beta$,
2. for every $a \in F$, $\sigma(a) = a$.

1.5.18 Conclusion Let F and K be any fields such that K is an extension of F. Let α, β be members of K. Let $p(x) \in F[x](\supset F)$. Let $p(x)$ be irreducible over F. Let

α, β be any roots of $p(x)$ in K. Then there exists an isomorphism σ from the field $F(\alpha)$ onto the field $F(\beta)$ such that

1. $\sigma(\alpha) = \beta$,
2. for every $a \in F$, $\sigma(a) = a$.

1.5.19 Example Let F be the field of all rational numbers, and let K be the field of all complex numbers. Let us take the polynomial $x^4 + x^2 + 1$ for $f(x)$ in $F[x]$. According to 1.5.9,

$$[(\text{splitting field over } F \text{ for } f(x)) : F] \le (\deg(f(x)))!,$$

so

$$\left[(\text{splitting field over } F \text{ for } x^4 + x^2 + 1) : F\right] \le (\deg(x^4 + x^2 + 1))! = 4! = 24,$$

and hence

$$1 \le \left[(\text{splitting field over } F \text{ for } x^4 + x^2 + 1) : F\right] \le 24.$$

Since

$$x^4 + x^2 + 1 = (x^2 + 1)^2 - x^2 = (x^2 + 1 + x)(x^2 + 1 - x)$$
$$= (x - \omega)(x - \omega^2) \cdot (x + \omega)(x + \omega^2),$$

where $\omega \equiv \frac{-1}{2} + i\frac{\sqrt{3}}{2}$, $F(\omega)$ is a splitting field over F for $x^4 + x^2 + 1$. Since the polynomial $1 + x + x^2$ is a member of $F[x]$, $1 + x + x^2$ is irreducible over F, $\deg(1 + x + x^2) = 2$, and ω is a root of $1 + x + x^2$ in K, by 1.5.12, ω is algebraic of degree 2 over F, and hence by 1.4.16, $[F(\omega) : F] = 2$. Thus

$$\left[(\text{splitting field over } F \text{ for } x^4 + x^2 + 1) : F\right] = 2.$$

1.5.20 Example Let F be the field of all rational numbers, and let K be the field of all complex numbers. Let us take the polynomial $x^3 - 2$ for $f(x)$ in $F[x]$. According to 1.5.9,

$$[(\text{splitting field over } F \text{ for } f(x)) : F] \le (\deg(f(x)))!,$$

so

$$\left[(\text{splitting field over } F \text{ for } x^3 - 2) : F\right] \le (\deg(x^3 - 2))! = 3! = 6,$$

and hence

$$1 \leq \left[(\text{splitting field over } F \text{ for } x^3 - 2) : F\right] \leq 6.$$

Observe that

$$x^3 - 2 = \left(x - \sqrt[3]{2}\right) \cdot \left(x - \sqrt[3]{2}\omega\right)\left(x - \sqrt[3]{2}\omega^2\right),$$

where $\omega \equiv \frac{-1}{2} + i\frac{\sqrt{3}}{2}$. Since the polynomial $x^3 - 2$ is a member of $F[x]$, $x^3 - 2$ is irreducible over F, $\deg(x^3 - 2) = 3$, and $\sqrt[3]{2}$ is a root of $x^3 - 2$ in K, by 1.5.12, $\sqrt[3]{2}$ is algebraic of degree 3 over F, and hence by 1.4.16, $\left[F(\sqrt[3]{2}) : F\right] = 3$. By 1.4.4,

$$\left[F\left(\sqrt[3]{2}\right) : F\right] \mid \left[(\text{splitting field over } F \text{ for } x^3 - 2) : F\right].$$

Now, since $\left[F(\sqrt[3]{2}) : F\right] = 3$, and $1 \leq \left[(\text{splitting field over } F \text{ for } x^3 - 2) : F\right]$ ≤ 6, we have

$$\left[(\text{splitting field over } F \text{ for } x^3 - 2) : F\right] = 3 \text{ or } 6. \quad (*)$$

Since members of $F(\sqrt[3]{2})$ are real numbers,

$$x^3 - 2 = \left(x - \sqrt[3]{2}\right) \cdot \left(x - \sqrt[3]{2}\omega\right)\left(x - \sqrt[3]{2}\omega^2\right),$$

and $\sqrt[3]{2}\omega, \sqrt[3]{2}\omega^2$ are not real numbers, $F(\sqrt[3]{2})$ is not a splitting field over F for $x^3 - 2$, and hence $3 < \left[(\text{splitting field over } F \text{ for } x^3 - 2) : F\right]$. It follows from $(*)$ that

$$\left[(\text{splitting field over } F \text{ for } x^3 - 2) : F\right] = 6.$$

1.5.21 Example Let F be the field of all rational numbers, and let K be the field of all complex numbers. Let $\alpha, \beta \in F$. Let us take the polynomial $x^2 + \alpha x + \beta$ for $f(x)$ in $F[x]$. Suppose that $a \in (K - F)$ such that a is a root of $x^2 + \alpha x + \beta$, that is, $a^2 + \alpha a + \beta = 0$.

According to 1.5.9,

$$\left[(\text{splitting field over } F \text{ for } f(x)) : F\right] \leq (\deg(f(x)))!,$$

so

$$\left[(\text{splitting field over } F \text{ for } x^2 + \alpha x + \beta) : F\right] \leq (\deg(x^2 + \alpha x + \beta))! = 2! = 2,$$

and hence

$$\left[\left(\text{splitting field over } F \text{ for } x^2 + \alpha x + \beta\right) : F\right] = 1 \text{ or } 2.$$

Since $a \in (K - F)$ such that a is a root of $x^2 + \alpha x + \beta$, we have $[(\text{splitting field over } F \text{ for } x^2 + \alpha x + \beta) : F] > 1$, and hence

$$\left[\left(\text{splitting field over } F \text{ for } x^2 + \alpha x + \beta\right) : F\right] = 2.$$

1.5.22 Theorem Let F and K be any fields such that K is an extension of F. Let F' and K' be any fields such that K' is an extension of F'. Let $\tau : a \mapsto a'$ be a ring isomorphism from F onto F'. For every $f(x) \in F[x]$, we shall denote $\tau^*(f(x))$ by $f'(t)$, where τ^* is the same as discussed in 1.5.10. Let $g(x) \in F[x]$. It follows that $g'(t) \in F'[t]$. Let E be a splitting field over F for $g(x)$. Let E' be a splitting field over F' for $g'(t)$. Suppose that $[E : F] = 1$. Then $E = F$.

Proof Suppose to the contrary that $E \neq F$. We seek a contradiction.

Since E is a splitting field over F for $g(x)$, E is a finite extension of F, and hence $F \subset E$. Now, since $E \neq F$, there exists a nonzero a in E such that $a \notin F$. Snce $1 \in F$, we have $a \neq 1$, and hence $\{1, a\}(\subset E)$ is a set of two elements.

Clearly, $\{1, a\}$ is a linearly independent subset of E.

Proof Suppose that $\lambda 1 + \mu a = 0$, where $\lambda, \mu \in F$. We have to show that $\lambda = 0$ and $\mu = 0$.

If $\mu \neq 0$, then $a = -\mu^{-1}\lambda \in F$, and hence $a \in F$. This is a contradiction. Hence $\mu = 0$. Since $\lambda 1 + \mu a = 0$,
we have $\lambda = 0$.   ∎

Thus we have shown that $\{1, a\}$ is a linearly independent subset of E, and the number of elements in $\{1, a\}$ is 2. It follows that $1 = [E : F] = \underbrace{\dim(E)}_{} \geq 2$. Thus

we get a contradiction.   ∎

1.5.23 Note Let F and K be any fields such that K is an extension of F. Let F' and K' be any fields such that K' is an extension of F'. Let $\tau : a \mapsto a'$ be a ring isomorphism from F onto F'. For every $f(x) \in F[x]$, we shall denote $\tau^*(f(x))$ by $f'(t)$, where τ^* is the same as discussed in 1.5.10. Let $g(x) \in F[x]$. It follows that $g'(t) \in F'[t]$. Let E be a splitting field over F for $g(x)$. Let E' be a splitting field over F' for $g'(t)$. Suppose that $[E : F] = 1$.

By 1.5.22, we have $E = F$. Since E is a splitting field over F for $g(x)$, F is a splitting field over F for $g(x)$, and hence $g(x)$ splits into a product of linear factors over F. By 1.5.10, $g'(t)$ splits into a product of linear factors over F'. Next, since E' is a splitting field over F' for $g'(t)$, we have $E' = F'$. Since τ is a ring isomorphism from F onto F', $E = F$, and $E' = F'$, τ is a ring isomorphism from E onto E'.

Let us take an arbitrary $\alpha \in F$. By the definition of τ, we have $\tau(\alpha) = \alpha'$.

1.5.24 Conclusion Let F and K be any fields such that K is an extension of F. Let F' and K' be any fields such that K' is an extension of F'. Let $\tau : a \mapsto a'$ be a ring isomorphism from F onto F'. For every $f(x) \in F[x]$, we shall denote $\tau^*(f(x))$ by $f'(t)$, where τ^* is the same as discussed in 1.5.10. Let $g(x) \in F[x]$. It follows that $g'(t) \in F'[t]$. Let E be a splitting field over F for $g(x)$. Let E' be a splitting field over F' for $g'(t)$. Suppose that $[E : F] = 1$. Then there exists a ring isomorphism φ from E onto E' such that for every $\alpha \in F$, $\varphi(\alpha) = \alpha'$.

1.5.25 Problem Let F and K be any fields such that K is an extension of F. Let F' and K' be any fields such that K' is an extension of F'. Let $\tau : a \mapsto a'$ be a ring isomorphism from F onto F'. For every $f(x) \in F[x]$, we shall denote $\tau^*(f(x))$ by $f'(t)$, where τ^* is the same as discussed in 1.5.10. Let $g(x) \in F[x]$. It follows that $g'(t) \in F'[t]$. Let E be a splitting field over F for $g(x)$. Let E' be a splitting field over F' for $g'(t)$. Suppose that $[E : F] = 2$. Then there exists a ring isomorphism φ from E onto E' such that for every $a \in F$, $\varphi(a) = a'$.

Proof Since $[E : F] = 2$, we have $[E : F] \neq 1$, and hence $E \neq F$. It follows that F is a proper subset of E. By 1.3.17, $g(x)$ can be expressed as a product of finitely many irreducible polynomials in $F[x]$. Since E is a splitting field over F for $g(x)$, and F is a proper subset of E, there exists $p(x) \in F[x]$ such that

1. $\deg(p(x)) > 1$,
2. $p(x)|g(x)$,
3. $p(x)$ is irreducible over F.

Since E is a splitting field over F for $g(x)$, $p(x)|g(x)$, and $p(x)$ is irreducible over F, all roots of $p(x)$ are members of E. Since $\deg(p(x)) > 1$, there exists $\alpha \in E$ such that

1. $\alpha \notin F$,
2. α is a root of $p(x)$ in E.

It follows, by 1.5.12, that α is algebraic of degree r over F, where $r \equiv \deg(p(x))$ (> 1). Now by 1.4.16, $[F(\alpha) : F] = r$. By 1.4.3,

$$2 = \underbrace{[E : F] = [E : F(\alpha)][F(\alpha) : F]} = [E : F(\alpha)]r,$$

and hence $[E : F(\alpha)] = \frac{2}{r} \leq 1$. Now, since $[E : F(\alpha)]$ is a positive integer, we have $[E : F(\alpha)] = 1$.

Since $p(x) \in F[x]$, we have $p'(t) \in F'[t]$. Since $\deg(p(x)) > 1$, we have $\deg(p'(t)) > 1$. Since $p(x)|g(x)$, we have $p'(t)|g'(t)$. Since $p(x)$ is irreducible over F, $p'(t)$ is irreducible over F'. It follows that there exists $\beta \in E'$ such that

1. $\beta \notin F'$,
2. β is a root of $p'(t)$ in E'.

Since $F \subset F(\alpha)$, we have $g(x) \in \underbrace{F[x] \subset (F(\alpha))[x]}$, and hence $g(x) \in (F(\alpha))[x]$.

Since $E (\supset F(\alpha) \supset F)$ is a splitting field over F for $g(x)$, E is a splitting field over $F(\alpha)$ for $g(x)$. Since $E' (\supset F'(\beta) \supset F')$ is a splitting field over F' for $g'(t)$, E' is a splitting field over $F'(\beta)$ for $g'(t)$. Now, since $[E : F(\alpha)] = 1$, by 1.5.24, there exists a ring isomorphism φ from E onto E' such that for every $a \in F(\alpha)(\supset F)$, $\varphi(a) = a'$. It follows that for every $a \in F$, $\varphi(a) = a'$.   ∎

1.5.26 Problem Let F and K be any fields such that K is an extension of F. Let F' and K' be any fields such that K' is an extension of F'. Let $\tau : a \mapsto a'$ be a ring isomorphism from F onto F'. For every $f(x) \in F[x]$, we shall denote $\tau^*(f(x))$ by $f'(t)$, where τ^* is the same as discussed in 1.5.10. Let $g(x) \in F[x]$. It follows that $g'(t) \in F'[t]$. Let E be a splitting field over F for $g(x)$. Let E' be a splitting field over F' for $g'(t)$. Suppose that $[E : F] = 3$. Then there exists a ring isomorphism φ from E onto E' such that for every $a \in F$, $\varphi(a) = a'$.

Proof Since $[E : F] = 3$, we have $[E : F] \neq 1$, and hence $E \neq F$. It follows that F is a proper subset of E. By 1.3.17, $g(x)$ can be expressed as a product of finitely many irreducible polynomials in $F[x]$. Now, since E is a splitting field over F for $g(x)$ and F is a proper subset of E, there exists $p(x) \in F[x]$ such that

1. $\deg(p(x)) > 1$,
2. $p(x) | g(x)$,
3. $p(x)$ is irreducible over F.

Since E is a splitting field over F for $g(x)$, $p(x) | g(x)$, and $p(x)$ is irreducible over F, all roots of $p(x)$ are members of E. Since $\deg(p(x)) > 1$, there exists $\alpha \in E$ such that

1. $\alpha \notin F$,
2. α is a root of $p(x)$ in E.

It follows, by 1.5.12, that α is algebraic of degree r over F, where $r \equiv \deg(p(x))(> 1)$. Now by 1.4.16, $[F(\alpha) : F] = r$. By 1.4.3,

$$3 = \underbrace{[E : F] = [E : F(\alpha)][F(\alpha) : F]} = [E : F(\alpha)]r,$$

and hence $[E : F(\alpha)] = \frac{3}{r} \leq 2$. Now, since $[E : F(\alpha)]$ is a positive integer, we have $[E : F(\alpha)] = 1$ or 2.

Since $p(x) \in F[x]$, we have $p'(t) \in F'[t]$. Since $\deg(p(x)) > 1$, we have $\deg(p'(t)) > 1$. Since $p(x) | g(x)$, we have $p'(t) | g'(t)$. Since $p(x)$ is irreducible over F, $p'(t)$ is irreducible over F'. It follows that there exists $\beta \in E'$ such that

1. $\beta \notin F'$,
2. β is a root of $p'(t)$ in E'.

Since $F \subset F(\alpha)$, we have $g(x) \in \underbrace{F[x] \subset (F(\alpha))[x]}$, and hence $g(x) \in (F(\alpha))[x]$.

Since $E(\supset F(\alpha) \supset F)$ is a splitting field over F for $g(x)$, E is a splitting field over $F(\alpha)$ for $g(x)$. Since $E'(\supset F'(\beta) \supset F')$ is a splitting field over F' for $g'(t)$, E' is a splitting field over $F'(\beta)$ for $g'(t)$. Now, since $[E : F(\alpha)] = 1$ or 2, by 1.5.24 and 1.5.25, there exists a ring isomorphism φ from E onto E' such that for every $a \in F(\alpha)(\supset F)$, $\varphi(a) = a'$. It follows that for every $a \in F$, $\varphi(a) = a'$. ∎

Similarly, we get the following.

1.5.27 Conclusion Let F and K be any fields such that K is an extension of F. Let F' and K' be any fields such that K' is an extension of F'. Let $\tau : a \mapsto a'$ be a ring isomorphism from F onto F'. For every $f(x) \in F[x]$, we shall denote $\tau^*(f(x))$ by $f'(t)$, where τ^* is the same as discussed in 1.5.10. Let $g(x) \in F[x]$. It follows that $g'(t) \in F'[t]$. Let E be a splitting field over F for $g(x)$. Let E' be a splitting field over F' for $g'(t)$. Then there exists a ring isomorphism φ from E onto E' such that for every $\alpha \in F$, $\varphi(\alpha) = \alpha'$.

1.5.28 Note In 1.5.27, let us take F for F', and the identity map $i : F \to F$ for τ. Thus for every $a \in F$, a' means a, and for every $f(x) \in F[x], f'(t)$ means $f(t)$. Let $g(x) \in F[x]$. It follows that $g(t) \in F'[t]$. Let E be a splitting field over F for $g(x)$. Let E' be a splitting field over F' for $g'(t)(= g(t))$. Then by 1.5.27, there exists a ring isomorphism φ from E onto E' such that for every $\alpha \in F$, $\varphi(\alpha) = \alpha'$.

1.5.29 Conclusion Let F and K be any fields such that $F \subset K$. Suppose that K is an extension of F. Let $g(x) \in F[x]$. Let E be a splitting field over F for $g(x)$. Let E' be a splitting field over F for $g(x)$. Then there exists a ring isomorphism φ from E onto E' such that for every $\alpha \in F$, $\varphi(\alpha) = \alpha'$.

Thus the splitting field over F for a polynomial is essentially unique, and hence it is justified in speaking about "the" splitting field.

Exercises

1. Find the greatest common divisor of

$$5 + 3\sqrt{-1} \text{ and } 3 - 4\sqrt{-1}$$

in $J[\sqrt{-1}]$.
(Hint: Observe that

$$\left. \begin{array}{l} 5 + i3 = (3 - i4)i + 1 \\ 3 - i4 = 1(3 - i4) + 0 \end{array} \right\}.$$

So the required gcd is 1.)

2. Suppose that p is a prime number, and a, b are integers such that $p|(a^2 + b^2)$, and $p^2 \nmid (a^2 + b^2)$. Show that p can be expressed as a sum of two perfect squares.

3. Show that $x^3 - 2$ is irreducible in the integral domain $\mathbb{Q}[x]$.

4. Show that if $4n - 3$ is a prime number, then $4n + 1$ can be expressed as a sum of two perfect squares.

5. Prove that $(72! + 1)$ is divisible by 73.

6. Suppose that R is a unique factorization domain with unit element 1. Show that $R[x, y]$ is also a unique factorization domain.

7. Let F and K be any fields such that K is an extension of F. Suppose that $\alpha \in K$. Show that α is algebraic of degree $[F(\alpha) : F]$ over F.

8. Prove that $\sqrt{2} + \sqrt{3}$ is algebraic of degree ≤ 4 over $\mathbb{Q}$.

9. Show that

$$\left[\left(\text{splitting field over } \mathbb{Q} \text{ for } x^2 + x + 1 \right) : \mathbb{Q} \right] = 2.$$

10. Show that $\sqrt{e}$ is a transcendental number.

Chapter 2
Galois Theory II

Roughly, a real number a is called a *constructible number* if by the application of straightedge and compass we can construct, given a line segment of unit length, a line segment of length a. In some familiar geometric situations, we shall apply the results of Galois theory. In our general development, we shall show that the general polynomial equation of degree five has no solution in radicals.

2.1 Simple Extensions

2.1.1 Definition Let D be an integral domain, that is, D is a commutative ring such that all products of nonzero members of D are nonzero. If for every positive integer m and every nonzero member a of D, $ma \left(\equiv \underbrace{a + a + \cdots + a}_{m \text{ terms}} \right)$ is nonzero, then we say that D *is of characteristic* 0.

Definition Let D be an integral domain. If there exists a positive integer m such that for every member a of D, $ma = 0$, then we say that D *is of finite characteristic*.

2.1.2 Problem Let D be an integral domain. Let a and b be any nonzero members of D. Then

$$\{m : m \text{ is a positive integer and } ma = 0\}$$
$$= \{m : m \text{ is a positive integer and } mb = 0\}.$$

© Springer Nature Singapore Pte Ltd. 2020
R. Sinha, *Galois Theory and Advanced Linear Algebra*,
https://doi.org/10.1007/978-981-13-9849-0_2

Proof Let us take an arbitrary positive integer m satisfying $ma = 0$. We shall show that $mb = 0$. Since $ma = 0$, we have

$$a(mb) = \underbrace{(ma)b = 0b} = 0,$$

and hence $a(mb) = 0$. Now, since a and mb are members of the integral domain D and a is nonzero, we have $mb = 0$. Thus

$$\{m : m \text{ is a positive integer and } ma = 0\}$$
$$\subset \{m : m \text{ is a positive integer and } mb = 0\}.$$

Similarly,

$$\{m : m \text{ is a positive integer and } mb = 0\}$$
$$\subset \{m : m \text{ is a positive integer and } ma = 0\}.$$

Hence

$$\{m : m \text{ is a positive integer and } ma = 0\}$$
$$= \{m : m \text{ is a positive integer and } mb = 0\}.$$

∎

2.1.3 Note Let D be an integral domain such that D is of finite characteristic. Let b be a nonzero member of D. It follows that

$$\{m : m \text{ is a positive integer such that for every member } a \text{ of } D, ma = 0\}$$

is a nonempty set of positive integers. Also, by 2.1.2,

$$\{m : m \text{ is a positive integer such that for every member } a \text{ of } D, ma = 0\}$$
$$= \{m : m \text{ is a positive integer and } mb = 0\}.$$

Since every set of positive integers has a least member, the smallest member n of

$$\{m : m \text{ is a positive integer such that for every member } a \text{ of } D, ma = 0\}$$

exists. Clearly, n is a prime number.

Also, for every nonzero member b of D,

$$\{m : m \text{ is a positive integer and } mb = 0\} = \{n, 2n, 3n, \ldots\}.$$

(The number n is called the *characteristic of D*.)

Proof Suppose to the contrary that n is not a prime number. We seek a contradiction.

Since n is not a prime number, there exist positive integers n_1, n_2 such that $n = n_1 n_2$ and $1 < n_1 \leq n_2 < n$. Here,

n is the smallest member of

$$\{m : m \text{ is a positive integer and } mb = 0\},$$

so $n_1 b \neq 0, n_2 b \neq 0$, and $nb = 0$. Since $n_1 b, n_2 b$ are nonzero members of the integral domain D, we have

$$0 = 0b = (nb)b = ((n_1 n_2)b)b = \underbrace{(n_1 b)(n_2 b) \neq 0},$$

and hence we get a contradiction. ∎

Definition Let F be a field. Let $f(x) \in F[x]$, where

$$f(x) \equiv a_0 x^n + a_1 x^{n-1} + \cdots + a_{n-1} x + a_n$$

and $a_i \in F \ (i = 0, 1, \ldots, n)$. It follows that $na_0, (n - 1)a_1, \ldots, 2a_{n-2}(= a_{n-2} + a_{n-2}), 1a_{n-1}(= a_{n-1})$ are members of F, and hence

$$(na_0)x^{n-1} + ((n - 1)a_1)x^{n-2} + \cdots + 1a_{n-1}$$

is a member of $F[x]$. The polynomial

$$(na_0)x^{n-1} + ((n - 1)a_1)x^{n-2} + \cdots + 1a_{n-1}$$

is denoted by $f'(x)$ and is called the *derivative of* $f(x)$.

2.1.4 Problem Let F be a field. Let $f(x), g(x) \in F[x]$. Let $\alpha \in F$. Suppose that $h(x) = f(x) + \alpha g(x) (\in F[x])$. Then

$$h'(x) = f'(x) + \alpha g'(x).$$

Proof Let

$$f(x) \equiv a_0 x^n + a_1 x^{n-1} + \cdots + a_{n-1} x + a_n,$$

where $a_i \in F \ (i = 0, 1, \ldots, n)$. Next, let

$$g(x) \equiv b_0 x^n + b_1 x^{n-1} + \cdots + b_{n-1} x + b_n,$$

where $b_i \in F$ $(i = 0, 1, \ldots, n)$. It follows that

$$h(x) = c_0 x^n + c_1 x^{n-1} + \cdots + c_{n-1} x + c_n,$$

where $c_i \equiv a_i + \alpha b_i$ $(i = 0, 1, \ldots, n)$. It follows that

$$f'(x) = (n a_0) x^{n-1} + ((n-1) a_1) x^{n-2} + \cdots + 1 a_{n-1},$$
$$g'(x) = (n b_0) x^{n-1} + ((n-1) b_1) x^{n-2} + \cdots + 1 b_{n-1},$$

and

$$h'(x) = (n c_0) x^{n-1} + ((n-1) c_1) x^{n-2} + \cdots + 1 c_{n-1}.$$

Here

$$
\begin{aligned}
\text{LHS} &= f'(x) + \alpha g'(x) \\
&= (n a_0 + \alpha(n b_0)) x^{n-1} + ((n-1) a_1 + \alpha((n-1) b_1)) x^{n-2} \\
&\quad + \cdots + (1 a_{n-1} + \alpha(1 b_{n-1})) \\
&= (n a_0 + n(\alpha b_0)) x^{n-1} + ((n-1) a_1 + (n-1)(\alpha b_1)) x^{n-2} \\
&\quad + \cdots + (1 a_{n-1} + 1(\alpha b_{n-1})) \\
&= (n(a_0 + (\alpha b_0))) x^{n-1} + ((n-1)(a_1 + \alpha b_1)) x^{n-2} \\
&\quad + \cdots + 1(a_{n-1} + \alpha b_{n-1}) \\
&= (n(c_0)) x^{n-1} + ((n-1)(c_1)) x^{n-2} \\
&\quad + \cdots + 1(c_{n-1}) \\
&= (n c_0) x^{n-1} + ((n-1) c_1) x^{n-2} + \cdots + 1 c_{n-1} = h'(x) = \text{RHS}
\end{aligned}
$$

∎

2.1.5 Problem Let F be a field. Let $f(x), g(x) \in F[x]$. Suppose that $h(x) = f(x) g(x) (\in F[x])$. Then

$$h'(x) = f'(x) g(x) + f(x) g'(x).$$

Proof Let

$$f(x) \equiv a_0 x^n + a_1 x^{n-1} + \cdots + a_{n-1} x + a_n,$$

where $a_i \in F$ $(i = 0, 1, \ldots, n)$. Next, let

$$g(x) \equiv b_0 x^n + b_1 x^{n-1} + \cdots + b_{n-1} x + b_n,$$

where $b_i \in F$ $(i = 0, 1, \ldots, n)$. It follows that

$$h(x) = c_0 x^{2n} + c_1 x^{2n-1} + \cdots + c_{2n-1} x + c_{2n},$$

where $c_0 \equiv a_0 b_0$, $c_1 \equiv a_0 b_1 + a_1 b_0$, etc. Here

$$
\begin{aligned}
\text{LHS} &= f'(x)g(x) + f(x)g'(x) \\
&= \big((na_0)x^{n-1} + ((n-1)a_1)x^{n-2} + \cdots\big)\big(b_0 x^n + b_1 x^{n-1} + \cdots + b_{n-1} x + b_n\big) \\
&\quad + \big(a_0 x^n + a_1 x^{n-1} + \cdots + a_{n-1} x + a_n\big)\big((nb_0)x^{n-1} \\
&\quad + ((n-1)b_1)x^{n-2} + \cdots + 1b_{n-1}\big) = \big(((na_0)b_0)x^{2n-1} \\
&\quad + ((na_0)b_1 + ((n-1)a_1)b_0)x^{2n-2} + \cdots\big) + \big((a_0(nb_0))x^{2n-1} \\
&\quad + (a_0((n-1)b_1) + a_1(nb_0))x^{2n-2} + \cdots\big) = ((na_0)b_0 + a_0(nb_0))x^{2n-1} \\
&\quad + ((na_0)b_1 + ((n-1)a_1)b_0 + a_0((n-1)b_1) + a_1(nb_0))x^{2n-2} + \cdots \\
&= (n(a_0 b_0) + n(a_0 b_0))x^{2n-1} + (n(a_0 b_1) + (n-1)(a_1 b_0) \\
&\quad + (n-1)(a_0 b_1) + n(a_1 b_0))x^{2n-2} + \cdots = ((2n)(a_0 b_0))x^{2n-1} \\
&\quad + (n(a_0 b_1 + a_1 b_0) + (n-1)(a_0 b_1 + a_1 b_0))x^{2n-2} + \cdots \\
&= ((2n)(a_0 b_0))x^{2n-1} + ((2n-1)(a_0 b_1 + a_1 b_0))x^{2n-2} + \cdots \\
&= ((2n)c_0)x^{2n-1} + ((2n-1)c_1)x^{2n-2} + \cdots = h'(x) = \text{RHS}.
\end{aligned}
$$

■

2.1.6 Note Let F and K be any fields such that K is an extension of F. Let α be a member of K. Suppose that $f(x) = (x - \alpha)^m g(x)$, where $m \in \{2, 3, \ldots\}$, $f(x) \in F[x]$, and $g(x) \in K[x]$.

It follows that $(x - \alpha)^m, f(x), g(x) \in K[x]$. Now by 2.1.5,

$f'(x)$

$$\underbrace{(x - \alpha)'(x - \alpha)(x - \alpha) \cdots (x - \alpha)}_{(m-1)\,\text{factors}}\, g(x)$$

$$= \underbrace{\underbrace{+ (x - \alpha)(x - \alpha)'(x - \alpha) \cdots (x - \alpha)}_{(m-2)\,\text{factors}}\, g(x) + \cdots}_{m\,\text{terms}}$$

$$+ \underbrace{(x - \alpha) \cdots (x - \alpha)}_{m\,\text{factors}}\, g'(x)$$

$$= \underbrace{1 \underbrace{(x - \alpha)(x - \alpha) \cdots (x - \alpha)}_{(m-1)\,\text{factors}}\, g(x) + (x - \alpha)1 \underbrace{(x - \alpha) \cdots (x - \alpha)}_{(m-2)\,\text{factors}}\, g(x) + \cdots}_{m\,\text{terms}}$$

$$+ \underbrace{(x - \alpha) \cdots (x - \alpha)}_{m\,\text{factors}}\, g'(x)$$

$$= \underbrace{\underbrace{(x - \alpha)(x - \alpha) \cdots (x - \alpha)}_{(m-1)\,\text{factors}}\, g(x) + \underbrace{(x - \alpha)(x - \alpha) \cdots (x - \alpha)}_{(m-1)\,\text{factors}}\, g(x) + \cdots}_{m\,\text{terms}}$$

$$+ \underbrace{(x - \alpha) \cdots (x - \alpha)}_{m\,\text{factors}}\, g'(x)$$

$$= \underbrace{(x - \alpha)^{m-1} g(x) + (x - \alpha)^{m-1} g(x) + \cdots}_{m\,\text{terms}} + (x - \alpha)^m g'(x)$$

$$= m(x - \alpha)^{m-1} g(x) + (x - \alpha)^m g'(x)$$

$$= (x - \alpha)\Big(m(x - \alpha)^{m-2} g(x) + (x - \alpha)^{m-1} g'(x)\Big) = (x - \alpha)r(x),$$

where $r(x) \equiv m(x - \alpha)^{m-2} g(x) + (x - \alpha)^{m-1} g'(x) \in K[x]$.   Since $f(x) = (x - \alpha)\Big((x - \alpha)^{m-1} g(x)\Big)$ and $f'(x) = (x - \alpha)r(x)$, $(x - \alpha)$ is a common factor of $f(x)$ and $f'(x)$.

2.1.7 Conclusion Let F and K be any fields such that K is an extension of F. Let α be a member of K. Let $f(x) \in F[x]$. Suppose that α is a multiple root of $f(x)$. Then $f(x)$ and $f'(x)$ have a nontrivial common factor in $K[x]$.

2.1.8 Note Let F and K be any fields such that K is an extension of F. Suppose that $f(x) \in F[x]$. It follows that $f'(x) \in F[x]$. Suppose that $f(x)$ and $f'(x)$ have a non-trivial common factor in $K[x]$, that is, $f(x)$ and $f'(x)$ have a common factor of degree ≥ 1 in $K[x]$.

It follows that there exists α such that $(x - \alpha)|f(x)$ and $(x - \alpha)|f'(x)$. We shall show that $(x - \alpha)^2|f(x)$.

Since $(x - \alpha)|f(x)$, there exist a positive integer m and a polynomial $r(x)$ such that

1. $f(x) = (x - \alpha)^m r(x)$,
2. $(x - \alpha)\!\!\not|\,r(x)$.

It suffices to show that $m \geq 2$. Suppose to the contrary that $m = 1$. We seek a contradiction. Here $f(x) = (x - \alpha)r(x)$, so by 2.1.5,

$$\underbrace{f'(x) = (x - \alpha)'r(x) + (x - \alpha)r'(x)}_{} = 1r(x) + (x - \alpha)r'(x) = r(x) + (x - \alpha)r'(x),$$

and hence

$$f'(x) = r(x) + (x - \alpha)r'(x).$$

It follows that

$$\underbrace{f'(\alpha) = r(\alpha) + (\alpha - \alpha)r'(\alpha)}_{} = r(\alpha) + 0r'(\alpha) = r(\alpha).$$

Thus $f'(\alpha) = r(\alpha)$. Since $(x - \alpha)\!\!\not|\,r(x)$, we have $r(\alpha) \neq 0$. Since $(x - \alpha)|f'(x)$, we have $f'(\alpha) = 0$. Since $f'(\alpha) = 0$ and $r(\alpha) \neq 0$, we have $f'(\alpha) \neq r(\alpha)$. This is a contradiction.

2.1.9 Conclusion Let F and K be any fields such that K is an extension of F. Suppose that $f(x) \in F[x]$. Suppose that $f(x)$ and $f'(x)$ have a nontrivial common factor in $K[x]$. Then $f(x)$ has a multiple root.

2.1.10 Problem Let F and K be any fields such that K is an extension of F. Let F be of characteristic 0. Suppose that $f(x) \in F[x]$. Let $f(x)$ be irreducible. Then $f(x)$ has no multiple root.

Proof Suppose to the contrary that $f(x)$ has a multiple root. We seek a contradiction.

Since $f(x)$ has a multiple root, by 2.1.6, $f(x)$ and $f'(x)$ have a nontrivial common factor in $K[x]$. Since $f(x)$ is irreducible, $f(x)$ is the only nontrivial factor of $f(x)$. Now, since $f(x)$ and $f'(x)$ have a nontrivial common factor in $K[x]$, $f(x)$ is a factor of $f'(x)$, and hence $\deg(f(x)) \leq \deg(f'(x))$. Suppose that

$$f(x) \equiv a_0 x^n + a_1 x^{n-1} + \cdots + a_{n-1}x + a_n,$$

where $a_i \in F$ $(i = 0, 1, \ldots, n)$, n is a positive integer, and $a_0 \neq 0$. Since F is of characteristic 0, we have na_0 is a nonzero member of F, and hence

$$\deg(f'(x)) = \underbrace{\deg\big((na_0)x^{n-1} + ((n-1)a_1)x^{n-2} + \cdots\big) = n-1}_{} < n$$
$$= \deg(a_0 x^n + a_1 x^{n-1} + \cdots + a_{n-1}x + a_n) = \deg(f(x)).$$

Thus $\deg(f'(x)) < \deg(f(x))$. This is a contradiction. ∎

2.1.11 Problem Let F and K be any fields such that K is an extension of F. Let F be of characteristic p. By 2.1.3, p is a prime number. Suppose that $f(x) \in F[x]$. Let $f(x)$ be irreducible. Suppose that $f(x)$ has a multiple root. Then $f(x)$ is of the form $g(x^p)$, where $g(x) \in F[x]$.

Proof Since $f(x)$ has a multiple root, by 2.1.6, $f(x)$ and $f'(x)$ have a nontrivial common factor in $K[x]$. Since $f(x)$ is irreducible, $f(x)$ is the only nontrivial factor of $f(x)$. Now, since $f(x)$ and $f'(x)$ have a nontrivial common factor in $K[x]$, $f(x)$ is a factor of $f'(x)$, and hence $\deg(f(x)) \le \deg(f'(x))$. But we know that if $f'(x)$ is nonzero, then $\deg(f'(x)) < \deg(f(x))$, hence $f'(x) = 0$. Suppose that

$$f(x) \equiv a_0 + a_1 x + \cdots + a_{p-1}x^{p-1} + a_p x^p + a_{p+1}x^{p+1} + \cdots$$
$$+ a_{2p-1}x^{2p-1} + a_{2p}x^{2p} + a_{2p+1}x^{2p+1} + \cdots + a_n x^n,$$

where $a_i \in F$ $(i = 0, 1, \ldots, n)$, and n is a positive integer. It suffices to show that $a_1 = 0, \ldots, a_{p-1} = 0, a_{p+1} = 0, a_{2p-1} = 0$, etc.
 Since $f'(x) = 0$, we have

$$a_1 + 2a_2 x + \cdots + (p-1)a_{p-1}x^{p-2} + pa_p x^{p-1} + (p+1)a_{p+1}x^p + \cdots$$
$$+ (2p-1)a_{2p-1}x^{2p-2} + 2pa_{2p}x^{2p-1} + (2p+1)a_{2p+1}x^{2p} + \cdots + na_n x^{n-1} = 0,$$

and hence $0 = a_1, 0 = 2a_2, 0 = (p-1)a_{p-1}, 0 = (p+1)a_{p+1}, 0 = (2p-1)a_{2p-1}$, $0 = (2p+1)a_{2p+1}$, etc.
 Since F is of characteristic p, $p \nmid (p-1)$, $0 = (p-1)a_{p-1}$, and $a_{p-1} \in F$, we have $a_{p-1} = 0$. Similarly, $a_{p+1} = 0$, $a_{2p-1} = 0$, etc. ∎

2.1.12 Note Let F and K be any fields such that K is an extension of F. Let F be of characteristic p.
 Observe that $1x^p + (-1)x$ is a member of $F[x]$. Also

$$(1x^p + (-1)x)' = (p1)x^{p-1} + (-1).$$

Since $1 \in F$, and F is of characteristic p, we have $p1 = 0$, and hence

$$(1x^p + (-1)x)' = \underbrace{(p1)x^{p-1} + (-1) = 0x^{p-1} + (-1)}_{} = -1.$$

Thus $(1x^p + (-1)x)' = -1$. Now, since $1x^p + (-1)x$ and 1 have no nontrivial common factor in $K[x]$, $1x^p + (-1)x$ and $(1x^p + (-1)x)'$ have no nontrivial common factor in $K[x]$, and hence by 2.1.6, $1x^p + (-1)x$ has no multiple root.

2.1.13 Conclusion Let F and K be any fields such that K is an extension of F. Let F be of characteristic p. Then $x^p - x$ has no multiple root. Similarly, $x^{p^2} - x$ has no multiple root, $x^{p^3} - x$ has no multiple root, etc.

2.1.14 Note Let F and K be any fields such that K is an extension of F. Let F be of characteristic 0. Let $a, b \in K$. Suppose that a, b are algebraic over F.

Since a is algebraic over F, there exists a nonzero polynomial $q(x) \in F[x]$ such that $(K \ni) q(a) = 0$. By 1.3.21, there exists an irreducible polynomial $f(x) \in F[x]$ such that $f(x) | q(x)$ in $F[x]$, $(K \ni) f(a) = 0$, and the leading coefficient of $f(x)$ is 1. Similarly, there exists an irreducible polynomial $g(x) \in F[x]$ such that $(K \ni)$ $g(b) = 0$, and the leading coefficient of $g(x)$ is 1. Suppose that $\deg(f(x)) = m (\geq 1)$ and $\deg(g(x)) = n (\geq 1)$. Let L be a field such that

1. $K \subset L$,
2. L is an extension of K,
3. $f(x)$ splits completely in L,
4. $g(x)$ splits completely in L.

Suppose that all the m roots of $f(x)$ in L are $a, a_2, \ldots, a_m$. Next, suppose that all the n roots of $g(x)$ in L are $b, b_2, \ldots, b_n$. Since $a, b \in L$ and $F \subset L$, we have $F(a, b) \subset L$. Also

$$f(x) = (x - a)(x - a_2)\ldots(x - a_m)$$

and

$$g(x) = (x - b)(x - b_2)\ldots(x - b_n)$$

in $L[x]$. Since F is of characteristic 0, $f(x) \in F[x]$, $f(x)$ is irreducible, and $a, a_2, \ldots, a_m$ are the roots of $f(x)$, by 2.1.10, $a, a_2, \ldots, a_m$ are distinct. Similarly, $b, b_2, \ldots, b_n$ are distinct. Also, $f(a) = 0$, $g(b) = 0$, $f(a_i) = 0$ $(i = 2, 3, \ldots, m)$, and $g(b_j) = 0$ $(j = 2, 3, \ldots, n)$. Since F is of characteristic 0 and $1 \in F$,

$$1, 1 + 1, 1 + 1 + 1, \ldots$$

are distinct members of F, and hence F is an infinite set.

Let us take arbitrary $i \in \{2, \ldots, m\}$ and $j \in \{2, \ldots, n\}$. Since $b, b_2, \ldots, b_n$ are distinct, $(b - b_j) \neq 0$, and hence $(b - b_j)^{-1}$ is a nonzero element of the field L.

Observe that there exists a unique $\lambda \in L$ such that

$$a_i + \lambda b_j = a + \lambda b.$$

Proof Existence: Since

$$a_i + \left((a_i - a)(b - b_j)^{-1}\right)b_j = \left(a_i(b - b_j) + (a_i - a)b_j\right)(b - b_j)^{-1}$$
$$= \left(a_ib - ab_j\right)(b - b_j)^{-1}$$

and

$$a + \left((a_i - a)(b - b_j)^{-1}\right)b = \left(a(b - b_j) + (a_i - a)b\right)(b - b_j)^{-1}$$
$$= \left(a_ib - ab_j\right)(b - b_j)^{-1},$$

$(a_i - a)(b - b_j)^{-1}$ $(\in L)$ is a solution of the λ-equation

$$a_i + \lambda b_j = a + \lambda b$$

in L.

Uniqueness: Suppose that

$$\left.\begin{aligned} a_i + \lambda_1 b_j &= a + \lambda_1 b \\ a_i + \lambda_2 b_j &= a + \lambda_2 b \end{aligned}\right\},$$

where $\lambda_1, \lambda_2 \in L$. We have to show that $\lambda_1 = \lambda_2$. Here,

$$(\lambda_1 - \lambda_2)b_j = \underbrace{\left(a_i + \lambda_1 b_j\right) - \left(a_i + \lambda_2 b_j\right) = (a + \lambda_1 b) - (a + \lambda_2 b)} = (\lambda_1 - \lambda_2)b,$$

so $(\lambda_1 - \lambda_2)b_j = (\lambda_1 - \lambda_2)b$, and hence $(\lambda_1 - \lambda_2)(b - b_j) = 0$. Since $(b - b_j) \neq 0$, we have $\lambda_1 - \lambda_2 = 0$, and hence $\lambda_1 = \lambda_2$. ∎

Thus we have shown that for every $i \in \{2, \ldots, m\}$ and $j \in \{2, \ldots, n\}$, there exists a unique $\lambda \in L \,(\supset F)$ such that $a_i + \lambda b_j = a + \lambda b$. It follows that the collection of all such λ is a finite set. Now, since F is an infinite set, there exists a nonzero $\gamma \in F \,(\subset F(a, b))$ such that for every $i \in \{2, \ldots, m\}$ and $j \in \{2, \ldots, n\}$, $a_i + \gamma b_j \neq a + \gamma b$, and hence for every $j \in \{2, \ldots, m\}$, $\left((a + \gamma b) - \gamma b_j\right)$ is different from $a, a_2, \ldots, a_m$. Since all the m distinct roots of $f(x)$ are $a, a_2, \ldots, a_m$, for every $j \in \{2, \ldots, m\}$, $(a + \gamma b) - \gamma b_j$ is not a root of $f(x)$, and hence for every $j \in \{2, \ldots, m\}$, we have $f\left((a + \gamma b) - \gamma b_j\right) \neq 0$.

Since a, b, γ are elements of the field $F(a, b)$, we have $(a + \gamma b) \in F(a, b)$, and hence $F(a + \gamma b) \subset F(a, b) \subset L$.

Now we shall show that $F(a, b) \subset F(a + \gamma b)$. To this end, put

$$h(x) \equiv f((a + \gamma b) - \gamma x)(\in (F(a + \gamma b))[x]).$$

Thus $h(x) \in (F(a + \gamma b))[x]$. Also

$$\underbrace{h(b) = f((a + \gamma b) - \gamma b)} = f(a) = 0,$$

so $h(b) = 0$, and hence $(x - b)$ is a factor of the polynomial $h(x)$ in $L[x]$. Since $g(b) = 0$, $(x - b)$ is a factor of the polynomial $g(x)$ in $L[x]$. Thus $(x - b)$ is a common factor of the polynomials $g(x)$ and $h(x)$ in $L[x]$. By 2.1.10, we have $(x - b)^2 \nmid g(x)$, and hence $(x - b)^2$ is not a common factor of the polynomials $g(x)$ and $h(x)$.

Clearly, for every $j \in \{2, \ldots, m\}$, $(x - b_j)$ is not a common divisor of the polynomials $g(x)$ and $h(x)$.

Proof Let us fix an arbitrary $j \in \{2, \ldots, m\}$. It follows that

$$\underbrace{h(b_j) = f((a + \gamma b) - \gamma b_j)} \neq 0,$$

and hence $h(b_j) \neq 0$. Thus $(x - b_j) \nmid h(x)$ in $L[x]$. It follows that for every $j \in \{2, \ldots, m\}$, $(x - b_j)$ is not a common divisor of the polynomials $g(x)$ and $h(x)$ in $L[x]$. ∎

Thus $(x - b)$ is a greatest common divisor of the polynomials $g(x)$ and $h(x)$ in $L[x](\supset (F(a + \gamma b))[x])$. It follows that $(x - b)$ $(\in (F(a, b))[x] \subset L[x])$ divides each member of the set

$$\{g(x)u(x) + h(x)v(x) : u(x), v(x) \in L[x]\}$$
$$(\supset \{g(x)u(x) + h(x)v(x) : u(x), v(x) \in (F(a + \gamma b))[x]\}),$$

and hence $(x - b)$ divides each member of

$$\{g(x)u(x) + h(x)v(x) : u(x), v(x) \in (F(a + \gamma b))[x]\}.$$

Since $g(x) \in F[x] (\subset (F(a + \gamma b))[x])$ and $h(x) \in (F(a + \gamma b))[x]$, a greatest common divisor of the polynomials $g(x)$ and $h(x)$ in $(F(a + \gamma b))[x]$ is a member of

$$\{g(x)u(x) + h(x)v(x) : u(x), v(x) \in (F(a + \gamma b))[x]\}.$$

Since $(x - b)$ divides each member of

$$\{g(x)u(x) + h(x)v(x) : u(x), v(x) \in (F(a + \gamma b))[x]\},$$

$(x - b)$ divides a greatest common divisor of the polynomials $g(x)$ and $h(x)$ in $(F(a + \gamma b))[x]$, and hence a greatest common divisor of the polynomials $g(x)$ and $h(x)$ in $(F(a + \gamma b))[x]$ is nontrivial. Since $(F(a + \gamma b))[x] \subset L[x]$, a greatest common divisor of the polynomials $g(x)$ and $h(x)$ in $(F(a + \gamma b))[x]$ divides a greatest common divisor of the polynomials $g(x)$ and $h(x)$ in $L[x]$, and hence a greatest common divisor of the polynomials $g(x)$ and $h(x)$ in $(F(a + \gamma b))[x]$ divides $(x - b)$. Now, since $(x - b)$ divides a greatest common divisor of the polynomials $g(x)$ and $h(x)$ in $(F(a + \gamma b))[x]$, $(x - b)$ is a greatest common divisor of the polynomials $g(x)$ and $h(x)$ in $(F(a + \gamma b))[x]$, and hence $(x - b) \in (F(a + \gamma b))[x]$. It follows that $(-b) \in F(a + \gamma b)$, and hence $b \in F(a + \gamma b)$. Since $(a + \gamma b) \in F(a + \gamma b)$ and $\gamma \in F(\subset F(a + \gamma b))$, we have

$$a = \underbrace{((a + \gamma b) - \gamma b)}_{} \in F(a + \gamma b),$$

and hence $a \in F(a + \gamma b)$. Thus $F \cup \{a, b\} \subset F(a + \gamma b)$, and hence $F(a, b) \subset F(a + \gamma b)$. Next, since $F(a + \gamma b) \subset F(a, b)$, we have $F(c) = F(a, b)$, where $c \equiv (a + \gamma b) \in F(a, b)$.

2.1.15 Conclusion I Let F and K be any fields such that K is an extension of F. Let F be of characteristic 0. Let $a, b \in K$. Suppose that a, b are algebraic over F. Then there exists $c \in F(a, b)$ such that $F(a, b) = F(c)$.

Similarly, we get the following.

2.1.16 Conclusion II Let F and K be any fields such that K is an extension of F. Let F be of characteristic 0. Let $a_1, a_2, \ldots, a_n \in K$. Suppose that $a_1, a_2, \ldots, a_n$ are algebraic over F. Then there exists $c \in F(a_1, a_2, \ldots, a_n)$ such that $F(a_1, a_2, \ldots, a_n) = F(c)$.

Definition Let F and K be any fields such that K is an extension of F. If there exists $c \in K$ such that $K = F(c)$, then we say that K *is a simple extension of F.*

Now Conclusion II can be stated as follows:

2.1.17 Conclusion III Let F and K be any fields such that K is an extension of F. Let F be of characteristic 0. Let $a_1, a_2, \ldots, a_n \in K$. Suppose that $a_1, a_2, \ldots, a_n$ are algebraic over F. Then $F(a_1, a_2, \ldots, a_n)$ is a simple extension of F.

Using 1.4.9, we get the following.

2.1.18 Conclusion IV Let F and K be any fields such that K is an extension of F. Let F be of characteristic 0. Let $a \in K$. Suppose that $F(a)$ is a finite extension of F. Then $F(a)$ is a simple extension of F. In short, every finite extension of a field of characteristic 0 is a simple extension.

2.2 Galois Groups

Caution: From henceforth, all our fields are of characteristic 0.

2.2.1 Definition Let K be any field. Let $\sigma : K \to K$ be a function. If

1. for every $a, b \in K$, $\sigma(a+b) = \sigma(a) + \sigma(b)$,
2. for every $a, b \in K$, $\sigma(ab) = \sigma(a)\sigma(b)$,
3. $\sigma : K \to K$ is onto,
4. $\sigma : K \to K$ is 1-1,

then we say that σ *is an automorphism of K.*

Here condition (4) is superfluous.

Proof Suppose to the contrary that there exist $a, b \in K$ such that $\sigma(a) = \sigma(b)$, and $a \neq b$. We seek a contradiction. Since $a \neq b$, $(a - b)$ is a nonzero member of K, and hence $(a - b)^{-1} \in K$. It follows that

$$1 = \underbrace{\sigma(1) = \sigma\Big((a-b)\big((a-b)^{-1}\big)\Big)} = \sigma(a-b)\sigma\Big((a-b)^{-1}\Big)$$

$$= \sigma(a + (-b))\sigma\Big((a-b)^{-1}\Big) = (\sigma(a) + \sigma(-b))\sigma\Big((a-b)^{-1}\Big)$$
$$= (\sigma(a) + (-\sigma(b)))\sigma\Big((a-b)^{-1}\Big) = (\sigma(a) - \sigma(b))\sigma\Big((a-b)^{-1}\Big)$$
$$= (\sigma(a) - \sigma(a))\sigma\Big((a-b)^{-1}\Big) = 0\,\sigma\Big((a-b)^{-1}\Big) = 0,$$

and hence $1 = 0$. This contradicts the fact that K is a field.   ∎

2.2.2 Note Let K be any field. Let $\sigma_1, \ldots, \sigma_n$ be n distinct automorphisms of K. Let $a_1, \ldots, a_n \in K$. Suppose that

1. for every $u \in K$, $a_1\sigma_1(u) + \cdots + a_n\sigma_n(u) = 0$,

2. not all a_i $(i = 1, \ldots, n)$ are 0.

We claim that this is impossible. We seek a contradiction.

In the case of $n = 1$, condition (1) becomes for every $u \in K$, $a_1\sigma_1(u) = 0$, and hence

$$a_1 = a_1 1 = \underbrace{a_1\sigma_1(1) = 0}.$$

Thus $a_1 = 0$. In this case of $n = 1$, condition (2) becomes $a_1 \neq 0$. This is a contradiction.

Now we consider the case $n = 2$.

Here condition (1) becomes for every $u \in K$, $a_1\sigma_1(u) + a_2\sigma_2(u) = 0$. Next, condition (2) becomes (either $a_1 \neq 0$ or $a_2 \neq 0$). For definiteness, suppose that $a_1 \neq 0$. We seek a contradiction.

Since σ_1, σ_2 are distinct, there exists a nonzero $c \in K$ such that $\sigma_1(c) \neq \sigma_2(c)$. Thus $\sigma_1(c), \sigma_2(c)$ are nonzero, and

$$a_1 \sigma_1(c) + a_2 \sigma_2(c) = 0.$$

Since $a_1 \neq 0$ and $\sigma_1(c) \neq 0$, we have

$$a_2 \sigma_2(c) = \underbrace{-a_1 \sigma_1(c) \neq 0},$$

and hence $a_2 \sigma_2(c) \neq 0$. This shows that $a_2 \neq 0$. Observe that for every $u \in K$, we have $cu \in K$, and hence

$$
\begin{aligned}
a_2 \sigma_2(c)(\sigma_2(u) - \sigma_1(u)) &= (-a_2 \sigma_2(c))\sigma_1(u) + a_2 \sigma_2(c)\sigma_2(u) \\
&= a_1 \sigma_1(c)\sigma_1(u) + a_2 \sigma_2(c)\sigma_2(u) \\
&= \underbrace{a_1 \sigma_1(cu) + a_2 \sigma_2(cu) = 0}.
\end{aligned}
$$

Thus for every $u \in K$,

$$a_2 \sigma_2(c)(\sigma_2(u) - \sigma_1(u)) = 0.$$

Now, since $a_2, \sigma_2(c)$ are nonzero members of K and $(\sigma_2(u) - \sigma_1(u))$ is a member of the field K, we have, for every $u \in K$, $\sigma_2(u) - \sigma_1(u) = 0$. Thus for every $u \in K$, $\sigma_1(u) = \sigma_2(u)$. Since $c \in K$, we have $\sigma_1(c) = \sigma_2(c)$. This is a contradiction.

Next we consider the case $n = 3$.

If $a_3 = 0$, then from the cases discussed above, we get a contradiction. Hence we have to deal with only the case $a_3 \neq 0$.

Since σ_1, σ_3 are distinct, there exists a nonzero $c \in K$ such that $\sigma_1(c) \neq \sigma_3(c)$. Thus $\sigma_1(c), \sigma_3(c)$ are nonzero, and

$$a_1 \sigma_1(c) + a_2 \sigma_2(c) + a_3 \sigma_3(c) = 0.$$

Here condition (1) becomes, for every $u \in K$, $a_1 \sigma_1(u) + a_2 \sigma_2(u) + a_3 \sigma_3(u) = 0$. Observe that for every $u \in K$, we have $cu \in K$, and hence

$$
\begin{aligned}
a_2(\sigma_2(c) &- \sigma_1(c)) \cdot \sigma_2(u) + a_3(\sigma_3(c) - \sigma_1(c)) \cdot \sigma_3(u) \\
&= (-a_2 \sigma_2(u) - a_3 \sigma_3(u))\sigma_1(c) + a_2 \sigma_2(c)\sigma_2(u) + a_3 \sigma_3(c)\sigma_3(u) \\
&= (a_1 \sigma_1(u))\sigma_1(c) + a_2 \sigma_2(c)\sigma_2(u) + a_3 \sigma_3(c)\sigma_3(u) \\
&= a_1 \sigma_1(c)\sigma_1(u) + a_2 \sigma_2(c)\sigma_2(u) + a_3 \sigma_3(c)\sigma_3(u) \\
&= \underbrace{a_1 \sigma_1(cu) + a_2 \sigma_2(cu) + a_3 \sigma_3(cu) = 0}
\end{aligned}
$$

Thus for every $u \in K$,

$$b_2 \sigma_2(u) + b_3 \sigma_3(u) = 0,$$

where $b_2 \equiv a_2(\sigma_2(c) - \sigma_1(c))$ and $b_3 \equiv a_3(\sigma_3(c) - \sigma_1(c))$. Since $a_3, (\sigma_3 (c) - \sigma_1(c))$ are nonzero members of the field K, $(b_3 =) a_3(\sigma_3(c) - \sigma_1(c))$ is a nonzero member of the field K, and hence not all b_i $(i = 2, 3)$ are 0. By our earlier case $n = 2$, we get a contradiction, etc.

2.2.3 Conclusion Let K be any field. Let $\sigma_1, \ldots, \sigma_n$ be n distinct automorphisms of K. There do not exist $a_1, \ldots, a_n \in K$ such that

1. for every $u \in K$, $a_1 \sigma_1(u) + \cdots + a_n \sigma_n(u) = 0$,
2. not all a_i $(i = 1, \ldots, n)$ are 0.

2.2.4 Problem Let K be any field. Let G be a nonempty collection automorphisms of K. Put

$$K_G \equiv \{a : a \in K, \text{and for every } \sigma \in G, \sigma(a) = a\}.$$

Then K_G is a subfield of K.
Here we say that K_G is the *fixed field of G*.

Proof Let us take an arbitrary $\sigma \in G$. It follows that $\sigma : K \to K$ is an automorphism of K, and hence $\sigma(0) = 0$ and $\sigma(1) = 1$. This shows that $0, 1 \in K_G$. Thus K_G is a subset of K, and K_G contains at least two elements.

Let $a, b \in K_G$. Let us take an arbitrary $\sigma \in G$. It follows by the definition of K_G that $\sigma(a) = a$ and $\sigma(b) = b$. Hence $\sigma(a+b) = \sigma(a) + \sigma(b) = a + b$ and $\sigma(ab) = \sigma(a)\sigma(b) = ab$. Thus $\sigma(a+b) = a+b$ and $\sigma(ab) = ab$. It follows that $(a+b), ab \in K_G$. Next, since $\sigma(-a) = -(\sigma(a)) = -a$, we have $\sigma(-a) = -a$. This shows that $(-a) \in K_G$. If $a \neq 0$, then $\sigma(a^{-1}) = (\sigma(a))^{-1} = a^{-1}$. Thus if a is a nonzero element of K_G, then $a^{-1} \in K_G$. Hence K_G is a subfield of K. ∎

2.2.5 Problem Let K be any field. The collection of all automorphisms of K is denoted by $\text{Aut}(K)$. Clearly, $\text{Aut}(K)$ is a group.

Proof The identity map $\text{Id} : a \mapsto a$ from K onto K is an automorphism of K, and hence $\text{Id} \in \text{Aut}(K)$.

a. Let $\sigma, \mu \in \text{Aut}(K)$. We have to show that $(\sigma\mu) \in \text{Aut}(K)$. Since $\sigma \in \text{Aut}(K)$, σ is a one-to-one map from K onto K. Similarly, μ is a one-to-one map from K onto K. It follows that the composite map $(\sigma\mu)$ is a one-to-one map from K onto K. Next, let us take arbitrary a, b in K. We have

$$(\sigma\mu)(a+b) = \sigma(\mu(a+b)) = \sigma(\mu(a) + \mu(b)) = \sigma(\mu(a)) + \sigma(\mu(b))$$
$$= (\sigma\mu)(a) + (\sigma\mu)(b),$$

and hence

$$(\sigma\mu)(a+b) = (\sigma\mu)(a) + (\sigma\mu)(b).$$

Similarly, $(\sigma\mu)(ab) = (\sigma\mu)(a) \cdot (\sigma\mu)(b)$. Thus $(\sigma\mu) \in \mathrm{Aut}(K)$.

b. Let $\sigma \in \mathrm{Aut}(K)$. We have to show that $\sigma^{-1} \in \mathrm{Aut}(K)$. Since $\sigma \in \mathrm{Aut}(K)$, σ is a one-to-one map from K onto K, and hence σ^{-1} is a one-to-one map from K onto K. Next, let us take arbitrary a, b in K. We have to show that

1. $\sigma^{-1}(a+b) = \sigma^{-1}(a) + \sigma^{-1}(b)$, that is, $a+b = \sigma(\sigma^{-1}(a) + \sigma^{-1}(b))$,
2. $\sigma^{-1}(ab) = \sigma^{-1}(a) \cdot \sigma^{-1}(b)$, that is, $ab = \sigma(\sigma^{-1}(a) \cdot \sigma^{-1}(b))$.

For 1:   RHS $= \sigma(\sigma^{-1}(a) + \sigma^{-1}(b)) = \sigma(\sigma^{-1}(a)) + \sigma(\sigma^{-1}(b)) = a+b =$ LHS.
For 2:   RHS $= \sigma(\sigma^{-1}(a) \cdot \sigma^{-1}(b)) = \sigma(\sigma^{-1}(a)) \cdot \sigma(\sigma^{-1}(b)) = ab =$ LHS.

∎

2.2.6 Problem Let K be any field. Let F be a subfield of K. Put

$$G(K,F) \equiv \{\sigma : \sigma \in \mathrm{Aut}(K), \text{and for every } a \in F, \sigma(a) = a\}.$$

Then $G(K,F)$ is a subgroup of $\mathrm{Aut}(K)$.
Here $G(K,F)$ is called the *group of automorphisms of K relative to F*.

Proof The identity map $\mathrm{Id} : a \mapsto a$ from K onto K is an automorphism of K, and hence $\mathrm{Id} \in \mathrm{Aut}(K)$. Also, for every $a \in F$, $\mathrm{Id}(a) = a$. Thus $\mathrm{Id} \in G(K,F)$.

Let $\sigma, \mu \in G(K,F)$. It suffices to show that $(\sigma\mu^{-1}) \in G(K,F)$. To this end, let us take an arbitrary $a \in F$. It suffices to show that $(\sigma\mu^{-1})(a) = a$. Since $\mu \in G(K,F)$ and $a \in F$, we have $\mu(a) = a$, and hence $\mu^{-1}(a) = a$.

$$\text{LHS} = (\sigma\mu^{-1})(a) = \sigma(\mu^{-1}(a)) = \sigma(a) = a = \text{RHS}.$$

∎

2.2.7 Note Let F and K be any fields such that K is a finite extension of F.

It follows that $[K : F] < \infty$. Hence there exists a basis $\{u_1, u_2, \ldots, u_n\}$ of the vector space K over F, where $n = [K : F]$. By 2.2.6, $G(K,F)$ is a group of automorphisms of K.

We claim that the number of elements of $G(K,F)$ is $\leq n$. Suppose to the contrary that the number of elements of $G(K,F)$ is $> n$. We seek a contradiction.

Since the number of elements of $G(K,F)$ is $> n$, there exist $(n+1)$ distinct automorphisms $\sigma_1, \sigma_2, \ldots, \sigma_{n+1}$ of K. It follows that for every $i \in \{1, 2, \ldots, n\}$ and for every $j \in \{1, 2, \ldots, n+1\}$, we have $\sigma_j(u_i) \in K$. It follows that the following system of n linear equations in $(n+1)$ variables $x_1, x_2, \ldots, x_n, x_{n+1}$,

$$\left.\begin{array}{l} \sigma_1(u_1)x_1 + \sigma_2(u_1)x_2 + \cdots + \sigma_n(u_1)x_n + \sigma_{n+1}(u_1)x_{n+1} = 0 \\ \sigma_1(u_2)x_1 + \sigma_2(u_2)x_2 + \cdots + \sigma_n(u_2)x_n + \sigma_{n+1}(u_2)x_{n+1} = 0 \\ \qquad\qquad\vdots \\ \sigma_1(u_n)x_1 + \sigma_2(u_n)x_2 + \cdots + \sigma_n(u_n)x_n + \sigma_{n+1}(u_n)x_{n+1} = 0 \end{array}\right\},$$

has   a   nontrivial   solution $(x_1, x_2, \ldots, x_n, x_{n+1}) = (a_1, a_2, \ldots, a_n, a_{n+1}) (\neq (0, 0, \ldots, 0, 0))$ in K. It follows that

$$\left.\begin{array}{l} \sigma_1(u_1)a_1 + \sigma_2(u_1)a_2 + \cdots + \sigma_{n+1}(u_1)a_{n+1} = 0 \\ \sigma_1(u_2)a_1 + \sigma_2(u_2)a_2 + \cdots + \sigma_{n+1}(u_2)a_{n+1} = 0 \\ \qquad\qquad\vdots \\ \sigma_1(u_n)a_1 + \sigma_2(u_n)a_2 + \cdots + \sigma_{n+1}(u_n)a_{n+1} = 0 \end{array}\right\},$$

that is,

$$\left.\begin{array}{l} a_1\sigma_1(u_1) + a_2\sigma_2(u_1) + \cdots + a_{n+1}\sigma_{n+1}(u_1) = 0 \\ a_1\sigma_1(u_2) + a_2\sigma_2(u_2) + \cdots + a_{n+1}\sigma_{n+1}(u_2) = 0 \\ \qquad\qquad\vdots \\ a_1\sigma_1(u_n) + a_2\sigma_2(u_n) + \cdots + a_{n+1}\sigma_{n+1}(u_n) = 0 \end{array}\right\},$$

that is,

$$\sum_{j=1}^{n+1} a_j \sigma_j(u_i) = 0 \quad (i = 1, \ldots, n).$$

Since $(a_1, a_2, \ldots, a_n, a_{n+1}) \neq (0, 0, \ldots, 0, 0)$, not all a_i $(i = 1, \ldots, n+1)$ are 0, and hence by 2.2.3, there exists $u \in K$ such that

$$a_1\sigma_1(u) + \cdots + a_{n+1}\sigma_{n+1}(u) \neq 0.$$

Since $u \in K$ and $\{u_1, u_2, \ldots, u_n\}$ is a basis of the vector space K over F, there exist $b_1, b_2, \ldots, b_n$ in F such that $u = \sum_{i=1}^{n} b_i u_i$. It follows that

$$a_1 \sigma_1(u) + \cdots + a_{n+1} \sigma_{n+1}(u) = a_1 \sigma_1 \left(\sum_{i=1}^{n} b_i u_i \right) + \cdots + a_{n+1} \sigma_{n+1} \left(\sum_{i=1}^{n} b_i u_i \right)$$

$$= a_1 \left(\sum_{i=1}^{n} b_i \sigma_1(u_i) \right) + \cdots + a_{n+1} \left(\sum_{i=1}^{n} b_i \sigma_{n+1}(u_i) \right)$$

$$= \sum_{i=1}^{n} a_1 b_i \sigma_1(u_i) + \cdots + \sum_{i=1}^{n} a_{n+1} b_i \sigma_{n+1}(u_i)$$

$$= \sum_{j=1}^{n+1} \left(\sum_{i=1}^{n} a_j b_i \sigma_j(u_i) \right) = \sum_{i=1}^{n} \left(\sum_{j=1}^{n+1} a_j b_i \sigma_j(u_i) \right)$$

$$= \sum_{i=1}^{n} \left(b_i \sum_{j=1}^{n+1} a_j \sigma_j(u_i) \right) = \sum_{i=1}^{n} (b_i \cdot 0) = 0,$$

and hence

$$a_1 \sigma_1(u) + \cdots + a_{n+1} \sigma_{n+1}(u) = 0.$$

This is a contradiction.

2.2.8 Conclusion Let F and K be any fields such that K is a finite extension of F. Then $o(G(K,F)) \leq [K : F]$.

Definition Let F be a field. By 1.3.7, $F[x_1, \ldots, x_n]$ is an integral domain. Its field of quotients is denoted by $F(x_1, \ldots, x_n)$. The members of $F(x_1, \ldots, x_n)$ are called rational functions in $x_1, \ldots, x_n$ over F. By S_n we shall mean the permutation group

$$\{\sigma : \sigma : \{1, 2, \ldots, n\} \rightarrow \{1, 2, \ldots, n\} \text{ is one-to-one and onto}\},$$

which is called the *symmetric group of degree n*.

2.2.9 Problem Observe that for every $\sigma \in S_n$, the mapping $\sigma^* : r(x_1, \ldots, x_n) \mapsto r(x_{\sigma(1)}, \ldots, x_{\sigma(n)})$ from $F(x_1, \ldots, x_n)$ to $F(x_1, \ldots, x_n)$ is an automorphism of the field $F(x_1, \ldots, x_n)$.

For simplicity, σ^* is also denoted by σ. Thus we can treat S_n as a group of automorphisms of the field $F(x_1, \ldots, x_n)$.

Proof Suppose that $\frac{p(x_1, \ldots, x_n)}{q(x_1, \ldots, x_n)}, \frac{r(x_1, \ldots, x_n)}{s(x_1, \ldots, x_n)} \in F(x_1, \ldots, x_n)$, where $p(x_1, \ldots, x_n)$, $q(x_1, \ldots, x_n), r(x_1, \ldots, x_n), s(x_1, \ldots, x_n) \in F[x_1, \ldots, x_n]$. Here

$$\sigma^*\left(\frac{p(x_1,\cdots,x_n)}{q(x_1,\cdots,x_n)} + \frac{r(x_1,\cdots,x_n)}{s(x_1,\cdots,x_n)}\right)$$

$$= \sigma^*\left(\frac{p(x_1,\cdots,x_n)s(x_1,\cdots,x_n) + r(x_1,\cdots,x_n)q(x_1,\cdots,x_n)}{q(x_1,\cdots,x_n)s(x_1,\cdots,x_n)}\right)$$

$$= \frac{p\big(x_{\sigma(1)},\cdots,x_{\sigma(n)}\big)s\big(x_{\sigma(1)},\cdots,x_{\sigma(n)}\big) + r\big(x_{\sigma(1)},\cdots,x_{\sigma(n)}\big)q\big(x_{\sigma(1)},\cdots,x_{\sigma(n)}\big)}{q\big(x_{\sigma(1)},\cdots,x_{\sigma(n)}\big)s\big(x_{\sigma(1)},\cdots,x_{\sigma(n)}\big)}$$

$$= \frac{p\big(x_{\sigma(1)},\cdots,x_{\sigma(n)}\big)}{q\big(x_{\sigma(1)},\cdots,x_{\sigma(n)}\big)} + \frac{r\big(x_{\sigma(1)},\cdots,x_{\sigma(n)}\big)}{s\big(x_{\sigma(1)},\cdots,x_{\sigma(n)}\big)}$$

$$= \sigma^*\left(\frac{p(x_1,\cdots,x_n)}{q(x_1,\cdots,x_n)}\right) + \sigma^*\left(\frac{r(x_1,\cdots,x_n)}{s(x_1,\cdots,x_n)}\right),$$

so

$$\sigma^*\left(\frac{p(x_1,\ldots,x_n)}{q(x_1,\ldots,x_n)} + \frac{r(x_1,\ldots,x_n)}{s(x_1,\ldots,x_n)}\right) = \sigma^*\left(\frac{p(x_1,\ldots,x_n)}{q(x_1,\ldots,x_n)}\right) + \sigma^*\left(\frac{r(x_1,\ldots,x_n)}{s(x_1,\ldots,x_n)}\right).$$

Next,

$$\sigma^*\left(\frac{p(x_1,\cdots,x_n)}{q(x_1,\cdots,x_n)}\frac{r(x_1,\cdots,x_n)}{s(x_1,\cdots,x_n)}\right) = \sigma^*\left(\frac{p(x_1,\cdots,x_n)r(x_1,\cdots,x_n)}{q(x_1,\cdots,x_n)s(x_1,\cdots,x_n)}\right)$$

$$= \frac{p\big(x_{\sigma(1)},\cdots,x_{\sigma(n)}\big)r\big(x_{\sigma(1)},\cdots,x_{\sigma(n)}\big)}{q\big(x_{\sigma(1)},\cdots,x_{\sigma(n)}\big)s\big(x_{\sigma(1)},\cdots,x_{\sigma(n)}\big)} = \frac{p\big(x_{\sigma(1)},\cdots,x_{\sigma(n)}\big)}{q\big(x_{\sigma(1)},\cdots,x_{\sigma(n)}\big)}\frac{r\big(x_{\sigma(1)},\cdots,x_{\sigma(n)}\big)}{s\big(x_{\sigma(1)},\cdots,x_{\sigma(n)}\big)}$$

$$= \sigma^*\left(\frac{p(x_1,\cdots,x_n)}{q(x_1,\cdots,x_n)}\right) \cdot \sigma^*\left(\frac{r(x_1,\cdots,x_n)}{s(x_1,\cdots,x_n)}\right),$$

so

$$\sigma^*\left(\frac{p(x_1,\ldots,x_n)}{q(x_1,\ldots,x_n)}\frac{r(x_1,\ldots,x_n)}{s(x_1,\ldots,x_n)}\right) = \sigma^*\left(\frac{p(x_1,\ldots,x_n)}{q(x_1,\ldots,x_n)}\right) \cdot \sigma^*\left(\frac{r(x_1,\ldots,x_n)}{s(x_1,\ldots,x_n)}\right).$$

Thus σ^* preserves addition and multiplication.

$\sigma^* : F(x_1,\ldots,x_n) \to F(x_1,\ldots,x_n)$ is one-to-one. To show this, let $\sigma^*(r(x_1,\ldots,x_n)) = \sigma^*(s(x_1,\ldots,x_n))$, where $r(x_1,\ldots,x_n), s(x_1,\ldots,x_n)$ $\in F(x_1,\ldots,x_n)$. We have to show that $r(x_1,\ldots,x_n) = s(x_1,\ldots,x_n)$. Since

$$r\big(x_{\sigma(1)},\ldots,x_{\sigma(n)}\big) = \underbrace{\sigma^*(r(x_1,\ldots,x_n)) = \sigma^*(s(x_1,\ldots,x_n))} = s\big(x_{\sigma(1)},\ldots,x_{\sigma(n)}\big),$$

we have $r\big(x_{\sigma(1)}, \ldots, x_{\sigma(n)}\big) = s\big(x_{\sigma(1)}, \ldots, x_{\sigma(n)}\big)$, and hence

$$
\begin{aligned}
r(x_1, \cdots, x_n) &= r\big(x_{\sigma^{-1}(\sigma(1))}, \cdots, x_{\sigma^{-1}(\sigma(n))}\big) \\
&= \underbrace{\big(\sigma^{-1}\big)^*\big(r\big(x_{\sigma(1)}, \cdots, x_{\sigma(n)}\big)\big) = \big(\sigma^{-1}\big)^*\big(s\big(x_{\sigma(1)}, \cdots, x_{\sigma(n)}\big)\big)} \\
&= s\big(x_{\sigma^{-1}(\sigma(1))}, \cdots, x_{\sigma^{-1}(\sigma(n))}\big) = s(x_1, \cdots, x_n).
\end{aligned}
$$

Thus $r(x_1, \ldots, x_n) = s(x_1, \ldots, x_n)$.

$\sigma^* : F(x_1, \ldots, x_n) \to F(x_1, \ldots, x_n)$ is onto. To show this, let us take an arbitrary $r(x_1, \ldots, x_n) \in F(x_1, \ldots, x_n)$. Put $s(x_1, \ldots, x_n) \equiv r\big(x_{\sigma^{-1}(1)}, \ldots, x_{\sigma^{-1}(n)}\big)$. Here

$$
\begin{aligned}
\sigma^*(s(x_1, \ldots, x_n)) &= \sigma^*\big(r\big(x_{\sigma^{-1}(1)}, \ldots, x_{\sigma^{-1}(n)}\big)\big) = r\big(x_{\sigma(\sigma^{-1}(1))}, \ldots, x_{\sigma(\sigma^{-1}(n))}\big) \\
&= r(x_1, \ldots, x_n),
\end{aligned}
$$

so

$$
\sigma^*(s(x_1, \ldots, x_n)) = r(x_1, \ldots, x_n).
$$

Thus $\sigma^* : F(x_1, \ldots, x_n) \to F(x_1, \ldots, x_n)$ is an automorphism of $F(x_1, \ldots, x_n)$. ∎

2.2.10 Definition Let F be a field. We know that $F(x_1, \ldots, x_n)$ is a field extension of F, and S_n is a group of automorphisms of $F(x_1, \ldots, x_n)$. Here the fixed field of S_n is denoted by S. Thus

$$
\begin{aligned}
S = \{r(x_1, \ldots, x_n) : r(x_1, \ldots, x_n) \in F(x_1, \ldots, x_n), \text{and for every} \\
\sigma \in S_n, \sigma(r(x_1, \ldots, x_n)) = r(x_1, \ldots, x_n)\},
\end{aligned}
$$

that is,

$$
\begin{aligned}
S = \{r(x_1, \ldots, x_n) : r(x_1, \ldots, x_n) \in F(x_1, \ldots, x_n), \text{and for every} \\
\sigma \in S_n, r\big(x_{\sigma(1)}, \ldots, x_{\sigma(n)}\big) = r(x_1, \ldots, x_n)\}.
\end{aligned}
$$

By 2.2.4, S is a subfield of $F(x_1, \ldots, x_n)$. Also $F \subset S$. Thus $F \subset S \subset F(x_1, \ldots, x_n)$.

The members of S are called *symmetric rational functions*. Thus S is the field of symmetric rational functions.

2.2.11 Example Suppose that $n = 3$. Here

$$
S_3 = \{\sigma_1, \sigma_2, \sigma_3, \sigma_4, \sigma_5, \sigma_6\},
$$

where

$$\sigma_1 \equiv \begin{pmatrix} 1\ 2\ 3 \\ 1\ 2\ 3 \end{pmatrix}, \sigma_2 \equiv \begin{pmatrix} 1\ 2\ 3 \\ 2\ 3\ 1 \end{pmatrix}, \sigma_3 \equiv \begin{pmatrix} 1\ 2\ 3 \\ 3\ 1\ 2 \end{pmatrix}, \sigma_4 \equiv \begin{pmatrix} 1\ 2\ 3 \\ 1\ 3\ 2 \end{pmatrix}, \sigma_5$$

$$\equiv \begin{pmatrix} 1\ 2\ 3 \\ 3\ 2\ 1 \end{pmatrix}, \sigma_6 \equiv \begin{pmatrix} 1\ 2\ 3 \\ 2\ 1\ 3 \end{pmatrix}.$$

Observe that $x_2 x_3 + x_3 x_1 + x_1 x_2$ is a symmetric rational function.
<u>Verification</u>: We must show that

$$x_{\sigma_i(2)} x_{\sigma_i(3)} + x_{\sigma_i(3)} x_{\sigma_i(1)} + x_{\sigma_i(1)} x_{\sigma_i(2)}$$
$$= x_2 x_3 + x_3 x_1 + x_1 x_2 \quad (i = 1, 2, 3, 4, 5, 6).$$

For $\quad i = 1$: LHS $= x_{\sigma_1(2)} x_{\sigma_1(3)} + x_{\sigma_1(3)} x_{\sigma_1(1)} + x_{\sigma_1(1)} x_{\sigma_1(2)}$
$$= x_2 x_3 + x_3 x_1 + x_1 x_2 = \text{RHS}.$$

For $\quad i = 2$: LHS $= x_{\sigma_2(2)} x_{\sigma_2(3)} + x_{\sigma_2(3)} x_{\sigma_2(1)} + x_{\sigma_2(1)} x_{\sigma_2(2)}$
$$= x_3 x_1 + x_1 x_2 + x_2 x_3 = x_2 x_3 + x_3 x_1 + x_1 x_2 = \text{RHS}.$$

For $\quad i = 3$: LHS $= x_{\sigma_3(2)} x_{\sigma_3(3)} + x_{\sigma_3(3)} x_{\sigma_3(1)} + x_{\sigma_3(1)} x_{\sigma_3(2)}$
$$= x_1 x_2 + x_2 x_3 + x_3 x_1 = x_2 x_3 + x_3 x_1 + x_1 x_2 = \text{RHS}.$$

For $\quad i = 4$: LHS $= x_{\sigma_4(2)} x_{\sigma_4(3)} + x_{\sigma_4(3)} x_{\sigma_4(1)} + x_{\sigma_4(1)} x_{\sigma_4(2)}$
$$= x_3 x_2 + x_2 x_1 + x_1 x_3 = x_2 x_3 + x_3 x_1 + x_1 x_2 = \text{RHS}.$$

For $\quad i = 5$: LHS $= x_{\sigma_5(2)} x_{\sigma_5(3)} + x_{\sigma_5(3)} x_{\sigma_5(1)} + x_{\sigma_5(1)} x_{\sigma_5(2)}$
$$= x_2 x_1 + x_1 x_3 + x_3 x_2 = x_2 x_3 + x_3 x_1 + x_1 x_2 = \text{RHS}.$$

For $\quad i = 6$: LHS $= x_{\sigma_6(2)} x_{\sigma_6(3)} + x_{\sigma_6(3)} x_{\sigma_6(1)} + x_{\sigma_6(1)} x_{\sigma_6(2)}$
$$= x_1 x_3 + x_3 x_2 + x_2 x_1 = x_2 x_3 + x_3 x_1 + x_1 x_2 = \text{RHS}.$$

Verified.

Definition Similarly, $x_1 x_2 x_3$ is a symmetric rational function, and $x_1 + x_2 + x_3$ is a symmetric rational function. The symmetric rational functions $x_1 + x_2 + x_3$, $x_2 x_3 + x_3 x_1 + x_1 x_2$, and $x_1 x_2 x_3$ are called the *elementary symmetric functions*.

The elementary symmetric function $x_1 + x_2 + x_3$ is denoted by a_1, the elementary symmetric function $x_2 x_3 + x_3 x_1 + x_1 x_2$ is denoted by a_2, and the elementary symmetric function $x_1 x_2 x_3$ is denoted by a_3. Thus $(\{a_1, a_2, a_3\} \cup F) \subset S$, and hence the smallest field $F(a_1, a_2, a_3)$ containing $\{a_1, a_2, a_3\} \cup F$ is contained in S. In short,

$$F \subset F(a_1, a_2, a_3) \subset S \subset F(x_1, x_2, x_3).$$

Since $F(x_1, x_2, x_3)$ is an extension of the field S, by 2.2.6,

$$
\begin{aligned}
&G(F(x_1, x_2, x_3), S)\\
&= \left\{ \begin{array}{c} \sigma : \sigma \in \mathrm{Aut}(F(x_1, x_2, x_3)), \text{and for every } r(x_1, x_2, x_3)\\ \in S, \sigma(r(x_1, x_2, x_3)) = r(x_1, x_2, x_3) \end{array} \right\}\\
&= \left\{ \begin{array}{c} \sigma : \sigma \in \mathrm{Aut}(F(x_1, x_2, x_3)), \text{and for every } r(x_1, x_2, x_3)\\ \in S, r\left(x_{\sigma(1)}, x_{\sigma(2)}, x_{\sigma(3)}\right) = r(x_1, x_2, x_3) \end{array} \right\}
\end{aligned}
$$

is a group of automorphisms of $F(x_1, x_2, x_3)$.

2.2.12 Problem Clearly, $S_3 \subset G(F(x_1, x_2, x_3), S)$.

Proof To show this, let us take an arbitrary $\sigma_i \in S_3$, where $i \in \{1, 2, 3, 4, 5, 6\}$. We have to show that

1. $\sigma_i \in \mathrm{Aut}(F(x_1, x_2, x_3))$,
2. for every $r(x_1, x_2, x_3) \in S, r\left(x_{\sigma_i(1)}, x_{\sigma_i(2)}, x_{\sigma_i(3)}\right) = r(x_1, x_2, x_3)$.

For 1: Since we can treat S_3 as a group of automorphisms of the field $F(x_1, x_2, x_3)$, and $\sigma_i \in S_3$, we have $\sigma_i \in \mathrm{Aut}(F(x_1, x_2, x_3))$.

For 2: Let us take an arbitrary $r(x_1, x_2, x_3) \in S$. Now, since $\sigma_i \in S_3$, by the definition of S, $r\left(x_{\sigma_i(1)}, x_{\sigma_i(2)}, x_{\sigma_i(3)}\right) = r(x_1, x_2, x_3)$. ∎

2.2.13 Note Since $S_3 \subset G(F(x_1, x_2, x_3), S)$, we have

$$3! = \underbrace{o(S_3) \leq o(G(F(x_1, x_2, x_3), S))},$$

and hence

$$3! \leq o(G(F(x_1, x_2, x_3), S)).$$

Since $F(a_1, a_2, a_3) \subset F(x_1, x_2, x_3)$, the field $F(x_1, x_2, x_3)$ is an extension of $F(a_1, a_2, a_3)$. Observe that

$$\left(t^3 + (-a_1)t^2 + a_2 t + (-a_3)\right) \in (F(a_1, a_2, a_3))[t]$$

and

$$
\begin{aligned}
t^3 + (-a_1)t^2 + a_2 t + (-a_3) &= t^3 - (x_1 + x_2 + x_3)t^2 + (x_2 x_3 + x_3 x_1 + x_1 x_2)t\\
&\quad - x_1 x_2 x_3\\
&= (t - x_1)(t - x_2)(t - x_3).
\end{aligned}
$$

So

$$t^3 + (-a_1)t^2 + a_2t + (-a_3) = (t - x_1)(t - x_2)(t - x_3).$$

It follows that $F(x_1, x_2, x_3)$ contains all the roots x_1, x_2, x_3 of $t^3 + (-a_1)t^2 + a_2t + (-a_3)$ in $F(x_1, x_2, x_3)$.

We claim that $F(x_1, x_2, x_3)$ is a splitting field over $F(a_1, a_2, a_3)$ for $t^3 + (-a_1)t^2 + a_2t + (-a_3)$.

Suppose to the contrary that G is a proper subfield of $F(x_1, x_2, x_3)$ that contains all the roots x_1, x_2, x_3 of $t^3 + (-a_1)t^2 + a_2t + (-a_3)$ in $F(x_1, x_2, x_3)$. We seek a contradiction. Since G contains $F \cup \{x_1, x_2, x_3\}$, and G is a field, G contains $F(x_1, x_2, x_3)$. This contradicts the fact that G is a proper subset of $F(x_1, x_2, x_3)$.

Hence our claim is substantiated, that is, $F(x_1, x_2, x_3)$ is a splitting field over $F(a_1, a_2, a_3)$ for $t^3 + (-a_1)t^2 + a_2t + (-a_3)$. By 1.5.9,

$$[(\text{splitting field over } F(a_1, a_2, a_3) \text{ for } t^3 + (-a_1)t^2 + a_2t + (-a_3)) : F(_1, a_2, a_3)]$$
$$\leq (\deg(t^3 + (-a_1)t^2 + a_2t + (-a_3)))!,$$

so

$$[F(x_1, x_2, x_3) : F(a_1, a_2, a_3)] \leq (\deg(t^3 + (-a_1)t^2 + a_2t + (-a_3)))!,$$

and hence

$$[F(x_1, x_2, x_3) : F(a_1, a_2, a_3)] \leq 3!.$$

Thus $F(x_1, x_2, x_3)$ is a finite extension of $F(a_1, a_2, a_3)$. Now, since $F(a_1, a_2, a_3) \subset S \subset F(x_1, x_2, x_3)$, by 1.4.4, S is a finite extension of $F(a_1, a_2, a_3)$, and $F(x_1, x_2, x_3)$ is a finite extension of S. Also, by 1.4.3,

$$[F(x_1, x_2, x_3) : F(a_1, a_2, a_3)] = [F(x_1, x_2, x_3) : S][S : F(a_1, a_2, a_3)].$$

Since $F(x_1, x_2, x_3)$ is a finite extension of S, by 2.2.8, we have

$$3! \leq \underbrace{o(G(F(x_1, x_2, x_3), S))}{} \leq [F(x_1, x_2, x_3) : S] \leq [F(x_1, x_2, x_3) : S][S : F(a_1, a_2, a_3)]$$
$$= [F(x_1, x_2, x_3) : F(a_1, a_2, a_3)] \leq 3!,$$

and hence

$$\boxed{[F(x_1, x_2, x_3) : S] = 3!.}$$

Also

$$\boxed{[F(x_1, x_2, x_3) : S] = 3!.}$$

$$o(G(F(x_1, x_2, x_3), S)) = 3!(= o(S_3))$$

and

$$[F(x_1, x_2, x_3) : S] = [F(x_1, x_2, x_3) : S][S : F(a_1, a_2, a_3)].$$

Since $[F(x_1, x_2, x_3) : S] = 3!$, we have $[S : F(a_1, a_2, a_3)] = 1$, and hence $S = F(a_1, a_2, a_3)$. Since $S_3 \subset G(F(x_1, x_2, x_3), S)$ and $o(G(F(x_1, x_2, x_3), S)) = 3!(= o(S_3))$, we have $S_3 = G(F(x_1, x_2, x_3), S)$.

2.2.14 Conclusion Let F be a field. Let n be a positive integer. Then

1. $[F(x_1, \ldots, x_n) : S] = n!$,
2. $G(F(x_1, \ldots, x_n), S) = S_n$,
3. $S = F(a_1, \ldots, a_n)$,
4. $F(x_1, \ldots, x_n)$ is a splitting field over S for $t^n - a_1 t^{n-1} + a_2 t^{n-2} - \ldots + (-1)^n a_n$,

where the symbols have their usual meanings.

2.2.15 Note Let F and K be any fields such K is a finite extension of F.

It follows that $[K : F] < \infty$. Hence there exists a basis $\{u_1, u_2, \ldots, u_n\}$ of the vector space K over F, where $n = [K : F]$. By 2.2.6, $G(K, F)$ is a group of automorphisms of K, where

$$G(K, F) \equiv \{\sigma : \sigma \in \mathrm{Aut}(K), \text{ and for every } a \in F, \sigma(a) = a\}.$$

Further, by 2.2.8, $\underbrace{o(G(K, F)) \leq [K : F]}= n$. Here the fixed field of $G(K, F)$ is

$$\{a : a \in K, \text{ and for every } \sigma \in G(K, F), \sigma(a) = a\} \; (\supset F),$$

so $F \subset (\text{fixed field of } G(K, F)) \subset K$. It follows that

$$(\text{fixed field of } G(F, F)) = F.$$

Definition Let F and K be any fields such that $F \subset K$. Let K be a finite extension of F. If $F = (\text{fixed field of } G(K, F))$, that is, $(\text{fixed field of } G(K, F)) \subset F$, then we say that K *is a normal extension of* F.

Since $(\text{fixed field of } G(F, F)) = F$, and $[F : F] = 1 < \infty$, F is a normal extension of F.

2.2.16 Note Let F and K be any fields such that K is a normal extension of F. Let H be a subgroup of the group $G(K, F)$ ($\subset \text{Aut}(K)$). Let K_H be the fixed field of H, that is,

$$(K \supset)K_H = \{a : a \in K, \text{and for every } \sigma \in H, \sigma(a) = a\} (\supset F).$$

Since K is a normal extension of F, K is a finite extension of F, and hence $[K : F] < \infty$. Next, by 2.2.7, $o(G(K, F)) \leq [K : F]$. Since H is a subgroup of the group $G(K, F)$, we have $\underbrace{o(H) \leq o(G(K, F))} \leq [K : F] < \infty$, and hence $o(H) < \infty$.

This shows that H is a finite subgroup of the finite group $G(K, F)$. Further, $F \subset K_H \subset K$.

Observe that

$$G(K, K_H) = \{\sigma : \sigma \in \text{Aut}(K), \text{and for every } a \in K_H, \sigma(a) = a\} (\supset H).$$

Since $F \subset K_H \subset K$, and K is a finite extension of F, by 1.4.4, K is a finite extension of K_H, and hence by 2.2.7, $o(G(K, K_H)) \leq [K : K_H] < \infty$. Now, since $H \subset G(K, K_H)$, we have

$$\underbrace{o(H) \leq o(G(K, K_H))} \leq [K : K_H] < \infty, \quad (*)$$

and hence $o(H) \leq [K : K_H]$.

Since $1 \leq [K : K_H] < \infty$, there exist $a_1, \ldots, a_m \in K$ such that $\{a_1, \ldots, a_m\}$ is a basis of the vector space K over the field K_H. It follows that for every $x \in K$, there exist $\alpha_1, \ldots, \alpha_m \in K_H$ such that

$$x = (\alpha_1 a_1 + \cdots + \alpha_m a_m) \in K_H(a_1, \ldots, a_m).$$

Thus $K \subset K_H(a_1, \ldots, a_m)$.   Since $K_H \cup \{a_1, \ldots, a_m\} \subset K$,   we   have $\underbrace{K_H(a_1, \ldots, a_m) \subset K} \subset K_H(a_1, \ldots, a_m)$, and hence

$$K = K_H(a_1, \ldots, a_m).$$

Since $F \subset K_H \subset K_H(a_1) \subset K_H(a_1, \ldots, a_m) = K$, we have $F \subset K_H(a_1) \subset K$. Since K is a normal extension of F, K is a finite extension of F. Since $F \subset K_H \subset K$, by 1.4.4, K is a finite extension of K_H. Next, since $K_H \subset K_H(a_1) \subset K$, by 1.4.4, $K_H(a_1)$ is a finite extension of K_H, and hence by 1.4.9, a_1 is algebraic over K_H. Similarly, a_2 is algebraic over K_H, etc. By 2.1.16, there exists $a \in K_H(a_1, \cdots, a_m)$ such that

$$K_H(a_1, \cdots, a_m) = K_H(a).$$

Since $K = K_H(a_1, \ldots, a_m)$, we have $a \in K$ and $K = K_H(a)$. Since K is a normal extension of F, $(K_H(a) =)K$ is a finite extension of F, and hence $K_H(a)$ is a finite extension of F. Since $F \subset K_H \subset K = K_H(a)$, we have $F \subset K_H \subset K_H(a)$. Now, since $K_H(a)$ is a finite extension of F, by 1.4.4, $K_H(a)$ is a finite extension of K_H, and hence by 1.4.9, a is algebraic over K_H.

Let a be algebraic of degree n over K_H.

By 1.4.11, there exists $q(x) \in K_H[x]$ such that $q(x)$ is the minimal polynomial of a over K_H, that is,

1. $(K \ni)q(a) = 0$,
2. $n = \deg(q(x)) \geq 1$,
3. the leading coefficient of $q(x)$ is 1.

Again by 1.4.12, $q(x)$ is irreducible over K_H. Also, since a is algebraic of degree n over K_H, by 1.4.16, we have

$$[K : K_H] = \underbrace{[K_H(a) : K_H] = n}_{} = \deg(q(x)),$$

and hence $[K : K_H] = \deg(q(x))$.

Since H is a finite subgroup of the finite group $G(K, F)$, we can suppose that $H = \{\sigma_1, \sigma_2, \ldots, \sigma_h\}(\subset G(K, F))$, $o(H) = h$, and σ_1 is the identity element of the group $G(K, F)$. It follows that each $\sigma_i(a) \in K$, and $\sigma_1(a) = a$. Put

$$\alpha_1 \equiv \sigma_1(a) + \sigma_2(a) + \cdots + \sigma_h(a) \ (\in K),$$
$$\alpha_2 \equiv \sum_{i<j} \big(\sigma_i(a)\sigma_j(a)\big) \ (\in K),$$
$$\alpha_3 \equiv \sum_{i<j<k} \big(\sigma_i(a)\sigma_j(a)\sigma_k(a)\big) \ (\in K),$$
$$\vdots$$

We want to show that

$$\alpha_1 \in K_H(= \{a : a \in K \text{ and for every } \sigma \in H, \sigma(a) = a\}).$$

To this end, let us take an arbitrary $\sigma_i \in H$, where $i \in \{1, 2, \ldots, h\}$. It suffices to show that $\sigma_i(\alpha_1) = \alpha_1$, that is,

$$\sigma_i(\sigma_1(a) + \sigma_2(a) + \cdots + \sigma_h(a)) = \sigma_1(a) + \sigma_2(a) + \cdots + \sigma_h(a).$$

Observe that

$$\sigma_i(\sigma_1(a) + \sigma_2(a) + \cdots + \sigma_h(a)) = \sigma_i(\sigma_1(a)) + \cdots + \sigma_i(\sigma_h(a))$$
$$= (\sigma_i\sigma_1)(a) + \cdots + (\sigma_i\sigma_h)(a).$$

Also, $\sigma_j \mapsto \sigma_i \sigma_j$ is a one-to-one mapping from $\{\sigma_1, \sigma_2, \ldots, \sigma_h\}$ onto $\{\sigma_1, \sigma_2, \ldots, \sigma_h\}$, so

$$\begin{aligned}
\text{LHS} &= \sigma_i(\sigma_1(a) + \sigma_2(a) + \cdots + \sigma_h(a)) \\
&= \underbrace{(\sigma_i\sigma_1)(a) + \cdots + (\sigma_i\sigma_h)(a) = \sigma_1(a) + \sigma_2(a) + \cdots + \sigma_h(a)}_{} = \text{RHS.}
\end{aligned}$$

Thus $\alpha_1 \in K_H$. Next, we want to show that $\alpha_2 \in K_H (= \{a : a \in K$ and for every $\sigma \in H, \sigma(a) = a\})$.

To this end, let us take an arbitrary $\sigma_i \in H$, where $i \in \{1, 2, \ldots, h\}$. It suffices to show that $\sigma_i(\alpha_2) = \alpha_2$, that is,

$$\begin{aligned}
&\sigma_i((\sigma_1(a)\sigma_2(a) + \sigma_1(a)\sigma_3(a) + \cdots) + (\sigma_2(a)\sigma_3(a) + \sigma_2(a)\sigma_4(a) + \cdots) + \cdots) \\
&= (\sigma_1(a)\sigma_2(a) + \sigma_1(a)\sigma_3(a) + \cdots) + (\sigma_2(a)\sigma_3(a) + \sigma_2(a)\sigma_4(a) + \cdots) + \cdots.
\end{aligned}$$

Observe that

$$\begin{aligned}
&\sigma_i((\sigma_1(a)\sigma_2(a) + \sigma_1(a)\sigma_3(a) + \cdots) + (\sigma_2(a)\sigma_3(a) + \sigma_2(a)\sigma_4(a) + \cdots) + \cdots) \\
&\quad = (\sigma_i(\sigma_1(a)\sigma_2(a)) + \sigma_i(\sigma_1(a)\sigma_3(a)) + \cdots) \\
&\qquad + (\sigma_i(\sigma_2(a)\sigma_3(a)) + \sigma_i(\sigma_2(a)\sigma_4(a)) \cdots) + \cdots \\
&\quad = (\sigma_i(\sigma_1(a))\sigma_i(\sigma_2(a)) + \sigma_i(\sigma_1(a))\sigma_i(\sigma_3(a)) + \cdots) \\
&\qquad + (\sigma_i(\sigma_2(a))\sigma_i(\sigma_3(a)) + \sigma_i(\sigma_2(a))\sigma_i(\sigma_4(a)) + \cdots) + \cdots \\
&\quad = ((\sigma_i\sigma_1)(a)(\sigma_i\sigma_2)(a) + (\sigma_i\sigma_1)(a)(\sigma_i\sigma_3)(a) + \cdots) \\
&\qquad + ((\sigma_i\sigma_2)(a)(\sigma_i\sigma_3)(a) + (\sigma_i\sigma_2)(a)(\sigma_i\sigma_4)(a) + \cdots) + \cdots \\
&\quad = (\sigma_i\sigma_1)(a)((\sigma_i\sigma_2)(a) + (\sigma_i\sigma_3)(a) + \cdots) \\
&\qquad + (\sigma_i\sigma_2)(a)((\sigma_i\sigma_3)(a) + (\sigma_i\sigma_4)(a) + \cdots) + \cdots.
\end{aligned}$$

Also, $\sigma_j \mapsto \sigma_i \sigma_j$ is a one-to-one mapping from $\{\sigma_1, \sigma_2, \ldots, \sigma_h\}$ onto $\{\sigma_1, \sigma_2, \ldots, \sigma_h\}$, so

$$\begin{aligned}
&(\sigma_i\sigma_1)(a)((\sigma_i\sigma_2)(a) + (\sigma_i\sigma_3)(a) + \cdots) \\
&\quad + (\sigma_i\sigma_2)(a)((\sigma_i\sigma_3)(a) + (\sigma_i\sigma_4)(a) + \cdots) + \cdots \\
&= (\sigma_1(a)\sigma_2(a) + \sigma_1(a)\sigma_3(a) + \cdots) \\
&\quad + (\sigma_2(a)\sigma_3(a) + \sigma_2(a)\sigma_4(a) + \cdots) + \cdots,
\end{aligned}$$

and hence LHS = RHS. Thus $\alpha_2 \in K_H$. Similarly, $\alpha_3 \in K_H$, etc. It follows that

$$\left(x^h - \alpha_1 x^{h-1} + \alpha_2 x^{h-2} - \cdots + (-1)^h \alpha_h\right) \in K_H[x].$$

Since

$$x^h - \alpha_1 x^{h-1} + \alpha_2 x^{h-2} - \cdots + (-1)^h \alpha_h = (x - \sigma_1(a))(x - \sigma_2(a)) \cdots (x - \sigma_h(a))$$
$$= (x - a)(x - \sigma_2(a)) \cdots (x - \sigma_h(a)),$$

it follows that a is a root of the polynomial $p(x)$ in $K_H[x]$, where

$$p(x) \equiv x^h - \alpha_1 x^{h-1} + \alpha_2 x^{h-2} - \cdots + (-1)^h \alpha_h.$$

Hence $p(x) \in K_H[x]$, $p(a) = 0$, $h = \deg(p(x)) \geq 1$, and the leading coefficient of $p(x)$ is 1. Now, since $q(x)$ is the minimal polynomial of a over K_H, we have

$$[K : K_H] = \underbrace{\deg(q(x)) \leq \deg(p(x))} = h = o(H),$$

and hence $[K : K_H] \leq o(H)$. Since $o(H) \leq [K : K_H]$, we have

$$\boxed{o(H) = [K : K_H]}$$

Next, from (*), $o(G(K, K_H)) = o(H)$. Now, since $H \subset G(K, K_H)$, we have $\boxed{H = G(K, K_H)}$.

We can substitute $G(K, F)$ for H in $o(H) = [K : K_H]$. We get $o(G(K, F)) = [K : K_{G(K,F)}]$. Now observe that

$$K_{G(K,F)} = \{b : b \in K, \text{and for every } \sigma \in G(K, F), \sigma(b) = b\}.$$

Since K is a normal extension of F, we have

$$\underbrace{F = (\text{fixed field of } G(K, F))} = \{b : b \in K \text{ and for every } \sigma \in G(K, F), \sigma(b) = b\}$$
$$= K_{G(K,F)},$$

and hence $K_{G(K,F)} = F$. Since $o(G(K, F)) = [K : K_{G(K,F)}]$, we have

$$\boxed{o(G(K, F)) = [K : F].}$$

We can substitute $G(K, F)$ for H in $K = K_H(a)$. We get $K = \underbrace{K_{G(K,F)}(a) = F(a)}$, and hence

$$\boxed{K = F(a),}$$

where $a \in K$. Further,

$$h = o(H) = o(G(K,F)) = [K : F] = [K : K_{G(K,F)}] = [K : K_H] = \deg(q(x)) = n,$$

so $h = n$. Now

$$p(x) = x^h - \alpha_1 x^{h-1} + \alpha_2 x^{h-2} - \cdots + (-1)^h \alpha_h$$

becomes

$$p(x) = x^n - \alpha_1 x^{n-1} + \alpha_2 x^{n-2} - \cdots + (-1)^n \alpha_n.$$

Next, $p(x) \in K_H[x]$ becomes $(x^n - \alpha_1 x^{n-1} + \alpha_2 x^{n-2} - \cdots + (-1)^n \alpha_n =)$ $p(x) \in F[x]$, and hence each α_i is in F. Also $p(a) = 0$, $h = \deg(p(x)) \geq 1$, and the leading coefficient of $p(x)$ is 1. Further,

$$x^h - \alpha_1 x^{h-1} + \alpha_2 x^{h-2} - \cdots + (-1)^h \alpha_h = (x - \sigma_1(a))(x - \sigma_2(a)) \cdots (x - \sigma_h(a))$$
$$= (x - a)(x - \sigma_2(a)) \cdots (x - \sigma_h(a))$$

becomes

$$x^n - \alpha_1 x^{n-1} + \alpha_2 x^{n-2} - \cdots + (-1)^n \alpha_n = (x - \sigma_1(a))(x - \sigma_2(a)) \cdots (x - \sigma_n(a))$$
$$= (x - a)(x - \sigma_2(a)) \cdots (x - \sigma_n(a)),$$

and $H = \{\sigma_1, \sigma_2, \cdots, \sigma_h\}$ becomes

$$G(K,F) = \{\sigma_1, \sigma_2, \cdots, \sigma_n\}.$$

Since each $\sigma_i(a)$ is in K and $(F[x] \ni) p(x) = (x - \sigma_1(a))(x - \sigma_2(a)) \ldots$ $(x - \sigma_n(a))$, K splits the polynomial $p(x)$ in $F[x]$ into a product of linear factors in $K[x]$.

We shall show that K is a splitting field over F for $p(x)$.

Assume to the contrary that G is a proper subfield of $K(= F(a))$ that contains F as well as all the roots of $p(x)$ in K. We seek a contradiction.

Since $\sigma_1(a)$ is a root of $p(x)$ in K, and G contains all the roots of $p(x)$ in K, we have $\sigma_1(a) \in G$. Now, since $\sigma_1(a) = a$, we have $a \in G$. Thus $F \cup \{a\} \subset G$, and hence $K = \underbrace{F(a) \subset G}$. Thus $K \subset G$. This contradicts the fact that G is a proper

subset of K.

Thus we have shown that K is a splitting field over F for $p(x)$.

2.2.17 Conclusion Let F and K be any fields such that K is a normal extension of F. Let H be a subgroup of the group $G(K,F)$ $(\subset \text{Aut}(K))$. Let K_H be the fixed field of H. Then

1. $o(H) = [K : K_H]$,
2. $H = G(K, K_H)$,
3. $o(G(K, F)) = [K : F]$,
4. there exists $a \in K$ such that $K = F(a)$, and K is a splitting field over F for $(x - \sigma_1(a))(x - \sigma_2(a)) \cdots (x - \sigma_n(a))$ in $F[x]$, where $G(K, F) = \{\sigma_1, \sigma_2, \ldots, \sigma_n\}$ and $\sigma_1(a) = a$.

2.2.18 Note Let F and K be any fields such that $F \subset K$. Let $f(x) \in F[x]$. Let K be a splitting field over F for $f(x)$. Suppose that $\deg(f(x)) \geq 1$. Let $p(x)$ be an irreducible factor of $f(x)$ in $F[x]$. Suppose that all the roots of $p(x)$ are $\alpha_1, \alpha_2, \ldots, \alpha_r$.

Since $p(x)$ is a factor of $f(x)$ in $F[x]$, all the roots of $p(x)$ are the roots of $f(x)$. Now, since $\alpha_1, \alpha_2, \ldots, \alpha_r$ are the roots of $p(x)$, $\alpha_1, \alpha_2, \ldots, \alpha_r$ are the roots of $f(x)$. Since K is a splitting field over F for $f(x)$, K contains all the roots of $f(x)$. Since $\alpha_1, \alpha_2, \ldots, \alpha_r$ are the roots of $f(x)$, K contains $\alpha_1, \alpha_2, \ldots, \alpha_r$.

Let us fix an arbitrary $i \in \{2, 3, \ldots, r\}$.

Since α_1, α_i are members of K, $p(x) \in F[x](\supset F)$, $p(x)$ is irreducible over F, and α_1, α_i are the roots of $p(x)$ in K, by 1.5.18, there exists an isomorphism τ_i from the field $F(\alpha_1)(\subset K)$ onto the field $F(\alpha_i)$ such that

1. $\tau_i(\alpha_1) = \alpha_i$,
2. for every $a \in F$, $\tau_i(a) = a$.

Since K is a splitting field over F for $f(x)$, K is a finite extension of F. Since $F \cup \{\alpha_1\} \subset K$, we have $F \subset F(\alpha_1) \subset K$. Since K is a finite extension of F, by 1.4.4, K is a finite extension of $F(\alpha_1)$. Since $f(x) \in F[x](\subset (F(\alpha_1))[x])$, we have $f(x) \in (F(\alpha_1))[x]$.

We want to show that K is a splitting field over $F(\alpha_1)$ for $f(x)$.

To this end, let us take a proper subfield G of K that contains $F(\alpha_1)(\supset F)$. It suffices to show that G does not contain all the roots of $f(x)$. Since G is a proper subfield of K that contains F, and K is a splitting field over F for $f(x)$, G does not contain all the roots of $f(x)$.

Thus K is a splitting field for $f(x)$ considered a polynomial over F_1, where $F_1 \equiv F(\alpha_1)$. Similarly, K is a splitting field for $f(x)$ considered a polynomial over $(F_1)'$, where $(F_1)' \equiv F(\alpha_i)$. Now, by 1.5.27, there exists a ring isomorphism σ_i from K onto K such that for every $a \in F_1(= F(\alpha_1) \supset F \cup \{\alpha_1\})$, $\sigma_i(a) = \tau_i(a)$.

Hence for every $a \in F$, $\underline{\sigma_i(a) = \tau_i(a)} = a$, and hence for every $a \in F$, $\sigma_i(a) = a$. Also, $\underline{\sigma_i(\alpha_1) = \tau_i(\alpha_1)} = \alpha_i$. Thus $\sigma_i(\alpha_1) = \alpha_i$. Since σ_i is a ring isomorphism from the field K onto K, we have $\sigma_i \in \mathrm{Aut}(K)$. Next, since for every $a \in F$, $\sigma_i(a) = a$, we have $\sigma_i \in G(K, F)$.

2.2.19 Conclusion Let F and K be any fields such that $F \subset K$. Let $f(x) \in F[x]$. Let K be a splitting field over F for $f(x)$. Suppose that $\deg(f(x)) \geq 1$. Let $p(x)$ be an irreducible factor of $f(x)$ in $F[x]$. Suppose that all the roots of $p(x)$ are $\alpha_1, \alpha_2, \ldots, \alpha_r$. Then for every $i \in \{1, 2, \ldots, r\}$, there exists $\sigma_i \in G(K, F)$ such that $\sigma_i(\alpha_1) = \alpha_i$.

2.2.20 Problem Let F and K be any fields such that K is an extension of F. Let $f(x) \in F[x]$. Let K be a splitting field over F for $f(x)$. Suppose that $\deg(f(x)) \geq 1$. Then K is a normal extension of F.

Proof Case I: $f(x)$ splits into linear factors over F.

Since $f(x)$ splits into linear factors over F, F is a splitting field over F for $f(x)$. Now, since K is a splitting field over F for $f(x)$, by 1.5.29, $K = F$. Since F is a normal extension of F, K is a normal extension of F.

Case II: $f(x)$ does not split into linear factors over F.

For induction on $[K : F]$, let us assume that for every pair of fields K_1, F_1 of degree $< [K : F]$,

$$(K_1 \text{ is a splitting field over } F_1 \text{ of some polynomial in } F_1[x])$$
$$\Rightarrow (K_1 \text{ is a normal extension of } F_1). \ (*)$$

Since $f(x)$ does not split into linear factors over F, by 1.2.21, there exists an irreducible factor $p(x)$ of $f(x)$ in $F[x]$ such that $\deg(p(x)) \geq 2$. Suppose that all the roots of $p(x)$ are $\alpha_1, \alpha_2, \ldots, \alpha_r$, where $r \equiv \deg(p(x))(\geq 2)$. Since K is a splitting field over F for $f(x)$, K is a finite extension of F. Since α_1 is a root of $p(x)$, and $p(x)$ is irreducible over F, α_1 is a nonzero.

Since $p(x)$ is a factor of $f(x)$ in $F[x]$, all the roots of $p(x)$ are roots of $f(x)$. Since $\alpha_1, \alpha_2, \ldots, \alpha_r$ are the roots of $p(x)$, $\alpha_1, \alpha_2, \ldots, \alpha_r$ are the roots of $f(x)$. Since K is a splitting field over F for $f(x)$, K contains all the roots of $f(x)$. Since $\alpha_1, \alpha_2, \ldots, \alpha_r$ are the roots of $f(x)$, K contains $\alpha_1, \alpha_2, \ldots, \alpha_r$. It follows that $F \cup \{\alpha_1\} \subset K$, and hence $F \subset F(\alpha_1) \subset K$. Now, since K is a finite extension of F, by 1.4.4, $F(\alpha_1)$ is a finite extension of F, and hence by 1.4.9, α_1 is algebraic over F.

By 1.4.3,

$$[K : F] = [K : F(\alpha_1)][F(\alpha_1) : F].$$

Since α_1 is a root of $p(x)$, we have $p(\alpha_1) = 0$. Further, since $p(x)$ is irreducible in $F[x]$ and $r = \deg(p(x)) \geq 2$, α_1 is algebraic of degree $r(\geq 2)$ over F. Now, by 1.4.16, $[F(\alpha_1) : F] = r$, and hence

$$[K : F] = \underbrace{[K : F(\alpha_1)][F(\alpha_1) : F] = [K : F(\alpha_1)]r} > [K : F(\alpha_1)]1 = [K : F(\alpha_1)].$$

Thus $[K : F(\alpha_1)] < [K : F]$.

Since $F \subset F(\alpha_1) \subset K$ and K is a finite extension of F, by 1.4.4, K is a finite extension of $F(\alpha_1)$. Since $f(x) \in F[x](\subset (F(\alpha_1))[x])$, we have $f(x) \in (F(\alpha_1))[x]$.

We want to show that K is a splitting field over $F(\alpha_1)$ for $f(x)$.

To this end, let us take a proper subfield G of K that contains $F(\alpha_1)(\supset F)$. It suffices to show that G does not contain all the roots of $f(x)$. Since G is a proper subfield of K that contains F, and K is a splitting field over F for $f(x)$, G does not contain all the roots of $f(x)$.

Thus K is a splitting field over $F(\alpha_1)$ for $f(x)$. Next, since $[K : F(\alpha_1)] < [K : F]$, by the induction hypothesis (*), K is a normal extension of $F(\alpha_1)$.

We claim that K is a normal extension of F.

Suppose to the contrary that K is not a normal extension of F. We seek a contradiction.

Since K is a finite extension of F, and K is not a normal extension of F, we have (fixed field of $G(K, F)$) $\not\subset F$. It follows that there exists $\theta \in$ (fixed field of $G(K, F)$) such that $\theta \notin F$. Since K is a normal extension of $F(\alpha_1)$, we have

$$\text{(fixed field of } G(K, F(\alpha_1))) = F(\alpha_1).$$

Since $F \subset F(\alpha_1)$, we have

$$G(K, F(\alpha_1)) = \underbrace{\begin{array}{l} \{\sigma : \sigma \in \text{Aut}(K), \text{ for every } a \in F(\alpha_1), \sigma(a) = a\} \\ \subset \{\sigma : \sigma \in \text{Aut}(K), \text{ for every } a \in F, \sigma(a) = a\} \end{array}}$$

$$= G(K, F),$$

and hence

$$G(K, F(\alpha_1)) \subset G(K, F).$$

It follows that

$$\theta \in \text{(fixed field of } G(K, F))$$

$$= \underbrace{\begin{array}{l} \{a : a \in K, \text{ for every } \sigma \in G(K, F), \sigma(a) = a\} \\ \subset \{a : a \in K, \text{ for ever } \sigma \in G(K, F(\alpha_1)), \sigma(a) = a\} \end{array}}$$

$$= \text{(fixed field of } G(K, F(\alpha_1))) = F(\alpha_1),$$

and hence $\theta \in F(\alpha_1)$. Also $\theta \notin F$.

Since α_1 is algebraic of degree $r(\geq 2)$ over F, $\left\{1, \alpha_1, (\alpha_1)^2, \ldots, (\alpha_1)^{r-1}\right\}$ is a linearly independent set of vectors in the vector space $F(\alpha_1)$ over the field F.

Proof Suppose to the contrary that there exist $\gamma_0, \gamma_1, \ldots, \gamma_{r-1}$ in F such that not all the γ_i are zero and

$$\gamma_0 1 + \gamma_1 \alpha_1 + \cdots + \gamma_{r-1}(\alpha_1)^{r-1} = 0.$$

We seek a contradiction. Here, it follows that $q(x) \equiv \gamma_0 + \gamma_1 x + \cdots + \gamma_{r-1} x^{r-1}$ is a nonzero polynomial in $F[x]$ such that $q(\alpha_1) = 0$, and $\deg(q(x)) \leq r - 1 < r$. This contradicts the fact that α_1 is algebraic of degree r over F. ∎

Thus, we have shown that $\left\{1, \alpha_1, (\alpha_1)^2, \ldots, (\alpha_1)^{r-1}\right\}$ is a linearly independent set of vectors in the vector space $F(\alpha_1)$ over the field F. Since $[F(\alpha_1) : F] = r$, the dimension of the vector space $F(\alpha_1)$ over the field F is r. Now, since $\left\{1, \alpha_1, (\alpha_1)^2, \ldots, (\alpha_1)^{r-1}\right\}$ is a linearly independent set of vectors in the vector space $F(\alpha_1)$ over the field F, $\left\{1, \alpha_1, (\alpha_1)^2, \ldots, (\alpha_1)^{r-1}\right\}$ constitutes a basis for the vector space $F(\alpha_1)$ over the field F. Now, since $\theta \in F(\alpha_1)$, there exist $\lambda_0, \lambda_1, \ldots, \lambda_{r-1}$ in F such that

$$\theta = \lambda_0 1 + \lambda_1 \alpha_1 + \cdots + \lambda_{r-1}(\alpha_1)^{r-1}.$$

Since K is a splitting field over F for $f(x)$, $\deg(f(x)) \geq 1$, $p(x)$ is an irreducible factor of $f(x)$ in $F[x]$, and all the roots of $p(x)$ are $\alpha_1, \alpha_2, \ldots, \alpha_r$, by 2.2.19, for every $i \in \{1, 2, \ldots, r\}$, there exists $\sigma_i \in G(K, F)$ such that $\sigma_i(\alpha_1) = \alpha_i$. It follows that for every $i \in \{1, 2, \ldots, r\}$,

$$\underbrace{\sigma_i(\theta) = \sigma_i\left(\lambda_0 1 + \lambda_1 \alpha_1 + \cdots + \lambda_{r-1}(\alpha_1)^{r-1}\right)}$$

$$= \sigma_i(\lambda_0 1) + \sigma_i(\lambda_1 \alpha_1) + \cdots + \sigma_i\left(\lambda_{r-1}(\alpha_1)^{r-1}\right)$$
$$= \sigma_i(\lambda_0)\sigma_i(1) + \sigma_i(\lambda_1)\sigma_i(\alpha_1) + \cdots + \sigma_i(\lambda_{r-1})(\sigma_i(\alpha_1))^{r-1}$$
$$= \sigma_i(\lambda_0)\sigma_i(1) + \sigma_i(\lambda_1)\alpha_i + \cdots + \sigma_i(\lambda_{r-1})(\alpha_i)^{r-1}$$
$$= \sigma_i(\lambda_0)1 + \sigma_i(\lambda_1)\alpha_i + \cdots + \sigma_i(\lambda_{r-1})(\alpha_i)^{r-1} = \lambda_0 1 + \lambda_1 \alpha_i + \cdots + \lambda_{r-1}(\alpha_i)^{r-1},$$

and hence

$$\sigma_i(\theta) = \lambda_0 1 + \lambda_1 \alpha_i + \cdots + \lambda_{r-1}(\alpha_i)^{r-1} \quad (i = 1, 2, \ldots, r).$$

Since

$$\theta \in (\text{fixed field of } G(K, F))$$
$$= \{a : a \in K, \text{ and for every } \sigma \in G(K, F), \sigma(a) = a\},$$

and each σ_i is in $G(K, F)$, we have $\sigma_i(\theta) = \theta$ $(i = 1, 2, \ldots, r)$. It follows, from (*) that

$$\theta = \lambda_0 1 + \lambda_1 \alpha_i + \cdots + \lambda_{r-1}(\alpha_i)^{r-1} \quad (i = 1, 2, \ldots, r).$$

This shows that $\underbrace{\alpha_1, \alpha_2, \ldots, \alpha_r}_{r \text{ in population}}$ are roots of the polynomial

$(\lambda_0 - \theta) + \lambda_1 x + \cdots + \lambda_{r-1} x^{r-1}$. Here $(\lambda_0 - \theta) + \lambda_1 x + \cdots + \lambda_{r-1} x^{r-1}$ is a polynomial of degree at most $(r-1)(<r)$, so $(\lambda_0 - \theta) + \lambda_1 x + \cdots + \lambda_{r-1} x^{r-1}$ is the

zero polynomial of $K[x]$. It follows that $(\lambda_0 - \theta) = 0$, and hence $\theta = \lambda_0$. Now, since $\lambda_0 \in F$, we have $\theta \in F$. This is a contradiction.

Thus our claim is substantiated, and hence K is a normal extension of F.

So in all cases, K is a normal extension of F.   ■

Definition Let F and K be any fields such that K is an extension of F. Let $f(x) \in F[x]$. Let K be a splitting field over F for $f(x)$. The group $G(K, F)(= \{\sigma : \sigma \in \mathrm{Aut}(K), \text{and for every } a \in F, \sigma(a) = a\})$ is called the *Galois group of* $f(x)$.

2.2.21 Note Let F and K be any fields such that K is an extension of F. Let $f(x)$ be a nonzero member of $F[x]$. Let K be a splitting field over F for $f(x)$. Suppose that $\deg(f(x)) \geq 1$. Let T be a subfield of K that contains F. Put

$$G(K, T) \equiv \{\sigma : \sigma \in G(K, F), \text{and for every } t \in T, \sigma(t) = t\}.$$

Clearly, $G(K, T)$ is a subgroup of the group $G(K, F)$. For every subgroup H of the group $G(K, F)$, put

$$K_H \equiv \{a : a \in K, \text{and for every } \sigma \in H, \sigma(a) = a\}.$$

Clearly, K is a splitting field over T for $f(x)$.

Proof Since K is a splitting field over F for $f(x)$, K is a finite extension of F. Now since $F \subset T \subset K$, by 1.4.4, K is a finite extension of T. Let G be a proper subfield of K which contains $T (\supset F)$. We have to show that G does not contain all the roots of $f(x)$. Since G is a proper subfield of K which contains F, and K is a splitting field over F for $f(x)$, G does not contain all the roots of $f(x)$.   ■

Thus we have shown that K is a splitting field over T for $f(x)$. Now, by 2.2.20, K is a normal extension of T, and hence, by definition of normal extension,

$$\underbrace{T = (\text{fixed field of } G(K, T))} = \{a : a \in K, \text{and for every } \sigma \in G(K, T), \sigma(a) = a\}$$

$$= K_{G(K,T)},$$

and hence $\boxed{T = K_{G(K,T)}}$. Since K is a normal extension of T, by 2.2.17, $\boxed{o(G(K, T)) = [K : T]}$, and $\boxed{H = G(K, K_H)}$. Since K is a splitting field over F for $f(x)$, by 2.2.20, K is a normal extension of F, and hence, by 2.2.17, $o(G(K, F)) = [K : F]$. Since K is a finite extension of F, and $F \subset T \subset K$, by 1.4.4, and 1.4.3, we have

$$o(G(K, F)) = \underbrace{[K : F] = [K : T][T : F]} = o(G(K, T))[T : F],$$

and hence,

$$[T : F] = \frac{o(G(K,F))}{o(G(K,T))} = (\text{index of subgroup } G(K,T) \text{of the group } G(K,F)).$$

2.2.22 Conclusion Let F and K be any fields such that $F \subset K$. Let K be an extension of F. Let $f(x)$ be a nonzero member of $F[x]$. Let K be a splitting field over F for $f(x)$. Suppose that $\deg(f(x)) \geq 1$. Let T be a subfield of K which contains F. Put

$$G(K,T) \equiv \{\sigma : \sigma \in G(K,F) \text{ and for every } t \in T, \sigma(t) = t\}.$$

Clearly, $G(K,T)$ is a subgroup of the group $G(K,F)$. For every subgroup H of the group $G(K,F)$, put

$$K_H \equiv \{a : a \in K, \text{ and for every } \sigma \in H, \sigma(a) = a\}.$$

Then:

1. $T = K_{G(K,T)}$, that is, the mapping $\Phi : H \mapsto K_H$ from the collection of all sub-groups of the group $G(K,F)$ to the collection of all subfields of K that contain F is onto.
2. $H = G(K,K_H)$, that is, the mapping $\Psi : T \mapsto G(K,T)$ from the collection of all subfields of K that contain F to the collection of all subgroups of the group $G(K,F)$ is onto. Also $(\Psi \circ \Phi)(H) = \Psi(\Phi(H)) = \Psi(K_H) = G(K,K_H) = H$, so $(\Psi \circ \Phi)(H) = H$. Next, $(\Phi \circ \Psi)(t) = \Phi(\Psi(t)) = \Phi(G(K,T)) = K_{G(K,T)} = T$, so $(\Phi \circ \Psi)(t) = T$. Thus $\Psi^{-1} = \Phi$.
3. It follows that $\Psi : T \mapsto G(K,T)$ is a one-to-one correspondence from the col-lection of all subfields of K that contain F onto the collection of all subgroups of the group $G(K,F)$.
4. $o(G(K,T)) = [K : T]$.
5. $[T : F]$ is equal to the index of the subgroup $G(K,T)$ in the group $G(K,F)$ that is, $[T : F] = \frac{o(G(K,F))}{o(G(K,T))}$.
6. $o(G(K,F)) = [K : F] = [K : T][T : F] = o(G(K,T))[T : F]$.

2.2.23 Problem Let F and K be any fields such that K is an extension of F. Let T be a subfield of K that contains F. Suppose that T is a normal extension of F. Let $\sigma \in G(K,F)$. Then $\sigma(t) \subset T$.

Proof Suppose to the contrary that there exists $\theta \in T$ such that $\sigma(\theta) \notin T$. We seek a contradiction.

Since T is a normal extension of F, by 2.2.17, there exists $a \in T$ such that $T = F(a)$ and T is a splitting field over F for

$$(x - \sigma_1(a))(x - \sigma_2(a))\ldots(x - \sigma_n(a))$$

in $F[x]$, where $G(T, F) = \{\sigma_1, \sigma_2, \ldots, \sigma_n\}$ and $\sigma_1(a) = a$. Thus T is a splitting field over F for $p(x) \in F[x]$, where

$$p(x) \equiv (x - \sigma_1(a))(x - \sigma_2(a))\ldots(x - \sigma_n(a))(= (x - a)(x - \sigma_2(a))\ldots(x - \sigma_n(a))).$$

Suppose that

$$p(x) \equiv x^n + b_1 x^{n-1} + \cdots + b_n,$$

where each b_i is in F. Clearly, $p(a) = 0$. Now,

$$p(\sigma(a)) \equiv (\sigma(a))^n + b_1(\sigma(a))^{n-1} + \cdots + b_n$$
$$= \sigma(a^n) + b_1\sigma(a^{n-1}) + \cdots + b_n\sigma(a^n) + \sigma(b_1)\sigma(a^{n-1}) + \cdots + \sigma(b_n)$$
$$= \sigma(a^n) + \sigma(b_1 a^{n-1}) + \cdots + \sigma(b_n)$$
$$= \sigma(a^n + b_1 a^{n-1} + \cdots + b_n) = \sigma(p(a)) = \sigma(0) = 0,$$

so $p(\sigma(a)) = 0$, and hence $\sigma(a)$ is a root of $p(x)$. Now, since T is a splitting field over F for $p(x) \in F[x]$, we have $\sigma(a) \in T(= F(a))$, and hence $\sigma(a) \in F(a)$.

Since T is a normal extension of F, $(F(a) =)T$ is a finite extension of F, and hence $F(a)$ is a finite extension of F. Since $\sigma(\theta) \notin T = F(a)$, we have $\sigma(\theta) \notin F(a)$. Since $\theta \in T = F(a)$, we have $\theta \in F(a)$. Since $F(a)$ is a finite extension of F, by 1.4.9, a is algebraic over F. Let $a(\in F(a))$ be algebraic of degree n over F. It follows, by 1.4.2, that $[F(a) : F] = n$, and hence n is the dimension of the vector space $F(a)$ over the field F.

Clearly, $\{1, a, a^2, \ldots, a^{n-1}\}$ is a linearly independent set of vectors for the vector space $F(a)$ over the field F.

Proof Suppose to the contrary that there exist $\lambda_0, \lambda_1, \ldots, \lambda_{n-1} \in F$ such that not all the λ_i are zero, and

$$\lambda_0 1 + \lambda_1 a + \cdots + \lambda_{n-1} a^{n-1} = 0.$$

We see, a contradiction. It follows that a is a root of the nonzero polynomial $\lambda_0 + \lambda_1 x + \cdots + \lambda_{n-1} x^{n-1}$ in $F[x]$. Further, the degree of this polynomial is strictly smaller than n. This contradicts the fact that a is algebraic of degree n over F.   ∎

Thus we have shown that $\{1, a, a^2, \ldots, a^{n-1}\}$ is a linearly independent set of vectors for the vector space $F(a)$ over the field F. Now, since n is the dimension of the vector space $F(a)$ over the field F, $\{1, a, a^2, \ldots, a^{n-1}\}$ is a basis of the vector space $F(a)$ over the field F. Next, since $\theta \in F(a)$, there exist $\gamma_0, \gamma_1, \ldots, \gamma_{n-1} \in F$ such that

$$\theta = \gamma_0 1 + \gamma_1 a + \cdots + \gamma_{n-1} a^{n-1}.$$

Hence

$$\underbrace{\sigma(\theta) = \sigma\left(\gamma_0 1 + \gamma_1 a + \cdots + \gamma_{n-1} a^{n-1}\right)}$$

$$= \sigma(\gamma_0 1) + \sigma(\gamma_1 a) + \cdots + \sigma(\gamma_{n-1} a^{n-1})$$
$$= \sigma(\gamma_0)\sigma(1) + \sigma(\gamma_1)\sigma(a) + \cdots + \sigma(\gamma_{n-1})\sigma(a^{n-1})$$
$$= \sigma(\gamma_0)\sigma(1) + \sigma(\gamma_1)\sigma(a) + \cdots + \sigma(\gamma_{n-1})(\sigma(a))^{n-1}$$
$$= \gamma_0 \sigma(1) + \gamma_1 \sigma(a) + \cdots + \gamma_{n-1}(\sigma(a))^{n-1} = \gamma_0 1 + \gamma_1 \sigma(a) + \cdots + \gamma_{n-1}(\sigma(a))^{n-1},$$

so

$$F(a) \not\ni \underbrace{\sigma(\theta) = \gamma_0 1 + \gamma_1 \sigma(a) + \cdots + \gamma_{n-1}(\sigma(a))^{n-1}},$$

and   hence $\left(\gamma_0 1 + \gamma_1 \sigma(a) + \cdots + \gamma_{n-1}(\sigma(a))^{n-1}\right) \notin F(a)$.   It   follows   that $\sigma(a) \notin F(a)$. This is a contradiction.   ∎

2.2.24 Problem Let F and K be any fields such that K is an extension of F. Let T be a subfield of K that contains F. Suppose that T is a normal extension of F. Clearly, $G(K,T)$ is a subgroup of the group $G(K,F)$. Also, $G(K,T)$ is a normal subgroup of the group $G(K,F)$.

Proof Let us take any $\tau \in G(K,T)$ and $\sigma \in G(K,F)$. We have to show that $\sigma^{-1}\tau\sigma \in G(K,T)$. To this end, let us take any $t \in T$. It suffices to show that $(\sigma^{-1}\tau\sigma)(t) = t$, that is, $\sigma^{-1}(\tau(\sigma(t))) = t$, that is, $\tau(\sigma(t)) = \sigma(t)$. Since $\tau \in G(K,T)$, it is enough to show that $\sigma(t) \in T$. By 2.2.23, $\sigma(t) \subset T$. Now, since $\sigma(t) \in \sigma(t)$, we have $\sigma(t) \in T$.   ∎

2.2.25 Note Let F and K be any fields such that K is an extension of F. Let T be a subfield of K that contains F. Suppose that T is a normal extension of F. By 2.2.24, $G(K,T)$ is a normal subgroup of the group $G(K,F)$. Hence $\frac{G(K,F)}{G(K,T)}$ is a quotient group. Also, $G(T,F)$ is a group.

Take an arbitrary $\sigma \in G(K,F)$. By 2.2.23, $\sigma(t) \subset T$. Since $\sigma \in G(K,F)$ and $G(K,F)$ is a group, the inverse function σ^{-1} is in $G(K,F)$, and hence by 2.2.23, $\sigma^{-1}(t) \subset T$. It follows that $T \subset \sigma(t)$. Thus $\sigma(t) = T$. Now, since $\sigma \in G(K,F)$ and $F \subset T \subset K$, we have $(\sigma|_T) \in G(T,F)$. Thus

$$\eta : \sigma \mapsto (\sigma|_T)$$

is a mapping from group $G(K,F)$ to the group $G(T,F)$.

η preserves the binary operation: To show this, let us take arbitrary $\sigma, \mu \in G(K,F)$. We have to show that $(\sigma\mu)|_T = (\sigma|_T)(\mu|_T)$.

Let us take an arbitrary $a \in T$. We have to show that $((\sigma\mu)|_T)(a)$ $= ((\sigma|_T)(\mu|_T))(a)$, that is, $(\sigma\mu)(a) = (\sigma|_T)((\mu|_T)(a))$, that is, $(\sigma\mu)(a) = (\sigma|_T)(\mu(a))$.

Since $\mu \in G(K, F)$, as above, we have $\mu(t) = T$. Since $a \in T$, we have $\mu(a) \in \mu(t) = T$, and hence $\mu(a) \in T$. It follows that $(\sigma|_T)(\mu(a)) = \sigma(\mu(a)) = (\sigma\mu)(a)$. Thus η preserves the binary operation.

$\ker(\eta) = G(K, T)$: Let us take an arbitrary $\sigma \in \ker(\eta)$, that is, $\sigma \in G(K, F)$ and $(\sigma|_T) = \text{Id}_T$. Since $\sigma \in G(K, F)$, we have $\sigma \in \text{Aut}(K)$. Next, since $(\sigma|_T) = \text{Id}_T$, for every $t \in T$, $\sigma(t) = t$. This shows that $\sigma \in G(K, T)$. Thus $\ker(\eta) \subset G(K, T)$.

Let us take an arbitrary $\sigma \in G(K, T)$, that is, $\sigma \in \text{Aut}(K)$ and $(\sigma|_T) = \text{Id}_T$. We have to show that $\sigma \in \ker(\eta)$, that is, $\sigma \in G(K, F)$ and $(\sigma|_T) = \text{Id}_T$. It remains to show that $(\sigma|_F) = \text{Id}_F$. Since $(\sigma|_T) = \text{Id}_T$ and $F \subset T$, we have $(\sigma|_F) = \text{Id}_F$. Thus $G(K, T) \subset \ker()$. Hence $\ker(\eta) = G(K, T)$.

Since $\eta : G(K, F) \rightarrow G(T, F)$ preserves the group binary operations, $\eta : G(K, F) \rightarrow G(T, F)$ is a homomorphism from $G(K, F)$ onto $\eta(G(K, F))$, and hence by the fundamental theorem of group homomorphisms, the quotient group $\frac{G(K,F)}{\ker(\eta)} \left(= \frac{G(K,F)}{G(K,T)} \right)$ is isomorphic to $\eta(G(K, F))$. It follows that $\frac{G(K,F)}{G(K,T)}$ is isomorphic to $\eta(G(K, F))$, and hence $\left(\frac{o(G(K,F))}{o(G(K,T))} = \right) o\left(\frac{G(K,F)}{G(K,T)} \right) = o(\eta(G(K, F)))$. Thus $\frac{o(G(K,F))}{o(G(K,T))} = o(\eta(G(K, F)))$. By 2.2.21,

$$\underbrace{[T : F] = \frac{o(G(K,F))}{o(G(K,T))}}= o(\eta(G(K, F))) \leq o(G(T, F)).$$

Since T is a normal extension of F, by 2.2.17, $o(G(T, F)) = [T : F]$, and hence

$$[T : F] = \frac{o(G(K,F))}{o(G(K,T))} = o(\eta(G(K, F))) \leq o(G(T, F)) = [T : F].$$

This shows that

$$o(\eta(G(K, F))) = o(G(T, F)).$$

Since $\eta : G(K, F) \rightarrow G(T, F)$, we have $\eta(G(K, F)) \subset G(T, F)$. Since $o(\eta(G(K, F))) = o(G(T, F))$, we have $\eta(G(K, F)) = G(T, F)$, and since $\frac{G(K,F)}{G(K,T)}$ is isomorphic to $\eta(G(K, F))$, $\frac{G(K,F)}{G(K,T)}$ is isomorphic to $G(T, F)$.

2.2.26 Conclusion Let F and K be any fields such that K is an extension of F. Let T be a subfield of K that contains F. Suppose that T is a normal extension of F. Then the quotient group $\frac{G(K,F)}{G(K,T)}$ is isomorphic to the group $G(T, F)$.

This result is known as the **fundamental theorem of Galois theory**.

2.3 Applications of Galois Theory

2.3.1 Definition Let G be a group. If there exists a finite collection $\{N_0, N_1, \ldots, N_k\}$ of subgroups of G such that

1. $G = N_0 \supset N_1 \supset \cdots \supset N_k = \{e\}$, where e denotes the identity element of G,
2. for every $i = 1, \ldots, k$, N_i is a normal subgroup of N_{i-1},
3. for every $i = 1, \ldots, k$, the quotient group $\frac{N_{i-1}}{N_i}$ is abelian,

then we say that G is *solvable*.

Definition Let G be a group. Let $a, b \in G$. By the *commutator of a and b* we mean $a^{-1}b^{-1}ab$.

Let C be the collection of all commutators in G. Then the subgroup G' of G generated by all the commutators in G is the smallest subgroup of G containing C. Clearly, G' is equal to the collection of all finite products of the members in C or their inverses. Observe that for every $a, b \in G$, $(a^{-1}b^{-1}ab)^{-1} = b^{-1}a^{-1}ba$, so the inverse of a member of C is also a member of C. Hence G' is equal to the collection of all finite products of the members in C.

Here G' is called the *commutator subgroup of G*.

2.3.2 Problem Let G be a group. Then the commutator subgroup G' of G is a normal subgroup of G.

Proof Let C be the collection of all commutators in G. Let $u \in G'$ and $g \in G$. We have to show that $g^{-1}ug \in G'$. Observe that

$$g^{-1}ug = u(u^{-1}g^{-1}ug).$$

Since $u, g \in G$, we have $u^{-1}g^{-1}ug \in C$. By the definition of G', we have $C \subset G'$. Now, since $u^{-1}g^{-1}ug \in C$, we have $u^{-1}g^{-1}ug \in G'$. Since $u, u^{-1}g^{-1}ug \in G'$ and G' is a group, we have $(g^{-1}ug =)u(u^{-1}g^{-1}ug) \in G'$, and hence $g^{-1}ug \in G'$. ∎

2.3.3 Problem Let G be a group. By 2.3.2, G' is a normal subgroup of G. Then the quotient group $\frac{G}{G'}$ is an abelian group.

Proof Let us take any $a, b \in G$. We have to show that $(aG')(bG') = (bG')(aG')$, that is, $(ab)G' = (ba)G'$, that is, $(ab)^{-1}(ba) \in G'$, that is, $b^{-1}a^{-1}ba \in G'$.

Since $b^{-1}a^{-1}ba \in C$, where C is the collection of all commutators in G and $C \subset G'$, we have $b^{-1}a^{-1}ba \in G'$. ∎

2.3.4 Problem Let G be a group. Let M be a normal subgroup of G. Suppose that the quotient group $\frac{G}{M}$ is an abelian group. Then $G' \subset M$.

Proof It suffices to show that $C \subset M$. To this end, let us take any $a, b \in G$. We have to show that $b^{-1}a^{-1}ba \in M$, that is, $(ab)^{-1}ba \in M$, that is, $(ab)M = (ba)M$, that is, $(aM)(bM) = (bM)(aM)$. This is known to be true, because, $\frac{G}{M}$ is an abelian group. ∎

Definition Let G be a group. Let C be a subgroup of G. If for every automorphism T of G, $T(C) \subset C$, then we say that C *is a characteristic subgroup of* G.

2.3.5 Problem Let G be a group. Then the commutator subgroup G' of G is a characteristic subgroup of G.

Proof To show this, let us take an arbitrary automorphism T of G. We have to show that $T(G') \subset G'$.

To this end, let us take arbitrary $(c_1 \ldots c_n) \in G'$, where each c_i is a commutator in G. We have to show that $(T(c_1) \ldots T(c_n) =)T(c_1 \ldots c_n) \in G'$. It suffices to show that each $T(c_i)$ is a commutator in G.

Since c_i is a commutator in G, there exist $a_i, b_i \in G$ such that $c_i = (a_i)^{-1}(b_i)^{-1}a_i b_i$. Now, since T is an automorphism of G, we have

$$T(c_i) = T\left((a_i)^{-1}(b_i)^{-1}a_i b_i\right) = T\left((a_i)^{-1}\right)T\left((b_i)^{-1}\right)T(a_i)T(b_i)$$
$$= (T(a_i))^{-1}(T(b_i))^{-1}T(a_i)T(b_i),$$

and hence $T(c_i) = (T(a_i))^{-1}(T(b_i))^{-1}T(a_i)T(b_i)$. Since $(T(a_i))^{-1}(T(b_i))^{-1}T(a_i)T(b_i)$ is a commutator in G, $T(c_i)$ is a commutator in G. ∎

2.3.6 Problem Let G be a group. Then the subgroup $(G')' (\equiv G^{(2)})$ of G is normal. Similarly, for every positive integer n, $G^{(n)}$ is a normal subgroup of G.

Proof To show this, let us take an arbitrary $g \in G$. We have to show that $g^{-1}(G')'g \subset (G')'$, that is, $T((G')') \subset (G')'$, where T is the automorphism $x \mapsto g^{-1}xg$ of G. By 2.3.5, G' is a characteristic subgroup of G. Again by 2.3.5, $(G')'$ is a characteristic subgroup of G'. It follows that $(G')'$ is a subgroup of G. Since G' is a characteristic subgroup of G, and T is an automorphism of G, we have $T(G') \subset G'$. Now, since $T : G \to G$ is an automorphism, its restriction $T|_{G'}$ is an automorphism of G'. Next, since $(G')'$ is a characteristic subgroup of G', we have

$$T((G')') = \underbrace{(T|_{G'})((G')')} \subset (G')',$$

and hence $T((G')') \subset (G')'$. ∎

2.3.7 Note Let G be a group. Let G be solvable.

It follows that there exists a finite collection $\{N_0, N_1, \ldots, N_k\}$ of subgroups of G such that

1. $G = N_0 \supset N_1 \supset \ldots \supset N_k = \{e\}$, where e denotes the identity element of G,
2. for every $i = 1, \ldots, k$, N_i is a normal subgroup of N_{i-1},
3. for every $i = 1, \ldots, k$, the quotient group $\frac{N_{i-1}}{N_i}$ is abelian.

Since N_1 is a normal subgroup of N_0, and the quotient group $\frac{N_0}{N_1}$ is abelian, by 2.3.4, $(N_0)' \subset N_1$. Since N_2 is a normal subgroup of N_1 and the quotient group $\frac{N_1}{N_2}$ is abelian, by 2.3.4, $(N_1)' \subset N_2$. Since $(N_0)' \subset N_1$, we have

$$G^{(2)} = (N_0)^{(2)} = \underbrace{\big((N_0)'\big)' \subset (N_1)'}_{} \subset N_2,$$

and hence $\boxed{G^{(2)} \subset N_2}$.

Since N_3 is a normal subgroup of N_2, and the quotient group $\frac{N_2}{N_3}$ is abelian, by 2.3.4, $(N_2)' \subset N_3$. Since $(N_1)' \subset N_2$, we have

$$G^{(3)} = (N_0)^{(3)} = \big((N_0)'\big)^{(2)} \subset (N_1)^{(2)} = \underbrace{\big((N_1)'\big)' \subset (N_2)'}_{} \subset N_3,$$

and hence $\boxed{G^{(3)} \subset N_3}$, etc. Hence $\{e\} = \underbrace{G^{(k)} \subset N_k}_{} = \{e\}$. Thus $G^{(k)} = \{e\}$.

2.3.8 Conclusion Let G be solvable. Then there exists a positive integer k such that $G^{(k)} = \{e\}$.

2.3.9 Problem Let G be a group. Let k be a positive integer. Suppose that $G^{(k)} = \{e\}$, where e denotes the identity element of G. Then G is solvable. It also follows that G' is solvable, etc.

Proof By 2.3.6, for every $i \in \{1, 2, \ldots, k\}$, $G^{(i)}$ is a normal subgroup of G. Now, since G is a normal subgroup of G, $\{G^{(0)}, G^{(1)}, \ldots, G^{(k)}\}$ is a collection of normal subgroups of G, where $G^{(0)} \equiv G$. Further,

$$G = G^{(0)} \supset G^{(1)} \supset \ldots \supset G^{(k)} = \{e\}.$$

By 2.3.2, for every $i = 1, \ldots, k$, $G^{(i)}$ is a normal subgroup of $G^{(i-1)}$. By 2.3.3, for every $i = 1, \ldots, k$, the quotient group $\frac{G^{(i-1)}}{G^{(i)}}$ is abelian. Thus G is solvable. ∎

2.3.10 Problem Let $G, \overline{G}$ be any groups. Let $f : G \to \overline{G}$ be a homomorphism from G onto $\overline{G}$. Thus $\overline{G}$ is the homomorphic f-image of G. Let G be solvable. Then $\overline{G}$ is solvable.

Proof Since G is solvable, by 2.3.8, there exists a positive integer k such that $G^{(k)} = \{e\}$, where e denotes the identity element of G. By 2.3.9, it suffices to show that $\overline{G}^{(k)} = \{\overline{e}\}$, where $\overline{e}$ denotes the identity element of $\overline{G}$.

Now, since $f : G \to \overline{G}$ is a homomorphism from G onto $\overline{G}$, we have $f(G) = \overline{G}$ and $f(e) = \overline{e}$. Thus it is enough to show that $(f(G))^{(k)} \subset \{f(e)\}$. Since $G^{(k)} = \{e\}$, we have $\{f(e)\} = f(G^{(k)})$, and hence it suffices to show that $(f(G))^{(k)} \subset f(G^{(k)})$.

Clearly, $(f(G))' \subset f(G')$.

Proof By 2.3.1, G' is a normal subgroup of G. We first show that $f(G')$ is a normal subgroup of $\overline{G}$.

To this end, let us take arbitrary $\overline{g} \in \overline{G}$ and $f(a) \in f(G')$, where $a \in G'$. We have to show that $(\overline{g})^{-1}(f(a))\overline{g} \in f(G')$.

Since $\overline{g} \in \overline{G} = f(G)$, there exists $\in G$ such that $\overline{g} = f(g)$. Now,

$$(\overline{g})^{-1}(f(a))\overline{g} = (f(g))^{-1}f(a)(f(g)) = f(g^{-1})f(a)f(g) = f(g^{-1}ag),$$

so $(\overline{g})^{-1}(f(a))\overline{g} = f(g^{-1}ag)$. Since G' is a normal subgroup of G, $a \in G'$, and $g \in G$, we have $g^{-1}ag \in G'$, and hence

$$(\overline{g})^{-1}(f(a))\overline{g} = \underbrace{f(g^{-1}ag)}_{} \in f(G').$$

Thus $(\overline{g})^{-1}(f(a))\overline{g} \in f(G')$.

Next we shall show that the quotient group $\frac{\overline{G}}{f(G')} \left(= \frac{f(G)}{f(G')} \right)$ is an abelian group.

To this end, let us take any $a, b \in G$. We have to show that $(f(a)f(G'))$ $(f(b)f(G')) = (f(b)f(G'))(f(a)f(G'))$, that is, $(f(a)f(b))f(G') = (f(b)f(a))$ $f(G')$, that is, $(f(ab))f(G') = (f(ba))f(G')$, that is, $(f(ba))^{-1}(f(ab)) \in f(G')$, that is, $f\left((ba)^{-1}\right)f(ab) \in f(G')$, that is, $f(a^{-1}b^{-1})f(ab) \in f(G')$, that is, $f(a^{-1}b^{-1}ab) \in f(G')$.

It suffices to show that $a^{-1}b^{-1}ab \in G'$. Since $a^{-1}b^{-1}ab$ is a commutator of a and b, we have $a^{-1}b^{-1}ab \in G'$.

Since $\frac{f(G)}{f(G')}$ is an abelian group, by 2.3.4, $\boxed{(f(G))' \subset f(G')}$. By 2.3.9, G' is solvable. Also $f|_{G'}$ is a homomorphism. So as above,

$$(f(G))'' = ((f(G))')' \subset \underbrace{(f|_{G'}(G'))' \subset f|_{G'}((G')')}_{} = f((G')') = f(G''),$$

so $\boxed{(f(G))' \subset f(G'')}$. Finally, we get $(f(G))^{(k)} \subset f(G^{(k)})$. ∎

2.3.11 Problem Let G be a group. Let N be a normal subgroup of G. Then N' is also a normal subgroup of G.

Proof Since N' is a subgroup of N, and N is a subgroup of G, N' is a subgroup of G. Next, let us take arbitrary $g \in G$ and $(c_1 \ldots c_n) \in N'$, where each c_i is a commutator in N. We have to show that

$$\left(g^{-1}c_1g\right)\left(g^{-1}c_2g\right)\ldots\left(g^{-1}c_ng\right) = \underbrace{g^{-1}(c_1\ldots c_n)g} \in N' .$$

It suffices to show that each $g^{-1}c_ig$ is in N', that is, $T(c_i) \in N'$, where T is the automorphism $x \mapsto g^{-1}xg$ of N.

By 2.3.5, N' is a characteristic subgroup of N, and T is the automorphism $x \mapsto g^{-1}xg$ of N, so $T(N') \subset N'$. Now, since c_i is a commutator in N, and N' contains all commutators in N, we have $c_i \in N'$, and hence $T(c_i) \in T(N') \subset N'$. Thus, $T(c_i) \in N'$. ∎

2.3.12 Problem Suppose that $n \in \{5, 6, 7, 8, \ldots\}$. Let S_n be the symmetric group of all permutations of n symbols $1, 2, \ldots, n$. Then

1. $(S_n)'$ contains all 3-cycles,
2. $(S_n)''$ contains all 3-cycles, etc.

In short, for every $n \geq 5$ and for every $k \geq 1$, $(S_n)^{(k)}$ contains all 3-cycles. It follows that for every $k \geq 1$, $(S_n)^{(k)} \neq \{e\}$, and hence by 2.3.9, S_n is not solvable when $n \geq 5$.

Proof 1 Let us take an arbitrary 3-cycle $(i_1i_2i_3)$ in S_n, where i_1, i_2, i_3 are three distinct members of $\{1, 2, \ldots, n\}$. We have to show that $(i_1i_2i_3) \in (S_n)'$.

Since $n \in \{5, 6, 7, 8, \ldots\}$, the 3-cycle $(1\,4\,5)\left(\equiv \begin{pmatrix} 1\,2\,3\,4\,5\ldots \\ 4\,2\,3\,5\,1\ldots \end{pmatrix}\right)$ is in S_n.

Observe that the 3-cycle (135) is a commutator in S_n.

Proof Since

$$(123)^{-1}(145)^{-1}(123)(145)$$

$$= \begin{pmatrix} 123456\cdots \\ 231456\cdots \end{pmatrix}^{-1} \begin{pmatrix} 123456\cdots \\ 423516\cdots \end{pmatrix}^{-1} \begin{pmatrix} 123456\cdots \\ 231456\cdots \end{pmatrix} \begin{pmatrix} 123456\cdots \\ 423516\cdots \end{pmatrix}$$

$$= \begin{pmatrix} 123456\cdots \\ 312456\cdots \end{pmatrix} \begin{pmatrix} 123456\cdots \\ 423516\cdots \end{pmatrix}^{-1} \begin{pmatrix} 123456\cdots \\ 231456\cdots \end{pmatrix} \begin{pmatrix} 123456\cdots \\ 423516\cdots \end{pmatrix}$$

$$= \begin{pmatrix} 123456\cdots \\ 312456\cdots \end{pmatrix} \begin{pmatrix} 123456\cdots \\ 523146\cdots \end{pmatrix} \begin{pmatrix} 123456\cdots \\ 231456\cdots \end{pmatrix} \begin{pmatrix} 123456\cdots \\ 423516\cdots \end{pmatrix}$$

$$= \begin{pmatrix} 123456\cdots \\ 312456\cdots \end{pmatrix} \begin{pmatrix} 123456\cdots \\ 523146\cdots \end{pmatrix} \begin{pmatrix} 123456\cdots \\ 431526\cdots \end{pmatrix}$$

$$= \begin{pmatrix} 123456\cdots \\ 312456\cdots \end{pmatrix} \begin{pmatrix} 123456\cdots \\ 135426\cdots \end{pmatrix} = \begin{pmatrix} 123456\cdots \\ 325416\cdots \end{pmatrix} = (135)$$

we have $(1\,3\,5) = (1\,2\,3)^{-1}(1\,4\,5)^{-1}(1\,2\,3)(1\,4\,5)$. Since $(1\,2\,3), (1\,4\,5) \in S_n$,

$$((1\,3\,5) =)(1\,2\,3)^{-1}(1\,4\,5)^{-1}(1\,2\,3)(1\,4\,5)$$

is a commutator in S_n, and hence $(1\,3\,5)$ is a commutator in S_n. ■

It follows that $(1\,3\,5) \in (S_n)'$. By 2.3.11, $(S_n)'$ is a normal subgroup of S_n. There exists a permutation κ of $1, 2, \ldots, n$ such that $\kappa(1) = i_1, \kappa(3) = i_2$, and $\kappa(5) = i_3$. Thus $\kappa \in S_n$. Since $(1\,3\,5) \in (S_n)'$, $\kappa \in S_n$, and $(S_n)'$ is a normal subgroup of S_n, we have $\kappa(1\,3\,5)\kappa^{-1} \in (S_n)'$. It suffices to show that

$$\kappa(1\,3\,5)\kappa^{-1} = (i_1 i_2 i_3).$$

For this we must prove that

$$\begin{cases} (\kappa(1\,3\,5)\kappa^{-1})(i_1) = i_2, \\ (\kappa(1\,3\,5)\kappa^{-1})(i_2) = i_3, \\ (\kappa(1\,3\,5)\kappa^{-1})(i_3) = i_1, \\ (\kappa(1\,3\,5)\kappa^{-1})(l) = l \text{ when } l \in \{1, 2, \ldots, n\} - \{i_1, i_2, i_3\}. \end{cases}$$

Here

$$\left(\kappa(1\,3\,5)\kappa^{-1}\right)(i_1) = \kappa(1\,3\,5)\left(\kappa^{-1}(i_1)\right) = \kappa((1\,3\,5)(1)) = \kappa(3) = i_2,$$
$$\left(\kappa(1\,3\,5)\kappa^{-1}\right)(i_2) = \kappa(1\,3\,5)\left(\kappa^{-1}(i_2)\right) = \kappa((1\,3\,5)(3)) = \kappa(5) = i_3,$$

and

$$\left(\kappa(1\,3\,5)\kappa^{-1}\right)(i_3) = \kappa(1\,3\,5)\left(\kappa^{-1}(i_3)\right) = \kappa((1\,3\,5)(5)) = \kappa(1) = i_1.$$

Suppose that $l \in \{1, 2, \ldots, n\} - \{i_1, i_2, i_3\}$. It suffices to show that $(\kappa(1\,3\,5)\kappa^{-1})(l) = l$. Here

$$\left(\kappa(1\,3\,5)\kappa^{-1}\right)(l) = (\kappa(1\,3\,5))\left(\kappa^{-1}(l)\right) = (\kappa(1\,3\,5))(m),$$

where $m \in \{1, 2, \ldots, n\} - \{1, 3, 5\}$, and $\kappa(m) = l$. It follows that $(1\,3\,5)(m) = m$, and hence

$$\text{LHS} = \left(\kappa(1\,3\,5)\kappa^{-1}\right)(l) = (\kappa(1\,3\,5))(m) = \underbrace{\kappa((1\,3\,5)(m))}_{} = \kappa(m) = l = \text{RHS}.$$

2: Let us take an arbitrary 3-cycle $(i_1 i_2 i_3)$ in $(S_n)'$, where i_1, i_2, i_3 are three distinct members of $\{1, 2, \ldots, n\}$. We have to show that $(i_1 i_2 i_3) \in ((S_n)')'$.

Since $n \in \{5, 6, 7, 8, \ldots\}$, by assumption the 3-cycle $(1\,4\,5)\left(\equiv \begin{pmatrix} 1\,2\,3\,4\,5\ldots \\ 4\,2\,3\,5\,1\ldots \end{pmatrix}\right)$ is in $(S_n)'$.

Observe that the 3-cycle $(1\,3\,5)$ is a commutator in $(S_n)'$.

Proof Since

$$(123)^{-1}(145)^{-1}(123)(145)$$

$$= \begin{pmatrix} 123456\cdots \\ 231456\cdots \end{pmatrix}^{-1} \begin{pmatrix} 123456\cdots \\ 423516\cdots \end{pmatrix}^{-1} \begin{pmatrix} 123456\cdots \\ 231456\cdots \end{pmatrix} \begin{pmatrix} \overline{123456\cdots} \\ \overline{423516\cdots} \end{pmatrix}$$

$$= \begin{pmatrix} 123456\cdots \\ 312456\cdots \end{pmatrix} \begin{pmatrix} 123456\cdots \\ 423516\cdots \end{pmatrix}^{-1} \begin{pmatrix} 123456\cdots \\ 231456\cdots \end{pmatrix} \begin{pmatrix} \overline{123456\cdots} \\ \overline{423516\cdots} \end{pmatrix}$$

$$= \begin{pmatrix} 123456\cdots \\ 312456\cdots \end{pmatrix} \begin{pmatrix} 123456\cdots \\ 523146\cdots \end{pmatrix} \begin{pmatrix} 123456\cdots \\ 231456\cdots \end{pmatrix} \begin{pmatrix} 123456\cdots \\ 423516\cdots \end{pmatrix}$$

$$= \begin{pmatrix} 123456\cdots \\ 312456\cdots \end{pmatrix} \begin{pmatrix} 123456\cdots \\ 523146\cdots \end{pmatrix} \begin{pmatrix} 123456\cdots \\ 431526\cdots \end{pmatrix}$$

$$= \begin{pmatrix} 123456\cdots \\ 312456\cdots \end{pmatrix} \begin{pmatrix} 123456\cdots \\ 135426\cdots \end{pmatrix} = \begin{pmatrix} 123456\cdots \\ 325416\cdots \end{pmatrix} = (135),$$

we have $(1\,3\,5) = (1\,2\,3)^{-1}(1\,4\,5)^{-1}(1\,2\,3)(1\,4\,5)$. Since $(1\,2\,3), (1\,4\,5) \in (S_n)'$,

$$((1\,3\,5) =)(1\,2\,3)^{-1}(1\,4\,5)^{-1}(1\,2\,3)(1\,4\,5)$$

is a commutator in $(S_n)'$, and hence $(1\,3\,5)$ is a commutator in $(S_n)'$.

It follows that $(1\,3\,5) \in ((S_n)')'$. By two applications of 2.3.11, $((S_n)')'$ is a normal subgroup of S_n. There exists a permutation κ of $1, 2, \ldots, n$ such that $\kappa(1) = i_1, \kappa(3) = i_2$, and $\kappa(5) = i_3$. Thus $\kappa \in S_n$. Since $(1\,3\,5) \in ((S_n)')'$, $\kappa \in S_n$, and $((S_n)')'$ is a normal subgroup of S_n, we have $\kappa(1\,3\,5)\kappa^{-1} \in ((S_n)')'$. It suffices to show that

$$\kappa(1\,3\,5)\kappa^{-1} = (i_1 i_2 i_3).$$

For this we must prove

$$\begin{cases} (\kappa(1\,3\,5)\kappa^{-1})(i_1) = i_2, \\ (\kappa(1\,3\,5)\kappa^{-1})(i_2) = i_3, \\ (\kappa(1\,3\,5)\kappa^{-1})(i_3) = i_1, \\ (\kappa(1\,3\,5)\kappa^{-1})(l) = l \text{ when } l \in \{1, 2, \ldots, n\} - \{i_1, i_2, i_3\}. \end{cases}$$

Here

$$\left(\kappa(1\,3\,5)\kappa^{-1}\right)(i_1) = \kappa(1\,3\,5)\left(\kappa^{-1}(i_1)\right) = \kappa((1\,3\,5)(1)) = \kappa(3) = i_2,$$
$$\left(\kappa(1\,3\,5)\kappa^{-1}\right)(i_2) = \kappa(1\,3\,5)\left(\kappa^{-1}(i_2)\right) = \kappa((1\,3\,5)(3)) = \kappa(5) = i_3,$$

and

$$\left(\kappa(1\,3\,5)\kappa^{-1}\right)(i_3) = \kappa(1\,3\,5)\left(\kappa^{-1}(i_3)\right) = \kappa((1\,3\,5)(5)) = \kappa(1) = i_1.$$

Suppose that $l \in \{1, 2, \ldots, n\} - \{i_1, i_2, i_3\}$. It suffices to show that $(\kappa(1\,3\,5)\kappa^{-1})(l) = l$. Here

$$\left(\kappa(1\,3\,5)\kappa^{-1}\right)(l) = (\kappa(1\,3\,5))\left(\kappa^{-1}(l)\right) = (\kappa(1\,3\,5))(m),$$

where $m \in \{1, 2, \ldots, n\} - \{1, 3, 5\}$ and $\kappa(m) = l$. It follows that $(1\,3\,5)(m) = m$, and hence

$$\text{LHS} = \left(\kappa(1\,3\,5)\kappa^{-1}\right)(l) = (\kappa(1\,3\,5))(m) = \underbrace{\kappa((1\,3\,5)(m))}_{} = \kappa(m) = l = \text{RHS}.$$

■

2.3.13 Example Let F be the field of all real numbers, and let K be the field of all complex numbers.

Clearly, F is a subfield of K. Next,

$$G(K, F) = \{\sigma : \sigma \in \text{Aut}(K), \text{ and for every } a \in F, \sigma(a) = a\}.$$

Suppose that $\sigma \in \text{Aut}(K)$. Here $1, i \in K$. Since $\sigma \in \text{Aut}(K)$, $\sigma : K \to K$ is an automorphism, and hence

$$-1 = -\sigma(1) = \sigma(-1) = \sigma(i^2) = \sigma(ii) = \sigma(i)\sigma(i).$$

Thus $\sigma(i)\sigma(i) = -1$, where $\sigma(i)$ is a complex number. It follows that $\sigma(i) = i$ or $\sigma(i) = -i$.

Case I: $\sigma(i) = i$. For every real a, b, we have $\sigma(a) = a$ and $\sigma(b) = b$. It follows that

$$\sigma(a + ib) = \sigma(a) + \sigma(ib) = a + \sigma(ib) = a + \sigma(i)\sigma(b) = a + \sigma(i)b = a + ib,$$

and hence $\sigma(a + ib) = a + ib$. This shows that in this case, σ is equal to the identity mapping of K. Let us denote this σ by σ_1.

Case II: $\sigma(i) = -i$. For every real a, b, we have $\sigma(a) = a$ and $\sigma(b) = b$. It follows that

$$\sigma(a + ib) = \sigma(a) + \sigma(ib) = a + \sigma(ib) = a + \sigma(i)\sigma(b) = a + \sigma(i)b = a + (-i)b$$
$$= a - ib,$$

and hence $\sigma(a + ib) = a - ib$. This shows that in this case, σ is equal to the complex-conjugation mapping of K. Let us denote this σ by σ_2.

Thus $G(K, F) = \{\sigma_1, \sigma_2\}$. Hence $o(G(K, F)) = o(\{\sigma_1, \sigma_2\}) = 2$. Further,

$$
\begin{aligned}
(\text{fixed field of } G(K, F)) &= \{a : a \in K, \text{and for every } \sigma \in G(K, F), \sigma(a) = a\} \\
&= \{a : a \in K, \text{and for every } \sigma \in \{\sigma_1, \sigma_2\}, \sigma(a) = a\} \\
&= \{a : a \in K, \sigma_1(a) = a \text{ and } \sigma_2(a) = a\} = \{a : a \in K, \sigma_2(a) = a\} \\
&= \{a : a \in K, \bar{a} = a\} = (\text{the set of all real numbers}) = F,
\end{aligned}
$$

so the fixed field of $G(K, F)$ is F.

2.3.14 Example Let F_0 be the field of all rational numbers and $K = F_0(\sqrt[3]{2})$, where $\sqrt[3]{2}$ is the real cube root of 2. By 1.5.20, we have $\left[F_0(\sqrt[3]{2}) : F_0\right] = 3$. Now, by 1.4.5, $\sqrt[3]{2}$ is algebraic of degree 3 over F_0. It follows that $\left\{1, \sqrt[3]{2}, \left(\sqrt[3]{2}\right)^2\right\}$ is a linearly independent set of vectors in the vector space $F_0(\sqrt[3]{2})$ over F_0. Since $\left[F_0(\sqrt[3]{2}) : F_0\right] = 3$, the dimension of the vector space $F_0(\sqrt[3]{2})$ over F_0 is 3. It follows that $\left\{1, \sqrt[3]{2}, \left(\sqrt[3]{2}\right)^2\right\}$ is a basis of the vector space $F_0(\sqrt[3]{2})$ over F_0. Hence

$$
F_0\left(\sqrt[3]{2}\right) = \left\{a_0 + a_1\sqrt[3]{2} + a_2\left(\sqrt[3]{2}\right)^2 : a_0, a_1, a_2 \in F_0\right\}.
$$

Next,

$$
G(K, F_0) = \{\sigma : \sigma \in \text{Aut}(K), \text{and for every } a \in F_0, \sigma(a) = a\}.
$$

Suppose that $\sigma \in G(K, F_0)$.

Here $1, \sqrt[3]{2}, \left(\sqrt[3]{2}\right)^2 \in K$. Since $\sigma \in G(K, F_0)$, we have $\sigma \in \text{Aut}(K)$, that is, $\sigma : K \to K$ is an automorphism, and hence

$$
2 = \sigma(2) = \sigma\left(\sqrt[3]{2}\sqrt[3]{2}\sqrt[3]{2}\right) = \sigma\left(\sqrt[3]{2}\right)\sigma\left(\sqrt[3]{2}\right)\sigma\left(\sqrt[3]{2}\right).
$$

Thus $\sigma(\sqrt[3]{2})\sigma(\sqrt[3]{2})\sigma(\sqrt[3]{2}) = 2$. Now, since $\sigma : F_0(\sqrt[3]{2}) \to F_0(\sqrt[3]{2})$ and $F_0(\sqrt[3]{2}) \subset \mathbb{R}$, we have $\sigma(\sqrt[3]{2}) = \sqrt[3]{2}$. Now, for every $a_0, a_1, a_2 \in F_0$,

$$
\begin{aligned}
\sigma\left(a_0 + a_1\sqrt[3]{2} + a_2\left(\sqrt[3]{2}\right)^2\right) &= \sigma(a_0) + \sigma\left(a_1\sqrt[3]{2}\right) + \sigma\left(a_2\left(\sqrt[3]{2}\right)^2\right) \\
&= \sigma(a_0) + \sigma(a_1)\sigma\left(\sqrt[3]{2}\right) + \sigma(a_2)\sigma\left(\sqrt[3]{2}\right)\sigma\left(\sqrt[3]{2}\right) \\
&= a_0 + a_1\sigma\left(\sqrt[3]{2}\right) + a_2\sigma\left(\sqrt[3]{2}\right)\sigma\left(\sqrt[3]{2}\right) \\
&= a_0 + a_1\sqrt[3]{2} + a_2\sqrt[3]{2}\sqrt[3]{2} = a_0 + a_1\sqrt[3]{2} + a_2\left(\sqrt[3]{2}\right)^2,
\end{aligned}
$$

that is, for every $a_0, a_1, a_2 \in F_0$, we have

$$\sigma\left(a_0 + a_1 \sqrt[3]{2} + a_2 \left(\sqrt[3]{2}\right)^2\right) = a_0 + a_1 \sqrt[3]{2} + a_2 \left(\sqrt[3]{2}\right)^2.$$

This shows that σ is equal to the identity mapping Id of K. Thus $G(K, F_0) = \{\mathrm{Id}\}$. Hence $o(G(K, F_0)) = o(\{\mathrm{Id}\}) = 1$.

Further,

$$\begin{aligned}
(\text{fixed field of } G(K, F_0)) &= \{a : a \in K, \text{and for every } \sigma \in G(K, F_0), \sigma(a) = a\} \\
&= \{a : a \in K, \text{and for every } \sigma \in \{\mathrm{Id}\}, \sigma(a) = a\} \\
&= \{a : a \in K, \mathrm{Id}(a) = a\} = K = F_0(\sqrt[3]{2}),
\end{aligned}$$

so the fixed field of $G\left(F_0(\sqrt[3]{2}), F_0\right)$ is $F_0(\sqrt[3]{2})$.

2.3.15 Example Let F_0 be the field of all rational numbers. Let us denote $e^{\frac{2\pi i}{5}}$ by α.

It follows that $\alpha^5 = 1$. Next, $\alpha^4 + \alpha^3 + \alpha^2 + \alpha + 1 = \frac{\alpha^5 - 1}{\alpha - 1} = \frac{1-1}{\alpha-1} = 0$, so α is a root of the polynomial $x^4 + x^3 + x^2 + x + 1 \in F_0[x]$.

Here $x^4 + x^3 + x^2 + x + 1$ is an irreducible polynomial over the field of rational numbers.

Proof Put $x \equiv y + 1$. It suffices to show that $(y+1)^4 + (y+1)^3 + (y+1)^2 + (y+1) + 1$ is an irreducible polynomial over the field of rational numbers.

Observe that

$$\begin{aligned}
&(y+1)^4 + (y+1)^3 + (y+1)^2 + (y+1) + 1 \\
&= (y^4 + 4y^3 + 6y^2 + 4y + 1) + (y^3 + 3y^2 + 3y + 1) + (y^2 + 2y + 1) + (y+1) + 1 \\
&= 5 + 10y + 10y^2 + 5y^3 + y^4 = 5(1 + 2y + 2y^2 + y^3) + y^4.
\end{aligned}$$

By 1.3.5, $5 + 10y + 10y^2 + 5y^3 + y^4$ is irreducible over the field of rational numbers, and

$$(y+1)^4 + (y+1)^3 + (y+1)^2 + (y+1) + 1 = 5 + 10y + 10y^2 + 5y^3 + y^4,$$

so

$$(y+1)^4 + (y+1)^3 + (y+1)^2 + (y+1) + 1$$

is irreducible over the field of rational numbers.   ■

Thus we have shown that $x^4 + x^3 + x^2 + x + 1$ is an irreducible polynomial over the field of rational numbers. Now, by 1.5.12, α is algebraic of degree 4 over F_0, and hence by 1.4.16, $[F_0(\alpha) : F_0] = 4$. Since α is algebraic of degree 4 over F_0, $\{1, \alpha, \alpha^2, \alpha^3\}$ is a linearly independent set of vectors in the vector space $F_0(\alpha)$ over F_0. Since $[F_0(\alpha) : F_0] = 4$, the dimension of the vector space $F_0(\alpha)$ over F_0 is 4. It follows that $\{1, \alpha, \alpha^2, \alpha^3\}$ is a basis of the vector space $F_0(\alpha)$ over F_0. Hence

$$F_0(\alpha) = \{a_0 + a_1\alpha + a_2\alpha^2 + a_3\alpha^3 : a_0, a_1, a_2 \in F_0\}.$$

Next,

$$G(K, F_0) = \{\sigma : \sigma \in \mathrm{Aut}(K), \text{and for every } a \in F_0, \sigma(a) = a\}.$$

Suppose that $\sigma \in G(K, F_0)$.

Here $1, \alpha, \alpha^2, \alpha^3 \in K$. Also $\boxed{\alpha^5 = 1}$. Since $\sigma \in G(K, F_0)$, we have $\sigma \in \mathrm{Aut}(K)$, that is, $\sigma : K \to K$ is an automorphism, and hence

$$1 = \sigma(1) = \sigma(\alpha^5) = \sigma(\alpha\alpha\alpha\alpha\alpha) = \sigma(\alpha)\sigma(\alpha)\sigma(\alpha)\sigma(\alpha)\sigma(\alpha) = (\sigma(\alpha))^5.$$

Thus $(\sigma(\alpha))^5 = 1$. Now, since $\sigma : F_0(\alpha) \to F_0(\alpha)$ and $F_0(\alpha) \subset \mathbb{C}$, we have $\sigma(\alpha) = 1$ or α or α^2 or α^3 or α^4. Since $\sigma : K \to K$ is an automorphism, σ is one-to-one. Since $\alpha \neq 1$, we have $\underbrace{\sigma(\alpha) \neq \sigma(1)}_{} = 1$, and hence $\sigma(\alpha) \neq 1$. Since $\sigma(\alpha) = 1$ or α or α^2 or α^3 or α^4, we have

$$\sigma(\alpha) = \alpha \text{ or } \alpha^2 \text{ or } \alpha^3 \text{ or } \alpha^4.$$

Case I: $\sigma(\alpha) = \alpha$. For every $a_0, a_1, a_2, a_3 \in F_0$,

$$\begin{aligned}
\sigma(a_0 + a_1\alpha + a_2\alpha^2 + a_3\alpha^3) &= \sigma(a_0) + \sigma(a_1\alpha) + \sigma(a_2\alpha^2) + \sigma(a_2\alpha^3) \\
&= \sigma(a_0) + \sigma(a_1)\sigma(\alpha) + \sigma(a_2)(\sigma(\alpha))^2 + \sigma(a_3)(\sigma(\alpha))^3 \\
&= a_0 + a_1\alpha + a_2(\alpha)^2 + a_3(\alpha)^3 = a_0 + a_1\alpha + a_2\alpha^2 + a_3\alpha^3,
\end{aligned}$$

that is, for every $a_0, a_1, a_2, a_3 \in F_0$, we have

$$\sigma(a_0 + a_1\alpha + a_2\alpha^2 + a_3\alpha^3) = a_0 + a_1\alpha + a_2\alpha^2 + a_3\alpha^3.$$

This shows that σ is equal to the identity mapping Id of K. Let us denote this σ by σ_1. Thus $\sigma_1(\alpha) = \alpha$ and $\sigma_1 \in G(K, F_0)$.

Case II: $\sigma(\alpha) = \alpha^2$. For every $a_0, a_1, a_2, a_3 \in F_0$,

$$\sigma\left(a_0 + a_1\alpha + a_2\alpha^2 + a_3\alpha^3\right)$$
$$= \sigma(a_0) + \sigma(a_1\alpha) + \sigma\left(a_2\alpha^2\right) + \sigma\left(a_2\alpha^3\right)$$
$$= \sigma(a_0) + \sigma(a_1)\sigma(\alpha) + \sigma(a_2)(\sigma(\alpha))^2 + \sigma(a_3)(\sigma(\alpha))^3$$
$$= a_0 + a_1\left(\alpha^2\right) + a_2\left(\alpha^2\right)^2 + a_3\left(\alpha^2\right)^3$$
$$= a_0 + a_1\alpha^2 + a_2\frac{1}{\alpha} + a_3\alpha$$
$$= a_0 + a_1\alpha^2 + a_2\left(-1 - \alpha - \alpha^2 - \alpha^3\right) + a_3\alpha$$
$$= (a_0 - a_2) + (a_3 - a_2)\alpha + (a_1 - a_2)\alpha^2 - a_2\alpha^3,$$

that is, for every $a_0, a_1, a_2, a_3 \in F_0$, we have

$$\sigma\left(a_0 + a_1\alpha + a_2\alpha^2 + a_3\alpha^3\right) = (a_0 - a_2) + (a_3 - a_2)\alpha + (a_1 - a_2)\alpha^2 - a_2\alpha^3.$$

Let us denote this σ by σ_2. Thus $\sigma_2(\alpha) = \alpha^2$, and for every $a_0, a_1, a_2, a_3 \in F_0$, we have

$$\sigma_2\left(a_0 + a_1\alpha + a_2\alpha^2 + a_3\alpha^3\right)$$
$$= (a_0 - a_2) + (a_3 - a_2)\alpha + (a_1 - a_2)\alpha^2 - a_2\alpha^3 (\in F_0(\alpha))$$
$$= a_0 + a_1\alpha^2 + a_2\frac{1}{\alpha} + a_3\alpha = \frac{1}{\alpha}\left(a_2 + a_0\alpha + a_3\alpha^2 + a_1\alpha^3\right)$$
$$= a_0 + a_3\alpha + a_1\alpha^2 + a_2\alpha^4.$$

Also $\sigma_2 \in G(K, F_0)$.

Proof $\sigma_2 : K \to K$ is one-to-one: Let

$$\sigma_2\left(a_0 + a_1\alpha + a_2\alpha^2 + a_3\alpha^3\right) = \sigma_2\left(b_0 + b_1\alpha + b_2\alpha^2 + b_3\alpha^3\right),$$

where each a_i, b_i is in F_0. We have to show that for every $i \in \{0, 1, 2, 3\}$, $a_i = b_i$. Here

$$(a_0 - a_2) + (a_3 - a_2)\alpha + (a_1 - a_2)\alpha^2 - a_2\alpha^3$$
$$= (b_0 - b_2) + (b_3 - b_2)\alpha + (b_1 - b_2)\alpha^2 - b_2\alpha^3.$$

Now, since $\left\{1, \alpha, \alpha^2, \alpha^3\right\}$ is a basis of the vector space $F_0(\alpha)$ over F_0, we have

$$\left.\begin{array}{r} a_0 - a_2 = b_0 - b_2 \\ a_3 - a_2 = b_3 - b_2 \\ a_1 - a_2 = b_1 - b_2 \\ -a_2 = -b_2 \end{array}\right\},$$

that is, for every $i \in \{0, 1, 2, 3\}$, $a_i = b_i$.   ∎

$\sigma_2 : K \to K$ is onto: Let us take an arbitrary sum $b_0 + b_1\alpha + b_2\alpha^2 + b_3\alpha^3 \in K$, where each $b_i \in F_0$. Since

$$\sigma_2\big((b_0 - b_3) + (b_2 - b_3)\alpha - b_3\alpha^2 + (b_1 - b_3)\alpha^3\big)$$
$$= ((b_0 - b_3) - (-b_3))$$
$$+ ((b_1 - b_3) - (-b_3))\alpha + ((b_2 - b_3) - (-b_3))\alpha^2 - (-b_3)\alpha^3$$
$$= b_0 + b_1\alpha + b_2\alpha^2 + b_3\alpha^3,$$

it follows that

$$\sigma_2\big((b_0 - b_3) + (b_2 - b_3)\alpha - b_3\alpha^2 + (b_1 - b_3)\alpha^3\big) = b_0 + b_1\alpha + b_2\alpha^2 + b_3\alpha^3,$$

where $(b_0 - b_3) + (b_2 - b_3)\alpha - b_3\alpha^2 + (b_1 - b_3)\alpha^3 \in K$.

It is clear that $\sigma_2\colon (a_0 + a_1\alpha + a_2\alpha^2 + a_3\alpha^3) \mapsto (a_0 + a_3\alpha + a_1\alpha^2 + a_2\alpha^4)$ preserves addition. We claim that

$\sigma_2\colon (a_0 + a_1\alpha + a_2\alpha^2 + a_3\alpha^3) \mapsto (a_0 + a_3\alpha + a_1\alpha^2 + a_2\alpha^4)$ preserves multiplication. We have to show that

$$\sigma_2\big((a_0 + a_1\alpha + a_2\alpha^2 + a_3\alpha^3)(b_0 + b_1\alpha + b_2\alpha^2 + b_3\alpha^3)\big)$$
$$= \left(a_0 + a_1\alpha^2 + a_2\frac{1}{\alpha} + a_3\alpha\right)\left(b_0 + b_1\alpha^2 + b_2\frac{1}{\alpha} + b_3\alpha\right).$$

Since

$$(a_0 + a_1\alpha + a_2\alpha^2 + a_3\alpha^3)(b_0 + b_1\alpha + b_2\alpha^2 + b_3\alpha^3)$$
$$= (a_0b_0 + a_2b_3 + a_3b_2) + (a_0b_1 + a_1b_0 + a_3b_3)\alpha + (a_0b_2 + a_1b_1 + a_2b_0)\alpha^2$$
$$+ (a_0b_3 + a_1b_2 + a_2b_1 + a_3b_0)\alpha^3 + (a_1b_3 + a_2b_2 + a_3b_1)(-1 - \alpha - \alpha^2 - \alpha^3)$$
$$= (a_0b_0 + a_2b_3 + a_3b_2 - (a_1b_3 + a_2b_2 + a_3b_1))$$
$$+ ((a_0b_1 + a_1b_0 + a_3b_3) - (a_1b_3 + a_2b_2 + a_3b_1))\alpha$$
$$+ ((a_0b_2 + a_1b_1 + a_2b_0) - (a_1b_3 + a_2b_2 + a_3b_1))\alpha^2$$
$$+ ((a_0b_3 + a_1b_2 + a_2b_1 + a_3b_0) - (a_1b_3 + a_2b_2 + a_3b_1))\alpha^3,$$

we have

$$\text{LHS} = (a_0b_0 + a_2b_3 + a_3b_2 - (a_1b_3 + a_2b_2 + a_3b_1))$$
$$+ ((a_0b_3 + a_1b_2 + a_2b_1 + a_3b_0) - (a_1b_3 + a_2b_2 + a_3b_1))\alpha$$
$$+ ((a_0b_1 + a_1b_0 + a_3b_3) - (a_1b_3 + a_2b_2 + a_3b_1))\alpha^2$$
$$+ ((a_0b_2 + a_1b_1 + a_2b_0) - (a_1b_3 + a_2b_2 + a_3b_1))\alpha^4$$
$$= (a_0b_0 + a_2b_3 + a_3b_2) + (a_0b_3 + a_1b_2 + a_2b_1 + a_3b_0)\alpha$$
$$(a_0b_1 + a_1b_0 + a_3b_3)\alpha^2 + (a_0b_2 + a_1b_1 + a_2b_0)\alpha^4$$
$$+ (a_1b_3 + a_2b_2 + a_3b_1)(-1 - \alpha - \alpha^2 - \alpha^4)$$
$$= (a_0b_0 + a_2b_3 + a_3b_2) + (a_0b_3 + a_1b_2 + a_2b_1 + a_3b_0)\alpha$$
$$+ (a_0b_1 + a_1b_0 + a_3b_3)\alpha^2 + (a_0b_2 + a_1b_1 + a_2b_0)\alpha^4 + (a_1b_3 + a_2b_2 + a_3b_1)\alpha^3$$
$$= (a_0b_0 + a_2b_3 + a_3b_2) + (a_0b_3 + a_1b_2 + a_2b_1 + a_3b_0)\alpha + (a_0b_1 + a_1b_0 + a_3b_3)\alpha^2$$
$$+ (a_1b_3 + a_2b_2 + a_3b_1)\alpha^3 + (a_0b_2 + a_1b_1 + a_2b_0)\alpha^4,$$

and

$$\text{RHS} = \left(a_0 + a_1\alpha^2 + a_2\tfrac{1}{\alpha} + a_3\alpha\right)\left(b_0 + b_1\alpha^2 + b_2\tfrac{1}{\alpha} + b_3\alpha\right)$$
$$= (a_0b_0 + a_2b_3 + a_3b_2) + (a_0b_3 + a_1b_2 + a_2b_1 + a_3b_0)\alpha + (a_0b_1 + a_1b_0 + a_3b_3)\alpha^2$$
$$+ (a_1b_3 + a_2b_2 + a_3b_1)\alpha^3 + (a_0b_2 + a_1b_1 + a_2b_0)\alpha^4,$$

so LHS = RHS.

Finally, let $a_0 \in F_0$. We have to show that $\sigma_2(a_0) = a_0$. Here,

$$\text{LHS} = \sigma_2(a_0) = \sigma_2\left(a_0 + 0\alpha + 0\alpha^2 + 0\alpha^3\right) = a_0 + 0\alpha + 0\alpha^2 + 0\alpha^4 = a_0$$
$$= \text{RHS}.$$

Thus we have shown that $\sigma_2 \in G(K, F_0)$.

Case III: $\sigma(\alpha) = \alpha^3$. For every $a_0, a_1, a_2, a_3 \in F_0$,

$$\sigma\left(a_0 + a_1\alpha + a_2\alpha^2 + a_3\alpha^3\right)$$
$$= \sigma(a_0) + \sigma(a_1\alpha) + \sigma\left(a_2\alpha^2\right) + \sigma\left(a_2\alpha^3\right)$$
$$= \sigma(a_0) + \sigma(a_1)\sigma(\alpha) + \sigma(a_2)(\sigma(\alpha))^2 + \sigma(a_3)(\sigma(\alpha))^3$$
$$= a_0 + a_1\left(\alpha^3\right) + a_2\left(\alpha^3\right)^2 + a_3\left(\alpha^3\right)^3 = a_0 + a_1\alpha^3 + a_2\alpha + a_3\frac{1}{\alpha}$$
$$= a_0 + a_1\alpha^3 + a_2\alpha + a_3\left(-1 - \alpha - \alpha^2 - \alpha^3\right)$$
$$= (a_0 - a_3) + (a_2 - a_3)\alpha - a_3\alpha^2 + (a_1 - a_3)\alpha^3,$$

that is, for every $a_0, a_1, a_2, a_3 \in F_0$, we have

$$\sigma\left(a_0 + a_1\alpha + a_2\alpha^2 + a_3\alpha^3\right) = (a_0 - a_3) + (a_2 - a_3)\alpha - a_3\alpha^2 + (a_1 - a_3)\alpha^3.$$

Let us denote this σ by σ_3. Thus $\sigma_3(\alpha) = \alpha^3$, and for every $a_0, a_1, a_2, a_3 \in F_0$, we have

$$\sigma_3\left(a_0 + a_1\alpha + a_2\alpha^2 + a_3\alpha^3\right)$$
$$= (a_0 - a_3) + (a_2 - a_3)\alpha - a_3\alpha^2 + (a_1 - a_3)\alpha^3 (\in F_0(\alpha))$$
$$= a_0 + a_1\alpha^3 + a_2\alpha + a_3\frac{1}{\alpha} = \frac{1}{\alpha^2}\left(a_1 + a_3\alpha + a_0\alpha^2 + a_2\alpha^3\right)$$
$$= a_0 + a_2\alpha + a_1\alpha^3 + a_3\alpha^4.$$

Also $\sigma_3 \in G(K, F_0)$.

Proof $\sigma_3 : K \to K$ is one-to-one: Let

$$\sigma_3\left(a_0 + a_1\alpha + a_2\alpha^2 + a_3\alpha^3\right) = \sigma_3\left(b_0 + b_1\alpha + b_2\alpha^2 + b_3\alpha^3\right),$$

where each a_i, b_i is in F_0. We have to show that for every $i \in \{0, 1, 2, 3\}$, $a_i = b_i$. Here

$$(a_0 - a_3) + (a_2 - a_3)\alpha - a_3\alpha^2 + (a_1 - a_3)\alpha^3$$
$$= (b_0 - b_3) + (b_2 - b_3)\alpha - b_3\alpha^2 + (b_1 - b_3)\alpha^3.$$

Now, since $\{1, \alpha, \alpha^2, \alpha^3\}$ is a basis of the vector space $F_0(\alpha)$ over F_0, we have

$$\left.\begin{array}{r} a_0 - a_3 = b_0 - b_3 \\ a_2 - a_3 = b_2 - b_3 \\ -a_3 = -b_3 \\ a_1 - a_3 = b_1 - b_3 \end{array}\right\},$$

that is, for every $i \in \{0, 1, 2, 3\}$, $a_i = b_i$.

$\sigma_3 : K \to K$ is onto: Let us take an arbitrary sum $b_0 + b_1\alpha + b_2\alpha^2 + b_3\alpha^3 \in K$, where each b_i is in F_0. Since
$$\sigma_3\left((b_0 - b_2) + (b_3 - b_2)\alpha + (b_1 - b_2)\alpha^2 - b_2\alpha^3\right)$$
$$= ((b_0 - b_2) - (-b_2)) + ((b_1 - b_2) - (-b_2))\alpha$$
$$- (-b_2)\alpha^2 + ((b_3 - b_2) - (-b_2))\alpha^3$$
$$= b_0 + b_1\alpha + b_2\alpha^2 + b_3\alpha^3,$$

it follows that

$$\sigma_3\big((b_0 - b_2) + (b_3 - b_2)\alpha + (b_1 - b_2)\alpha^2 - b_2\alpha^3\big) = b_0 + b_1\alpha + b_2\alpha^2 + b_3\alpha^3,$$

where $(b_0 - b_2) + (b_3 - b_2)\alpha + (b_1 - b_2)\alpha^2 - b_2\alpha^3 \in K$.

It is clear that $\sigma_3 \colon (a_0 + a_1\alpha + a_2\alpha^2 + a_3\alpha^3) \mapsto (a_0 + a_2\alpha + a_1\alpha^3 + a_3\alpha^4)$ preserves addition. We claim that

$$\sigma_3 \colon \quad (a_0 + a_1\alpha + a_2\alpha^2 + a_3\alpha^3) \mapsto (a_0 + a_2\alpha + a_1\alpha^3 + a_3\alpha^4) \qquad \text{preserves}$$

multiplication:

We have to show that

$$\sigma_3\big((_0 + a_1\alpha + a_2\alpha^2 + a_3\alpha^3)(b_0 + b_1\alpha + b_2\alpha^2 + b_3\alpha^3)\big)$$
$$= \Big(a_0 + a_1\alpha^3 + a_2\alpha + a_3\frac{1}{\alpha}\Big)\Big(b_0 + b_1\alpha^3 + b_2\alpha + b_3\frac{1}{\alpha}\Big).$$

Since

$$(a_0 + a_1\alpha + a_2\alpha^2 + a_3\alpha^3)(b_0 + b_1\alpha + b_2\alpha^2 + b_3\alpha^3)$$
$$= (a_0b_0 + a_2b_3 + a_3b_2) + (a_0b_1 + a_1b_0 + a_3b_3)\alpha + (a_0b_2 + a_1b_1 + a_2b_0)\alpha^2$$
$$+ (a_0b_3 + a_1b_2 + a_2b_1 + a_3b_0)\alpha^3 + (a_1b_3 + a_2b_2 + a_3b_1)(-1 - \alpha - \alpha^2 - \alpha^3)$$
$$= (a_0b_0 + a_2b_3 + a_3b_2 - (a_1b_3 + a_2b_2 + a_3b_1))$$
$$+ ((a_0b_1 + a_1b_0 + a_3b_3) - (a_1b_3 + a_2b_2 + a_3b_1))\alpha$$
$$+ ((a_0b_2 + a_1b_1 + a_2b_0) - (a_1b_3 + a_2b_2 + a_3b_1))\alpha^2$$
$$+ ((a_0b_3 + a_1b_2 + a_2b_1 + a_3b_0) - (a_1b_3 + a_2b_2 + a_3b_1))\alpha^3,$$

we have

$$\text{LHS} = (a_0b_0 + a_2b_3 + a_3b_2 - (a_1b_3 + a_2b_2 + a_3b_1))$$
$$+ ((a_0b_2 + a_1b_1 + a_2b_0) - (a_1b_3 + a_2b_2 + a_3b_1))\alpha$$
$$+ ((a_0b_1 + a_1b_0 + a_3b_3) - (a_1b_3 + a_2b_2 + a_3b_1))\alpha^3$$
$$+ ((a_0b_3 + a_1b_2 + a_2b_1 + a_3b_0) - (a_1b_3 + a_2b_2 + a_3b_1))\alpha^4$$
$$= (a_0b_0 + a_2b_3 + a_3b_2) + (a_0b_2 + a_1b_1 + a_2b_0)\alpha + (a_0b_1 + a_1b_0 + a_3b_3)\alpha^3$$
$$+ (a_0b_3 + a_1b_2 + a_2b_1 + a_3b_0)\alpha^4 + (a_1b_3 + a_2b_2 + a_3b_1)(-1 - \alpha - \alpha^3 - \alpha^4)$$
$$= (a_0b_0 + a_2b_3 + a_3b_2) + (a_0b_2 + a_1b_1 + a_2b_0)\alpha + (a_0b_1 + a_1b_0 + a_3b_3)\alpha^3$$
$$+ (a_0b_3 + a_1b_2 + a_2b_1 + a_3b_0)\alpha^4 + (a_1b_3 + a_2b_2 + a_3b_1)\alpha^2$$
$$= (a_0b_0 + a_2b_3 + a_3b_2) + (a_0b_2 + a_1b_1 + a_2b_0)\alpha + (a_1b_3 + a_2b_2 + a_3b_1)\alpha^2$$
$$+ (a_0b_1 + a_1b_0 + a_3b_3)\alpha^3 + (a_0b_3 + a_1b_2 + a_2b_1 + a_3b_0)\alpha^4,$$

and

$$\text{RHS} = \left(a_0 + a_1\alpha^3 + a_2\alpha + a_3\tfrac{1}{\alpha}\right)\left(b_0 + b_1\alpha^3 + b_2\alpha + b_3\tfrac{1}{\alpha}\right)$$
$$= (a_0b_0 + a_2b_3 + a_3b_2) + (a_0b_2 + a_1b_1 + a_2b_0)\alpha + (a_1b_3 + a_2b_2 + a_3b_1)\alpha^2$$
$$+ (a_0b_1 + a_1b_0 + a_3b_3)\alpha^3 + (a_0b_3 + a_1b_2 + a_2b_1 + a_3b_0)\alpha^4,$$

so LHS = RHS.

Finally, let $a_0 \in F_0$. We have to show that $\sigma_3(a_0) = a_0$. Here

$$\text{LHS} = \sigma_3(a_0) = \sigma_3\left(a_0 + 0\alpha + 0\alpha^2 + 0\alpha^3\right) = a_0 + 0\alpha + 0\alpha^3 + 0\alpha^4 = a_0$$
$$= \text{RHS}.$$

Thus we have shown that $\sigma_3 \in G(K, F_0)$. ∎

Case IV: $\sigma(\alpha) = \alpha^4$. For every $a_0, a_1, a_2, a_3 \in F_0$,

$$\sigma\left(a_0 + a_1\alpha + a_2\alpha^2 + a_3\alpha^3\right)$$
$$= \sigma(a_0) + \sigma(a_1\alpha) + \sigma\left(a_2\alpha^2\right) + \sigma\left(a_2\alpha^3\right)$$
$$= \sigma(a_0) + \sigma(a_1)\sigma(\alpha) + \sigma(a_2)(\sigma(\alpha))^2 + \sigma(a_3)(\sigma(\alpha))^3$$
$$= a_0 + a_1\left(\alpha^4\right) + a_2\left(\alpha^4\right)^2 + a_3\left(\alpha^4\right)^3 = a_0 + a_1\frac{1}{\alpha} + a_2\alpha^3 + a_3\alpha^2$$
$$= a_0 + a_1\left(-1 - \alpha - \alpha^2 - \alpha^3\right) + a_2\alpha^3 + a_3\alpha^2$$
$$= (a_0 - a_1) - a_1\alpha + (a_3 - a_1)\alpha^2 + (a_2 - a_1)\alpha^3,$$

that is, for every $a_0, a_1, a_2, a_3 \in F_0$, we have

$$\sigma\left(a_0 + a_1\alpha + a_2\alpha^2 + a_3\alpha^3\right) = (a_0 - a_1) - a_1\alpha + (a_3 - a_1)\alpha^2 + (a_2 - a_1)\alpha^3.$$

Let us denote this σ by σ_4. Thus $\sigma_4(\alpha) = \alpha^4$, and for every $a_0, a_1, a_2, a_3 \in F_0$, we have

$$\sigma_4\left(a_0 + a_1\alpha + a_2\alpha^2 + a_3\alpha^3\right)$$
$$= (a_0 - a_1) - a_1\alpha + (a_3 - a_1)\alpha^2 + (a_2 - a_1)\alpha^3 (\in F_0(\alpha))$$
$$= a_0 + a_1\frac{1}{\alpha} + a_2\alpha^3 + a_3\alpha^2 = \frac{1}{\alpha^3}\left(a_3 + a_2\alpha + a_1\alpha^2 + a_0\alpha^3\right)$$
$$= a_0 + a_3\alpha^2 + a_2\alpha^3 + a_1\alpha^4.$$

Also $\sigma_4 \in G(K, F_0)$.

Proof $\sigma_4 : K \to K$ is one-to-one: Let

$$\sigma_4\left(a_0 + a_1\alpha + a_2\alpha^2 + a_3\alpha^3\right) = \sigma_4\left(b_0 + b_1\alpha + b_2\alpha^2 + b_3\alpha^3\right),$$

where each a_i, b_i is in F_0. We have to show that for every $i \in \{0, 1, 2, 3\}$, $a_i = b_i$. Here

$$(a_0 - a_1) - a_1\alpha + (a_3 - a_1)\alpha^2 + (a_2 - a_1)\alpha^3$$
$$= (b_0 - b_1) - b_1\alpha + (b_3 - b_1)\alpha^2 + (b_2 - b_1)\alpha^3.$$

Now, since $\{1, \alpha, \alpha^2, \alpha^3\}$ is a basis of the vector space $F_0(\alpha)$ over F_0, we have

$$\left.\begin{array}{r} a_0 - a_1 = b_0 - b_1 \\ -a_1 = -b_1 \\ a_3 - a_1 = b_3 - b_1 \\ a_2 - a_1 = b_2 - b_1 \end{array}\right\},$$

that is, for every $i \in \{0, 1, 2, 3\}$, $a_i = b_i$.

$\sigma_4 : K \to K$ is onto: Let us take an arbitrary sum $b_0 + b_1\alpha + b_2\alpha^2 + b_3\alpha^3 \in K$, where each b_i is in F_0. Since
$$\sigma_4\big((b_0 - b_1) - b_1\alpha + (b_3 - b_1)\alpha^2 + (b_2 - b_1)\alpha^3\big)$$
$$= ((b_0 - b_1) - (-b_1)) - (-b_1)\alpha + ((b_2 - b_1) - (-b_1))\alpha^2$$
$$+ ((b_3 - b_1) - (-b_1))\alpha^3$$
$$= b_0 + b_1\alpha + b_2\alpha^2 + b_3\alpha^3,$$

it follows that

$$\sigma_4\big((b_0 - b_1) - b_1\alpha + (b_3 - b_1)\alpha^2 + (b_2 - b_1)\alpha^3\big) = b_0 + b_1\alpha + b_2\alpha^2 + b_3\alpha^3,$$

where $(b_0 - b_1) - b_1\alpha + (b_3 - b_1)\alpha^2 + (b_2 - b_1)\alpha^3 \in K$.

It is clear that $\sigma_4 \colon (a_0 + a_1\alpha + a_2\alpha^2 + a_3\alpha^3) \mapsto (a_0 + a_3\alpha^2 + a_2\alpha^3 + a_1\alpha^4)$ preserves addition.

$\sigma_4 \colon (a_0 + a_1\alpha + a_2\alpha^2 + a_3\alpha^3) \mapsto (a_0 + a_3\alpha^2 + a_2\alpha^3 + a_1\alpha^4)$ preserves multiplication: We have to show that
$$\sigma_4\big((a_0 + a_1\alpha + a_2\alpha^2 + a_3\alpha^3)(b_0 + b_1\alpha + b_2\alpha^2 + b_3\alpha^3)\big)$$
$$= \left(a_0 + a_1\frac{1}{\alpha} + a_2\alpha^3 + a_3\alpha^2\right)\left(b_0 + b_1\frac{1}{\alpha} + b_2\alpha^3 + b_3\alpha^2\right).$$

Since

$$(a_0 + a_1\alpha + a_2\alpha^2 + a_3\alpha^3)(b_0 + b_1\alpha + b_2\alpha^2 + b_3\alpha^3)$$
$$= (a_0b_0 + a_2b_3 + a_3b_2) + (a_0b_1 + a_1b_0 + a_3b_3)\alpha + (a_0b_2 + a_1b_1 + a_2b_0)\alpha^2$$
$$+ (a_0b_3 + a_1b_2 + a_2b_1 + a_3b_0)\alpha^3 + (a_1b_3 + a_2b_2 + a_3b_1)(-1 - \alpha - \alpha^2 - \alpha^3)$$
$$= (a_0b_0 + a_2b_3 + a_3b_2 - (a_1b_3 + a_2b_2 + a_3b_1))$$
$$+ ((a_0b_1 + a_1b_0 + a_3b_3) - (a_1b_3 + a_2b_2 + a_3b_1))\alpha$$
$$+ ((a_0b_2 + a_1b_1 + a_2b_0) - (a_1b_3 + a_2b_2 + a_3b_1))\alpha^2$$
$$+ ((a_0b_3 + a_1b_2 + a_2b_1 + a_3b_0) - (a_1b_3 + a_2b_2 + a_3b_1))\alpha^3,$$

we have

$$\textbf{LHS} = (a_0b_0 + a_2b_3 + a_3b_2 - (a_1b_3 + a_2b_2 + a_3b_1))$$
$$+ ((a_0b_3 + a_1b_2 + a_2b_1 + a_3b_0) - (a_1b_3 + a_2b_2 + a_3b_1))\alpha^2$$
$$+ ((a_0b_2 + a_1b_1 + a_2b_0) - (a_1b_3 + a_2b_2 + a_3b_1))\alpha^3$$
$$+ ((a_0b_1 + a_1b_0 + a_3b_3) - (a_1b_3 + a_2b_2 + a_3b_1))\alpha^4$$
$$= (a_0b_0 + a_2b_3 + a_3b_2) + (a_0b_3 + a_1b_2 + a_2b_1 + a_3b_0)\alpha^2$$
$$+ (a_0b_2 + a_1b_1 + a_2b_0)\alpha^3 + (a_0b_1 + a_1b_0 + a_3b_3)\alpha^4 + (a_1b_3 + a_2b_2 + a_3b_1)$$
$$(-1 - \alpha^2 - \alpha^3 - \alpha^4) = (a_0b_0 + a_2b_3 + a_3b_2) + (a_0b_3 + a_1b_2 + a_2b_1 + a_3b_0)\alpha^2$$
$$+ (a_0b_2 + a_1b_1 + a_2b_0)\alpha^3 + (a_0b_1 + a_1b_0 + a_3b_3)\alpha^4 + (a_1b_3 + a_2b_2 + a_3b_1)\alpha$$
$$= (a_0b_0 + a_2b_3 + a_3b_2) + (a_1b_3 + a_2b_2 + a_3b_1)\alpha + (a_0b_3 + a_1b_2 + a_2b_1 + a_3b_0)\alpha^2$$
$$+ (a_0b_2 + a_1b_1 + a_2b_0)\alpha^3 + (a_0b_1 + a_1b_0 + a_3b_3)\alpha^4,$$

and

$$\textbf{RHS} = \left(a_0 + a_1\tfrac{1}{\alpha} + a_2\alpha^3 + a_3\alpha^2\right)\left(b_0 + b_1\tfrac{1}{\alpha} + b_2\alpha^3 + b_3\alpha^2\right)$$
$$= (a_0b_0 + a_2b_3 + a_3b_2) + (a_1b_3 + a_3b_1 + a_2b_2)\alpha + (a_0b_3 + a_1b_2 + a_2b_1 + a_3b_0)\alpha^2$$
$$+ (a_0b_2 + a_1b_1 + a_2b_0)\alpha^3 + (a_0b_1 + a_1b_0 + a_3b_3)\alpha^4,$$

so LHS = RHS.

Finally, let $a_0 \in F_0$. We have to show that $\sigma_4(a_0) = a_0$. Here

$$\textbf{LHS} = \sigma_4(a_0) = \sigma_4\left(a_0 + 0\alpha + 0\alpha^2 + 0\alpha^3\right) = \left(a_0 + 0\alpha^2 + 0\alpha^3 + 0\alpha^4\right) = a_0$$
$$= \textbf{RHS}.$$

Thus we have shown that $\sigma_4 \in G(K, F_0)$.   ∎

Hence

$$G(K, F_0) = \{\sigma_1, \sigma_2, \sigma_3, \sigma_4\}.$$

Next, $o(G(K, F_0)) = o(\{\sigma_1, \sigma_2, \sigma_3, \sigma_4\}) = 4$. Further,

(fixed field of $G(K, F_0)$)
$$= \{a : a \in K, \text{and for every } \sigma \in G(K, F_0), \sigma(a) = a\}$$
$$= \{a : a \in K, \text{and for every } \sigma \in \{\sigma_1, \sigma_2, \sigma_3, \sigma_4\}, \sigma(a) = a\}$$
$$= \{a : a \in K, \sigma_1(a) = a, \sigma_2(a) = a, \sigma_3(a) = a, \sigma_4(a) = a\}$$
$$= \{a : a \in K, \sigma_2(a) = a, \sigma_3(a) = a, \sigma_4(a) = a\}$$
$$= \{a_0 + a_1\alpha + a_2\alpha^2 + a_3\alpha^3 : a_0, a_1a_2, a_3 \in F_0, (a_0 - a_2)$$
$$+ (a_3 - a_2)\alpha + (a_1 - a_2)\alpha^2 - a_2\alpha^3$$
$$= a_0 + a_1\alpha + a_2\alpha^2 + a_3\alpha^3, (a_0 - a_3) + (a_2 - a_3)\alpha - a_3\alpha^2 + (a_1 - a_3)\alpha^3$$
$$= a_0 + a_1\alpha + a_2\alpha^2 + a_3\alpha^3, (a_0 - a_1) - a_1\alpha + (a_3 - a_1)\alpha^2 + (a_2 - a_1)\alpha^3$$
$$= a_0 + a_1\alpha + a_2\alpha^2 + a_3\alpha^3\}$$
$$= \{a_0 + a_1\alpha + a_2\alpha^2 + a_3\alpha^3 : a_0 \in F_0, a_1 = a_2 = a_3 = 0\} = F_0,$$

so the fixed field of $G(F_0(\alpha), F_0)$ is F_0.

Observe that

$$
\begin{aligned}
(\sigma_2)^2 & \left(a_0 + a_1\alpha + a_2\alpha^2 + a_3\alpha^3\right) \\
&= \sigma_2\left(\sigma_2\left(a_0 + a_1\alpha + a_2\alpha^2 + a_3\alpha^3\right)\right) \\
&= \sigma_2\left((a_0 - a_2) + (a_3 - a_2)\alpha + (a_1 - a_2)\alpha^2 - a_2\alpha^3\right) \\
&= ((a_0 - a_2) - (a_1 - a_2)) + ((-a_2) - (a_1 - a_2))\alpha \\
&\quad + ((a_3 - a_2) - (a_1 - a_2))\alpha^2 - (a_1 - a_2)\alpha^3 \\
&= (a_0 - a_1) - a_1\alpha + (a_3 - a_1)\alpha^2 + (a_2 - a_1)\alpha^3 \\
&= \sigma_4\left(a_0 + a_1\alpha + a_2\alpha^2 + a_3\alpha^3\right),
\end{aligned}
$$

so $(\sigma_2)^2 = \sigma_4$. Next,

$$
\begin{aligned}
(\sigma_2)^3 & \left(a_0 + a_1\alpha + a_2\alpha^2 + a_3\alpha^3\right) \\
&= \sigma_2\left((a_0 - a_1) - a_1\alpha + (a_3 - a_1)\alpha^2 + (a_2 - a_1)\alpha^3\right) \\
&= ((a_0 - a_1) - (a_3 - a_1)) + ((a_2 - a_1) - (a_3 - a_1))\alpha \\
&\quad + ((-a_1) - (a_3 - a_1))\alpha^2 - (a_3 - a_1)\alpha^3 \\
&= (a_0 - a_3) + (a_2 - a_3)\alpha - a_3\alpha^2 + (a_1 - a_3)\alpha^3 \\
&= \sigma_3\left(a_0 + a_1\alpha + a_2\alpha^2 + a_3\alpha^3\right),
\end{aligned}
$$

so $(\sigma_2)^3 = \sigma_3$. Finally,

$$
\begin{aligned}
(\sigma_2)^4 & \left(a_0 + a_1\alpha + a_2\alpha^2 + a_3\alpha^3\right) \\
&= \sigma_2\left((a_0 - a_3) + (a_2 - a_3)\alpha - a_3\alpha^2 + (a_1 - a_3)\alpha^3\right) \\
&= ((a_0 - a_3) - (-a_3)) + ((a_1 - a_3) - (-a_3))\alpha \\
&\quad + ((a_2 - a_3) - (-a_3))\alpha^2 - (-a_3)\alpha^3 \\
&= a_0 + a_1\alpha + a_2\alpha^2 + a_3\alpha^3 = \sigma_1\left(a_0 + a_1\alpha + a_2\alpha^2 + a_3\alpha^3\right),
\end{aligned}
$$

so $(\sigma_2)^4 = \sigma_1$. Thus

$$
G(K, F_0) = \{\sigma_1, \sigma_2, \sigma_3, \sigma_4\} = \left\{\sigma_2, (\sigma_2)^2, (\sigma_2)^3, (\sigma_2)^4\right\}.
$$

It follows that $G(K, F_0)$ is a cyclic group generated by

$$
\sigma_2 : \left(a_0 + a_1\alpha + a_2\alpha^2 + a_3\alpha^3\right) \mapsto \left(a_0 + a_3\alpha + a_1\alpha^2 + a_2\alpha^4\right).
$$

Since

$$(\sigma_4)^2(a_0 + a_1\alpha + a_2\alpha^2 + a_3\alpha^3)$$
$$= \sigma_4((a_0 - a_1) - a_1\alpha + (a_3 - a_1)\alpha^2 + (a_2 - a_1)\alpha^3)$$
$$= ((a_0 - a_1) - (-a_1)) - (-a_1)\alpha + ((a_2 - a_1) - (-a_1))\alpha^2$$
$$+ ((a_3 - a_1) - (-a_1))\alpha^3$$
$$= a_0 + a_1\alpha + a_2\alpha^2 + a_3\alpha^3$$
$$= \sigma_1(a_0 + a_1\alpha + a_2\alpha^2 + a_3\alpha^3),$$

we have $(\sigma_4)^2 = \sigma_1$, and hence $\{\sigma_1, \sigma_4\}$ is a subgroup of $G(K, F_0)$. Here

(fixed field of $\{\sigma_1, \sigma_4\}$)
$$= \{a : a \in K, \text{and for every } \sigma \in \{\sigma_1, \sigma_4\}, \sigma(a) = a\}$$
$$= \{a : a \in K, \sigma_1(a) = a, \sigma_4(a) = a\} = \{a : a \in K, \sigma_4(a) = a\}$$
$$= \{a_0 + a_1\alpha + a_2\alpha^2 + a_3\alpha^3 : a_0, a_1a_2, a_3 \in F_0, (a_0 - a_1)$$
$$- a_1\alpha + (a_3 - a_1)\alpha^2 + (a_2 - a_1)\alpha^3 = a_0 + a_1\alpha + a_2\alpha^2 + a_3\alpha^3\}$$
$$= \{a_0 + a_1\alpha + a_2\alpha^2 + a_3\alpha^3 : a_0 \in F_0, a_1 = 0, a_2 = a_3\}$$
$$= \{a_0 + a_2(\alpha^2 + \alpha^3) : a_0, a_2 \in F_0\},$$

so the fixed field of $\{\sigma_1, \sigma_4\}$ is $\{a_0 + a_2(\alpha^2 + \alpha^3) : a_0, a_2 \in F_0\}$.

2.3.16 Problem Let n be a positive integer ≥ 2. Let F be a field such that $F \subset \mathbb{C}$. Suppose that F contains all the nth roots of unity, that is, F contains all the roots of the polynomial $x^n - 1$, that is, $\left\{1, e^{\frac{2\pi i}{n}}, \left(e^{\frac{2\pi i}{n}}\right)^2, \ldots, \left(e^{\frac{2\pi i}{n}}\right)^{n-1}\right\} \subset F$. Let a be a nonzero member of F. Let u be a root of the polynomial $x^n - a$ in $\mathbb{C}$, that is, $u^n = a$. Then $F(u)$ is the splitting field over F for $x^n - a$.

Proof We must prove:

1. $F(u)$ is a finite extension of F,
2. $F(u)$ contains all the roots of $x^n - a$ in $\mathbb{C}$,
3. if G is a proper subfield of $F(u)$ that contains F, then G does not contain all the roots of $x^n - a$ in $\mathbb{C}$.

For 1: Since $u(\in \mathbb{C})$ is a root of the polynomial $(x^n - a) \in F[x]$, u is algebraic over F, and hence by 1.4.17, $F(u)$ is a finite extension of F.

For 2: Here we have to show that $\left\{u1, ue^{\frac{2\pi i}{n}}, u\left(e^{\frac{2\pi i}{n}}\right)^2, \ldots, u\left(e^{\frac{2\pi i}{n}}\right)^{n-1}\right\} \subset F(u)$.

Since $\left\{1, e^{\frac{2\pi i}{n}}, \left(e^{\frac{2\pi i}{n}}\right)^2, \ldots, \left(e^{\frac{2\pi i}{n}}\right)^{n-1}\right\} \subset F \subset F(u)$, $u \in F(u)$, and $F(u)$ is a field, we have

$$\left\{u1, ue^{\frac{2\pi i}{n}}, u\left(e^{\frac{2\pi i}{n}}\right)^2, \ldots, u\left(e^{\frac{2\pi i}{n}}\right)^{n-1}\right\} \subset F(u).$$

For 3: Suppose to the contrary that G is a subfield of $F(u)$ such that $G \neq F(u)$, G contains F, and G contains

$$\left\{u1, ue^{\frac{2\pi i}{n}}, u\left(e^{\frac{2\pi i}{n}}\right)^2, \ldots, u\left(e^{\frac{2\pi i}{n}}\right)^{n-1}\right\} (\ni u).$$

We seek a contradiction. Since G is a subfield of $F(u)$, we have $G \subset F(u)$. Since G is a field that contains $F \cup \{u\}$, we have $F(u) \subset G$, and hence $G = F(u)$. This is a contradiction.   ∎

2.3.17 Problem Let n be a positive integer ≥ 2. Let F be a field such that $F \subset \mathbb{C}$. Suppose that F contains all the nth roots of unity, that is, F contains all the roots of the polynomial $x^n - 1$, that is, $\left\{1, e^{\frac{2\pi i}{n}}, \left(e^{\frac{2\pi i}{n}}\right)^2, \ldots, \left(e^{\frac{2\pi i}{n}}\right)^{n-1}\right\} \subset F$. Let a be a nonzero member of F. Hence $(x^n - a) \in F[x]$. Then the Galois group of $x^n - a$ over F is abelian.

Proof By 2.3.16, $F(u)$ is the splitting field over F for $x^n - a$, where $u^n = a$ and $u \in \mathbb{C}$, and hence $F(u)$ contains all the roots of $x^n - a$ in $\mathbb{C}$. Further, the set of all the roots of $x^n - a$ is $\left\{u1, ue^{\frac{2\pi i}{n}}, u\left(e^{\frac{2\pi i}{n}}\right)^2, \ldots, u\left(e^{\frac{2\pi i}{n}}\right)^{n-1}\right\}(\subset F(u))$.

The Galois group of $x^n - a$ is the group $G(F(u), F)(= \{\sigma : \sigma \in \text{Aut}(F(u)), \text{and for every } a \in F, \sigma(a) = a\})$. We have to show that $G(F(u), F)$ is abelian.

To this end, let us take an automorphism $\sigma : F(u) \to F(u)$ such that for every $a \in F, \sigma(a) = a$. Next let us take an automorphism $\tau : F(u) \to F(u)$ such that for every $a \in F, \tau(a) = a$. We have to show that for every $b \in F(u)$, $\tau(\sigma(b)) = \sigma(\tau(b))$.

Observe that $\{t : t \in F(u), \tau(\sigma(t)) = \sigma(\tau(t))\}$ is a subfield of $F(u)$.

Proof Let $s, t \in F(u)$, where $\tau(\sigma(s)) = \sigma(\tau(s))$, and $\tau(\sigma(t)) = \sigma(\tau(t))$. It suffices to show that

1. $\tau(\sigma(s - t)) = \sigma(\tau(s - t))$,
2. $\tau(\sigma(st)) = \sigma(\tau(st))$,
3. if s, t are nonzero, then $\tau(\sigma(st^{-1})) = \sigma(\tau(st^{-1}))$.

For 1:

$$\text{LHS} = \tau(\sigma(s - t)) = \tau(\sigma(s) - \sigma(t)) = \tau(\sigma(s)) - \tau(\sigma(t))$$
$$= \sigma(\tau(s)) - \tau(\sigma(t)) = \sigma(\tau(s)) - \sigma(\tau(t)) = \sigma(\tau(s) - \tau(t)) = \sigma(\tau(s - t)) = \text{RHS}.$$

For 2:

$$\text{LHS} = \tau(\sigma(st)) = \tau(\sigma(s)\sigma(t)) = \tau(\sigma(s))\tau(\sigma(t))$$
$$= \sigma(\tau(s))\tau(\sigma(t)) = \sigma(\tau(s))\sigma(\tau(t)) = \sigma(\tau(s)\tau(t)) = \sigma((st)) = \text{RHS}.$$

For 3: Let s, t be nonzero. We have

$$\text{LHS} = \tau(\sigma(st^{-1})) = \tau\left(\sigma(s)(\sigma(t))^{-1}\right) = \tau(\sigma(s))(\tau(\sigma(t)))^{-1} = \sigma(\tau(s))(\tau(\sigma(t)))^{-1}$$
$$= \sigma(\tau(s))(\sigma(\tau(t)))^{-1} = \sigma\left(\tau(s)(\tau(t))^{-1}\right) = \sigma(\tau(st^{-1})) = \text{RHS}.$$

∎

We have shown that $\{t : t \in F(u), \tau(\sigma(t)) = \sigma(\tau(t))\}$ is a subfield of $F(u)$. It is clear that

$$F \subset \{t : t \in F(u), \tau(\sigma(t)) = \sigma(\tau(t))\}.$$

Since $u^n = a$, we have

$$(\sigma(u))^n = \underbrace{\sigma(u^n) = \sigma(a)}_{} = a,$$

and hence $(\sigma(u))^n - a = 0$. Thus $\sigma(u)$ is a root of $x^n - a$ in $\mathbb{C}$. Now, since the set of all the roots of $x^n - a$ is $\left\{u1, ue^{\frac{2\pi i}{n}}, u\left(e^{\frac{2\pi i}{n}}\right)^2, \ldots, u\left(e^{\frac{2\pi i}{n}}\right)^{n-1}\right\}$, there exists an integer $k \in \{0, 1, \ldots, n-1\}$ such that $\sigma(u) = u\left(e^{\frac{2\pi i}{n}}\right)^k$. Similarly, there exists an integer $l \in \{0, 1, \ldots, n-1\}$ such that $\tau(u) = u\left(e^{\frac{2\pi i}{n}}\right)^l$.

Clearly, $\tau(\sigma(u)) = \sigma(\tau(u))$.

Proof

$$\text{LHS} = \tau(\sigma(u)) = \tau\left(u\left(e^{\frac{2\pi i}{n}}\right)^k\right) = \tau(u)\tau\left(\left(e^{\frac{2\pi i}{n}}\right)^k\right) = \tau(u)\left(e^{\frac{2\pi i}{n}}\right)^k = u\left(e^{\frac{2\pi i}{n}}\right)^l.$$
$$\left(e^{\frac{2\pi i}{n}}\right)^k = u\left(e^{\frac{2\pi i}{n}}\right)^{k+l},$$

and

$$\text{RHS} = \sigma(\tau(u)) = \sigma\left(u\left(e^{\frac{2\pi i}{n}}\right)^l\right) = \sigma(u)\sigma\left(\left(e^{\frac{2\pi i}{n}}\right)^l\right) = \sigma(u)\left(e^{\frac{2\pi i}{n}}\right)^l$$
$$= u\left(e^{\frac{2\pi i}{n}}\right)^k \cdot \left(e^{\frac{2\pi i}{n}}\right)^l = u\left(e^{\frac{2\pi i}{n}}\right)^{k+l}.$$

Thus LHS = RHS.   ∎

We have shown that $\tau(\sigma(u)) = \sigma(\tau(u))$, and hence $u \in \{t : t \in F(u),$ $\tau(\sigma(t)) = \sigma(\tau(t))\}$. Thus $\{t : t \in F(u), \tau(\sigma(t)) = \sigma(\tau(t))\}$ is a field containing $F \cup \{u\}$, and hence

$$F(u) \subset \{t : t \in F(u), \tau(\sigma(t)) = \sigma(\tau(t))\}(\subset F(u)).$$

Thus $F(u) = \{t : t \in F(u), \tau(\sigma(t)) = \sigma(\tau(t))\}$. Thus for every $b \in F(u)$, $\tau(\sigma(b)) = \sigma(\tau(b))$.   ∎

2.4   Solvability By Radicals

2.4.1 Definition Let F, K be any fields such that K is an extension of F. Let $p(x) \in F[x]$. Let $\deg(p(x)) = n$. Suppose that K contains all the n roots of $p(x)$ in K. If there exists a finite sequence

$$(\omega_1, r_1), (\omega_2, r_2), \ldots, (\omega_k, r_k)$$

such that

1. each ω_i is a member of K,
2. each r_i is an integer ≥ 2,
3. $(\omega_1)^{r_1} \in F, (\omega_2)^{r_2} \in F(\omega_1), (\omega_3)^{r_3} \in F(\omega_1, \omega_2), \ldots, (\omega_k)^{r_k} \in F(\omega_1, \omega_2, \ldots, \omega_{k-1})$,
4. $F(\omega_1, \omega_2, \ldots, \omega_{k-1}, \omega_k)$ contains all the n roots of $p(x)$ in K,

then we say that $p(x)$ *is solvable by radicals over* F.

2.4.2 Example Let us take $\mathbb{Q}$ for F, $\mathbb{C}$ for K, and $x^5 - 2x^3 - x^2 + 2$ for $p(x)$.
    Since

$$
\begin{aligned}
& x^5 - 2x^3 - x^2 + 2 \\
&= x^3(x^2 - 2) - (x^2 - 2) \\
&= (x^2 - 2)(x^3 - 1) = \left(x - \sqrt{2}\right)\left(x + \sqrt{2}\right)(x^3 - 1) \\
&= \left(x - \sqrt{2}\right)\left(x + \sqrt{2}\right)(x - 1)(x^2 + x + 1) \\
&= \left(x - \sqrt{2}\right)\left(x + \sqrt{2}\right)(x - 1)\left(\left(x + \frac{1}{2}\right)^2 - \left(i\frac{\sqrt{3}}{2}\right)^2\right) \\
&= (x - 1)\left(x - \sqrt{2}\right)\left(x + \sqrt{2}\right)\left(x - \left(\frac{-1}{2} + i\frac{\sqrt{3}}{2}\right)\right)\left(x - \left(\frac{-1}{2} - i\frac{\sqrt{3}}{2}\right)\right),
\end{aligned}
$$

we have

$$x^5 - 2x^3 - x^2 + 2 = (x - 1)\left(x - \sqrt{2}\right)\left(x + \sqrt{2}\right)(x - \omega)(x - \omega^2),$$

where $\omega \equiv \frac{-1}{2} + i\frac{\sqrt{3}}{2}$. It follows that all the roots of $p(x)$ are $1, \sqrt{2}, -\sqrt{2}, \omega, \omega^2$. They all are members of $\mathbb{C}$. Let us take $\sqrt{2}$ for ω_1 and ω for ω_2. Let us take 2 for r_1 and 3 for r_2. Put $F_1 \equiv \mathbb{Q}(\sqrt{2})$ and $F_2 \equiv F_1(\omega)$. Let us take $k = 2$. Now all six conditions of the above definition are satisfied, so $x^5 - 2x^3 - x^2 + 2$ is solvable by radicals over $\mathbb{Q}$.

2.4.3 Example Let us take the general cubic polynomial

$$x^3 + a_1 x^2 + a_2 x + a_3$$

over the field F_0 of all rational numbers. By $F_0(a_1, a_2, a_3)$ we shall mean the field of rational functions in a_1, a_2, a_3 over F_0.
     Since

$$x^3 + a_1 x^2 + a_2 x + a_3$$
$$= \left(x^3 + 3x^2 \frac{a_1}{3} + 3x\left(\frac{a_1}{3}\right)^2 + \left(\frac{a_1}{3}\right)^3\right)$$
$$- 3x\left(\frac{a_1}{3}\right)^2 - \left(\frac{a_1}{3}\right)^3 + a_2 x + a_3$$
$$= \left(x + \frac{a_1}{3}\right)^3 + x\left(a_2 - \frac{(a_1)^2}{3}\right) + \left(a_3 - \left(\frac{a_1}{3}\right)^3\right)$$
$$= \left(x + \frac{a_1}{3}\right)^3 + \left(a_2 - \frac{(a_1)^2}{3}\right)\left(x + \frac{a_1}{3}\right) - \left(a_2 - \frac{(a_1)^2}{3}\right)\frac{a_1}{3} + \left(a_3 - \left(\frac{a_1}{3}\right)^3\right)$$
$$= y^3 + py + q,$$

where $y \equiv x + \frac{a_1}{3}$, $p \equiv a_2 - \frac{(a_1)^2}{3} (\in F_0(a_1, a_2, a_3))$, $q \equiv -\left(a_2 - \frac{(a_1)^2}{3}\right)$
$\frac{a_1}{3} + \left(a_3 - \left(\frac{a_1}{3}\right)^3\right) = \frac{2(a_1)^3}{27} - \frac{a_1 a_2}{3} + a_3 (\in F_0(a_1, a_2, a_3))$, we have

$$x^3 + a_1 x^2 + a_2 x + a_3 = y^3 + py + q.$$

Put $y \equiv u + v$, where $uv = \frac{-p}{3}$. It follows that

$$y^3 + py + q = \left(u^3 + v^3 + 3uvy\right) + py + q = u^3 + v^3 + q.$$

Hence

$$x^3 + a_1x^2 + a_2x + a_3 = 0$$

is equivalent to

$$y^3 + py + q = 0,$$

which, in turn, is equivalent to the simultaneous equations

$$\left.\begin{array}{r} u^3 + v^3 + q = 0 \\ uv = \frac{-p}{3} \end{array}\right\},$$

that is,

$$\left.\begin{array}{r} u^3 + v^3 = -q \\ u^3v^3 = \frac{-p^3}{27} \end{array}\right\},$$

that is,

$$\left.\begin{array}{r} u^3 + v^3 = -q \\ (u^3 - v^3)^2 = q^2 - 4\left(\frac{-p^3}{27}\right) \end{array}\right\}.$$

Now we can take

$$\left.\begin{array}{l} u^3 = \frac{1}{2}\left(-q + \sqrt{q^2 + \frac{4p^3}{27}}\right) \\ v^3 = \frac{1}{2}\left(-q - \sqrt{q^2 + \frac{4p^3}{27}}\right) \end{array}\right\}.$$

It follows that we can take

$$\left.\begin{array}{l} u = \sqrt[3]{-\frac{1}{2}q + \sqrt{\frac{1}{27}p^3 + \frac{1}{4}q^2}} \\ v = \sqrt[3]{-\frac{1}{2}q - \sqrt{\frac{1}{27}p^3 + \frac{1}{4}q^2}} \end{array}\right\}.$$

Hence all the roots of $x^3 + a_1x^2 + a_2x + a_3$ are

$$-\frac{a_1}{3} + (u + v), \quad -\frac{a_1}{3} + \left(u\omega + v\omega^2\right), \quad -\frac{a_1}{3} + \left(u\omega^2 + v\omega\right).$$

This is known as **Cardan's formula**.

2.4.4 Example Let us take the general biquadratic polynomial

$$x^4 + a_1 x^3 + a_2 x^2 + a_3 x + a_4$$

over the field F_0 of all rational numbers. By $F_0(a_1, a_2, a_3, a_4)$ we shall mean the field of rational functions in a_1, a_2, a_3, a_3 over F_0. Since

$$x^4 + a_1 x^3 + a_2 x^2 + a_3 x + a_4$$

$$= \left(x^4 + 4x^3 \frac{a_1}{4} + 6x^2 \left(\frac{a_1}{4}\right)^2 + 4x \left(\frac{a_1}{4}\right)^3 + \left(\frac{a_1}{4}\right)^4 \right)$$

$$- 6x^2 \left(\frac{a_1}{4}\right)^2 - 4x \left(\frac{a_1}{4}\right)^3 - \left(\frac{a_1}{4}\right)^4 + a_2 x^2 + a_3 x + a_4$$

$$= \left(x + \frac{a_1}{4} \right)^4 + x^2 \left(a_2 - 6 \left(\frac{a_1}{4}\right)^2 \right) + x \left(a_3 - 4 \left(\frac{a_1}{4}\right)^3 \right) + \left(a_4 - \left(\frac{a_1}{4}\right)^4 \right)$$

$$= y^4 + \left(y - \frac{a_1}{4} \right)^2 (\cdots) + \left(y - \frac{a_1}{4} \right)(\cdots) + (\cdots) = y^4 + py^2 + qy + r,$$

where $y \equiv x + \frac{a_1}{4}$, and $p, q, r \in F_0(a_1, a_2, a_3, a_4)$. It suffices to solve the equation

$$y^4 + py^2 + qy + r = 0,$$

that is,

$$\left(y^2 + \frac{p}{2} \right)^2 = \frac{p^2}{4} - qy - r,$$

or

$$\left(y^2 + \frac{p}{2} + m \right)^2 = m^2 + \left(2y^2 + p \right)m + \frac{p^2}{4} - qy - r,$$

where m is to be determined. Here

$$\left(y^2 + \frac{p}{2} + m \right)^2 = 2my^2 - qy + \left(m^2 + pm + \frac{p^2}{4} - r \right).$$

The equation is solved when the quadratic expression on the right-hand side of the above equation is a perfect square, that is, if its discriminant is zero, that is,

$$(-q)^2 - 4(2m)\left(m^2 + pm + \frac{p^2}{4} - r \right) = 0,$$

that is,

$$m^3 + pm^2 + \left(\frac{p^2}{4} - r\right)m - \frac{q^2}{8} = 0.$$

This is a cubic equation in m, so all its roots can be found as in Example 2.4.3. This is known as **Ferrari's method**.

2.4.5 Note Let F, K be any fields such that K is an extension of F. Suppose that for every positive integer l, F contains all the lth roots of unity. Let $p(x) \in F[x]$. Let $\deg(p(x)) = n$. Suppose that K contains all the n roots of $p(x)$. Suppose that $p(x)$ is solvable by radicals over F.

Hence there exists a finite sequence

$$(\omega_1, r_1), (\omega_2, r_2), \ldots, (\omega_k, r_k)$$

such that

1. each ω_i is a nonzero member of K,
2. each r_i is an integer ≥ 2,

3. $(\omega_1)^{r_1} \in F, (\omega_2)^{r_2} \in F(\omega_1), (\omega_3)^{r_3} \in F(\omega_1, \omega_2), \ldots, (\omega_k)^{r_k}$
        $\in F(\omega_1, \omega_2, \ldots, \omega_{k-1}),$
4. $F(\omega_1, \omega_2, \ldots, \omega_{k-1}, \omega_k)$ contains all the n roots of $p(x)$ in K.

Let L be the splitting field over F for $p(x)$. It follows that L is the smallest field containing all the n roots of $p(x)$ in K. Now since $F(\omega_1, \omega_2, \ldots, \omega_{k-1}, \omega_k)$ is a field containing all the n roots of $p(x)$ in K, we have $L \subset F(\omega_1, \omega_2, \ldots, \omega_{k-1}, \omega_k)$.

Since $(\omega_1)^{r_1} \in F$, there exists a nonzero $a \in F$ such that ω_1 is a root of the polynomial $(x^{r_1} - a) \in F[x]$. By assumption, F contains all the r_1 th roots of unity. So by 2.3.16, $F(\omega_1)$ is the splitting field over F for $x^{r_1} - a$, and hence by 2.2.20, $F(\omega_1)$ is a normal extension of F. Thus $F(\omega_1)$ is a normal extension of F.

Since $(\omega_2)^{r_2} \in F(\omega_1)$, there exists $b \in F(\omega_1)$ such that ω_2 is a root of the polynomial equation $(x^{r_2} - b) \in (F(\omega_1))[x]$. By assumption, $F(\subset F(\omega_1))$ contains all the r_2 th roots of unity, so $F(\omega_1)$ contains all the r_2 th roots of unity. Now by 2.3.16, $(F(\omega_1))(\omega_2)(= F(\omega_1, \omega_2))$ is the splitting field over $(F(\omega_1))$ for $x^{r_2} - b$, and hence by 2.2.20, $F(\omega_1, \omega_2)$ is a normal extension of $F(\omega_1)$.

Similarly, $F(\omega_1, \omega_2, \omega_3)$ is a normal extension of $F(\omega_1, \omega_2)$, etc. Since $F(\omega_1)$ is a normal extension of F, by 2.2.24, $G(F(\omega_1, \omega_2, \ldots, \omega_k), F(\omega_1))$ is a normal subgroup of the group $G(F(\omega_1, \omega_2, \ldots, \omega_k), F)$.

Similarly, $G(F(\omega_1, \omega_2, \ldots, \omega_k), F(\omega_1, \omega_2))$ is a normal subgroup of the group $G(F(\omega_1, \omega_2, \ldots, \omega_k), F(\omega_1))$, $G(F(\omega_1, \omega_2, \ldots, \omega_k), F(\omega_1, \omega_2, \omega_3))$ is a normal subgroup of the group $G(F(\omega_1, \omega_2, \ldots, \omega_k), F(\omega_1, \omega_2))$, etc. Thus

$$\{G(F(\omega_1, \omega_2, \ldots, \omega_k), F(\omega_1, \omega_1, \omega_2, \ldots, \omega_{k-1})),$$
$$\ldots, G(F(\omega_1, \omega_2, \ldots, \omega_k), F(\omega_1)), G(F(\omega_1, \omega_2, \ldots, \omega_k), F)\}$$

is a collection of subgroups of $G(F(\omega_1, \omega_2, \ldots, \omega_k), F)$ such that

1. $G(F(\omega_1, \omega_2, \ldots, \omega_k), F) \supset G(F(\omega_1, \omega_2, \ldots, \omega_k), F(\omega_1)) \supset \quad \cdots \supset$ $G(F(\omega_1, \omega_2, \ldots, \omega_k), F(\omega_1, \omega_1, \omega_2, \ldots, \omega_{k-1})) \supset \{\text{Id}\}$, where Id denotes the identity automorphism of $F(\omega_1, \omega_2, \ldots, \omega_k)$.

2. for every $i = 1, \ldots, k$, $G(F(\omega_1, \omega_2, \ldots, \omega_k), F(\omega_1, \omega_2, \ldots, \omega_i))$ is a normal subgroup of $G(F(\omega_1, \omega_2, \ldots, \omega_k), F(\omega_1, \omega_2, \ldots, \omega_{i-1}))$.

By 2.2.26, for every $i = 1, \ldots, k$, the quotient group

$$\frac{G(F(\omega_1, \omega_2, \ldots, \omega_k), F(\omega_1, \omega_2, \ldots, \omega_{i-1}))}{G(F(\omega_1, \omega_2, \ldots, \omega_k), F(\omega_1, \omega_2, \ldots, \omega_i))}$$

is isomorphic onto the group

$$G(F(\omega_1, \omega_2, \cdots, \omega_i), F(\omega_1, \omega_2, \cdots, \omega_{i-1}))$$
$$(= G((F(\omega_1, \omega_2, \cdots, \omega_{i-1}))(\omega_i), F(\omega_1, \omega_2, \cdots, \omega_{i-1}))).$$

We want to show that, for every $i = 1, \ldots, k$, the quotient group

$$\frac{G(F(\omega_1, \omega_2, \ldots, \omega_k), F(\omega_1, \omega_2, \ldots, \omega_{i-1}))}{G(F(\omega_1, \omega_2, \ldots, \omega_k), F(\omega_1, \omega_2, \ldots, \omega_i))}$$

is abelian. It suffices to show that for every $i = 1, \ldots, k$, $G((F(\omega_1, \omega_2, \ldots, \omega_{i-1}))(\omega_i), F(\omega_1, \omega_2, \ldots, \omega_{i-1}))$ is abelian, that is, each $G(F(\omega_1, \omega_2, \ldots, \omega_i), F(\omega_1, \omega_2, \ldots, \omega_{i-1}))$ is abelian.

To this end, we shall apply 2.3.17.

Since $(\omega_i)^{r_i} \in F(\omega_1, \omega_2, \ldots, \omega_{i-1})$, there exists a nonzero $a \in F(\omega_1, \omega_2, \ldots, \omega_{i-1})$ such that ω_i is a root of the polynomial $(x^{r_i} - a) \in (F(\omega_1, \omega_2, \ldots, \omega_{i-1}))[x]$. By assumption, $F(\omega_1, \omega_2, \ldots, \omega_{i-1})$ contains all the r_i th roots of unity. So by 2.3.16, $(F(\omega_1, \omega_2, \ldots, \omega_{i-1}))(\omega_i)(= F(\omega_1, \omega_2, \ldots, \omega_i))$ is the splitting field over $F(\omega_1, \omega_2, \ldots, \omega_{i-1})$ for $(x^{r_i} - a)$.

By 2.3.17, the Galois group of $x^{r_i} - a$ over $F(\omega_1, \omega_2, \ldots, \omega_{i-1})$ is abelian. Here the Galois group of $x^{r_i} - a$ is

$$G(F(\omega_1, \omega_2, \ldots, \omega_i), F(\omega_1, \omega_2, \ldots, \omega_{i-1})),$$

so $G(F(\omega_1, \omega_2, \ldots, \omega_i), F(\omega_1, \omega_2, \ldots, \omega_{i-1}))$ is abelian.

Thus we have shown that $G(F(\omega_1, \omega_2, \ldots, \omega_k), F)$ is a solvable group.

We want to show that the Galois group over F of $p(x)$ is a solvable group, that is, $G(L, F)$ is a solvable group.

Since L is the splitting field over F for $p(x)(\in F[x])$, by 2.2.20, L is a normal extension of F. Now, since $L \subset F(\omega_1, \omega_2, \ldots, \omega_{k-1}, \omega_k)$, by 2.2.24, $G(F(\omega_1, \omega_2, \ldots, \omega_k), L)$ is a normal subgroup of the group $G(F(\omega_1, \omega_2, \ldots, \omega_k), F)$. Further, by 2.2.25, the quotient group $\frac{G(F(\omega_1, \omega_2, \ldots, \omega_k), F)}{G(F(\omega_1, \omega_2, \ldots, \omega_k), L)}$ is isomorphic to the group $G(L, F)$. Since the quotient group $\frac{G(F(\omega_1, \omega_2, \ldots, \omega_k), F)}{G(F(\omega_1, \omega_2, \ldots, \omega_k), L)}$ is a homomorphic image of $G(F(\omega_1, \omega_2, \ldots, \omega_k), F)$, the group $G(L, F)$ is a homomorphic image of $G(F(\omega_1, \omega_2, \ldots, \omega_k), F)$. Next, since $G(F(\omega_1, \omega_2, \ldots, \omega_k), F)$ is a solvable group, the group $G(L, F)$ is a homomorphic image of a solvable group. It follows, by 2.3.10, that the group $G(L, F)$ is solvable.

2.4.6 Conclusion Let F, K be any fields such that K is an extension of F. Suppose that for every positive integer l, F contains all the lth roots of unity. Let $p(x) \in F[x]$. Let $\deg(p(x)) = n$. Suppose that K contains all the n roots of $p(x)$. Suppose that $p(x)$ is solvable by radicals over F. Then the Galois group over F of $p(x)$ is a solvable group.

2.4.7 Note Let F be any field. By the *general polynomial $x^n + a_1 x^{n-1} + \cdots + a_n$ of degree n over F*, we mean the following:

$F(a_1, \ldots, a_n)$ is the field of all rational functions in n variables $a_1, \ldots, a_n$ over F, and $x^n + a_1 x^{n-1} + \cdots + a_n$ is a polynomial in x over the field $F(a_1, \ldots, a_n)$ $(\ni a_1, \ldots, a_n)$.

If $x^n + a_1 x^{n-1} + \cdots + a_n$ is solvable by radicals over $F(a_1, \ldots, a_n)$, then we say that the *general polynomial $x^n + a_1 x^{n-1} + \cdots + a_n$ of degree n over F is solvable by radicals*.

By 2.2.14, the splitting field over $F(-a_1, \ldots, (-1)^n a_n)(= F(a_1, \ldots, a_n))$ for $t^n + a_1 t^{n-1} + a_2 t^{n-2} + \cdots + a_n$ is $F(x_1, \ldots, x_n)$, and

$$\underbrace{S_n = G(F(x_1, \cdots, x_n), F(a_1, \cdots, a_n))}$$

$$= G \left(\begin{matrix} \text{splitting field over } F(a_1, \cdots, a_n) \text{for } t^n + a_1 t^{n-1} + a_2 t^{n-2} \\ + \cdots + a_n, F(a_1, \cdots, a_n) \end{matrix} \right)$$

$$= \left(\text{Galois group of } t^n + a_1 t^{n-1} + a_2 t^{n-2} + \cdots + a_n \right).$$

It follows that the Galois group of $t^n + a_1 t^{n-1} + a_2 t^{n-2} + \cdots + a_n$ is S_n. Next, by 2.3.12, S_n is not solvable for $n \geq 5$, so the Galois group of $t^n + a_1 t^{n-1} + a_2 t^{n-2} + \cdots + a_n$ is not solvable for $n \geq 5$, and hence by 2.4.6, $t^n + a_1 t^{n-1} + a_2 t^{n-2} + \cdots + a_n$ is not solvable by radicals over F for $n \geq 5$.

2.4.8 Conclusion Let F be any field. Suppose that for every positive integer l, F contains all the lth roots of unity. Let $n \geq 5$. Let $x^n + a_1 x^{n-1} + \cdots + a_n$ be the general polynomial of degree n over F. Then $t^n + a_1 t^{n-1} + a_2 t^{n-2} + \cdots + a_n$ is not solvable by radicals over F.

Roughly speaking, for $n \geq 5$, there exists no formula for the roots of $t^n + a_1 t^{n-1} + a_2 t^{n-2} + \cdots + a_n$ involving only a combination of mth roots of rational functions of $a_1, \cdots, a_n$, for various values of m.

This result is due to **Niels Henrik Abel** (1802–1829).

2.4.9 Note Here we shall recapitulate something about high-school "construction geometry," We shall assume that we have a straightedge (that is, an ungraduated scale), a compass (that is, an instrument with two arms, one with a metallic needle end, and another with a pencil's lead end). We also assume that we are given the measure of a "unit distance."

Some of the well-known constructions are sketched below:

1. In Fig. 2.1, the perpendicular bisector of a given line segment is constructed.
2. In Fig. 2.2, the perpendicular at given point of a line is constructed.
3. In Fig. 2.3, the perpendicular line from a given point on a given line is constructed.
4. In Fig. 2.4, a line parallel to a given line and passing through a given point is constructed. Here, construction (3) and then construction (2) are made.
5. In Fig. 2.5, a given line segment is divided into three equal parts. Similarly, a given line segment can be divided into any number of equal parts.

It follows that every rational number is a "constructible number." In other words, $\mathbb{Q} \subset W$, where W denotes the collection of all constructible numbers. It is easy to

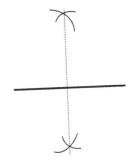

Fig. 2.1 Perpendicular bisector of a given line segment

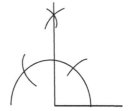

Fig. 2.2 Perpendicular at a given point of a line

Fig. 2.3 Perpendicular
line drawn from a given
point on a given line

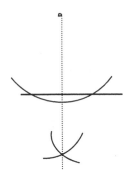

Fig. 2.4 Line parallel to a
given line and passing
through a given point

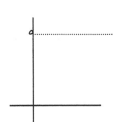

Fig. 2.5 A given line segment to be divided into
three equal parts

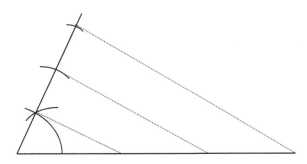

Fig. 2.6 Construction for
squaring the size of a line
segment

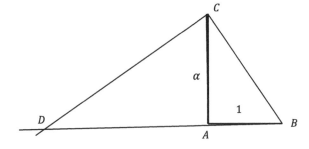

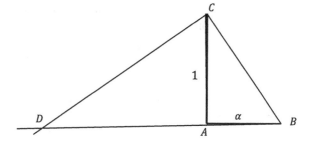

Fig. 2.7 Construction for reciprocal of the size of a line segment

observe that if α and β are constructible numbers, then $\alpha + \beta$ and $\alpha - \beta$ are constructible numbers.

Now we shall show that if α is a constructible number, then so is α^2 (Fig. 2.6).

We first construct a right triangle ABC one of whose legs, AB, is of length 1, and the other leg, AC, is of length α. Now we draw a line perpendicular to BC at C. Suppose that this line meets AB at D. Thus we get two similar triangles, ACB and ADC. It follows that

$$\frac{\alpha}{1} = \underbrace{\frac{AC}{AB} = \frac{AD}{AC}}_{} = \frac{AD}{\alpha},$$

and hence $AD = \alpha^2$. Thus α^2 is a constructible number.

Now, since $\alpha\beta = \frac{(\alpha+\beta)^2 + (\alpha-\beta)^2}{4}$, we get the following result: if α and β are constructible numbers, then so is $\alpha\beta$. We want to show that W is a field. For this, it suffices to show that if α is a positive constructible number, then so is $\frac{1}{\alpha}$ (Fig. 2.7).

We first construct a right triangle ABC one of whose legs, AB, is of length α, and the other leg, AC, is of length 1. Now we draw a line perpendicular to BC at C. Suppose that this line meets AB at D. Thus we get two similar triangles, ACB and ADC. It follows that

$$\frac{1}{\alpha} = \underbrace{\frac{AC}{AB} = \frac{AD}{AC}}_{} = \frac{AD}{1},$$

and hence $AD = \frac{1}{\alpha}$. Thus $\frac{1}{\alpha}$ is a constructible number.

Thus W is a subfield of $\mathbb{R}$ that contains $\mathbb{Q}$.

2.4.10 Definition Let F be a subfield of $\mathbb{R}$. By the *plane of F* we mean the Cartesian product $F \times F (\subset \mathbb{R}^2)$.

Observe that the straight line joining the point $(x_1, y_1)(\in F \times F)$ and the point $(x_2, y_2)(\in F \times F)$ is

$$y - y_1 = \frac{y_2 - y_1}{x_2 - x_1}(x - x_1),$$

which is of the form $ax + by + c = 0$, where $a, b, c \in F$. Similarly, every equation of the form $ax + by + c = 0$ represents a straight line passing through two points of the plane of F. Such straight lines are called *straight lines in F*.

It is clear that if two straight lines in F intersect in the real plane $\mathbb{R}^2$, then their point of intersection is a point in the plane of F.

Every circle having center at a point of the plane of F and radius an element of F is of the form $x^2 + y^2 + ax + by + c = 0$, where $a, b, c \in F$. Such circles are called *circles in F*.

It is clear that if a circle in F and a straight line in F intersect in the real plane $\mathbb{R}^2$, then their points of intersection are either points in the plane of F or points in the plane of the field extension $F(\sqrt{\gamma})$ of F, for some positive $\gamma \in F$.

Similarly, it is clear that if two circles in F intersect in the real plane $\mathbb{R}^2$, then their points of intersection are either points in the plane of F or points in the plane of the field extension $F(\sqrt{\gamma})$ of F, for some positive $\gamma \in F$.

Thus, if a straight line or a circle in the field F intersects another straight line or a circle in the field F in the real plane $\mathbb{R}^2$, then there exists a positive real number γ_1 such that their point(s) of intersection are points in the plane of the field extension $F(\gamma_1)$ of F, where $(\gamma_1)^2 \in F$.

As above, if a straight line or a circle in the field $F(\gamma_1)$ intersects another straight line or a circle in the field $F(\gamma_1)$ in the real plane $\mathbb{R}^2$, then there exists a positive real number γ_2 such that their point(s) of intersection are points in the plane of the field extension $(F(\gamma_1))(\gamma_2)(= F(\gamma_1, \gamma_2))$ of $F(\gamma_1)$, where $(\gamma_2)^2 \in F(\gamma_1)$, etc.

Hence, if a point is constructible from F, then there exists a finite sequence $\gamma_1, \ldots, \gamma_n$ of real numbers such that

1. $(\gamma_1)^2 \in F, (\gamma_2)^2 \in F(\gamma_1), (\gamma_3)^2 \in F(\gamma_1, \gamma_2), \ldots, (\gamma_n)^2 \in F(\gamma_1, \ldots, \gamma_{n-1})$,
2. the point is in the plane of $F(\gamma_1, \ldots, \gamma_n)$.

Since $(\gamma_1)^2 \in F$, there exists $\alpha \in F$ such that γ_1 is a root of the polynomial $(x^2 - \alpha) \in F(x)$. Here $\deg(x^2 - \alpha) = 2$, so γ_1 is algebraic of degree 1 or 2, and hence by 1.4.16, $[F(\gamma_1), F] = 1$ or 2. Similarly , $[F(\gamma_1, \gamma_2), F(\gamma_1)] = 1$ or 2, and hence by 1.4.3, $[F(\gamma_1, \gamma_2), F] = [F(\gamma_1, \gamma_2), F(\gamma_1)][F(\gamma_1), F](= 1\text{or}2\text{or}2^2)$. Thus $[F(\gamma_1, \gamma_2), F] = 1$ or 2 or 2^2. Similarly, $[F(\gamma_1, \gamma_2, \gamma_3), F] = 1$ or 2 or 2^2 or 2^3, etc.

2.4.11 Conclusion Suppose that a real number α is constructible. Then there exist an extension K of $\mathbb{Q}$ and a nonnegative integer k such that $\alpha \in K$ and α is algebraic of degree 2^k.

2.4.12 Theorem It is impossible, by straightedge and a compass alone, to trisect the angle $60°$.

Proof Suppose to the contrary that the angle $20°$ is constructible. We seek a contradiction.

Fig. 2.8 Construction   for the size of cos $20°$, provided $20°$ angle is constructible

Let us draw a circle of radius 1 with center at the vertex of the 20° angle. Now draw the foot of perpendicular as shown in Fig. 2.8:

Thus $\cos 20°$ is constructible. On using the formula $\cos 3\theta = 4\cos^3 \theta - 3\cos \theta$, we get $\frac{1}{2} = 4\cos^3 20° - 3\cos 20°$, and hence $8x^3 - 6x - 1 = 0$, where $x \equiv \cos 20°$. Thus $\cos 20°$ is a root of the polynomial $8x^3 - 6x - 1(\in \mathbb{Q}[x])$.

We claim that $8x^3 - 6x - 1$ is an irreducible polynomial over the field of rational numbers.

Proof Put $x \equiv y - 1$. It suffices to show that $8(y-1)^3 - 6(y-1) - 1$ is irreducible over the field of rational numbers.

Observe that

$$8(y-1)^3 - 6(y-1) - 1 = 8(y^3 - 3y^2 + 3y - 1) - 6y + 6 - 1$$
$$= -3 + 18y - 24y^2 + 8y^3$$
$$= 3(-1 + 6y - 8y^2) + 8y^3,$$

so

$$8(y-1)^3 - 6(y-1) - 1 = 3(-1 + 6y - 8y^2) + 8y^3.$$

By 1.3.5, $3(-1 + 6y - 8y^2) + 8y^3$ is irreducible over the field of rational numbers, and hence
$8(y-1)^3 - 6(y-1) - 1$ is irreducible over the field of rational numbers.   ∎

Thus we have shown that $8x^3 - 6x - 1$ is an irreducible polynomial over the field of rational numbers. Then by 1.5.12, $\cos 20°$ is algebraic of degree 3 over $\mathbb{Q}$. Since $\cos 20°$ is constructible, by 2.4.12, $\cos 20°$ is algebraic of a degree of the form 2^k. This contradicts the fact that $\cos 20°$ is algebraic of degree 3 over $\mathbb{Q}$.   ∎

2.4.13 Theorem It is impossible by straightedge and a compass alone to duplicate the cube in the sense of constructing an edge of a cube whose volume is twice the volume of a given cube.

Proof For simplicity, suppose that the volume of the given cube is 1. We have to construct a length α such that $\alpha^3 = 2$.

Suppose that α is constructible. We seek a contradiction.
Here α is a root of the polynomial $x^3 - 2(\in \mathbb{Q}[x])$.
We claim that $x^3 - 2$ is an irreducible polynomial over the field of rational numbers.

Proof Put $x \equiv y - 1$. It suffices to show that $(y-1)^3 - 2$ is an irreducible polynomial over the field of rational numbers.

Observe that

$$(y-1)^3-2 = \left(y^3 - 3y^2 + 3y - 1\right) - 2 = -3 + 3y - 3y^2 + y^3$$
$$= 3\left(-1+y-y^2\right) + y^3,$$

so

$$(y-1)^3-2 = 3\left(-1+y-y^2\right) + y^3.$$

By 1.3.5, $3(-1+y-y^2)+y^3$ is irreducible over the field of rational numbers, and hence

$(y-1)^3-2$ is irreducible over the field of rational numbers.   ∎

Thus we have shown that $x^3 - 2$ is an irreducible polynomial over the field of rational numbers. So by 1.5.12, α is algebraic of degree 3 over $\mathbb{Q}$. Since α is constructible, by 2.4.12, α is algebraic of a degree of the form 2^k. This contradicts the fact that α is algebraic of degree 3 over $\mathbb{Q}$.   ∎

2.4.14 Theorem It is impossible, by straightedge and a compass alone, to construct a regular septagon.

Proof Construction of a regular septagon requires the construction of an angle $\frac{2\pi}{7}$. Suppose that the angle $\frac{2\pi}{7}$ is constructible, and hence $2\cos\frac{2\pi}{7}$ is constructible. We seek a contradiction. Put $\theta \equiv \frac{2\pi}{7}$.

It follows that

$$4\theta = 2\pi - 3\theta,$$

and hence

$$2(2\sin\theta\cos\theta)\cos 2\theta = 2\sin 2\theta\cos 2\theta = \underbrace{\sin 4\theta = \sin(2\pi - 3\theta)} = -\sin 3\theta$$
$$= -3\sin\theta + 4(\sin\theta)^3 = \sin\theta\left(-3 + 4\sin^2\theta\right).$$

This shows that

$$2\cos\theta\left((2\cos\theta)^2 - 2\right) = 4\cos\theta(2\cos^2\theta - 1) = \underbrace{4\cos\theta\cos 2\theta = -3 + 4\sin^2\theta}$$
$$= -3 + 4(1 - \cos^2\theta) = -4\cos^2\theta + 1,$$

that is,

$$y\left(y^2 - 2\right) = -y^2 + 1,$$

or

$$y^3 + y^2 - 2y - 1 = 0,$$

where $y \equiv 2\cos\theta$. Thus $2\cos\theta$ is a root of the polynomial $x^3 + x^2 - 2x - 1 (\in \mathbb{Q}[x])$.

Next, we claim that $x^3 + x^2 - 2x - 1$ is an irreducible polynomial over the field of rational numbers.

Proof Put $x \equiv y + 2$. It suffices to show that $(y+2)^3 + (y+2)^2 - 2(y+2) - 1$ is an irreducible polynomial over the field of rational numbers. Observe that

$$(y+2)^3 + (y+2)^2 - 2(y+2) - 1 = 7 + 14y + 7y^2 + y^3 = 7(1 + 2y + y^2) + y^3,$$

so

$$(y+2)^3 + (y+2)^2 - 2(y+2) - 1 = 7(1 + 2y + y^2) + y^3.$$

By 1.3.5, $7(1 + 2y + y^2) + y^3$ is irreducible over the field of rational numbers, and hence

$(y+2)^3 + (y+2)^2 - 2(y+2) - 1$ is irreducible over the field of rational numbers.

Thus we have shown that $x^3 + x^2 - 2x - 1$ is an irreducible polynomial over the field of rational numbers.

So by 1.5.12, $2\cos\theta$ is algebraic of degree 3 over $\mathbb{Q}$. Since $2\cos\theta$ is constructible, by 2.4.11, $2\cos\theta$ is algebraic of a degree of the form 2^k. This contradicts the fact that $2\cos\theta$ is algebraic of degree 3 over $\mathbb{Q}$. ∎

Exercises

1. Let F be a field. Let $f(x), g(x), k(x) \in F[x]$. Let $\alpha \in F$. Suppose that $h(x) = f(x) + k(\alpha)g(x)(\in F[x])$. Then

$$h'(x) = f'(x) + k(\alpha)g'(x).$$

2. Show that every finite extension of a field of characteristic 0 is a simple extension.

3. Let F and K be any fields such that K is an extension of F. Let F be of characteristic p. Suppose that α is a member of K such that $\alpha^{(p^2-1)} = 1$. Show that $p^2\alpha^{(p^2-1)} \neq 1$.

4. Suppose that K is a finite extension of F. Show that the order of the group of automorphisms of K relative to F cannot be greater than $[K : F]$.

5. Suppose that K is a finite extension of F. Suppose that T is a subfield of K that contains F. Suppose that T is a normal extension of F. Show that $G(K, T)$ is a normal subgroup of $G(K, F)$.

6. Show that for all positive integers m, n, the group $(S_m)^{(n)}$ is a normal subgroup of the symmetric group S_m.

7. Show that $\mathbb{Q}(\sqrt[3]{2})$ is the fixed field of $G(\mathbb{Q}(\sqrt[3]{2}), \mathbb{Q})$.

8. Let F be a field such that $F \subset \mathbb{C}$. Suppose that F contains all the nth roots of unity. Show that the Galois group of $x^5 - 5$ over F is abelian.

9. Show that it is impossible, by straightedge and a compass alone, to construct the angle $10°$.

10. Suppose that α, β are nonzero constructible numbers. Show that

$$\frac{\alpha^2 - \beta^2}{\alpha^2 + \beta^2}$$

is a constructible number.

Chapter 3
Linear Transformations

The subject matter in this chapter is also known as *linear algebra*. As we shall see, the *theory of matrices* is intimately related to linear algebra. Its applications to other branches of knowledge is overwhelming. That is why it is considered an independent subfield of mathematics, exciting on its own.

3.1 Eigenvalues

3.1.1 Theorem Let V be an n-dimensional inner product space. Let $T : V \rightarrow V$ be a linear transformation. Suppose that for every $v \in V$, $\langle T(v), v \rangle = 0$. Then $T = 0$.

Proof We have to show that $T = 0$. To this end, let us fix an arbitrary $w \in V$. We have to show that $T(w) = 0$, that is, $\langle T(w), T(w) \rangle = 0$.

Let us take arbitrary $u, v \in V$. By the given condition, we have

$$
\begin{aligned}
\langle T(u), v \rangle + \langle T(v), u \rangle &= 0 + \langle T(u), v \rangle + \langle T(v), u \rangle + 0 \\
&= \langle T(u), u \rangle + \langle T(u), v \rangle + \langle T(v), u \rangle + \langle T(v), v \rangle \\
&= \langle T(u) + T(v), u + v \rangle = \underbrace{\langle T(u+v), u+v \rangle = 0}.
\end{aligned}
$$

Thus for every $u, v \in V$,

$$
\langle T(u), v \rangle + \langle T(v), u \rangle = 0.
$$

It follows that for every $u, v \in V$,

© Springer Nature Singapore Pte Ltd. 2020
R. Sinha, *Galois Theory and Advanced Linear Algebra*,
https://doi.org/10.1007/978-981-13-9849-0_3

$$2i\langle T(v), u\rangle = i\langle T(v), u\rangle + i\langle T(v), u\rangle$$
$$= -i\langle T(u), v\rangle + i\langle T(v), u\rangle = \bar{i}\langle T(u), v\rangle + i\langle T(v), u\rangle$$
$$= \langle T(u), iv\rangle + i\langle T(v), u\rangle = \langle T(u), iv\rangle + \langle iT(v), u\rangle$$
$$= \underbrace{\langle T(u), iv\rangle + \langle T(iv), u\rangle = 0},$$

and hence for every $u, v \in V$,

$$\langle T(v), u\rangle = 0.$$

It follows that

$$\langle T(w), T(w)\rangle = 0.$$

∎

Definition Let V be an n-dimensional inner product space. Let $T : V \to V$ be a linear transformation. If for every $u, v \in V$, $\langle T(u), T(v)\rangle = \langle u, v\rangle$, then we say that T is *unitary*.

3.1.2 Theorem Let V be an n-dimensional inner product space. Let $T : V \to V$ be a unitary linear transformation. Suppose that for every $v \in V$, $\langle T(v), T(v)\rangle = \langle v, v\rangle$. Then T is unitary.

Proof Let us take any $u, v \in V$. By the given condition, we have

$$\langle u, u\rangle + \langle T(u), T(v)\rangle + \langle T(v), T(u)\rangle + \langle v, v\rangle$$
$$= \langle T(u), T(u)\rangle + \langle T(u), T(v)\rangle + \langle T(v), T(u)\rangle + \langle T(v), T(v)\rangle$$
$$= \langle T(u) + T(v), T(u) + T(v)\rangle = \underbrace{\langle T(u+v), T(u+v)\rangle = \langle u+v, u+v\rangle}$$

$$= \langle u, u\rangle + \langle u, v\rangle + \langle v, u\rangle + \langle v, v\rangle.$$

Thus for every $u, v \in V$,

$$\langle T(u), T(v)\rangle + \langle T(v), T(u)\rangle = \langle u, v\rangle + \langle v, u\rangle.$$

It follows that for every $u, v \in V$,

$$2i\langle T(v), T(u)\rangle - i\langle u, v\rangle - i\langle v, u\rangle = i(\langle T(v), T(u)\rangle - \langle u, v\rangle - \langle v, u\rangle) + i\langle T(v), T(u)\rangle$$
$$= -i\langle T(u), T(v)\rangle + i\langle T(v), T(u)\rangle$$
$$= -i\langle T(u), T(v)\rangle + \langle iT(v), T(u)\rangle = \langle T(u), iT(v)\rangle + \langle iT(v), T(u)\rangle$$
$$= \underbrace{\langle T(u), T(iv)\rangle + \langle T(iv), T(u)\rangle = \langle u, iv\rangle + \langle iv, u\rangle}$$

$$= -i\langle u, v\rangle + \langle iv, u\rangle = -i\langle u, v\rangle + i\langle v, u\rangle,$$

and hence for every $u, v \in V$,

$$\langle T(v), T(u)\rangle = \langle v, u\rangle.$$

Thus T is unitary.   ∎

3.1.3 Theorem Let V be an n-dimensional inner product space. Let $T : V \to V$ be a unitary linear transformation. Let $\{v_1, \ldots, v_n\}$ be any orthonormal basis of V. Then $\{T(v_1), \ldots, T(v_n)\}$ is an orthonormal basis of V.

Proof Let

$$\alpha_1 T(v_1) + \cdots + \alpha_n T(v_n) = 0.$$

It follows that

$$\alpha_1 = \alpha_1 1 + \alpha_2 0 + \cdots + \alpha_n 0 = \alpha_1\langle v_1, v_1\rangle + \alpha_2\langle v_2, v_1\rangle + \cdots + \alpha_n\langle v_n, v_1\rangle$$
$$= \alpha_1\langle T(v_1), T(v_1)\rangle + \alpha_2\langle T(v_2), T(v_1)\rangle + \cdots + \alpha_n\langle T(v_n), T(v_1)\rangle$$
$$= \underbrace{\langle \alpha_1 T(v_1) + \cdots + \alpha_n T(v_n), T(v_1)\rangle = \langle 0, T(v_1)\rangle} = 0,$$

and hence $\alpha_1 = 0$. Similarly, $\alpha_2 = 0, \ldots, \alpha_n = 0$. Thus we have shown that $T(v_1), \ldots, T(v_n)$ are linearly independent. Since $T(v_1), \ldots, T(v_n)$ are linearly independent, $T(v_1), \ldots, T(v_n)$ are distinct members of V. It follows that $\{T(v_1), \ldots, T(v_n)\}$ is a basis of V. Next, for distinct indices $i, j \in \{1, \ldots, n\}$, $\langle T(v_i), T(v_j)\rangle = \langle v_i, v_j\rangle = 0$. Also, for every index $i \in \{1, \ldots, n\}$, $\langle T(v_i), T(v_i)\rangle = \langle v_i, v_i\rangle = 1$. Thus $\{T(v_1), \ldots, T(v_n)\}$ is an orthonormal basis of V.   ∎

3.1.4 Theorem Let V be an n-dimensional inner product space. Let $T : V \to V$ be a linear transformation. Suppose that T sends every orthonormal basis of V to an orthonormal basis of V. Then T is unitary.

Proof Since V is an n-dimensional inner product space, there exists an orthonormal basis $\{v_1, \ldots, v_n\}$ of V. By assumption, $\{T(v_1), \ldots, T(v_n)\}$ is also an orthonormal basis of V. Let $u \equiv \sum_{i=1}^{n} \alpha_i v_i$ and $w \equiv \sum_{j=1}^{n} \beta_j v_j$ be any members of V. We have to show that $\langle T(u), T(w) \rangle = \langle u, w \rangle$. Here,

$$\text{LHS} = \langle T(u), T(w) \rangle = \left\langle T\left(\sum_{i=1}^{n} \alpha_i v_i\right), T\left(\sum_{j=1}^{n} \beta_j v_j\right) \right\rangle$$

$$= \left\langle \left(\sum_{i=1}^{n} \alpha_i T(v_i)\right), \left(\sum_{j=1}^{n} \beta_j T(v_j)\right) \right\rangle$$

$$= \sum_{i,j} \alpha_i \overline{\beta_j} \langle T(v_i), T(v_j) \rangle$$

$$= \sum_{i,j} \alpha_i \overline{\beta_j} \delta^{ij} = \sum_{i=1}^{n} \alpha_i \overline{\beta_i},$$

and

$$\text{RHS} = \langle u, w \rangle = \left\langle \sum_{i=1}^{n} \alpha_i v_i, \sum_{j=1}^{n} \beta_j v_j \right\rangle = \sum_{i,j} \alpha_i \overline{\beta_j} \langle v_i, v_j \rangle = \sum_{i,j} \alpha_i \overline{\beta_j} \delta^{ij} = \sum_{i=1}^{n} \alpha_i \overline{\beta_i}$$

so LHS = RHS.   ∎

3.1.5 Theorem Let V be an n-dimensional inner product space. Let $T : V \to V$ be a linear transformation. Let $v \in V$. Then there exists a unique $w \in V$ such that

$$u \in V \Rightarrow \langle u, w \rangle = \langle T(u), v \rangle.$$

We denote w by $T^*(v)$. Thus $T^* : V \to V$, and for every $u, v \in V$, $\langle u, T^*(v) \rangle = \langle T(u), v \rangle$. Also, $T^* : V \to V$ is linear.

Proof Existence: Since V is an n-dimensional inner product space, there exists an orthonormal basis $\{u_1, \ldots, u_n\}$ of V. Put

$$w \equiv \overline{\langle T(u_1), v \rangle} u_1 + \cdots + \overline{\langle T(u_n), v \rangle} u_n.$$

Let us fix an arbitrary $u \equiv \sum_{i=1}^{n} \alpha_i u_i$. We have to show that

$$\left\langle \sum_{i=1}^{n} \alpha_i u_i, \sum_{j=1}^{n} \overline{\langle T(u_j), v \rangle} u_j \right\rangle = \left\langle T\left(\sum_{i=1}^{n} \alpha_i u_i \right), v \right\rangle,$$

$$\text{LHS} = \left\langle \sum_{i=1}^{n} \alpha_i u_i, \sum_{j=1}^{n} \overline{\langle T(u_j), v \rangle} u_j \right\rangle$$

$$= \sum_{i,j} \alpha_i \langle T(u_j), v \rangle \langle u_i, u_j \rangle = \sum_{i,j} \alpha_i \langle T(u_j), v \rangle \delta^{ij}$$

$$= \sum_{i=1}^{n} \alpha_i \langle T(u_i), v \rangle = \left\langle \sum_{i=1}^{n} \alpha_i T(u_i), v \right\rangle$$

$$= \left\langle T\left(\sum_{i=1}^{n} \alpha_i u_i \right), v \right\rangle = \text{RHS}.$$

Uniqueness: Suppose that there exist $w_1, w_2 \in V$ such that

$$u \in V \Rightarrow \langle u, w_1 \rangle = \langle T(u), v \rangle, \text{ and } \langle u, w_2 \rangle = \langle T(u), v \rangle.$$

We have to show that $w_1 = w_2$, that is, $\langle w_1 - w_2, w_1 - w_2 \rangle = 0$. Here

$$u \in V \Rightarrow \langle u, w_1 \rangle = \langle u, w_2 \rangle,$$

so for every $u \in V$, $\langle u, w_1 - w_2 \rangle = 0$. It follows that $\langle w_1 - w_2, w_1 - w_2 \rangle = 0$.

Linearity: Let us take arbitrary $v_1, v_2 \in V$. Let α, β be arbitrary complex numbers. We have to show that

$$T^*(\alpha v_1 + \beta v_2) = \alpha T^*(v_1) + \beta T^*(v_2).$$

It suffices to show that for every $u \in V$,

$$\langle u, T^*(\alpha v_1 + \beta v_2) \rangle = \langle u, \alpha T^*(v_1) + \beta T^*(v_2) \rangle.$$

To this end, let us fix an arbitrary $u \in V$. We have to show that

$$\langle u, T^*(\alpha v_1 + \beta v_2) \rangle = \langle u, \alpha T^*(v_1) + \beta T^*(v_2) \rangle,$$
$$\text{LHS} = \langle u, T^*(\alpha v_1 + \beta v_2) \rangle = \langle T(u), \alpha v_1 + \beta v_2 \rangle = \bar{\alpha} \langle T(u), v_1 \rangle + \bar{\beta} \langle T(u), v_2 \rangle$$
$$= \bar{\alpha} \langle u, T^*(v_1) \rangle + \bar{\beta} \langle u, T^*(v_2) \rangle = \langle u, \alpha T^*(v_1) + \beta T^*(v_2) \rangle = \text{RHS}.$$

∎

Definition Let V be an n-dimensional inner product space. Let $T : V \to V$ be a linear transformation. By 3.1.5, $T^* : V \to V$ is a linear transformation such that for every $u, v \in V$, $\langle u, T^*(v) \rangle = \langle T(u), v \rangle$. Here T^* is called the *Hermitian adjoint of T*.

3.1.6 Problem Let V be an n-dimensional inner product space. Let $T : V \to V$ be a linear transformation. Then $(T^*)^* = T$.

Proof Let us take an arbitrary $v \in V$. We have to show that

$$(T^*)^*(v) = T(v).$$

To this end, let us take an arbitrary $u \in V$. It suffices to show that

$$\langle u, (T^*)^*(v) \rangle = \langle u, T(v) \rangle,$$
$$\text{LHS} = \langle u, (T^*)^*(v) \rangle = \langle T^*(u), v \rangle$$
$$= \overline{\langle v, T^*(u) \rangle} = \overline{\langle T(v), u \rangle} = \langle u, T(v) \rangle = \text{RHS}.$$

∎

3.1.7 Problem Let V be an n-dimensional inner product space. Let $S : V \to V$ and $T : V \to V$ be linear transformations. Let λ, μ be any complex numbers. Then $(\lambda S + \mu T)^* = \bar{\lambda} S^* + \bar{\mu} T^*$.

Proof Let us take an arbitrary $v \in V$. We have to show that

$$(\lambda S + \mu T)^*(v) = \left(\bar{\lambda} S^* + \bar{\mu} T^* \right)(v),$$

that is,

$$(\lambda S + \mu T)^*(v) = \bar{\lambda} S^*(v) + \bar{\mu} T^*(v).$$

To this end, let us take an arbitrary $u \in V$. It suffices to show that

$$\langle u, (\lambda S + \mu T)^*(v) \rangle = \langle u, \bar{\lambda} S^*(v) + \bar{\mu} T^*(v) \rangle :$$
$$\text{LHS} = \langle u, (\lambda S + \mu T)^*(v) \rangle = \langle (\lambda S + \mu T)(u), v \rangle$$
$$= \langle \lambda S(u) + \mu T(u), v \rangle = \lambda \langle S(u), v \rangle + \mu \langle T(u), v \rangle$$
$$= \lambda \langle u, S^*(v) \rangle + \mu \langle u, T^*(v) \rangle$$
$$= \langle u, \bar{\lambda} S^*(v) + \bar{\mu} T^*(v) \rangle = \text{RHS}.$$

∎

3.1.8 Problem Let V be an n-dimensional inner product space. Let $S : V \to V$ and $T : V \to V$ be linear transformations. Then $(ST)^* = T^* S^*$.

Proof Let us take an arbitrary $v \in V$. We have to show that

$$(ST)^*(v) = (T^* S^*)(v),$$

that is,

$$(ST)^*(v) = T^*(S^*(v)).$$

To this end, let us take an arbitrary $u \in V$. It suffices to show that

$$\langle u, (ST)^*(v) \rangle = \langle u, T^*(S^*(v)) \rangle:$$
$$\text{LHS} = \langle u, (ST)^*(v) \rangle = \langle (ST)(u), v \rangle$$
$$= \langle S(T(u)), v \rangle = \langle T(u), S^*(v) \rangle = \langle u, T^*(S^*(v)) \rangle = \text{RHS}.$$

∎

3.1.9 Problem Let V be an n-dimensional inner product space. Let $T : V \to V$ be a unitary linear transformation. Then $T^*T = I$.

Proof Let us take an arbitrary $v \in V$. We have to show that $T^*(T(v)) = v$. To this end, let us take an arbitrary $u \in V$. It suffices to show that $\langle u, T^*(T(v)) \rangle = \langle u, v \rangle$:

$$\text{RHS} = \langle u, v \rangle = \langle T(u), T(v) \rangle = \langle u, T^*(T(v)) \rangle = \text{LHS}.$$

∎

3.1.10 Problem Let V be an n-dimensional inner product space. Let $T : V \to V$ be a linear transformation such that $T^*T = I$. Then T is unitary.

Proof Let us take arbitrary $u, v \in V$. We have to show that $\langle T(u), T(v) \rangle = \langle u, v \rangle$:

$$\text{LHS} = \langle T(u), T(v) \rangle = \langle u, T^*(T(v)) \rangle = \langle u, (T^*T)(v) \rangle = \langle u, I(v) \rangle = \langle u, v \rangle = \text{RHS}.$$

∎

3.1.11 Theorem Let V be an n-dimensional inner product space. Let $T : V \to V$ be a linear transformation. Let $\{v_1, \ldots, v_n\}$ be an orthonormal basis of V. Let $\left[\alpha_{ij}\right]$ be the matrix of T relative to the basis $\{v_1, \ldots, v_n\}$, in the sense that

$$T(v_1) = a_{11}v_1 + a_{21}v_2 + \cdots + a_{n1}v_n \left(= \sum_{i=1}^{n} a_{i1}v_i\right),$$
$$T(v_2) = a_{12}v_1 + a_{22}v_2 + \cdots + a_{n2}v_n,$$
$$\vdots$$
$$T(v_n) = a_{1n}v_1 + a_{2n}v_2 + \cdots + a_{nn}v_n.$$

In short, $T(v_j) = \sum_{i=1}^{n} a_{ij}v_i$.

Then the matrix of T^* relative to the basis $\{v_1, \ldots, v_n\}$ is $\left[\beta_{ij}\right]$, where $\beta_{ij} = \overline{\alpha_{ji}}$. In short, $T^*(v_j) = \sum_{i=1}^{n} \beta_{ij}v_i$.

Proof By the proof of 3.1.5,

$$T^*(v_1) = \overline{\langle T(v_1), v_1 \rangle}\, v_1 + \cdots + \overline{\langle T(v_n), v_1 \rangle}\, v_n,$$
$$T^*(v_2) = \overline{\langle T(v_1), v_2 \rangle}\, v_1 + \cdots + \overline{\langle T(v_n), v_2 \rangle}\, v_n,$$
$$\vdots$$
$$T^*(v_n) = \overline{\langle T(v_1), v_n \rangle}\, v_1 + \cdots + \overline{\langle T(v_n), v_n \rangle}\, v_n.$$

Since

$$T^*(v_1) = \sum_{i=1}^{n} \overline{\langle T(v_i), v_1 \rangle}\, v_i$$

$$= \sum_{i=1}^{n} \overline{\langle a_{1i} v_1 + a_{2i} v_2 + \cdots + a_{ni} v_n, v_1 \rangle}\, v_i$$

$$= \sum_{i=1}^{n} \overline{a_{1i} \langle v_1, v_1 \rangle + a_{2i} \langle v_2, v_1 \rangle + \cdots + a_{ni} \langle v_n, v_1 \rangle}\, v_i$$

$$= \sum_{i=1}^{n} \overline{a_{1i} 1 + a_{i2} 0 + \cdots + a_{in} 0}\, v_i$$

$$= \sum_{i=1}^{n} \overline{a_{1i}}\, v_i = \overline{a_{11}} v_1 + \overline{a_{12}} v_2 + \cdots + \overline{a_{1n}} v_n,$$

we have

$$T^*(v_1) = \overline{a_{11}}\, v_1 + \overline{a_{12}}\, v_2 + \cdots + \overline{a_{1n}}\, v_n \left(= \sum_{i=1}^{n} \overline{a_{1i}}\, v_i \right),$$

Similarly,

$$T^*(v_2) = \overline{a_{21}}\, v_1 + \overline{a_{22}}\, v_2 + \cdots + \overline{a_{2n}}\, v_n,$$

etc. In short, $T^*(v_j) = \sum_{i=1}^{n} \overline{a_{ji}}\, v_i$. If the matrix of T^* relative to the basis $\{v_1, \ldots, v_n\}$ is $[\beta_{ij}]$, then $\beta_{ij} = \overline{\alpha_{ji}}$. ∎

3.1.12 Problem Let V be an n-dimensional inner product space. Let $T : V \to V$ be a unitary linear transformation. Let $\{v_1, \ldots, v_n\}$ be an orthonormal basis of V. Let $[\alpha_{ij}]$ be the matrix of T relative to the basis $\{v_1, \ldots, v_n\}$. Then $\sum_{j=1}^{n} \alpha_{ji} \overline{\alpha_{jk}} = \delta_{ik}$.

Proof It is given that

$$T(v_i) = \sum_{j=1}^{n} \alpha_{ji} v_j.$$

By 3.1.11,

$$T^*(v_j) = \sum_{k=1}^{n} \overline{\alpha_{jk}}\, v_k.$$

Since $T : V \to V$ is unitary, by 3.1.9, $T^*T = I$. It follows that

$$\sum_{k=1}^{n} \delta_{ki} v_k = v_i = \underbrace{I(v_i) = (T^*T)(v_i)}$$

$$= T^*(T(v_i)) = T^*\left(\sum_{j=1}^{n} \alpha_{ji} v_j\right) = \sum_{j=1}^{n} \alpha_{ji} T^*(v_j)$$

$$= \sum_{j=1}^{n} \alpha_{ji}\left(\sum_{k=1}^{n} \overline{\alpha_{jk}} v_k\right) = \sum_{j=1}^{n}\left(\sum_{k=1}^{n} \alpha_{ji}\overline{\alpha_{jk}} v_k\right)$$

$$= \sum_{k=1}^{n}\left(\sum_{j=1}^{n} \alpha_{ji}\overline{\alpha_{jk}} v_k\right) = \sum_{j=1}^{n}\left(\sum_{j=1}^{n} \alpha_{ji}\overline{\alpha_{jk}}\right) v_k,$$

and hence

$$\sum_{j=1}^{n} \alpha_{ji}\overline{\alpha_{jk}} = \delta_{ki}.$$

■

Definition Let V be an n-dimensional inner product space. Let $T : V \to V$ be a linear transformation. If $T^* = T$, then we say that T *is Hermitian*.

3.1.13 Problem Let V be an n-dimensional inner product space. Let $T : V \to V$ be a Hermitian linear transformation. Then all its eigenvalues are real.

Proof Let λ be an eigenvalue of T. We have to show that λ is real, that is, $\bar{\lambda} = \lambda$.

Since λ is an eigenvalue of T, there exists a nonzero $v \in V$ such that $T(v) = \lambda v$. Since $T : V \to V$ is Hermitian, we have $T^* = T$. Now,

$$\lambda\langle v, v\rangle = \langle \lambda v, v\rangle = \langle T(v), v\rangle = \underbrace{\langle v, T^*(v)\rangle = \langle v, T(v)\rangle} = \langle v, \lambda v\rangle = \bar{\lambda}\langle v, v\rangle,$$

so

$$(\bar{\lambda} - \lambda)\langle v, v\rangle = 0,$$

and hence, $\bar{\lambda} = \lambda$ or $\langle v, v\rangle = 0$. Since v is nonzero, $\langle v, v\rangle \neq 0$, and hence $\bar{\lambda} = \lambda$. ■

3.1.14 Problem Let V be an n-dimensional inner product space. Let $T : V \to V$ be a linear transformation. Then

$$(T^*T)(v) = 0 \Rightarrow T(v) = 0.$$

Proof Let $v \in V$ be such that $(T^*T)(v) = 0$. We have to show that $T(v) = 0$, that is, $\langle T(v), T(v) \rangle = 0$. Since $(T^*T)(v) = 0$, we have

$$\langle T(v), T(v) \rangle = \langle v, T^*(T(v)) \rangle = \underbrace{\langle v, (T^*T)(v) \rangle = 0},$$

and hence $\langle T(v), T(v) \rangle = 0$.   ■

3.1.15 Problem Let V be an n-dimensional inner product space. Let $T : V \to V$ be a Hermitian linear transformation. Let k be a positive integer. Then

$$T^k(v) = 0 \Rightarrow T(v) = 0.$$

Proof For $k = 1$, the theorem is trivial. So we consider the case $k = 2$. Since $T : V \to V$ is Hermitian, we have $T^* = T$. Now,

$$0 = \underbrace{T^k(v) = T^2(v)} = (TT)(v) = (T^*T)(v),$$

so $(T^*T)(v) = 0$. It follows from 3.1.14 that $T(v) = 0$.

Next, we consider the case $k = 3$. Here

$$0 = \underbrace{T^k(v) = T^3(v)} = (TTT)(v) = (T^*TT)(v) = (T^*T)(T(v)),$$

so $(T^*T)(T(v)) = 0$. It follows from 3.1.14 that $T(T(v)) = 0$, that is, $T^2(v) = 0$. Since the theorem has been proved for $k = 2$, we have $T(v) = 0$, etc.   ■

Definition Let V be an n-dimensional inner product space. Let $T : V \to V$ be a linear transformation. If $T^*T = TT^*$, then we say that *T is normal*.

3.1.16 Theorem Let V be an n-dimensional inner product space. Let $T : V \to V$ be a unitary linear transformation. Then T is normal.

Proof We have to show that $T^*T = TT^*$. Since $T : V \to V$ is unitary, by 3.1.9, $T^*T = I$. It follows that $T^{-1} = T^*$:

$$\text{LHS} = T^*T = T^{-1}T = I = TT^{-1} = TT^* = \text{RHS}.$$

                                                                              ■

3.1.17 Problem Let V be an n-dimensional inner product space. Let $T : V \to V$ be a Hermitian linear transformation. Then T is normal.

Proof We have to show that $T^*T = TT^*$. Since $T : V \rightarrow V$ is Hermitian, $T^* = T$:

$$\text{LHS} = T^*T = TT = TT^* = \text{RHS}.$$

∎

Definition Let V be an n-dimensional inner product space. Let $T : V \rightarrow V$ be a linear transformation. If $T^* = -T$, then we say that T *is skew-Hermitian*.

3.1.18 Problem Let V be an n-dimensional inner product space. Let $T : V \rightarrow V$ be a skew-Hermitian linear transformation. Then T is normal.

Proof We have to show that $T^*T = TT^*$. Since $T : V \rightarrow V$ is skew-Hermitian, $T^* = -T$:

$$\text{LHS} = T^*T = (-T)T = T(-T) = TT^* = \text{RHS}.$$

∎

3.1.19 Problem Let V be an n-dimensional inner product space. Let $T : V \rightarrow V$ be a normal linear transformation. Then

$$T(v) = 0 \Rightarrow T^*(v) = 0.$$

Proof Let $v \in V$ be such that $T(v) = 0$. We have to show that $T^*(v) = 0$, that is, $\langle T^*(v), T^*(v) \rangle = 0$, that is, $\langle T(T^*(v)), v \rangle = 0$, that is, $\langle (TT^*)(v), v \rangle = 0$. Since $T : V \rightarrow V$ is normal, we have $T^*T = TT^*$. It suffices to show that $\langle (T^*T)(v), v \rangle = 0$:

$$\text{LHS} = \langle (T^*T)(v), v \rangle = \langle T^*(T(v)), v \rangle = \langle T^*(0), v \rangle = \langle 0, v \rangle = 0 = \text{RHS}.$$

∎

3.1.20 Problem Let V be an n-dimensional inner product space. Let $T : V \rightarrow V$ be a normal linear transformation. Let λ be any complex number. Then $(T - \lambda I) : V \rightarrow V$ is a normal linear transformation.

Proof By 3.1.7, $(T - \lambda I)^* = T^* - \bar{\lambda}I^*$. Since for every $u, v \in V$, $\langle u, I^*(v) \rangle = \langle I(u), v \rangle = \langle u, v \rangle = \langle u, I(v) \rangle$, we have, for every $u, v \in V$, $\langle u, I^*(v) \rangle = \langle u, I(v) \rangle$. It follows that $I^* = I$. Thus $(T - \lambda I)^* = T^* - \bar{\lambda}I$. It suffices to show that

$$(T^* - \bar{\lambda}I)(T - \lambda I) = (T - \lambda I)(T^* - \bar{\lambda}I),$$

that is,

$$T^*T - \lambda T^* - \bar{\lambda}T + |\lambda|^2 I = TT^* - \bar{\lambda}T - \lambda T^* + |\lambda|^2 I,$$

that is,

$$T^*T = TT^*.$$

This is known to be true, because $T : V \to V$ is normal. ∎

3.1.21 Problem Let V be an n-dimensional inner product space. Let $T : V \to V$ be a normal linear transformation. Let λ be an eigenvalue of T. Let v be an eigenvector belonging to λ, in the sense that v is nonzero, and $T(v) = \lambda v$. Then $T^*(v) = \bar{\lambda}v$.

Proof It suffices to show that $(T^* - \bar{\lambda}I)(v) = 0$, that is, $(T^* - \bar{\lambda}I^*)(v) = 0$, that is, $(T - \lambda I)^*(v) = 0$.

Since $T(v) = \lambda v$, we have $(T - \lambda I)(v) = 0$. By 3.1.20, $(T - \lambda I) : V \to V$ is a normal linear transformation. Now, since $(T - \lambda I)(v) = 0$, by 3.1.19, $(T - \lambda I)^*(v) = 0$. ∎

3.1.22 Problem Let V be an n-dimensional inner product space. Let $T : V \to V$ be a unitary linear transformation. Let λ be an eigenvalue of T. Then $|\lambda| = 1$.

Proof Since T is unitary, by 3.1.16, T is normal. Since λ is an eigenvalue of T, there exists a nonzero $v \in V$ such that $T(v) = \lambda v$. Now, by 3.1.21, $T^*(v) = \bar{\lambda}v$. Since T is unitary, by 3.1.9, $T^*T = I$, and hence

$$|\lambda|^2 v = (\lambda\bar{\lambda})v = \lambda(\bar{\lambda}v) = \lambda(T^*(v)) = T^*(\lambda v) = T^*(T(v)) = \underbrace{(T^*T)(v) = I(v)} = v.$$

It follows that $\left(|\lambda|^2 - 1\right)v = 0$. Now since v is nonzero, we have $|\lambda|^2 - 1 = 0$, that is, $|\lambda| = 1$. ∎

3.1.23 Problem Let V be an n-dimensional inner product space. Let $T : V \to V$ be a normal linear transformation. Let k be a positive integer. Then

$$T^k(v) = 0 \Rightarrow T(v) = 0.$$

Proof For $k = 1$, the theorem is trivial. So we consider the case $k = 2$. Since $T : V \to V$ is normal, we have $TT^* = T^*T$. Since

$$(T^*T)^* = T^*(T^*)^* = T^*T,$$

we have $(T^*T)^* = T^*T$, and hence T^*T is a Hermitian linear transformation. Now suppose that $T^k(v) = 0$, that is, $T^2(v) = 0$. We have to show that $T(v) = 0$. Since

$$(T^*T)^2(v) = ((T^*T)(T^*T))(v)$$
$$= (T^*(TT^*)T)(v) = (T^*(T^*T)T)(v)$$
$$= (T^*T^*)((TT)(v)) = (T^*T^*)(T^2(v))$$
$$= (T^*T^*)(0) = 0,$$

we have $(T^*T)^2(v) = 0$. Now, since T^*T is Hermitian, by 3.1.15, $(T^*T)(v) = 0$, and hence by 3.1.14, $T(v) = 0$.

Next, we consider the case $k = 3$. Suppose that $T^k(v) = 0$, that is, $T^3(v) = 0$. We have to show that $T(v) = 0$. Since T is normal, we have

$$(T^*T)^3(v) = ((T^*T)(T^*T)(T^*T))(v) = \left((T^*)^3 T^3\right)(v) = (T^*)^3(T^3(v)) = (T^*)^3(0)$$
$$= 0,$$

and hence $(T^*T)^3(v) = 0$. Now, since T^*T is Hermitian, by 3.1.15, $(T^*T)(v) = 0$, and hence by 3.1.14, $T(v) = 0$, etc. ∎

3.1.24 Problem Let V be an n-dimensional inner product space. Let $T : V \to V$ be a normal linear transformation. Let λ be any complex number. Let k be a positive integer. Then

$$(T - \lambda I)^k(v) = 0 \Rightarrow T(v) = \lambda v.$$

Proof By 3.1.20, $(T - \lambda I) : V \to V$ is a normal linear transformation. Now, by 3.1.23,

$$(T - \lambda I)^k(v) = 0 \Rightarrow (T - \lambda I)(v) = 0,$$

and hence

$$(T - \lambda I)^k(v) = 0 \Rightarrow T(v) = \lambda v.$$

∎

3.1.25 Problem Let V be an n-dimensional inner product space. Let $T : V \to V$ be a normal linear transformation. Let λ, μ be two distinct eigenvalues of T. Let $v, w \in V$ be such that $T(v) = \lambda v$ and $T(w) = \mu w$. Then $\langle v, w \rangle = 0$.

Proof If $v = 0$ or $w = 0$, then the theorem is trivial. So we consider the case that v and w both are nonzero. Since μ is an eigenvalue of T and $T(w) = \mu w$, w is an eigenvector belonging to μ. It follows, from 3.1.21, that $T^*(w) = \bar{\mu} w$. Hence

$$\mu\langle v, w \rangle = \underbrace{\langle v, \bar{\mu} w \rangle = \langle v, T^*(w) \rangle} = \langle T(v), w \rangle = \langle \lambda v, w \rangle = \lambda\langle v, w \rangle.$$

Thus

$$(\mu - \lambda)\langle v, w \rangle = 0.$$

Now, since $\mu \neq \lambda$, we have $\langle v, w \rangle = 0$.   ∎

Definition Let A be a ring. Let A be a vector space over the field $\mathbb{C}$ of complex numbers. If for every $a, b \in A$, and, for every complex number α,

$$\alpha(ab) = (\alpha a)b = a(\alpha b),$$

then we say that A *is an algebra.*

Let V be any vector space over the field $\mathbb{C}$ of complex numbers. Let $A(V)$ be the collection of all linear transformations from V to V. We know that $A(V)$ is an algebra with unit element I. If dim $V = n$, then dim $A(V) = n^2$.

Definition Let A be an algebra with unit element e. Let

$$p(x) \equiv \alpha_0 + \alpha_1 x + \cdots + \alpha_n x^n$$

be any polynomial in x with complex coefficients α_i. Let $a \in A$. By *a satisfies $p(x)$*, we mean

$$\alpha_0 e + \alpha_1 a + \cdots + \alpha_n a^n = 0. \text{ (In short, } p(a) = 0.)$$

3.1.26 Problem Let A be an algebra with unit element e. Let m be the dimension of A. Let $a \in A$. Then there exists a nontrivial polynomial $p(x)$ such that a satisfies $p(x)$. Also, the degree of $p(x)$ is not greater than m.

Proof If any two of $e, a, a^2, \ldots, a^m$ are equal, say $a^2 = a^5$, then the polynomial $1x^2 + (-1)x^5$ serves the purpose of $p(x)$.

Finally, we consider the case that $e, a, a^2, \ldots, a^m$ are $(m+1)$ distinct members of A. Since m is the dimension of A, $e, a, a^2, \ldots, a^m$ are linearly dependent, and hence there exist complex numbers $\alpha_0, \alpha_1, \ldots, \alpha_m$, not all zero, such that

$$\alpha_0 e + \alpha_1 a + \cdots + \alpha_m a^m = 0.$$

It follows that the polynomial $\alpha_0 + \alpha_1 x + \cdots + \alpha_m x^m$ serves the purpose of $p(x)$. Here, the degree of $p(x)$ is not greater than m.   ∎

3.1.27 Problem Let V be any n-dimensional vector space. Let $T \in A(V)$. Then there exists a nontrivial polynomial $p(x)$ of degree $\leq n^2$ such that $p(T) = 0$.

Proof We know that dim $A(V) = n^2$, so by 3.1.26, there exists a nontrivial polynomial $p(x)$ of degree $\leq n^2$ such that $p(T) = 0$.   ∎

Definition Let V be any n-dimensional vector space. Let $T \in A(V)$. A nontrivial polynomial $p(x)$ of lowest degree such that $p(T) = 0$ is called a *minimal polynomial of T.*

If $p(x)$ is a minimal polynomial of T, and T satisfies another polynomial $h(x)$, then $p(x)$ divides $h(x)$.

3.1.28 Problem Let V be any n-dimensional vector space. Let $T \in A(V)$. Let T be invertible. Suppose that $\alpha_0 + \alpha_1 x + \cdots + \alpha_m x^m$ is a minimal polynomial of T, where $\alpha_m \neq 0$. Then $a_0 \neq 0$.

Proof Suppose to the contrary that $a_0 = 0$. We seek a contradiction.

Since $\alpha_1 x + \cdots + \alpha_m x^m$ is a minimal polynomial of T, we have

$$\alpha_1 T + \cdots + \alpha_m T^m = 0,$$

and hence

$$\alpha_1 I + \alpha_2 T \cdots + \alpha_m T^{m-1} = \underbrace{T(\alpha_1 T + \cdots + \alpha_m T^m) = 0T = 0.}$$

Thus

$$\alpha_1 I + \alpha_2 T \cdots + \alpha_m T^{m-1} = 0.$$

Hence $h(T) = 0$, where $h(x) \equiv \alpha_1 + \alpha_2 x \cdots + \alpha_m x^{m-1}$. Now, since $\alpha_m \neq 0$,

$$\underbrace{\deg h(x) = m - 1} < m = \deg(\alpha_1 x + \cdots + \alpha_m x^m).$$

Since $\deg h(x) < \deg(\alpha_1 x + \cdots + \alpha_m x^m)$ and $\alpha_1 x + \cdots + \alpha_m x^m$ is a minimal polynomial of T, we have $h(T) \neq 0$. This is a contradiction. ∎

3.1.29 Problem Let V be any n-dimensional vector space. Let $T \in A(V)$. Suppose that $\alpha_0 + \alpha_1 x + \cdots + \alpha_m x^m$ is a minimal polynomial of T, where $\alpha_m \neq 0$ and $a_0 \neq 0$. Then T^{-1} exists.

Proof Since $\alpha_0 + \alpha_1 x + \cdots + \alpha_m x^m$ is a minimal polynomial of T, we have

$$\alpha_0 I + \alpha_1 T + \cdots + \alpha_m T^m = 0.$$

It follows that

$$I = \frac{-\alpha_1}{\alpha_0} T + \frac{-\alpha_2}{\alpha_0} T^2 + \cdots + \frac{-\alpha_m}{\alpha_0} T^m,$$

or

$$T\left(\frac{-\alpha_1}{\alpha_0} + \frac{-\alpha_2}{\alpha_0}T + \cdots + \frac{-\alpha_m}{\alpha_0}T^{m-1}\right) = I.$$

This shows that $\frac{-\alpha_1}{\alpha_0} + \frac{-\alpha_2}{\alpha_0}T + \cdots + \frac{-\alpha_m}{\alpha_0}T^{m-1}$ is the inverse of T.   ∎

3.1.30 Problem Let V be any n-dimensional vector space. Let $T \in A(V)$. Suppose that T^{-1} does not exist. Then there exists a nonzero $S \in A(V)$ such that $ST = TS = 0$.

Proof Suppose that $\alpha_0 + \alpha_1 x + \cdots + \alpha_m x^m$ is a minimal polynomial of T, where $\alpha_m \neq 0$. By 3.1.29, $a_0 = 0$. Hence

$$\alpha_1 T + \cdots + \alpha_m T^m = 0,$$

or

$$\left(\alpha_1 I + \alpha_2 T + \cdots + \alpha_m T^{m-1}\right)T = T\left(\alpha_1 I + \alpha_2 T + \cdots + \alpha_m T^{m-1}\right) = 0.$$

Thus $ST = TS = 0$, where $S \equiv \alpha_1 I + \alpha_2 T + \cdots + \alpha_m T^{m-1} (\in A(V))$. Since $\alpha_0 + \alpha_1 x + \cdots + \alpha_m x^m$ is a minimal polynomial of T and $\alpha_m \neq 0$, we have $\alpha_1 I + \alpha_2 T + \cdots + \alpha_m T^{m-1} \neq 0$, and hence $S \neq 0$.   ∎

3.1.31 Problem Let V be any n-dimensional vector space. Let $T \in A(V)$. Suppose that T^{-1} does not exist. Then there exists a nonzero $v \in V$ such that $T(v) = 0$.

Proof By 3.1.30, there exists a nonzero $S \in A(V)$ such that $TS = 0$. Since S is nonzero, there exists $u \in V$ such that $S(u) \neq 0$. Now, since $TS = 0$, we have

$$T(S(u)) = \underbrace{(TS)(u)}_{} = 0(u) = 0,$$

and hence $T(v) = 0$, where $v \equiv S(u) (\neq 0)$.   ∎

3.1.32 Problem Let V be any n-dimensional vector space. Let $T \in A(V)$. Suppose that there exists a nonzero $v \in V$ such that $T(v) = 0$. Then T^{-1} does not exist.

Proof Suppose to the contrary that T^{-1} exists. We seek a contradiction.

Suppose that $\alpha_0 + \alpha_1 x + \cdots + \alpha_m x^m$ is a minimal polynomial of T, where $\alpha_m \neq 0$. Now, by 3.1.28, $\alpha_0 \neq 0$. Since $\alpha_0 + \alpha_1 x + \cdots + \alpha_m x^m$ is a minimal polynomial of T and $\alpha_m \neq 0$, we have

$$\alpha_0 I + \alpha_1 T + \cdots + \alpha_m T^m = 0.$$

It follows that

$$\alpha_0 v = \alpha_0 v + \alpha_1 0 + \cdots + \alpha_m 0 = \alpha_0 I(v) + \alpha_1 T(v) + \cdots + \alpha_m T^m(v)$$
$$= \underbrace{(\alpha_0 I + \alpha_1 T + \cdots + \alpha_m T^m)(v) = 0(v)} = 0,$$

and hence $\alpha_0 v = 0$. Now, since v is nonzero, $\alpha_0 = 0$. This is a contradiction. ∎

3.1.33 Problem Let V be any n-dimensional vector space. Let $S, T \in A(V)$. Let $v_1, \ldots, v_n$ be any basis of V. Let $m(S)$ be the matrix of S relative to the basis $v_1, \ldots, v_n$, in the sense that $m(S) = [a_{ij}]_{n \times n}$, where $S(v_j) \equiv \sum_{i=1}^{n} a_{ij} v_i$. Let $m(T)$ be the matrix of T relative to the basis $v_1, \ldots, v_n$. Then

$$m(ST) = m(S)m(T).$$

Proof Let $m(T) = [b_{ij}]_{n \times n}$, where $T(v_j) \equiv \sum_{i=1}^{n} b_{ij} v_i$. It suffices to show that

$$(ST)(v_j) = \sum_{k=1}^{n} \left(\sum_{i=1}^{n} a_{ki} b_{ij} \right) v_k :$$

$$\text{LHS} = (ST)(v_j) = S(T(v_j)) = S\left(\sum_{i=1}^{n} b_{ij} v_i \right)$$

$$= \sum_{i=1}^{n} b_{ij} S(v_i) = \sum_{i=1}^{n} b_{ij} \left(\sum_{k=1}^{n} a_{ki} v_k \right)$$

$$= \sum_{i=1}^{n} \left(\sum_{k=1}^{n} b_{ij} a_{ki} v_k \right) = \sum_{k=1}^{n} \left(\sum_{i=1}^{n} b_{ij} a_{ki} v_k \right) = \sum_{k=1}^{n} \left(\sum_{i=1}^{n} b_{ij} a_{ki} \right) v_k$$

$$= \sum_{k=1}^{n} \left(\sum_{i=1}^{n} a_{ki} b_{ij} \right) v_k = \text{RHS}. \qquad ∎$$

3.1.34 Problem Let V be any n-dimensional vector space. Let $T \in A(V)$. Let $v_1, \ldots, v_n$ be any basis of V. Suppose that T^{-1} exists. Let $m(T)$ be the matrix of T relative to the basis $v_1, \ldots, v_n$. Then

$$m(T^{-1}) = (m(T))^{-1}.$$

Proof Here, it suffices to show that $m(T^{-1})m(T) = [\delta_{ij}]_{n \times n}$. Since $T^{-1}T = I$, by Problem 3.1.33, we have

$$[\delta_{ij}]_{n \times n} = m(I) = \underbrace{m(T^{-1}T) = m(T^{-1})m(T)},$$

and hence $m(T^{-1})m(T) = [\delta_{ij}]_{n \times n}$. ∎

3.1.35 Theorem

a. Let V be any n-dimensional vector space. Let $T \in A(V)$. Let $v_1, \ldots, v_n$ and $w_1, \ldots, w_n$ be any two bases of V. Let $m_1(T)$ be the matrix of T relative to the basis $v_1, \ldots, v_n$. Let $m_2(T)$ be the matrix of T relative to the basis $w_1, \ldots, w_n$. Let $S : V \to V$ be the linear transformation such that for every $i \in \{1, \ldots, n\}$, $S(v_i) = w_i$. Then

$$m_2(T) = (m_1(S))^{-1} m_1(T) m_1(S).$$

b. Let V be any n-dimensional vector space. Let $T \in A(V)$. Let $v_1, \ldots, v_n$ be any basis of V. Let $A \equiv [a_{ij}]_{n \times n}$ be the matrix of T relative to the basis $v_1, \ldots, v_n$. Let $P \equiv [p_{ij}]_{n \times n}$ be any invertible matrix. Then there exists a basis $w_1, \ldots, w_n$ of V such that $P^{-1}AP$ is the matrix of T relative to the basis $w_1, \ldots, w_n$.

Proof (a) Let $m_1(T) = [a_{ij}]_{n \times n}$, where $T(v_j) \equiv \sum_{i=1}^{n} a_{ij} v_i$. Let $m_2(T) = [b_{ij}]_{n \times n}$, where $T(w_j) \equiv \sum_{i=1}^{n} b_{ij} w_i$.

Clearly, S is invertible, that is, S^{-1} exists.

Proof Suppose that $S(v) = 0$. It suffices to show that $v = 0$. There exist scalars $\alpha_1, \ldots, \alpha_n$ such that
$v = \alpha_1 v_1 + \cdots + \alpha_n v_n$. It follows that

$$0 = \underbrace{S(v) = S(\alpha_1 v_1 + \cdots + \alpha_n v_n)}_{} = \alpha_1 S(v_1) + \cdots + \alpha_n S(v_n)$$

$$= \alpha_1 w_1 + \cdots + \alpha_n w_n,$$

so $\alpha_1 w_1 + \cdots + \alpha_n w_n = 0$. Now, since $w_1, \ldots, w_n$ is a basis of V, each α_i is 0, and hence
$v = \underbrace{\alpha_1 v_1 + \cdots + \alpha_n v_n = 0}_{}$. Thus $v = 0$. ∎

Since for every $j \in \{1, \ldots, n\}$,

$$T(w_j) = \sum_{i=1}^{n} b_{ij} w_i,$$

we have, for every $j \in \{1, \ldots, n\}$,

$$(TS)(v_j) = T(S(v_j)) = \underbrace{T(w_j) = \sum_{i=1}^{n} b_{ij} w_i}_{} = \sum_{i=1}^{n} b_{ij} S(v_i) = S\left(\sum_{i=1}^{n} b_{ij} v_i\right),$$

and hence for every $j \in \{1, \ldots, n\}$,

$$(TS)(v_j) = S\left(\sum_{i=1}^{n} b_{ij}v_i\right).$$

It follows that for every $j \in \{1, \ldots, n\}$,

$$\left(S^{-1}TS\right)(v_j) = S^{-1}\left((TS)(v_j)\right) = \underbrace{\sum_{i=1}^{n} b_{ij}v_i}_{}.$$

Thus for every $j \in \{1, \ldots, n\}$,

$$\left(S^{-1}TS\right)(v_j) = \sum_{i=1}^{n} b_{ij}v_i.$$

It follows that

$$\underbrace{m_1\left(S^{-1}TS\right) = \left[b_{ij}\right]}_{} = m_2(T),$$

where $m_1\left(S^{-1}TS\right)$ is the matrix of $S^{-1}TS$ relative to the basis $v_1, \ldots, v_n$. Thus

$$m_2(T) = m_1\left(S^{-1}TS\right).$$

By 3.1.33, and 3.1.34,

$$m_2(T) = \underbrace{m_1\left(S^{-1}TS\right) = m_1\left(S^{-1}\right)m_1(T)m_1(S)}_{} = (m_1(S))^{-1}m_1(T)m_1(S),$$

and hence

$$m_2(T) = (m_1(S))^{-1}m_1(T)m_1(S).$$

Proof **(b)** Put $w_j \equiv \sum_{i=1}^{n} p_{ij}v_i$ $(j = 1, \ldots, n)$.
  Clearly, $\{w_1, \ldots, w_n\}$ is linearly independent.

  Proof To show this, suppose that $\alpha_1 w_1 + \cdots + \alpha_n w_n = 0$. We have to show that each α_i equals 0. Since

$$\sum_{k=1}^{n}\left(\sum_{i=1}^{n}p_{ki}\alpha_i\right)v_k = \sum_{k=1}^{n}\left(\sum_{i=1}^{n}\alpha_i p_{ki}\right)v_k = \sum_{k=1}^{n}\left(\sum_{i=1}^{n}\alpha_i p_{ki}v_k\right)$$

$$= \sum_{i=1}^{n}\left(\sum_{k=1}^{n}\alpha_i p_{ki}v_k\right) = \sum_{i=1}^{n}\alpha_i\left(\sum_{k=1}^{n}p_{ki}v_k\right) = \underbrace{\sum_{i=1}^{n}\alpha_i w_i}_{} = 0,$$

we have

$$\sum_{k=1}^{n}\left(\sum_{i=1}^{n}p_{ki}\alpha_i\right)v_k = 0.$$

Since $v_1, \ldots, v_n$ is a basis of V, we have, for every $k \in \{1, \ldots, n\}$,

$$\sum_{i=1}^{n}p_{ki}\alpha_i = 0,$$

and hence $\left[p_{ij}\right]_{n\times n}\begin{bmatrix}\alpha_1 \\ \vdots \\ \alpha_n\end{bmatrix} = \begin{bmatrix}0 \\ \vdots \\ 0\end{bmatrix}$, that is, $P[\alpha_1, \ldots, \alpha_n]^T = [0, \ldots, 0]^T$. It follows that

$$[\alpha_1, \ldots, \alpha_n]^T = I[\alpha_1, \ldots, \alpha_n]^T = \left(P^{-1}P\right)[\alpha_1, \ldots, \alpha_n]^T$$

$$= \underbrace{P^{-1}\left(P[\alpha_1, \ldots, \alpha_n]^T\right) = P^{-1}[0, \ldots, 0]^T = [0, \ldots, 0]^T}_{},$$

and hence $[\alpha_1, \ldots, \alpha_n]^T = [0, \ldots, 0]^T$. This shows that each α_i is 0. ∎

Thus, we have shown that $\{w_1, \ldots, w_n\}$ is linearly independent. Now, since V is an n-dimensional vector space, $w_1, \ldots, w_n$ is a basis of V. It remains to show that $P^{-1}AP$ is the matrix of T relative to the basis $w_1, \ldots, w_n$.

Let $S : V \to V$ be the linear transformation defined as follows: for every $i \in \{1, \ldots, n\}$, $S(v_i) = w_i$. Since

$$\underbrace{S(v_i) = w_i}_{} = \sum_{k=1}^{n}p_{ki}v_k,$$

we have $S(v_i) = \sum_{k=1}^{n}p_{ki}v_k$, and hence the matrix of S relative to the basis $v_1, \ldots, v_n$ is $[p_{ki}] (= P)$. Now, by 3.1.35(a), $P^{-1}AP$ is the matrix of T relative to the basis $w_1, \ldots, w_n$. ∎

3.2   Canonical Forms

Definition Let V be any n-dimensional vector space. Let $T \in A(V)$. Let V_1 be any subspace of V. If $T(V_1) \subset V_1$, then we say that V_1 is *invariant under T*.

3.2.1 Theorem Let V be any n-dimensional vector space. Let $T \in A(V)$. Let V_1 and V_2 be any subspaces of V. Suppose that $V = V_1 \oplus V_2$, in the sense that every $v \in V$ can be expressed uniquely as $v_1 + v_2$, where $v_1 \in V_1$ and $v_2 \in V_2$. Suppose that V_1 is invariant under T, and V_2 is invariant under T, in the sense that the restriction $T|_{V_1}$ is in $A(V_1)$ and the restriction $T|_{V_2}$ is in $A(V_2)$. Let $p_1(x)$ be a minimal polynomial of $T|_{V_1}$, and $p_2(x)$ a minimal polynomial of $T|_{V_2}$. Then the least common multiple of $p_1(x)$ and $p_2(x)$ is a minimal polynomial of T.

Proof Let $p(x)$ be a minimal polynomial of T. It suffices to show that

1. $p_1(x)$ divides $p(x)$, that is, $p\left(T|_{V_1}\right) = 0$,
2. $p_2(x)$ divides $p(x)$, that is, $p\left(T|_{V_2}\right) = 0$,
3. if $p_1(x)$ divides $q(x)$, and $p_2(x)$ divides $q(x)$, then $p(x)$ divides $q(x)$, that is,

$$(p_1(x)|q(x) \text{ and } p_2(x)|q(x)) \Rightarrow q(T) = 0.$$

For 1: Since $p_1(x)$ is a minimal polynomial of T_1, we have $p_1(T_1) = 0$.
Clearly, for every polynomial $q(x)$, $q(T)|_{V_1} = q\left(T|_{V_1}\right)$.

Proof Suppose that

$$q(x) \equiv \alpha_0 + \alpha_1 x + \cdots + \alpha_m x^m$$

and $v \in V_1$. We have to show that

$$\alpha_0 I(v) + \alpha_1 T(v) + \alpha_2 T(T(v)) \cdots + \alpha_m T^m(v)$$
$$= \alpha_0 \left(I|_{V_1}\right)(v) + \alpha_1 \left(T|_{V_1}\right)(v) + \alpha_2 \left(T|_{V_1}\right)\left(\left(T|_{V_1}\right)(v)\right) \cdots + \alpha_m \left(T|_{V_1}\right)^m(v):$$
$$\text{RHS} = \alpha_0 \left(I|_{V_1}\right)(v) + \alpha_1 \left(T|_{V_1}\right)(v) + \alpha_2 \left(T|_{V_1}\right)\left(\left(T|_{V_1}\right)(v)\right) \cdots + \alpha_m \left(T|_{V_1}\right)^m(v)$$
$$= \alpha_0 v + \alpha_1 T(v) + \alpha_2 \left(T|_{V_1}\right)(T(v)) \cdots + \alpha_m \left(T|_{V_1}\right)^m(v)$$
$$= \alpha_0 v + \alpha_1 T(v) + \alpha_2 T(T(v)) \cdots + \alpha_m \left(T|_{V_1}\right)^m(v)$$
$$= \alpha_0 I(v) + \alpha_1 T(v) + \alpha_2 T(T(v)) \cdots + \alpha_m \left(T|_{V_1}\right)^m(v)$$
$$\vdots$$
$$= \alpha_0 I(v) + \alpha_1 T(v) + \alpha_2 T(T(v)) \cdots + \alpha_m T^m(v) = \text{LHS}.$$

∎

Similarly, for every polynomial $q(x)$, we have $q(T)|_{V_2} = q\left(T|_{V_2}\right)$. Since $p(x)$ is a minimal polynomial of T, we have $p(T) = 0$, and hence

$$p\left(T|_{V_1}\right) = \underbrace{p(T)|_{V_1} = 0}.$$

Thus $p\left(T|_{V_1}\right) = 0$.

For 2: The proof is similar to (1).

For 3: Suppose that $p_1(x)$ divides $q(x)$ and $p_2(x)$ divides $q(x)$. We have to show that $q(T) = 0$.

To this end, let us take an arbitrary $v \in V$. We have to show that $(q(T))(v) = 0$. Since $v \in V$ and $V = V_1 \oplus V_2$, there exist $v_1 \in V_1$ and $v_2 \in V_2$ such that $v = v_1 + v_2$. We have to show that

$$(q(T))(v_1) + (q(T))(v_2) = \underbrace{(q(T))(v_1 + v_2) = 0},$$

that is,

$$(q(T))(v_1) + (q(T))(v_2) = 0.$$

It suffices to show that

$$\left.\begin{array}{l} q(T)|_{V_1} = 0 \\ q(T)|_{V_2} = 0 \end{array}\right\},$$

that is,

$$\left.\begin{array}{l} q\left(T|_{V_1}\right) = 0 \\ q\left(T|_{V_2}\right) = 0 \end{array}\right\}.$$

Since $p_1(x)$ is a minimal polynomial of $T|_{V_1}$, we have $p_1\left(T|_{V_1}\right) = 0$. Now, since $p_1(x)$ divides $q(x)$, we have $q\left(T|_{V_1}\right) = 0$. Similarly, $q\left(T|_{V_2}\right) = 0$. ∎

3.2.2 Note Let V be any n-dimensional vector space over the field F. Let $T \in A(V)$. Let $p(x) \in F[x]$. Let $p(x)$ be the minimal polynomial of T over F.

By 1.3.19, we can write

$$p(x) = (q_1(x))^{l_1}(q_2(x))^{l_2} \ldots (q_k(x))^{l_k},$$

where $q_1(x), q_2(x), \ldots, q_k(x)$ are distinct irreducible polynomials (of degree ≥ 1) in $F[x]$, and $l_1, l_2, \ldots, l_k$ are positive integers.

Put

$$V_1 \equiv \left\{ v : v \in V \text{ and } \left((q_1(T))^{l_1} \right)(v) = 0 \right\},$$
$$V_2 \equiv \left\{ v : v \in V \text{ and } \left((q_2(T))^{l_2} \right)(v) = 0 \right\}, \text{ etc.}$$

It is clear that each V_i is a linear subspace of V. It follows that

$$V_1 + V_2 + \cdots + V_k \subset V. \quad (*)$$

Next suppose that $v \in V_1$. It follows that $\left((q_1(T))^{l_1} \right)(v) = 0$. Now

$$\left((q_1(T))^{l_1} \right)(T(v)) = \left((q_1(T))^{l_1} \circ T \right)(v) = \left(T \circ (q_1(T))^{l_1} \right)(v)$$
$$= \underbrace{T \left((q_1(T))^{l_1}(v) \right) = T(0) = 0,}$$

so

$$\left((q_1(T))^{l_1} \right)(T(v)) = 0,$$

and hence $T(v) \in V_1$. This shows that V_1 is invariant under T. Similarly, V_2 is invariant under T, etc.

We assume that $k > 1$. Put

$$h_1(x) \equiv (q_2(x))^{l_2} (q_3(x))^{l_3} \cdots (q_k(x))^{l_k},$$
$$h_2(x) \equiv (q_1(x))^{l_1} (q_3(x))^{l_3} \cdots (q_k(x))^{l_k},$$

$$\vdots$$

$$h_k(x) \equiv (q_1(x))^{l_1} (q_2(x))^{l_2} \cdots (q_{k-1}(x))^{l_{k-1}}.$$

Observe that the greatest common divisor of $h_1(x), h_2(x), \ldots, h_k(x)$ is 1. So by 1.1.5, there exist $a_1(x), a_2(x), \ldots, a_k(x) \in F[x]$ such that

$$1 = h_1(x)a_1(x) + h_2(x)a_2(x) + \cdots + h_k(x)a_k(x),$$

and hence

$$I = (h_1(T)) \circ (a_1(T)) + (h_2(T)) \circ (a_2(T)) + \cdots + (h_k(T)) \circ (a_k(T)).$$

It follows that for every $v \in V$, we have

$$v = I(v) = ((h_1(T)) \circ (a_1(T)) + (h_2(T)) \circ (a_2(T)) + \cdots + (h_k(T)) \circ (a_k(T)))(v)$$

$$= (h_1(T))((a_1(T))(v)) + (h_2(T))((a_2(T))(v)) + \cdots + (h_k(T))((a_k(T))(v)),$$

and hence for every $v \in V$, there exist $w_1, w_2, \ldots, w_k \in V$ such that

$$v = (h_1(T))(w_1) + (h_2(T))(w_2) + \cdots + (h_k(T))(w_k). \quad (**)$$

Since $q_1(x)$ is a polynomial of degree ≥ 1 in $F[x]$ and l_1 is a positive integer, we have

$$\deg(h_1(x)) = \deg\left((q_2(x))^{l_2}(q_3(x))^{l_3}\cdots(q_k(x))^{l_k}\right)$$
$$= \underbrace{l_2\deg(q_2(x)) + \cdots + l_k\deg(q_k(x))}_{} < \underbrace{l_1\deg(q_1(x)) + \cdots + l_k\deg(q_k(x))}_{}$$

$$= \deg\left((q_1(x))^{l_1}(q_2(x))^{l_2}\cdots(q_k(x))^{l_k}\right) = \deg(p(x)),$$

and hence $\deg(h_1(x)) < \deg(p(x))$. Now since $p(x)$ is the minimal polynomial of T over F, we have $p(T) = 0$ and $h_1(T) \neq 0$. Since $h_1(T) \neq 0$ and $h_1(T) : V \to V$, there exists $v_1 \in V$ such that $(h_1(T))(v_1) \neq 0$. Since $p(x) = (q_1(x))^{l_1}h_1(x)$, we have $0 = \underbrace{p(T) = (q_1(T))^{l_1} \circ h_1(T)}_{}$, and hence $(q_1(T))^{l_1} \circ h_1(T) = 0$. It follows that

$$(q_1(T))^{l_1}((h_1(T))(v_1)) = \underbrace{\left((q_1(T))^{l_1} \circ h_1(T)\right)(v_1) = 0(v_1)}_{} = 0,$$

and hence $(q_1(T))^{l_1}(w_1) = 0$, where $w_1 \equiv (h_1(T))(v_1) (\neq 0)$. Hence by the definition of V_1, we have $w_1 \in V_1$. Also $w_1 \neq 0$. This shows that $V_1 \neq \{0\}$. Similarly, $V_2 \neq \{0\}$, etc.

Since $h_1(T) \neq 0$ and $h_1(T) : V \to V$, we have $(h_1(T))(V) \neq \{0\}$. Similarly, $(h_2(T))(V) \neq \{0\}$, etc.

Clearly, $(h_1(T))(V) \subset V_1$.

Proof To show this, let us take an arbitrary $u_1 \in V$. We have to show that

$$(h_1(T))(u_1) \in V_1 \left(= \left\{v : v \in V \text{ and } \left((q_1(T))^{l_1}\right)(v) = 0\right\}\right),$$

that is, $\left((q_1(T))^{l_1}\right)((h_1(T))(u_1)) = 0$, that is, $\left((q_1(T))^{l_1} \circ (h_1(T))\right)(u_1) = 0$. Since $p(x) = (q_1(x))^{l_1} h_1(x)$, we have

$$0 = \underbrace{p(T) = (q_1(T))^{l_1} \circ h_1(T)},$$

and hence $(q_1(T))^{l_1} \circ h_1(T) = 0$. It follows that

$$\text{LHS} = \left((q_1(T))^{l_1} \circ (h_1(T))\right)(u_1) = (0)(u_1) = 0 = \text{RHS}.$$

∎

Similarly, $(h_2(T))(V) \subset V_2$, etc. Now, since $w_1, w_2, \ldots, w_k \in V$, we have $(h_1(T))(w_1) \in V_1$, $(h_2(T))(w_2) \in V_2, \ldots, (h_k(T))(w_k) \in V_k$. Hence from (**), for every $v \in V$, there exist $v_1 \in V_1, v_2 \in V_2, \ldots, v_k \in V_k$ such that

$$v = v_1 + v_2 + \cdots + v_k.$$

This proves that

$$V \subset V_1 + V_2 + \cdots + V_k.$$

Now, from (*), we have

$$V = V_1 + V_2 + \cdots + V_k. \quad (* * *)$$

Observe that if $v \in V_2$, then $\left((q_2(T))^{l_2}\right)(v) = 0$, and hence

$(h_1(T))(v)$
$$= \underbrace{\left((q_3(T))^{l_3} \cdots (q_k(T))^{l_k}\right)\left(\left((q_2(T))^{l_2}\right)(v)\right) = \left((q_3(T))^{l_3} \cdots (q_k(T))^{l_k}\right)(0)}$$
$$= 0.$$

Thus

$$v_2 \in V_2 \Rightarrow (h_1(T))(v_2) = 0.$$

Similarly,

$$v_3 \in V_3 \Rightarrow (h_1(T))(v_3) = 0,$$

etc. In short, for any distinct $i, j \in \{1, 2, \ldots, k\}$, we have $(h_i(T))(V_j) = \{0\}$.

We claim that

$$V = V_1 \oplus V_2 \oplus \cdots \oplus V_k.$$

In view of (***), it suffices to show that if each v_i is in V_i, and $v_1 + v_2 + \cdots + v_k = 0$, then each v_i equals 0.

Suppose to the contrary that there exist $v_i \in V_i$ $(i = 1, 2, \ldots, k)$ such that $v_1 + v_2 + \cdots + v_k = 0$ and $v_1 \neq 0$. We seek a contradiction.

Since $v_1 + v_2 + \cdots + v_k = 0$, we have

$$
\begin{aligned}
\left((q_2(T))^{l_2} (q_3(T))^{l_3} \cdots (q_k(T))^{l_k} \right)(v_1) &= (h_1(T))(v_1) + 0 + \cdots + 0 \\
&= (h_1(T))(v_1) + (h_1(T))(v_2) + \cdots \\
&\quad + (h_1(T))(v_k) \\
&= \underbrace{(h_1(T))(v_1 + v_2 + \cdots + v_k)}_{} = (h_1(T))(0) \\
&= 0,
\end{aligned}
$$

and hence

$$\left((q_2(T))^{l_2} \circ (q_3(T))^{l_3} \circ \cdots \circ (q_k(T))^{l_k} \right)(v_1) = 0.$$

Since $v_1 \in V_1 = \left\{ v : v \in V \text{ and } \left((q_1(T))^{l_1} \right)(v) = 0 \right\}$, we have $\left((q_1(T))^{l_1} \right)(v_1) = 0$.

Observe that the greatest common divisor of $(q_1(x))^{l_1}, (q_2(x))^{l_2}(q_3(x))^{l_3} \cdots (q_k(x))^{l_k}$ is 1. So by 1.1.5, there exist $b_1(x), b_2(x) \in F[x]$ such that

$$1 = b_1(x) \cdot (q_1(x))^{l_1} + b_2(x) \cdot (q_2(x))^{l_2}(q_3(x))^{l_3} \cdots (q_k(x))^{l_k},$$

and hence

$$I = (b_1(T)) \circ \left((q_1(T))^{l_1} \right) + (b_2(T)) \circ \left((q_2(T))^{l_2} \circ (q_3(T))^{l_3} \circ \cdots \circ (q_k(T))^{l_k} \right).$$

It follows that

$$v_1 = I(v_1) = \begin{aligned} &\Big((b_1(T)) \circ \Big((q_1(T))^{l_1} \Big) + (b_2(T)) \circ \\ &\underbrace{\Big((q_2(T))^{l_2} \circ (q_3(T))^{l_3} \circ \cdots \circ (q_k(T))^{l_k} \Big) \Big)}_{}(v_1) \end{aligned}$$

$$= (b_1(T)) \Big(\Big((q_1(T))^{l_1} \Big)(v_1) \Big)$$

$$+ (b_2(T)) \Big(\Big((q_2(T))^{l_2} \circ (q_3(T))^{l_3} \circ \cdots \circ (q_k(T))^{l_k} \Big)(v_1) \Big)$$

$$= (b_1(T)) \Big(\Big((q_1(T))^{l_1} \Big)(v_1) \Big) + (b_2(T))(0)$$

$$= (b_1(T))(0) + (b_2(T))(0) = 0 + 0 = 0,$$

and hence $v_1 = 0$. This is a contradiction.

Thus we have shown that $V = V_1 \oplus V_2 \oplus \cdots \oplus V_k$.

Observe that $\Big(q_1 \big(T|_{V_1} \big) \Big)^{l_1} = 0$.

Proof To show this, let us take an arbitrary $v_1 \in V_1$. We have to show that $\Big(\big(q_1 \big(T|_{V_1} \big) \big)^{l_1} \Big)(v_1) = 0$.

Since

$$v_1 \in V_1 = \Big\{ v : v \in V \text{ and } \Big((q_1(T))^{l_1} \Big)(v) = 0 \Big\},$$

we have $\Big((q_1(T))^{l_1} \Big)(v_1) = 0$. Now

$$\text{LHS} = \Big(\big(q_1 \big(T|_{V_1} \big) \big)^{l_1} \Big)(v_1) = \Big(q_1 \big(\big(T|_{V_1} \big)(v_1) \big) \Big)^{l_1} = (q_1(T(v_1)))^{l_1}$$

$$= \Big((q_1(T))^{l_1} \Big)(v_1) = 0 = \text{RHS}.$$

∎

It follows that the minimal polynomial of $T|_{V_1}$ is of the form $(q_1(x))^{m_1}$, where m_1 is a positive integer $\leq l_1$. Similarly, the minimal polynomial of $T|_{V_2}$ is of the form $(q_2(x))^{m_2}$, where m_2 is a positive integer $\leq l_2$, etc. Now, since $q_1(x), q_2(x), \ldots, q_k(x)$ are distinct irreducible polynomials, the least common multiple of $(q_1(x))^{m_1}, (q_2(x))^{m_2}, \ldots, (q_k(x))^{m_k}$ is $(q_1(x))^{m_1}(q_2(x))^{m_2} \ldots (q_k(x))^{m_k}$.

By 3.2.1, the minimal polynomial of T is the least common multiple of $(q_1(x))^{m_1}, (q_2(x))^{m_2}, \ldots, (q_k(x))^{m_k}$, and hence the minimal polynomial of T is $(q_1(x))^{m_1}(q_2(x))^{m_2} \ldots (q_k(x))^{m_k}$. Since

$$(q_1(x))^{l_1}(q_2(x))^{l_2} \ldots (q_k(x))^{l_k}$$

is the minimal polynomial of T over F, and each $m_i \leq l_i$, we have $m_1 = l_1, \ldots, m_k = l_k$. Since the minimal polynomial of $T|_{V_1}$ is of the form $(q_1(x))^{m_1}$, and $m_1 = l_1$, the minimal polynomial of $T|_{V_1}$ is $(q_1(x))^{l_1}$. Similarly, the minimal polynomial of $T|_{V_2}$ is $(q_2(x))^{l_2}$, etc.

3.2.3 Conclusion Let V be any n-dimensional vector space over the field F. Let $T \in A(V)$. Let $p(x) \in F[x]$. Let $p(x)$ be the minimal polynomial of T over F. Suppose that

$$p(x) = (q_1(x))^{l_1}(q_2(x))^{l_2}\ldots(q_k(x))^{l_k},$$

where $q_1(x), q_2(x), \ldots, q_k(x)$ are distinct irreducible polynomials (of degree ≥ 1) in $F[x]$, and $l_1, l_2, \ldots, l_k$ are positive integers. Put

$$V_1 \equiv \left\{ v : v \in V \text{ and } \left((q_1(T))^{l_1}\right)(v) = 0 \right\}, V_2$$
$$\equiv \left\{ v : v \in V \text{ and } \left((q_2(T))^{l_2}\right)(v) = 0 \right\}, \text{etc.}$$

Then,

1. each V_i is a nontrivial linear subspace of V,
2. each V_i is invariant under T,
3. $V = V_1 \oplus V_2 \oplus \cdots \oplus V_k$,
4. for each $i = 1, 2, \ldots, k$, the minimal polynomial of $T|_{V_1}$ is $(q_i(x))^{l_i}$.

3.2.4 Problem Let V be any n-dimensional vector space over the field F. Let $T \in A(V)$. Let $\lambda \in F$. Suppose that λ is an eigenvalue of T. Then $(\lambda I - T) : V \to V$ is not invertible.

Proof Since λ is an eigenvalue of T, there exists a nonzero $v \in V$ such that $T(v) = \lambda v$, and hence $(\lambda I - T)(v) = 0 \, (= (\lambda I - T)(0))$. Here $v \neq 0$, and $(\lambda I - T)(v) = (\lambda I - T)(0)$, $(\lambda I - T) : V \to V$ is not one-to-one, and hence $(\lambda I - T)$ is not invertible. ∎

3.2.5 Problem Let V be any n-dimensional vector space over the field F. Let $T \in A(V)$. Let $\lambda \in F$. Suppose that $(\lambda I - T) : V \to V$ is not invertible, that is, $(\lambda I - T)$ is singular. Then λ is an eigenvalue of T.

Proof Here $(\lambda I - T) : V \to V$ is not invertible, that is, $(\lambda I - T)^{-1}$ does not exist, so by 3.1.31, there exists a nonzero $v \in V$ such that $(\lambda I - T)(v) = 0$. It follows that $T(v) = \lambda v$, where $v \neq 0$. Hence λ is an eigenvalue of T. ∎

3.2.6 Problem Let V be any n-dimensional vector space over the field F. Let $T \in A(V)$. Let $\lambda \in F$. Suppose that λ is an eigenvalue of T. Let $q(x) \in F[x]$. (It follows that $q(\lambda) \in F$ and $q(T) \in A(V)$.) Then $q(\lambda)$ is an eigenvalue of $q(T)$.

Proof Suppose that

$$q(x) = a_0 + a_1 x + \cdots + a_n x^n,$$

where each a_i is in F. It follows that

$$q(\lambda) = a_0 + a_1 \lambda + \cdots + a_n \lambda^n$$

and

$$q(T) = a_0 I + a_1 T + \cdots + a_n T^n.$$

Since λ is an eigenvalue of T, there exists a nonzero $v \in V$ such that $T(v) = \lambda v$. It suffices to show that

$$a_0 I(v) + a_1 T(v) + a_2 T(T(v)) + a_3 T(T(v)) + \cdots + a_n T^n(v)$$
$$= (a_0 I + a_1 T + \cdots + a_n T^n)(v) = \underbrace{(q(T))(v) = (q(\lambda))v}$$

$$= (a_0 + a_1 \lambda + \cdots + a_n \lambda^n)v,$$

that is,

$$a_0 I(v) + a_1 T(v) + a_2 T(T(v)) + a_3 T(T(T(v))) + \cdots + a_n T^n(v)$$
$$= (a_0 + a_1 \lambda + \cdots + a_n \lambda^n)v.$$
$$\text{LHS} = a_0 I(v) + a_1 T(v) + a_2 T(T(v)) + a_3 T(T(T(v))) + \cdots + a_n T^n(v)$$
$$= a_0 v + a_1 T(v) + a_2 T(T(v)) + a_3 T(T(T(v))) + \cdots + a_n T^n(v)$$
$$= a_0 v + a_1 (\lambda v) + a_2 T(T(v)) + a_3 T(T(T(v))) + \cdots + a_n T^{n-1}(T(v))$$
$$= a_0 v + a_1 (\lambda v) + a_2 T(\lambda v) + a_3 T(T(\lambda v)) + \cdots + a_n T^{n-1}(\lambda v)$$
$$= a_0 v + a_1 (\lambda v) + a_2 \lambda T(v) + a_3 \lambda T(T(v)) + \cdots + a_n \lambda T^{n-1}(v)$$
$$= a_0 v + a_1 (\lambda v) + a_2 \lambda (\lambda v) + a_3 \lambda T(\lambda v) + \cdots + a_n \lambda T^{n-2}(\lambda v)$$
$$= a_0 v + a_1 (\lambda v) + a_2 (\lambda^2 v) + a_3 \lambda^2 T(v) + \cdots + a_n \lambda^2 T^{n-2}(v)$$
$$\vdots$$
$$= a_0 v + a_1 (\lambda v) + a_2 (\lambda^2 v) + a_3 (\lambda^3 v) + \cdots + a_n (\lambda^n v)$$
$$= (a_0 + a_1 \lambda + a_2 \lambda^2 + a_3 \lambda^3 + \cdots + a_n \lambda^n)v = \text{RHS}.$$

∎

3.2.7 Problem Let V be any n-dimensional vector space over the field F. Let $T \in A(V)$. Let $p(x)$ be the minimal polynomial of T over F. Let $\lambda \in F$. Suppose that λ is an eigenvalue of T. Then λ is a root of $p(x)$, that is, $p(\lambda) = 0$.

(Since the number roots of $p(x)$ is finite, the number of eigenvalues of T is finite.)

Proof Since $p(x)$ is the minimal polynomial of T over F, we have $p(x) \in F[x]$ and $p(T) = 0$. Suppose that

$$p(x) = a_0 + a_1 x + \cdots + a_n x^n,$$

where each a_i is in F. It follows that

$$p(\lambda) = a_0 + a_1 \lambda + \cdots + a_n \lambda^n$$

and

$$p(T) = a_0 I + a_1 T + \cdots + a_n T^n.$$

Since λ is an eigenvalue of T, there exists a nonzero $v \in V$ such that $T(v) = \lambda v$. We claim that $(p(T))(v) = (p(\lambda))v$, that is,

$$a_0 I(v) + a_1 T(v) + a_2 T(T(v)) + a_3 T(T(v)) + \cdots + a_n T^n(v)$$
$$= (a_0 + a_1 \lambda + \cdots + a_n \lambda^n)v,$$
$$\text{LHS} = a_0 I(v) + a_1 T(v) + a_2 T(T(v)) + a_3 T(T(T(v))) + \cdots + a_n T^n(v)$$
$$= a_0 v + a_1 T(v) + a_2 T(T(v)) + a_3 T(T(T(v))) + \cdots + a_n T^n(v)$$
$$= a_0 v + a_1 (\lambda v) + a_2 T(T(v)) + a_3 T(T(T(v))) + \cdots + a_n T^{n-1}(T(v))$$
$$= a_0 v + a_1 (\lambda v) + a_2 T(\lambda v) + a_3 T(T(\lambda v)) + \cdots + a_n T^{n-1}(\lambda v)$$
$$= a_0 v + a_1 (\lambda v) + a_2 \lambda T(v) + a_3 \lambda T(T(v)) + \cdots + a_n \lambda T^{n-1}(v)$$
$$= a_0 v + a_1 (\lambda v) + a_2 \lambda (\lambda v) + a_3 \lambda T(\lambda v) + \cdots + a_n \lambda T^{n-2}(\lambda v)$$
$$= a_0 v + a_1 (\lambda v) + a_2 (\lambda^2 v) + a_3 \lambda^2 T(v) + \cdots + a_n \lambda^2 T^{n-2}(v)$$
$$\vdots$$
$$= a_0 v + a_1 (\lambda v) + a_2 (\lambda^2 v) + a_3 (\lambda^3 v) + \cdots + a_n (\lambda^n v)$$
$$= (a_0 + a_1 \lambda + a_2 \lambda^2 + a_3 \lambda^3 + \cdots + a_n \lambda^n)v = \text{RHS}.$$

Thus, $0 = 0(v) = \underbrace{(p(T))(v) = (p(\lambda))v}$. Now, since $(p(\lambda))v = 0$ and v is non-

zero, we have $p(\lambda) = 0$.   ∎

3.2.8 Problem Let V be any n-dimensional vector space over the field F. Let $S, T \in A(V)$. Let $S : V \to V$ be invertible. Then T and $S^{-1} \circ T \circ S$ have the same minimal polynomial.

Proof Let $p(x)$ be the minimal polynomial of T, and $q(x)$ the minimal polynomial of $S^{-1} \circ T \circ S$. It suffices to show that

1. $q(x)|p(x)$, that is, $p(S^{-1} \circ T \circ S) = 0$,
2. $p(x)|q(x)$, that is, $q(T) = 0$,

For 1: Since $p(x)$ is the minimal polynomial of T, we have $p(T) = 0$. Suppose that

$$p(x) = a_0 + a_1 x + \cdots + a_n x^n,$$

where each a_i is in F. It follows that

$$0 = \underbrace{p(T) = a_0 I + a_1 T + \cdots + a_n T^n}$$

and

$$p\left(S^{-1} \circ T \circ S\right) = a_0 I + a_1 \left(S^{-1} \circ T \circ S\right) + a_2 \left(S^{-1} \circ T \circ S\right) \circ \left(S^{-1} \circ T \circ S\right)$$
$$+ a_3 \left(S^{-1} \circ T \circ S\right) \circ \left(S^{-1} \circ T \circ S\right) \circ \left(S^{-1} \circ T \circ S\right)$$
$$+ \cdots + a_n \left(S^{-1} \circ T \circ S\right)^n.$$

Hence

$$p\left(S^{-1} \circ T \circ S\right) = a_0 I + a_1 \left(S^{-1} \circ T \circ S\right) + a_2 \left(S^{-1} \circ T^2 \circ S\right) + a_3 \left(S^{-1} \circ T^3 \circ S\right)$$
$$+ \cdots + a_n \left(S^{-1} \circ T^n \circ S\right).$$

Now, since

$$a_0 I + a_1 \left(S^{-1} \circ T \circ S\right) + a_2 \left(S^{-1} \circ T^2 \circ S\right) + a_3 \left(S^{-1} \circ T^3 \circ S\right)$$
$$+ \cdots + a_n \left(S^{-1} \circ T^n \circ S\right) = S^{-1} \circ (a_0 I) \circ S + S^{-1} \circ (a_1 T) \circ S$$
$$+ S^{-1} \circ \left(a_2 T^2\right) \circ S + S^{-1} \circ \left(a_3 T^3\right) \circ S$$
$$+ \cdots + S^{-1} \circ (a_n T^n) \circ S$$
$$= S^{-1} \circ \left(a_0 I + a_1 T + a_2 T^2 + a_3 T^3 + \cdots + a_n T^n\right) \circ S$$
$$= S^{-1} \circ (p(T)) \circ S = S^{-1} \circ 0 \circ S = 0,$$

we have $p(S^{-1} \circ T \circ S) = 0$.

For 2: Since $q(x)$ is the minimal polynomial of $S^{-1} \circ T \circ S$, we have $q(S^{-1} \circ T \circ S) = 0$. Suppose that

$$q(x) = b_0 + b_1 x + \cdots + b_m x^m,$$

where each b_i is in F. It follows that

$$q(T) = b_0 I + b_1 T + b_2 T^2 + \cdots + b_m T^m$$

and

$$0 = q\left(S^{-1} \circ T \circ S\right) = b_0 I + b_1 \left(S^{-1} \circ T \circ S\right) + \cdots + b_m \underbrace{\left(S^{-1} \circ T \circ S\right)^m}$$

$$\begin{aligned} &= b_0 (S^{-1} \circ I \circ S) + b_1 (S^{-1} \circ T \circ S) + \cdots + b_m (S^{-1} \circ T^m \circ S) \\ &= S^{-1} \circ (b_0 I) \circ S + S^{-1} \circ (b_1 T) \circ S + \cdots + S^{-1} \circ (b_m T^m) \circ S \\ &= S^{-1} \circ (b_0 I + b_1 T + b_2 T^2 + \cdots + b_m T^m) \circ S \end{aligned}$$

and hence

$$0 = S^{-1} \circ \left(b_0 I + b_1 T + b_2 T^2 + \cdots + b_m T^m\right) \circ S.$$

It follows that

$$q(T) = b_0 I + b_1 T + b_2 T^2 + \cdots + b_m T^m = \underbrace{S \circ 0 \circ S^{-1}} = 0,$$

and hence $q(T) = 0$. ∎

Definition Let V be any n-dimensional vector space over the field F. Let $T \in A(V)$. Let $\lambda \in F$. Suppose that λ is an eigenvalue of T. Let v be a nonzero vector in V. If $T(x) = \lambda x$, then we say that v *is an eigenvector of T belonging to the eigenvalue λ.*

3.2.9 Problem Let V be any n-dimensional vector space over the field F. Let $T \in A(V)$. Let $\lambda_1, \lambda_2, \ldots, \lambda_k$ be distinct eigenvalues of T. Suppose that for every $i \in \{1, 2, \ldots, k\}$, v_i is an eigenvector of T belonging to the eigenvalue λ_i. Then $\{v_1, v_2, \ldots, v_k\}$ is a linearly independent set of vectors over F.

Proof Suppose to the contrary (after suitably rearranging the indices) that

$$v_1 = \underbrace{a_2 v_2 + a_3 v_3 + \ldots + a_l v_l}_{(l-1)\,\text{terms}}$$

is the shortest linear relation, where $a_2, a_3, \ldots, a_l$ are nonzero members of F. We seek a contradiction.

Since $v_1 = a_2 v_2 + a_3 v_3 + \cdots + a_l v_l$, we have

$$\begin{aligned} \lambda_1 a_2 v_2 + \lambda_1 a_3 v_3 + \cdots + \lambda_1 a_l v_l = \lambda_1 (a_2 v_2 + a_3 v_3 + \cdots + a_l v_l) &= \lambda_1 v_1 \\ &= \underbrace{T(v_1) = T(a_2 v_2 + a_3 v_3 + \cdots + a_l v_l)} \end{aligned}$$

$$= a_2 T(v_2) + a_3 T(v_3) + \cdots + a_l T(v_l) = a_2 \cdot \lambda_2 v_2 + a_3 \cdot \lambda_3 v_3 + \cdots + a_l \cdot \lambda_l v_l,$$

and hence

$$\lambda_1 a_2 v_2 + \lambda_1 a_3 v_3 + \cdots + \lambda_1 a_l v_l = a_2 \lambda_2 v_2 + a_3 \lambda_3 v_3 + \cdots + a_l \lambda_l v_l.$$

This shows that

$$(\lambda_1 - \lambda_2) a_2 v_2 + (\lambda_1 - \lambda_3) a_3 v_3 + \cdots + (\lambda_1 - \lambda_l) a_l v_l = 0.$$

Since $\lambda_1 \neq \lambda_2$ and $a_2 \neq 0$, we have

$$v_2 = \underbrace{\frac{-(\lambda_1 - \lambda_3) a_3}{(\lambda_1 - \lambda_2) a_2} v_3 + \cdots + \frac{-(\lambda_1 - \lambda_l) a_l}{(\lambda_1 - \lambda_2) a_2} v_l}_{(l-2)\,\text{terms}}.$$

This contradicts the fact that

$$v_1 = \underbrace{a_2 v_2 + a_3 v_3 + \cdots + a_l v_l}_{(l-1)\,\text{terms}}$$

is the shortest linear relation.   ∎

3.2.10 Problem Let V be any n-dimensional vector space over the field F. Let $T \in A(V)$. Then the number of distinct eigenvalues of T is $\leq n$.

Proof Let $\lambda_1, \lambda_2, \ldots, \lambda_k$ be distinct eigenvalues of T. We have to show that $k \leq n$, that is, $k \leq \dim(V)$.

Suppose that for every $i \in \{1, 2, \ldots, k\}$, v_i is an eigenvector of T belonging to the eigenvalue λ_i. By 3.2.9, $\{v_1, v_2, \ldots, v_k\}$ is a linearly independent set of vectors over F. It follows that the number of elements in $\{v_1, v_2, \ldots, v_k\}$ is $\leq \dim(V)$. Since $\{v_1, v_2, \ldots, v_k\}$ is a linearly independent set of vectors, the number of elements in $\{v_1, v_2, \ldots, v_k\}$ is k, and hence $k \leq \dim(V)$.   ∎

3.2.11 Problem Let V be any n-dimensional vector space over the field F. Let $T \in A(V)$. Suppose that T has n distinct eigenvalues in F. Then there exists a basis of V over F such that each member of the basis is an eigenvector of T.

Proof Let $\lambda_1, \lambda_2, \ldots, \lambda_n$ be distinct eigenvalues of T. Suppose that for every $i \in \{1, 2, \ldots, n\}$, v_i is an eigenvector of T belonging to the eigenvalue λ_i. By 3.2.9, $\{v_1, v_2, \ldots, v_n\}$ is a linearly independent set of vectors over F. Now, since the number of elements in $\{v_1, v_2, \ldots, v_n\}$ is equal to $\dim(V)$, $\{v_1, v_2, \ldots, v_n\}$ constitutes a basis of V.   ∎

Definition Let V be any n-dimensional vector space over the field F. Let $S, T \in A(V)$. If there exists $C \in A(V)$ such that C^{-1} exists, and $C^{-1} \circ S \circ C = T$, then we say that S *is similar to* T, and we write $S \sim T$.

3.2.12 Problem $\sim$ is an equivalence relation over $A(V)$.

And hence $A(V)$ is partitioned into equivalence classes. Each equivalence class is called a *similarity class*.

Proof

a. Let us take an arbitrary $T \in A(V)$. Since $I^{-1} \circ T \circ I = T$, we have $T \sim T$. Thus $\sim$ is reflexive over $A(V)$.
b. Let us take arbitrary $S, T \in A(V)$ satisfying $S \sim T$. It follows that there exists $C \in A(V)$ such that C^{-1} exists, and $C^{-1} \circ S \circ C = T$. It follows that $(C^{-1})^{-1}(= C \in A(V))$ and $C \circ T \circ C^{-1} = S$. Thus $S \sim T$. Hence $\sim$ is symmetric.
c. Let us take arbitrary $R, S, T \in A(V)$ satisfying $R \sim S$ and $S \sim T$. We have to show that $R \sim T$. Since $R \sim S$, there exists $C \in A(V)$ such that C^{-1} exists, and $C^{-1} \circ R \circ C = S$. Again there exists $D \in A(V)$ such that D^{-1} exists, and

$$(C \circ D)^{-1} \circ R \circ (C \circ D) = (D^{-1} \circ C^{-1}) \circ R \circ (C \circ D)$$
$$= D^{-1} \circ (C^{-1} \circ R \circ C) \circ D = \underbrace{D^{-1} \circ S \circ D = T}.$$

Thus $E^{-1} \circ R \circ E = T$, where $E \equiv (C \circ D) \in A(V)$.

This proves that $\sim$ is an equivalence relation over $A(V)$. ∎

3.2.13 Problem Let V be any n-dimensional vector space over F. Let $T \in A(V)$. Let W be any subspace of V. Suppose that W is invariant under T. Then $\widehat{T}$: $(v + W) \mapsto (T(v) + W)$ from the quotient space $\frac{V}{W}$ to $\frac{V}{W}$ is a linear transformation.

Proof $\widehat{T}$ is a well-defined function from the quotient space $\frac{V}{W}$ to $\frac{V}{W}$: To show this, let us take arbitrary $u, v \in V$ such that $(u + W) = (v + W)$, that is, $(u - v) \in W$. We have to show that $(T(u) + W) = (T(v) + W)$, that is, $(T(u) - T(v)) \in W$, that is, $T(u - v) \in W$. Since $(u - v) \in W$ and W is invariant under T, we have $T(u - v) \in W$.

$\widehat{T} : \frac{V}{W} \to \frac{V}{W}$ is linear: To show this, let us take arbitrary $u, v \in V$ and $\alpha, \beta \in F$. We have to show that

$$T(\alpha u + \beta v) + W = \alpha(T(u) + W) + \beta(T(v) + W) :$$
$$\text{LHS} = T(\alpha u + \beta v) + W = (\alpha T(u) + \beta T(v)) + W$$
$$= \alpha(T(u) + W) + \beta(T(v) + W) = \text{RHS}.$$

∎

3.2.14 Problem Let V be any n-dimensional vector space over F. Let $T \in A(V)$. Let W be any subspace of V. Suppose that W is invariant under T. Let $p(x) \in F[x]$. Here $\widehat{T}$, as defined in 3.2.13, is a member of $A\left(\frac{V}{W}\right)$, and hence $p\left(\widehat{T}\right) \in A\left(\frac{V}{W}\right)$. Also $p(T) \in A(V)$. Suppose that $p(T)$ is the zero element of $A(V)$. Then $p\left(\widehat{T}\right)$ is the zero element of $A\left(\frac{V}{W}\right)$.

Proof Suppose that

$$p(x) = a_0 + a_1 x + \cdots + a_n x^n,$$

where each a_i is in F. It follows that

$$A(V) \ni 0 = \underbrace{p(T) = a_0 I + a_1 T + \cdots + a_n T^n}$$

and

$$p\left(\widehat{T}\right) = a_0 \widehat{I} + a_1 \widehat{T} + a_2 \widehat{T} \circ \widehat{T} + a_3 \widehat{T} \circ \widehat{T} \circ \widehat{T} + \cdots + a_n T^n.$$

We have to show that $a_0 \widehat{I} + a_1 \widehat{T} + a_2 \widehat{T} \circ \widehat{T} + a_3 \widehat{T} \circ \widehat{T} \circ \widehat{T} + \cdots + a_n T^n$ is the zero element of $A\left(\frac{V}{W}\right)$. To this end, let us take an arbitrary $v \in V$. We have to show that

$$\left(a_0 \widehat{I} + a_1 \widehat{T} + a_2 \widehat{T} \circ \widehat{T} + a_3 \widehat{T} \circ \widehat{T} \circ \widehat{T} + \cdots + a_n T^n\right)(v + W) = (0 + W),$$

that is,

$$a_0 \widehat{I}(v + W) + a_1 \widehat{T}(v + W) + a_2 \left(\widehat{T} \circ \widehat{T}\right)(v + W)$$
$$+ a_3 \left(\widehat{T} \circ \widehat{T} \circ \widehat{T}\right)(v + W) + \cdots + a_n \left(\widehat{T^n}\right)(v + W) = (0 + W):$$
$$\text{LHS} = a_0 \widehat{I}(v + W) + a_1 \widehat{T}(v + W) + a_2 \left(\widehat{T} \circ \widehat{T}\right)(v + W)$$
$$+ a_3 \left(\widehat{T} \circ \widehat{T} \circ \widehat{T}\right)(v + W) + \cdots + a_n \left(\widehat{T^n}\right)(v + W)$$
$$= a_0 (I(v) + W) + a_1 (T(v) + W) + a_2 \widehat{T} \left(\widehat{T}(v + W)\right)$$
$$+ a_3 \left(\widehat{T} \circ \widehat{T}\right)\left(\widehat{T}(v + W)\right) + \cdots + a_n \left(\widehat{T^{n-1}}\right)\left(\widehat{T}(v + W)\right)$$
$$= a_0 (v + W) + a_1 (T(v) + W) + a_2 \widehat{T}(T(v) + W)$$
$$+ a_3 \left(\widehat{T} \circ \widehat{T}\right)(T(v) + W) + \cdots + a_n \left(\widehat{T^{n-1}}\right)(T(v) + W)$$
$$= a_0 (v + W) + a_1 (T(v) + W) + a_2 (T(T(v)) + W)$$
$$+ a_3 \widehat{T}(T(T(v)) + W) + \cdots + a_n \left(\widehat{T^{n-2}}\right)(T(T(v)) + W)$$
$$= a_0 (v + W) + a_1 (T(v) + W) + a_2 \left(T^2(v) + W\right)$$
$$+ a_3 \widehat{T}\left(T^2(v) + W\right) + \cdots + a_n \left(\widehat{T^{n-2}}\right)\left(T^2(v) + W\right)$$
$$= a_0 (v + W) + a_1 (T(v) + W) + a_2 \left(T^2(v) + W\right)$$
$$+ a_3 \left(T^3(v) + W\right) + \cdots + a_n (T^n(v) + W)$$
$$= (a_0 v + W) + (a_1 T(v) + W) + \left(a_2 T^2(v) + W\right)$$
$$+ \left(a_3 T^3(v) + W\right) + \cdots + (a_n T^n(v) + W)$$
$$= \left(a_0 v + a_1 T(v) + a_2 T^2(v) + a_3 T^3(v) + \cdots + a_n T^n(v)\right) + W$$
$$= (a_0 I + a_1 T + \cdots + a_n T^n)(v) + W = 0(v) + W = 0 + W = \text{RHS}.$$

3.2.15 Problem Let V be any n-dimensional vector space over F. Let $T \in A(V)$. Let W be any subspace of V. Suppose that W is invariant under T. Let $p(x), q(x) \in F[x]$. Here $\widehat{T}$, as defined in 3.2.13, is a member of $A\left(\frac{V}{W}\right)$, and hence $p\left(\widehat{T}\right) \in A\left(\frac{V}{W}\right)$. Also $p(T) \in A(V)$. Similarly, $q\left(\widehat{T}\right) \in A\left(\frac{V}{W}\right)$ and $q(T) \in A(V)$. Suppose that $p(x)$ is the minimal polynomial of $T : V \to V$ over F, and $q(x)$ is the minimal polynomial of $\widehat{T} : \frac{V}{W} \to \frac{V}{W}$ over F. Then $q(x)|p(x)$.

Proof Since $p(x)$ is the minimal polynomial of $T : V \to V$ over F, $p(T)$ is the zero element of $A(V)$, and hence by 3.2.14, $p\left(\widehat{T}\right)$ is the zero element of $A\left(\frac{V}{W}\right)$. Now, since $q(x)$ is the minimal polynomial of $\widehat{T} : \frac{V}{W} \to \frac{V}{W}$ over F, we have $q(x)|p(x)$. ∎

3.2.16 Theorem Let V be any n-dimensional vector space over F. Let $T \in A(V)$. Suppose that all the roots of the minimal polynomial of T over F are in F. Then there exists a basis of V in which the matrix of T is such that all its entries above the diagonal are zero. In short, there exists a basis of V in which the matrix of T is triangular.

Proof (Induction on n): The theorem is trivially true when $n = 1$.

Now let us assume that the theorem is true for all vector spaces of dimension $(n - 1)$.

Let $\lambda_1 \in F$. Let λ_1 be an eigenvalue of T.

Since λ_1 is an eigenvalue of T, there exists a nonzero vector $v_1 \in V$ such that $T(v_1) = \lambda_1 v_1$. Put

$$W \equiv \{\alpha v_1 : \alpha \in F\}.$$

Clearly, W is a linear subspace of V. Also, since v_1 is nonzero, W is a one-dimensional space. Next, since $T(v_1) = \lambda_1 v_1$, W is invariant under T. It follows that

$$\dim\left(\frac{V}{W}\right) = \dim(V) - \dim(W) = \dim(V) - 1 = n - 1.$$

Thus $\frac{V}{W}$ is an $(n - 1)$-dimensional vector space over F.

Suppose that $p(x)$ is the minimal polynomial of $T : V \to V$ over F, and $q(x)$ is the minimal polynomial of $\widehat{T} : \frac{V}{W} \to \frac{V}{W}$ over F, where $\widehat{T} : (v + W) \mapsto (T(v) + W)$ from the quotient space $\frac{V}{W}$ to $\frac{V}{W}$ is a linear transformation. By 3.2.15, $q(x)|p(x)$, and hence every root of $q(x)$ is a root of $p(x)$. Now, since all the roots of the minimal polynomial $p(x)$ of T over F are in F, all the roots of $q(x)$ are in F.

Since $\frac{V}{W}$ is an $(n - 1)$-dimensional vector space, $\widehat{T} \in A\left(\frac{V}{W}\right)$, and all the roots of the minimal polynomial $q(x)$ of $\widehat{T}$ over F are in F, it follows by our induction hypothesis that there exists a basis $\{v_2 + W, v_3 + W, \dots, v_n + W\}$ of $\frac{V}{W}$ in which the matrix of $\widehat{T}$ is such that all its entries above the diagonal are zero. So there exists a matrix

$$\begin{bmatrix} \alpha_{22} & \alpha_{23} & \alpha_{24} & \cdots & \alpha_{2n} \\ \alpha_{32} & \alpha_{33} & \alpha_{34} & \cdots & \alpha_{3n} \\ & & \ddots & & \\ \alpha_{n2} & \alpha_{n3} & \alpha_{n4} & \cdots & \alpha_{nn} \end{bmatrix}_{(n-1)\times(n-1)}$$

such that all its entries above the diagonal are zero, and

$$\widehat{T}(v_2 + W) = \alpha_{22}(v_2 + W),$$
$$\widehat{T}(v_3 + W) = \alpha_{32}(v_2 + W) + \alpha_{33}(v_3 + W),$$
$$\vdots$$
$$\widehat{T}(v_n + W) = \alpha_{n2}(v_2 + W) + \alpha_{n3}(v_3 + W) + \cdots + \alpha_{nn}(v_n + W).$$

It follows that

$$T(v_2) + W = \alpha_{22}v_2 + W,$$
$$T(v_3) + W = (\alpha_{32}v_2 + \alpha_{33}v_3) + W,$$
$$\vdots$$
$$T(v_n) + W = (\alpha_{n2}v_2 + \alpha_{n3}v_3 + \cdots + \alpha_{nn}v_n) + W,$$

and hence

$$\left. \begin{array}{c} T(v_2) - \alpha_{22}v_2 \in W \\ T(v_3) - (\alpha_{32}v_2 + \alpha_{33}v_3) \in W \\ \vdots \\ T(v_n) - (\alpha_{n2}v_2 + \alpha_{n3}v_3 + \cdots + \alpha_{nn}v_n) \in W \end{array} \right\}.$$

Thus

$$\left. \begin{array}{c} T(v_2) - \alpha_{22}v_2 \in \{\alpha v_1 : \alpha \in F\} \\ T(v_3) - (\alpha_{32}v_2 + \alpha_{33}v_3) \in \{\alpha v_1 : \alpha \in F\} \\ \vdots \\ T(v_n) - (\alpha_{n2}v_2 + \alpha_{n3}v_3 + \cdots + \alpha_{nn}v_n) \in \{\alpha v_1 : \alpha \in F\} \end{array} \right\}.$$

It follows that there exists $\alpha_{21} \in F$ such that $T(v_2) - \alpha_{22}v_2 = \alpha_{21}v_1$, and hence $T(v_2) = \alpha_{21}v_1 + \alpha_{22}v_2$. Similarly, there exists $\alpha_{31} \in F$ such that $T(v_3) = \alpha_{31}v_1 + \alpha_{32}v_2 + \alpha_{33}v_3$, etc. Also $T(v_1) = \lambda_1 v_1$. Thus

$$\left.\begin{array}{l} T(v_1) = \lambda_1 v_1 \\ T(v_2) = \alpha_{21} v_1 + \alpha_{22} v_2 \\ T(v_3) = \alpha_{31} v_1 + \alpha_{32} v_2 + \alpha_{33} v_3 \\ \qquad\vdots \\ T(v_n) = \alpha_{n1} v_1 + \alpha_{n2} v_2 + \cdots + \alpha_{n3} v_n \end{array}\right\}. \quad (*)$$

Clearly, $\{v_1, v_2, \ldots, v_n\}$ is a linearly independent set of vectors in V.

Proof To show this, suppose that

$$\alpha_1 v_1 + \alpha_2 v_2 + \cdots + \alpha_n v_n = 0. \quad (*)$$

We have to show that each α_i equals 0. Here

$$\alpha_2 v_2 + \cdots + \alpha_n v_n = (-\alpha_1) v_1 \in \{\alpha v_1 : \alpha \in F\} = W,$$

so

$$\alpha_2 (v_2 + W) + \cdots + \alpha_n (v_n + W) = 0 + W.$$

Now, since $\{v_2 + W, v_3 + W, \ldots, v_n + W\}$ is a basis of $\frac{V}{W}$, $\{v_2 + W, v_3 + W, \ldots, v_n + W\}$ is a linearly independent set of vectors in $\frac{V}{W}$, and hence $\alpha_2 = 0, \alpha_3 = 0$, and $\alpha_n = 0$. It remains to show that $\alpha_1 = 0$.

From (*),

$$\alpha_1 v_1 = \underbrace{\alpha_1 v_1 + 0 v_2 + \cdots + 0 v_n = 0},$$

and hence $\alpha_1 v_1 = 0$. Now, since v_1 is nonzero, we have $\alpha_1 = 0$. ∎

Thus we have shown that $\{v_1, v_2, \ldots, v_n\}$ is a linearly independent set of vectors in the n-dimensional vector space V. It follows that $\{v_1, v_2, \ldots, v_n\}$ is a basis of V. Next, from (*), the matrix of T relative to the basis $\{v_1, v_2, \ldots, v_n\}$ is triangular. ∎

3.2.17 Problem Let F be a field. Let A be an $n \times n$ matrix with entries in F. Suppose that all its eigenvalues are in F. Then there exists an invertible $n \times n$ matrix C with entries in F such that $C^{-1}AC$ is a triangular matrix.

Such matrices of a particularly nice form are called *canonical forms*. In short, we say that A *can be brought to triangular form over* F *by similarity*.

Proof We know that F^n constitutes a vector space over F under pointwise addition and pointwise scalar multiplication. Put

$$v_1 \equiv \left(\underbrace{1,0,0,\ldots,0}_{n}\right), v_2 \equiv \left(\underbrace{0,1,0,\ldots,0}_{n}\right), \text{ etc.}$$

We know that $\{v_1, v_2, \ldots, v_n\}$ is a basis of F^n. Suppose that $A \equiv [a_{ij}]_{n \times n}$, where each a_{ij} is in F. Suppose that $T : F^n \to F^n$ is a linear transformation such that

$$T(v_1) = a_{11}v_1 + a_{12}v_2 + \cdots + a_{1n}v_n,$$
$$T(v_2) = a_{21}v_1 + a_{22}v_2 + \cdots + a_{2n}v_n, \text{ etc.}$$

It follows that $m_1(T) = [a_{ij}]_{n \times n}$, where $m_1(T)$ is the matrix of T relative to the basis $v_1, \ldots, v_n$. Since all the eigenvalues of the matrix $[a_{ij}]_{n \times n}$ are in F, all the eigenvalues of T are in F, and hence by 3.2.7, all the roots of the minimal polynomial of T over F are in F. Hence by 3.2.16, there exists a basis $w_1, \ldots, w_n$ of F^n such that the matrix $m_2(T)$ of T relative to the basis $w_1, \ldots, w_n$ is triangular.

Let $S : F^n \to F^n$ be the linear transformation such that for every $i \in \{1, \ldots, n\}$, $S(v_i) = w_i$. By 3.1.35,

$$m_2(T) = (m_1(S))^{-1} m_1(T) m_1(S).$$

Thus

$$m_2(T) = C^{-1}AC,$$

where $C \equiv m_1(S)$. Now, since the matrix $m_2(T)$ of T is triangular, $C^{-1}AC$ is a triangular matrix. ∎

3.2.18 Problem Let A be a triangular matrix with entries in the field F. Suppose that no entry on the diagonal is 0. Then A is invertible.

Proof Let $A \equiv [a_{ij}]_{n \times n}$, where $i < j \Rightarrow \alpha_{ij} = 0$. Next, suppose that each a_{ii} is nonzero. We have to show that $[a_{ij}]_{n \times n}$ is invertible.

It suffices to show that $\det[a_{ij}]_{n \times n} \neq 0$. Since $[a_{ij}]_{n \times n}$ is a triangular matrix, we have $\det[a_{ij}]_{n \times n} = a_{11}a_{22}\ldots a_{nn}$, and since each a_{ii} is nonzero, we have $\det[a_{ij}]_{n \times n} \neq 0$. ∎

3.2.19 Problem Let A be a triangular matrix with entries in the field F. Suppose that some entry on the diagonal is 0. Then A^{-1} does not exist.

Proof Let $A \equiv [a_{ij}]_{n \times n}$, where $i < j \Rightarrow \alpha_{ij} = 0$. Next, there exists $i \in \{1, 2, \ldots, n\}$ such that $a_{ii} = 0$. We have to show that $[a_{ij}]_{n \times n}$ is not invertible.

It suffices to show that $\det[a_{ij}]_{n \times n} = 0$. Since $[a_{ij}]_{n \times n}$ is a triangular matrix, we have $\det[a_{ij}]_{n \times n} = a_{11}a_{22}\ldots a_{nn}$, and since $a_{ii} = 0$, we have $\det[a_{ij}]_{n \times n} = 0$. ∎

3.2.20 Problem Let $\left[a_{ij}\right]_{n\times n}$ be a triangular matrix with entries in the field F. Then the set of all eigenvalues of $\left[a_{ij}\right]_{n\times n}$ is $\{a_{11}, a_{22}, \ldots, a_{nn}\}$.

Proof Let λ be an eigenvalue of $\left[a_{ij}\right]_{n\times n}$. It follows that $\left(\left[a_{ij}\right]_{n\times n} - \lambda I\right)$ is singular, that is, $\det\left(\left[a_{ij}\right]_{n\times n} - \lambda I\right) = 0$. Since $\left[a_{ij}\right]_{n\times n}$ is a triangular matrix, we have $\det\left(\left[a_{ij}\right]_{n\times n} - \lambda I\right) = (a_{11} - \lambda)(a_{22} - \lambda)\cdots(a_{nn} - \lambda)$. Since $\det\left(\left[a_{ij}\right]_{n\times n} - \lambda I\right) = 0$, we have

$$(a_{11} - \lambda)(a_{22} - \lambda)\cdots(a_{nn} - \lambda) = 0.$$

This shows that $\lambda \in \{a_{11}, a_{22}, \ldots, a_{nn}\}$.

Conversely, since $\det\left(\left[a_{ij}\right]_{n\times n} - a_{11}I\right) = (a_{11} - a_{11})(a_{22} - a_{11})\cdots(a_{nn} - a_{11})$ $= 0$, it follows that $\left(\left[a_{ij}\right]_{n\times n} - a_{11}I\right)$ is singular, and hence a_{11} is an eigenvalue of $\left[a_{ij}\right]_{n\times n}$. Similarly, a_{22} is an eigenvalue of $\left[a_{ij}\right]_{n\times n}$, etc. ∎

3.2.21 Theorem Let V be any n-dimensional vector space over F. Let $T \in A(V)$. Suppose that all the roots of the minimal polynomial of T over F are in F. Then there exists a polynomial $p(x)$ in $F[x]$ such that $p(x)$ is of degree n, and $p(T) = 0$.

Proof By 3.2.16, there exists a basis $\{v_1, v_2, \ldots, v_n\}$ of V such that the matrix $\left[\alpha_{ij}\right]_{n\times n}$ of T relative to the basis $\{v_1, v_2, \ldots, v_n\}$ is triangular. It follows that

$$i < j \Rightarrow \alpha_{ij} = 0,$$

and

$$
\begin{aligned}
T(v_1) &= \alpha_{11}v_1, \\
T(v_2) &= \alpha_{21}v_1 + \alpha_{22}v_2, \\
T(v_3) &= \alpha_{31}v_1 + \alpha_{32}v_2 + \alpha_{33}v_3, \\
&\;\;\vdots
\end{aligned}
$$

For every $i \in \{1, 2, \ldots, n\}$, put $\lambda_i \equiv \alpha_{ii}$. By 3.2.20, the set of all eigenvalues of $\left[a_{ij}\right]_{n\times n}$ is $\{\lambda_1, \lambda_2, \ldots, \lambda_n\}$. Further,

$$
\begin{aligned}
(T - \lambda_1 I)(v_1) &= 0, \\
(T - \lambda_2 I)(v_2) &= \alpha_{21}v_1, \\
(T - \lambda_3 I)(v_3) &= \alpha_{31}v_1 + \alpha_{32}v_2, \\
&\;\;\vdots
\end{aligned}
$$

Observe that

$$((T - \lambda_1 I) \circ (T - \lambda_2 I))(v_2) = \underbrace{(T - \lambda_1 I)((T - \lambda_2 I)(v_2))}_{} = (T - \lambda_1 I)(\alpha_{21} v_1)$$

$$= \alpha_{21}(T - \lambda_1 I)(v_1) = \alpha_{21} 0 = 0,$$

$$((T - \lambda_1 I) \circ (T - \lambda_2 I) \circ (T - \lambda_3 I))(v_3)$$

$$= \underbrace{(((T - \lambda_1 I) \circ (T - \lambda_2 I)))((T - \lambda_3 I)(v_3)) = (((T - \lambda_1 I) \circ (T - \lambda_2 I)))}_{}$$

$$(\alpha_{31} v_1 + \alpha_{32} v_2)$$

$$= \alpha_{31}((T - \lambda_1 I) \circ (T - \lambda_2 I))(v_1) + \alpha_{32}((T - \lambda_1 I) \circ (T - \lambda_2 I))(v_2)$$

$$= \alpha_{31}((T - \lambda_1 I) \circ (T - \lambda_2 I))(v_1) + \alpha_{32} 0$$

$$= \alpha_{31}((T - \lambda_1 I) \circ (T - \lambda_2 I))(v_1) = \alpha_{31}(T \circ T - \lambda_1 T - \lambda_2 T + \lambda_1 \lambda_2 I)(v_1)$$

$$= \alpha_{31}((T - \lambda_2 I) \circ (T - \lambda_1 I))(v_1)$$

$$= \alpha_{31}(T - \lambda_2 I)((T - \lambda_1 I)(v_1)) = \alpha_{31}(T - \lambda_2 I)(0) = 0, \text{ etc.}$$

Thus

$$(T - \lambda_1 I)(v_1) = 0,$$
$$((T - \lambda_1 I) \circ (T - \lambda_2 I))(v_2) = 0,$$
$$((T - \lambda_1 I) \circ (T - \lambda_2 I) \circ (T - \lambda_3 I))(v_3) = 0,$$
$$\vdots$$

Now, since each $(T - \lambda_i I)$ commutes with each $(T - \lambda_j I)$, we have

$$\left.\begin{array}{r}
((T - \lambda_1 I) \circ (T - \lambda_2 I) \circ \cdots \circ (T - \lambda_n I))(v_1) = 0 \\
((T - \lambda_1 I) \circ (T - \lambda_2 I) \circ \cdots \circ (T - \lambda_n I))(v_2) = 0 \\
\vdots \\
((T - \lambda_1 I) \circ (T - \lambda_2 I) \circ \cdots \circ (T - \lambda_n I))(v_n) = 0
\end{array}\right\}.$$

Next, since $\{v_1, v_2, \ldots, v_n\}$ is a basis of V, for every $v \in V$, we have

$$((T - \lambda_1 I) \circ (T - \lambda_2 I) \circ \cdots \circ (T - \lambda_n I))(v) = 0.$$

This shows that $(T - \lambda_1 I) \circ (T - \lambda_2 I) \circ \cdots \circ (T - \lambda_n I) = 0$, and hence $p(T) = 0$, where

$$p(x) \equiv (x - \lambda_1)(x - \lambda_2) \cdots (x - \lambda_n).$$

Since each λ_i is in F, $(p(x) \equiv)(x - \lambda_1)(x - \lambda_2) \cdots (x - \lambda_n)$ is a polynomial in $F[x]$, and it is of degree n. Thus $p(x)$ is a polynomial in $F[x]$ of degree n. ∎

Definition Let V be any n-dimensional vector space over the field F. Let $T \in A(V)$. If there exists a positive integer m such that $T^m = 0$, then we say that T is nilpotent.

3.2.22 Problem Let V be any n-dimensional vector space over the field F. Let $T \in A(V)$. Suppose that T is nilpotent. Let λ be an eigenvalue of T. Then $\lambda = 0$.

Proof Suppose to the contrary that $\lambda \neq 0$. We seek a contradiction.

Since λ is an eigenvalue of T, there exists a nonzero v in V such that $T(v) = \lambda v$. Since T is nilpotent, there exists a positive integer m such that $T^m = 0$. It follows that $T^m(v) = 0$. Hence

$$\{k : k \text{ is a positive integer, and } T^k(v) = 0\}$$

is a nonempty set of positive integers. It follows that

$$\min\{k : k \text{ is a positive integer, and } T^k(v) = 0\}$$

exists. Put

$$n \equiv \min\{k : k \text{ is a positive integer, and } T^k(v) = 0\}.$$

It follows that $T^{n-1}(v) \neq 0$ and $T^n(v) = 0$. Now,

$$\lambda T^{n-1}(v) = T^{n-1}(\lambda v) = T^{n-1}(T(v)) = \underbrace{T^n(v) = 0}.$$

Thus $\lambda T^{n-1}(v) = 0$. Since $\lambda \neq 0$, we have $T^{n-1}(v) = 0$. This is a contradiction. ∎

3.2.23 Problem Let V be any n-dimensional vector space over the field F. Let $T \in A(V)$. Suppose that T is nilpotent. Let $\alpha_0, \alpha_1, \ldots, \alpha_n \in F$. Let $\alpha_0 \neq 0$. Then

$$\alpha_0 I + \alpha_1 T + \cdots + \alpha_n T^n$$

is invertible, and $(\alpha_0 I + \alpha_1 T + \cdots + \alpha_n T^n)^{-1}$ is a polynomial in the linear transformation S, where $S \equiv \alpha_1 T + \cdots + \alpha_n T^n$.

Proof We have to show that $\alpha_0 I + S$ is invertible.

Since T is nilpotent, there exists a positive integer m such that $T^m = 0$. It follows that

$$r \geq m \Rightarrow T^r = 0,$$

and hence

$$S^m = (\alpha_1 T + \cdots + \alpha_n T^n)^m$$
$$= \underbrace{(\alpha_1 T + \cdots + \alpha_n T^n) \circ (\alpha_1 T + \cdots + \alpha_n T^n) \circ \cdots \circ (\alpha_1 T + \cdots + \alpha_n T^n)}_{m \text{ factors}} = 0.$$

Thus $S^m = 0$. Observe that

$$\left(I + \tfrac{1}{\alpha_0}S\right) \circ \left(\tfrac{1}{\alpha_0}I - \tfrac{1}{(\alpha_0)^2}S + \tfrac{1}{(\alpha_0)^3}S^2 - \cdots + (-1)^{m-1}\tfrac{1}{(\alpha_0)^m}S^{m-1}\right).$$

$$= \left(\tfrac{1}{\alpha_0}I - \tfrac{1}{(\alpha_0)^2}S + \tfrac{1}{(\alpha_0)^3}S^2 - \cdots + (-1)^{m-1}\tfrac{1}{(\alpha_0)^m}S^{m-1}\right)$$

$$+ \left(\tfrac{1}{(\alpha_0)^2}S - \tfrac{1}{(\alpha_0)^3}S^2 + \tfrac{1}{(\alpha_0)^4}S^3 - \cdots - (-1)^{m-1}\tfrac{1}{(\alpha_0)^m}S^{m-1} + (-1)^{m-1}\tfrac{1}{(\alpha_0)^{m+1}}S^m\right)$$

$$= \tfrac{1}{\alpha_0}I + (-1)^{m-1}\tfrac{1}{(\alpha_0)^{m+1}}S^m = \tfrac{1}{\alpha_0}I + (-1)^{m-1}\tfrac{1}{(\alpha_0)^{m+1}}0 = \tfrac{1}{\alpha_0}I,$$

so

$$\left(I + \frac{1}{\alpha_0}S\right) \circ \left(\frac{1}{\alpha_0}I - \frac{1}{(\alpha_0)^2}S + \frac{1}{(\alpha_0)^3}S^2 - \cdots + (-1)^{m-1}\frac{1}{(\alpha_0)^m}S^{m-1}\right) = \frac{1}{\alpha_0}I.$$

This shows that

$$(\alpha_0 I + S)^{-1} = \left(\frac{1}{\alpha_0}I - \frac{1}{(\alpha_0)^2}S + \frac{1}{(\alpha_0)^3}S^2 - \cdots + (-1)^{m-1}\frac{1}{(\alpha_0)^m}S^{m-1}\right),$$

and hence $\alpha_0 I + S$ is invertible.   ∎

Definition Let V be any n-dimensional vector space over the field F. Let $T \in A(V)$. Suppose that T is nilpotent.

Since T is nilpotent, there exists a positive integer m such that $T^m = 0$. Hence

$$\{k : k \text{ is a positive integer, and } T^k = 0\}$$

is a nonempty set of positive integers. It follows that

$$\min\{k : k \text{ is a positive integer, and } T^k = 0\}$$

exists. Put

$$n \equiv \min\{k : k \text{ is a positive integer, and } T^k = 0\}.$$

It follows that $T^{n-1} \neq 0$ and $T^n = 0$. Here n is called the *index of nilpotence* of T.

3.2.24 Note Let V be any n-dimensional vector space over the field F. Let $T \in A(V)$. Suppose that T is nilpotent. Let n_1 be a positive integer. Let n_1 be the index of nilpotence of T.

It follows that $T^{n_1-1} \neq 0$ and $T^{n_1} = 0$. Since $T^{n_1-1} : V \to V$ is a nonzero function, there exists a nonzero $v \in V$ such that $T^{n_1-1}(v) \neq 0$ and $T^{n_1}(v) = 0$.

Clearly, $v, T(v), T^2(v), \ldots, T^{n_1-1}(v)$ are linearly independent over F.

Proof Suppose to the contrary that $v, T(v), T^2(v), \ldots, T^{n_1-1}(v)$ are linearly dependent over F. We seek a contradiction.

It follows that there exist $\alpha_1, \alpha_2, \ldots, \alpha_{n_1} \in F$ such that not all the α_i are zero, and

$$\alpha_1 v + \alpha_2 T(v) + \alpha_3 T^2(v) + \cdots + \alpha_{n_1} T^{n_1-1}(v) = 0.$$

Suppose that α_s is the first nonzero α_i. It follows that $s \leq n_1$, $\alpha_s \neq 0$, and $(r < s \Rightarrow \alpha_r = 0)$. Hence

$$\left(\alpha_s I + \alpha_{s+1} T + \alpha_{s+2} T^2 + \cdots + \alpha_{n_1} T^{n_1-s}\right)\left(T^{s-1}(v)\right)$$
$$= \underbrace{\alpha_s T^{s-1}(v) + \alpha_{s+1} T^s(v) + \cdots + \alpha_{n_1} T^{n_1-1}(v) = 0}.$$

This shows that

$$\left(\alpha_s I + \alpha_{s+1} T + \alpha_{s+2} T^2 + \cdots + \alpha_{n_1} T^{n_1-s}\right)\left(T^{s-1}(v)\right) = 0. \quad (*)$$

Since $\alpha_s \neq 0$ and T is nilpotent, by 3.2.23,

$$\left(\alpha_s I + \alpha_{s+1} T + \alpha_{s+2} T^2 + \cdots + \alpha_{n_1} T^{n_1-s}\right)^{-1}$$

exists. Now, from $(*)$, $T^{s-1}(v) = 0$. Since $s \leq n_1$, we have $(s-1) \leq (n_1 - 1)$. Since $T^{n_1-1}(v) \neq 0$, we have $T^{s-1}(v) \neq 0$. This is a contradiction. ∎

Put

$$v_1 \equiv v, v_2 \equiv T(v), v_3 \equiv T^2(v), \ldots, v_{n_1} \equiv T^{n_1-1}(v).$$

Since $v, T(v), T^2(v), \ldots, T^{n_1-1}(v)$ are linearly independent over F, it follows that $\{v_1, v_2, \ldots, v_{n_1}\}$ is a linearly independent set of vectors in V.

Suppose that V_1 is the linear span of $\{v_1, v_2, \ldots, v_{n_1}\}$.

It follows that V_1 is an n_1-dimensional linear subspace of V, and $\{v_1, v_2, \ldots, v_{n_1}\}$ is a basis of V_1. Observe that

$$T(v_1) = T(v) = v_2 \in V_1,$$
$$T(v_2) = T(T(v)) = T^2(v) = v_3 \in V_1,$$
$$\vdots$$
$$T(v_{n_1}) = T(T^{n_1-1}(v)) = T^{n_1}(v) = 0 \in V_1,$$

and hence $\{T(v_1), T(v_2), \ldots, T(v_{n_1})\} \subset V_1$. Now, since $\{v_1, v_2, \ldots, v_{n_1}\}$ is a basis of V_1, we have $(w \in V_1 \Rightarrow T(w) \in V_1)$.

Thus we have shown that V_1 is invariant under T.

3.2.24.1 There exists a linear subspace W of V, of largest possible dimension, such that

1. $V_1 \cap W = \{0\}$,
2. W is invariant under T.

Proof Since V_1 is an n_1-dimensional linear subspace of the n-dimensional vector space V, and $\{v_1, v_2, \ldots, v_{n_1}\}$ is a basis of V_1, there exists a basis $\{v_1, v_2, \ldots, v_{n_1}, w_{n_1+1}, w_{n_1+2}, \ldots, w_n\}$ of V.

Let W_1 be the linear span of $\{w_{n_1+1}, w_{n_1+2}, \ldots, w_n\}$.

It is clear that W_1 is a linear subspace of V, $\dim(W_1) = n - n_1 (\geq 1)$, and $V = V_1 \oplus W_1$. Thus $V_1 \cap W_1 = \{0\}$. Since $\{v_1, v_2, \ldots, v_{n_1}, w_{n_1+1}, w_{n_1+2}, \ldots, w_n\}$ is a basis of V, we have $w_{n_1+1} \neq 0$. Now, since $T^{n_1} = 0$, we have that

$$\{k : k \text{ is a positive integer, and } T^k(w_{n_1+1}) = 0\}$$

is a nonempty set of positive integers, and it follows that

$$\min\{k : k \text{ is a positive integer, and } T^k(w_{n_1+1}) = 0\}$$

exists. Put

$$n_2 \equiv \min\{k : k \text{ is a positive integer, and } T^k(w_{n_1+1}) = 0\}.$$

It follows that $T^{n_2-1}(w_{n_1+1}) \neq 0$ and $T^{n_2}(w_{n_1+1}) = 0$. Put $w \equiv w_{n_1+1}$. We have $w \in W_1, T^{n_2-1}(w) \neq 0$ and $T^{n_2}(w) = 0$.

Clearly, $w, T(w), T^2(w), \ldots, T^{n_2-1}(w)$ are linearly independent over F.

Proof Suppose to the contrary that $w, T(w), T^2(w), \ldots, T^{n_2-1}(w)$ are linearly dependent over F. We seek a contradiction.

It follows that there exist $\alpha_1, \alpha_2, \ldots, \alpha_{n_2} \in F$ such that not all the α_i are zero and

$$\alpha_1 w + \alpha_2 T(w) + \alpha_3 T^2(w) + \cdots + \alpha_{n_2} T^{n_2-1}(w) = 0.$$

Suppose that α_s is the first nonzero α_i. It follows that $s \leq n_2$, $\alpha_s \neq 0$, and $(r < s \Rightarrow \alpha_r = 0)$. Hence

$$\left(\alpha_s I + \alpha_{s+1} T + \alpha_{s+2} T^2 + \cdots + \alpha_{n_2} T^{n_2-s}\right)\left(T^{s-1}(w)\right)$$
$$= \underbrace{\alpha_s T^{s-1}(w) + \alpha_{s+1} T^s(w) + \cdots + \alpha_{n_2} T^{n_2-1}(w) = 0}.$$

This shows that

$$\left(\alpha_s I + \alpha_{s+1} T + \alpha_{s+2} T^2 + \cdots + \alpha_{n_2} T^{n_2-s}\right)\left(T^{s-1}(w)\right) = 0. \quad (*)$$

Since $\alpha_s \neq 0$ and T is nilpotent, by 3.2.23,

$$\left(\alpha_s I + \alpha_{s+1} T + \alpha_{s+2} T^2 + \cdots + \alpha_{n_2} T^{n_2-s}\right)^{-1}$$

exists. Now, from (*), $T^{s-1}(w) = 0$. Since $s \leq n_2$, we have $(s-1) \leq (n_2 - 1)$. Since
$T^{n_2-1}(w) \neq 0$, we have $T^{s-1}(w) \neq 0$. This is a contradiction. ∎

Put

$$w_1 \equiv w, w_2 \equiv T(w), w_3 \equiv T^2(w), \ldots, w_{n_2} \equiv T^{n_2-1}(w).$$

Since $w, T(w), T^2(w), \ldots, T^{n_2-1}(w)$ are linearly independent over F, it follows that $\{w_1, w_2, \ldots, w_{n_2}\}$ is a linearly independent set of vectors in V.
Suppose that W_1 is the linear span of $\{w_1, w_2, \ldots, w_{n_2}\}$.
It follows that W_1 is an n_2-dimensional linear subspace of V, and $\{w_1, w_2, \ldots, w_{n_2}\}$ is a basis of W_1. Observe that

$$T(w_1) = T(w) = w_2 \in W_1,$$
$$T(w_2) = T(T(w)) = T^2(w) = w_3 \in W_1,$$
$$\vdots$$
$$T(w_{n_2}) = T(T^{n_2-1}(w)) = T^{n_2}(w) = 0 \in W_1,$$

and hence $\{T(w_1), T(w_2), \ldots, T(w_{n_2})\} \subset W_1$. Now, since $\{w_1, w_2, \ldots, w_{n_2}\}$ is a basis of W_1, we have

$(z \in W_1 \Rightarrow T(z) \in W_1)$. Thus we have shown that W_1 is invariant under T. Hence there exists a linear subspace W_1 of V such that

1. $V_1 \cap W_1 = \{0\}$,
2. W_1 is invariant under T.

It follows that there exists a linear subspace W of V, of largest possible dimension, such that

1. $V_1 \cap W = \{0\}$,
2. W is invariant under T. ∎

3.2.24.2 Suppose that $u \in V_1$. Let k be a positive integer such that $k \leq n_1$. Suppose that $T^{(n_1-k)}(u) = 0$. Then there exists $u_0 \in V_1$ such that $T^k(u_0) = u$.

Proof Since $u \in V_1$ and V_1 is the linear span of $\{v, T(v), T^2(v), \ldots, T^{n_1-1}(v)\}$, there exist $\alpha_1, \alpha_2, \ldots, \alpha_{n_1} \in F$ such that

$$u = \alpha_1 v + \alpha_2 T(v) + \alpha_3 T^2(v) + \cdots + \alpha_{n_1} T^{n_1-1}(v).$$

Since

$$\underbrace{0 = T^{(n_1-k)}(u)}$$

$$= T^{(n_1-k)}\left(\alpha_1 v + \alpha_2 T(v) + \alpha_3 T^2(v) + \cdots + \alpha_k T^{k-1}(v)\right.$$
$$\left. + \alpha_{k+1} T^k(v) + \cdots + \alpha_{n_1} T^{n_1-1}(v)\right)$$
$$= \alpha_1 T^{(n_1-k)}(v) + \alpha_2 T^{(n_1-k)+1}(v) + \alpha_3 T^{(n_1-k)+2}(v) + \cdots$$
$$+ \alpha_k T^{n_1-1}(v) + \alpha_{k+1} T^{n_1}(v)$$
$$+ \alpha_{k+2} T^{n_1+1}(v) + \cdots + \alpha_{n_1} T^{(n_1-k)+(n_1-1)}(v)$$
$$= \alpha_1 T^{(n_1-k)}(v) + \alpha_2 T^{(n_1-k)+1}(v) + \alpha_3 T^{(n_1-k)+2}(v) + \cdots + \alpha_k T^{n_1-1}(v)$$
$$+ \alpha_{k+1} 0(v) + \alpha_{k+2} 0(v) + \cdots + \alpha_{n_1} 0(v)$$
$$= \alpha_1 T^{(n_1-k)}(v) + \alpha_2 T^{(n_1-k)+1}(v) + \alpha_3 T^{(n_1-k)+2}(v) + \cdots + \alpha_k T^{n_1-1}(v),$$

we have

$$\alpha_1 T^{(n_1-k)}(v) + \alpha_2 T^{(n_1-k)+1}(v) + \alpha_3 T^{(n_1-k)+2}(v) + \cdots + \alpha_k T^{n_1-1}(v) = 0.$$

Next, since $\{v, T(v), T^2(v), \ldots, T^{n_1-1}(v)\}$ is a basis of V_1, we have $\alpha_1 = 0, \alpha_2 = 0, \ldots, \alpha_k = 0$. It follows that

$$\underbrace{u = \alpha_{k+1} T^k(v) + \alpha_{k+2} T^{k+1}(v) + \cdots + \alpha_{n_1} T^{n_1-1}(v)}$$

$$= T^k\left(\alpha_{k+1} v + \alpha_{k+2} T(v) + \cdots + \alpha_{n_1} T^{n_1-k-1}(v)\right)$$
$$= T^k(\alpha_{k+1} v_1 + \alpha_{k+2} v_2 + \cdots + \alpha_{n_1} v_{n_1-k}) = T^k(u_0),$$

where $u_0 \equiv (\alpha_{k+1} v_1 + \alpha_{k+2} v_2 + \cdots + \alpha_{n_1} v_{n_1-k}) \in [v_1, v_2, \ldots, v_{n_1}] = V_1$. Thus $u_0 \in V_1$ and $T^k(u_0) = u$. ∎

3.2.24.3 There exists a linear subspace W of V, of largest possible dimension, such that

1. $V = V_1 \oplus W$,
2. W is invariant under T.

Proof By 3.2.24.1, there exists a linear subspace W of V such that

1. $V_1 \cap W = \{0\}$,
2. W is invariant under T.

It suffices to show that $V \subset V_1 + W$.

Suppose to the contrary that there exists $z \in V$ such that $z \notin (V_1 + W)$. We seek a contradiction.

Clearly, $(V_1 + W) \subset T^{-1}(V_1 + W)$.

Proof To show this, let us take arbitrary $x \in V_1$ and $y \in W$. We have to show that

$$(T(x) + T(y) =) T(x + y) \in (V_1 + W),$$

that is, $(T(x) + T(y)) \in (V_1 + W)$. It suffices to show that $T(x) \in V_1$ and $T(y) \in W$. Since V_1 is invariant under T, and $x \in V_1$, we have $T(x) \in V_1$. Since W is invariant under T, and $y \in W$, we have $T(y) \in W$.   ∎

Clearly, $T^{-1}(V_1 + W) \subset (T^2)^{-1}(V_1 + W)$.

Proof To show this, let us take arbitrary $x \in V_1$ and $y \in W$ such that $T(x + y) \in (V_1 + W)$. We have to show that $(T^2(x) + T^2(y) =) T^2(x + y) \in (V_1 + W)$, that is, $(T^2(x) + T^2(y)) \in (V_1 + W)$. It suffices to show that $T^2(x) \in V_1$ and $T^2(y) \in W$. Since V_1 is invariant under T, and $x \in V_1$, we have $T(x) \in V_1$, and hence $(T^2(x) =) T(T(x)) \in V_1$. Since W is invariant under T, and $y \in W$, we have $T(y) \in W$, and hence $(T^2(y) =) T(T(Y)) \in W$.   ∎

Clearly, $(T^2)^{-1}(V_1 + W) \subset (T^3)^{-1}(V_1 + W)$.

Proof To show this, let us take arbitrary $x \in V_1$ and $y \in W$ such that $T^2(x + y) \in (V_1 + W)$. We have to show that

$$\left(T^3(x) + T^3(y) =\right) T^3(x + y) \in (V_1 + W),$$

that is, $(T^3(x) + T^3(y)) \in (V_1 + W)$. It suffices to show that $T^3(x) \in V_1$ and $T^3(y) \in W$. Since V_1 is invariant under T, and $x \in V_1$, we have $T(x) \in V_1$, and hence $(T^2(x) =) T(T(x)) \in V_1$. Now, $(T^3(x) =) T(T^2(x)) \in V_1$, so $T^3(x) \in V_1$. Similarly, $T^3(y) \in W$.

Thus we have shown that $(T^2)^{-1}(V_1 + W) \subset (T^3)^{-1}(V_1 + W)$, etc.   ∎

Hence

$$z \notin (V_1 + W) \subset T^{-1}(V_1 + W) \subset (T^2)^{-1}(V_1 + W) \subset (T^3)^{-1}(V_1 + W) \subset \cdots$$
$$\subset (T^{n_1})^{-1}(V_1 + W) = V.$$

Now, since $z \in V$, there exists a positive integer k such that

1. $k \leq n_1$,
2. $z \in (T^k)^{-1}(V_1 + W)$,
3. $r < k \Rightarrow z \notin (T^r)^{-1}(V_1 + W)$.

Since $z \in (T^k)^{-1}(V_1 + W)$, we have $(T^k)(z) \in (V_1 + W)$, and hence there exist $u \in V_1$ and $w \in W$ such that $(T^k)(z) = u + w$. It follows that

$$0 = 0(z) = (T^{n_1})(z) = \underbrace{T^{n_1-k}\big((T^k)(z)\big) = T^{n_1-k}(u+w)} = T^{n_1-k}(u) + T^{n_1-k}(w),$$

and hence $T^{n_1-k}(u) = -T^{n_1-k}(w)$. Since $u \in V_1$, and V_1 is invariant under T, we have $T^{n_1-k}(u) \in V_1$. Since $w \in W$, and W is invariant under T, we have $T^{n_1-k}(w) \in W$, and hence $\big(T^{n_1-k}(u) =\big) - T^{n_1-k}(w) \in W$. Thus

$$\underbrace{T^{n_1-k}(u) \in V_1 \cap W} = \{0\}.$$

It follows that $T^{n_1-k}(u) = 0$. Now by 3.2.24.2, there exists $u_0 \in V_1$ such that $T^k(u_0) = u \left(= (T^k)(z) - w\right)$, and hence $T^k(z - u_0) = w (\in W)$. Thus $T^k(z - u_0) \in W$. Next, since W is invariant under T, we have

$$\underbrace{m \geq k \Rightarrow T^{(m-k)}\big(T^k(z - u_0)\big) \in W,}$$

and hence

$$m \geq k \Rightarrow T^m(z - u_0) \in W.$$

It follows that $T^k(z - u_0) \in W$.

Let us take an arbitrary $r < k$. Clearly, $T^r(z - u_0) \notin (V_1 + W)$. (*)

Proof Suppose to the contrary that $(T^r(z) - T^r(u_0) =)T^r(z - u_0) \in (V_1 + W)$. We seek a contradiction.

Since $u_0 \in V_1$, and V_1 is invariant under T, we have $T^r(u_0) \in V_1$. Now, since $(T^r(z) - T^r(u_0)) \in (V_1 + W)$, we have

$$T^r(z) = \underbrace{T^r(u_0) + (T^r(z) - T^r(u_0)) = V_1 + (V_1 + W)} = (V_1 + V_1) + W$$

$$= V_1 + W,$$

and hence $(T^r)(z) \in (V_1 + W)$. Since $r < k$, we have by point 3 above that $(T^r)(z) \notin (V_1 + W)$. This is a contradiction. ∎

Thus we have shown that

$$r < k \Rightarrow T^r(z - u_0) \notin (V_1 + W). \quad \text{(A)}$$

Clearly, $(z - u_0) \notin (V_1 + W)$.

Proof Suppose to the contrary that $(z - u_0) \in (V_1 + W)$. We seek a contradiction. Since $u_0 \in V_1$ and $(z - u_0) \in (V_1 + W)$, we have

$$z \in u_0 + (V_1 + W) = (u_0 + V_1) + W \subset (V_1 + V_1) + W = V_1 + W,$$

and hence $z \in (V_1 + W)$. This is a contradiction.   ∎

Thus we have shown that $(z - u_0) \notin (V_1 + W) (\supset \{0\} + W = W)$. Hence $(z - u_0) \notin W$. Similarly, from (A),

$$T(z - u_0), T^2(z - u_0), \ldots, T^{k-1}(z - u_0) \notin (V_1 + W) (\supset \{0\} + W = W).$$

Thus

$$\left\{ (z - u_0), T(z - u_0), T^2(z - u_0), \ldots, T^{k-1}(z - u_0) \right\} \subset W^c.$$

Suppose that W_2 is the linear span of

$$\left\{ (z - u_0), T(z - u_0), T^2(z - u_0), \ldots, T^{k-1}(z - u_0) \right\} \cup W.$$

Since $(z - u_0) \notin W$, the dimension of the linear span of

$$\left\{ (z - u_0), T(z - u_0), T^2(z - u_0), \ldots, T^{k-1}(z - u_0) \right\} \cup W$$

is strictly greater than $\dim(W)$, and hence

$$\dim(W_2) > \dim(W).$$

It follows that

either $(V_1 \cap W_2 \neq \{0\})$ or W_2 is not invariant under T.   (∗∗)

Observe that

$$T(z - u_0) \in \left(\left\{ (z - u_0), T(z - u_0), T^2(z - u_0), \ldots, T^{k-1}(z - u_0) \right\} \cup W \right) \subset W_2,$$

so $T(z - u_0) \in W_2$. Next,

$$T(T(z - u_0)) \in \left(\left\{ (z - u_0), T(z - u_0), T^2(z - u_0), \ldots, T^{k-1}(z - u_0) \right\} \cup W \right) \subset W_2,$$

so $T(T(z - u_0)) \in W_2$. Similarly, $T(T^2(z - u_0)) \in W_2, \ldots$. Next, $T(T^{k-1}(z - u_0)) = T^k(z - u_0) \in W \subset W_2$, so $T(T^{k-1}(z - u_0)) \in W_2$. Thus

$$T\left(\left\{ (z - u_0), T(z - u_0), T^2(z - u_0), \ldots, T^{k-1}(z - u_0) \right\} \right) \subset W_2.$$

Since W is invariant under T, we have $T(W) \subset W \subset W_2$. Thus

$$
\begin{aligned}
&T\big(\{(z-u_0), T(z-u_0), T^2(z-u_0), \ldots, T^{k-1}(z-u_0)\} \cup W\big) \\
&= \underbrace{T\big(\{(z-u_0), T(z-u_0), T^2(z-u_0), \ldots, T^{k-1}(z-u_0)\}\big) \cup T(W) \subset W_2}.
\end{aligned}
$$

Hence

$$
T\big(\{(z-u_0), T(z-u_0), T^2(z-u_0), \ldots, T^{k-1}(z-u_0)\} \cup W\big) \subset W_2.
$$

Now, since T is a linear transformation and W_2 is a linear space, we have

$$
T(W_2) = \underbrace{\dfrac{T \text{ (linear span of } (\{(z-u_0), T(z-u_0),}{T^2(z-u_0), \ldots, T^{k-1}(z-u_0)\} \cup W)) \subset W_2}},
$$

and hence $T(W_2) \subset W_2$. This shows that W_2 is invariant under T. It follows from
(**) that $V_1 \cap W_2 \neq \{0\}$.

Since $V_1 \cap W_2 \neq \{0\}$, there exists a nonzero

$$
z_0 \in W_2 \left(= \text{linear span of} \left(\left\{ \begin{array}{c} (z-u_0), T(z-u_0), T^2(z-u_0), \ldots, \\ T^{k-1}(z-u_0) \end{array} \right\} \cup W \right) \right)
$$

such that $z_0 \in V_1$. It follows that there exist $\alpha_1, \alpha_2, \ldots, \alpha_k \in F$ and $w^* \in W$ such
that

$$
(0 \neq) z_0 = \alpha_1(z-u_0) + \alpha_2 T(z-u_0) + \cdots + \alpha_k T^{k-1}(z-u_0) + w^*.
$$

Clearly, not all of $\alpha_1, \alpha_2, \ldots, \alpha_k$ are zero,

Proof Suppose to the contrary that each α_i is zero. We seek a contradiction.
Since each α_i is zero, we have

$$
V_1 \ni z_0 = \underbrace{0(z-u_0) + 0T(z-u_0) + \cdots + 0T^{k-1}(z-u_0) + w^*}_{} = w^*,
$$

and hence $w^* \in V_1$. Also $w^* \in W$. It follows that $(z_0 =) w^*$
$\in V_1 \cap W (= \{0\})$, and hence $z_0 = 0$. This is a contradiction. ∎

Suppose that α_s is the first nonzero α_i, where $s \leq k$. It follows from (*) that
$T^{s-1}(z-u_0) \notin (V_1 + W)$. Also

$$(0 \neq) \underbrace{z_0 = \alpha_s T^{s-1}(z - u_0) + \alpha_{s+1} T^s (z - u_0) + \cdots + \alpha_k T^{k-1}(z - u_0) + w^*}$$

$$= \left(\alpha_s I + \alpha_{s+1} T + \cdots + \alpha_k T^{k-s} \right) \left(T^{s-1}(z - u_0) \right) + w^*,$$

so

$$\left(\alpha_s I + \alpha_{s+1} T + \cdots + \alpha_k T^{k-s} \right) \left(T^{s-1}(z - u_0) \right) = z_0 - w^*.$$

By 3.2.23, $\left(\alpha_s I + \alpha_{s+1} T + \cdots + \alpha_k T^{k-s} \right)^{-1}$ exists and is a polynomial $p(S)$ in S, where

$$S \equiv \alpha_{s+1} T + \cdots + \alpha_k T^{k-s}.$$

Thus

$$(V_1 + W) \not\ni \underbrace{T^{s-1}(z - u_0) = (p(S))(z_0 - w^*)} = (p(S))(z_0) - (p(S))(w^*).$$

Since V_1 is invariant under T, V_1 is invariant under $(\alpha_{s+1} T + \cdots + \alpha_k T^{k-s}) (= S)$, and hence V_1 is invariant under S. Now, since $p(S)$ is a polynomial in S, V_1 is invariant under $p(S)$. Next, since $z_0 \in V_1$, we have $(p(S))(z_0) \in V_1$. Since $((p(S))(z_0) - (p(S))(w^*)) \notin (V_1 + W)$, we have $(p(S))(w^*) \notin W$.

Since W is invariant under T, W is invariant under $(\alpha_{s+1} T + \cdots + \alpha_k T^{k-s}) (= S)$, and hence W is invariant under S. Since $p(S)$ is a polynomial in S, W is invariant under $p(S)$. Next, since $w^* \in W$, we have

$$(p(S))(w^*) \in W.$$

This is a contradiction.   ∎

3.2.24.4 Note Let $V_1, V_2, \ldots, V_k$ be linear subspaces of V such that

1. $V = V_1 \oplus V_2 \oplus \cdots \oplus V_k$,
2. each V_i is invariant under T.

Suppose that $\dim(V_1) = n_1$, and $\{v_1^1, v_1^2, \ldots, v_1^{n_1}\}$ is a basis of V_1. Suppose that $\dim(V_2) = n_2$, and $\{v_2^1, v_2^2, \ldots, v_2^{n_2}\}$ is a basis of V_2, etc.

Since $V = V_1 \oplus V_2 \oplus \cdots \oplus V_k$, we have

$$n = \dim(V) = n_1 + n_2 + \cdots + n_k,$$

and

$$\{v_1^1, v_1^2, \ldots, v_1^{n_1}; v_2^1, v_2^2, \ldots, v_2^{n_2}; \ldots; v_k^1, v_k^2, \ldots, v_k^{n_k}\}$$

is a basis of V. Since $v_1^1 \in V_1$, and V_1 is invariant under T, we have $T(v_1^1) \in V_1$. Now, since $\{v_1^1, v_1^2, \ldots, v_1^{n_1}\}$ is a basis of V_1, there exist $\alpha_1^1, \alpha_1^2, \ldots, \alpha_1^{n_1} \in F$ such that

$$T(v_1^1) = \alpha_1^1 v_1^1 + \alpha_1^2 v_1^2 + \cdots + \alpha_1^{n_1} v_1^{n_1}.$$

Hence

$$\begin{aligned}
T(v_1^1) = &\left(\alpha_1^1 v_1^1 + \alpha_1^2 v_1^2 + \cdots + \alpha_1^{n_1} v_1^{n_1}\right) \\
&+ \left(0 v_2^1 + 0 v_2^2 + \cdots + 0 v_2^{n_2}\right) + \cdots \\
&+ \left(0 v_k^1 + 0 v_k^2 + \cdots + 0 v_k^{n_k}\right).
\end{aligned}$$

Similarly,

$$T(v_1^2) = \left(\alpha_2^1 v_1^1 + \cdots + \alpha_2^{n_1} v_1^{n_1}\right) + \left(0 v_2^1 + \cdots + 0 v_2^{n_2}\right) + \cdots + \left(0 v_k^1 + \cdots + 0 v_k^{n_k}\right),$$

$$\vdots$$

$$T(v_1^{n_1}) = \left(\alpha_{n_1}^1 v_1^1 + \cdots + \alpha_{n_1}^{n_1} v_1^{n_1}\right) + \left(0 v_2^1 + \cdots + 0 v_2^{n_2}\right) + \cdots + \left(0 v_k^1 + \cdots + 0 v_k^{n_k}\right),$$

$$T(v_2^1) = \left(0 v_1^1 + \cdots + 0 v_1^{n_1}\right) + \left(\beta_1^1 v_2^1 + \cdots + \beta_1^{n_2} v_2^{n_2}\right) + \cdots + \left(0 v_k^1 + \cdots + 0 v_k^{n_k}\right),$$

$$T(v_2^2) = \left(0 v_1^1 + \cdots + 0 v_1^{n_1}\right) + \left(\beta_2^1 v_2^1 + \cdots + \beta_2^{n_2} v_2^{n_2}\right) + \cdots + \left(0 v_k^1 + \cdots + 0 v_k^{n_k}\right),$$

$$\vdots$$

$$T(v_2^{n_2}) = \left(0 v_1^1 + \cdots + 0 v_1^{n_1}\right) + \left(\beta_{n_2}^1 v_2^1 + \cdots + \beta_{n_2}^{n_2} v_2^{n_2}\right) + \cdots + \left(0 v_k^1 + \cdots + 0 v_k^{n_k}\right),$$

$$\vdots$$

Thus the matrix of $T(\in A(V))$ relative to the basis

$$\{v_1^1, v_1^2, \ldots, v_1^{n_1}; v_2^1, v_2^2, \ldots, v_2^{n_2}; \ldots; v_k^1, v_k^2, \ldots, v_k^{n_k}\}$$

is the $n \times n$ matrix in the canonical form

$$\begin{bmatrix} A_1 & 0 & 0 \\ 0 & A_2 & 0 \\ 0 & 0 & \ddots \end{bmatrix},$$

where

$$A_1 \equiv \begin{bmatrix} \alpha_1^1 & \alpha_1^2 & \cdots \\ \alpha_2^1 & \alpha_2^2 & \cdots \\ \vdots & \vdots & \ddots \end{bmatrix}_{n_1 \times n_1},$$

$$A_2 \equiv \begin{bmatrix} \beta_1^1 & \beta_1^2 & \cdots \\ \beta_2^1 & \beta_2^2 & \cdots \\ \vdots & \vdots & \ddots \end{bmatrix}_{n_2 \times n_2} , \text{ etc.}$$

Since

$$\left(T|_{V_1}\right)\left(v_1^1\right) = T\left(v_1^1\right) = \alpha_1^1 v_1^1 + \alpha_1^2 v_1^2 + \cdots + \alpha_1^{n_1} v_1^{n_1},$$
$$\left(T|_{V_1}\right)\left(v_1^2\right) = T\left(v_1^2\right) = \alpha_2^1 v_1^1 + \alpha_2^2 v_1^2 + \cdots + \alpha_2^{n_1} v_1^{n_1},$$
$$\vdots$$
$$\left(T|_{V_1}\right)\left(v_1^{n_1}\right) = T\left(v_1^{n_1}\right) = \alpha_{n_1}^1 v_1^1 + \alpha_{n_1}^2 v_1^2 + \cdots + \alpha_{n_1}^{n_1} v_1^{n_1},$$

A_1 is the $n_1 \times n_1$ matrix of the linear transformation $T|_{V_1}$ induced by T on V_1. Similarly, A_2 is the $n_2 \times n_2$ matrix of the linear transformation $T|_{V_2}$ induced by T on V_2, etc.

3.2.24.5 Conclusion Let $V_1, V_2, \ldots, V_k$ be linear subspaces of V such that

1. $V = V_1 \oplus V_2 \oplus \cdots \oplus V_k$,
2. each V_i is invariant under T.

Then there exist a basis $\{v_1^1, v_1^2, \ldots, v_1^{n_1}\}$ of V_1, a basis $\{v_2^1, v_2^2, \ldots, v_2^{n_2}\}$ of $V_2, \ldots$, a basis $\{v_k^1, v_k^2, \ldots, v_k^{n_k}\}$ of V_k such that the matrix of $T \ (\in A(V))$ relative to the basis $\{v_1^1, v_1^2, \ldots, v_1^{n_1}; v_2^1, v_2^2, \ldots, v_2^{n_2}; \ldots; v_k^1, v_k^2, \ldots, v_k^{n_k}\}$ has the canonical form

$$\begin{bmatrix} A_1 & 0 & 0 \\ 0 & A_2 & 0 \\ 0 & 0 & \ddots \end{bmatrix}_{n \times n},$$

where A_1 is the matrix of the linear transformation $T|_{V_1}$ relative to $\{v_1^1, v_1^2, \ldots, v_1^{n_1}\}$, A_2 is the matrix of the linear transformation $T|_{V_2}$ relative to $\{v_2^1, v_2^2, \ldots, v_2^{n_2}\}$, etc. Also $v_1^2 = T\left(v_1^1\right)$, $v_1^3 = T^2\left(v_1^1\right), \ldots$; $v_2^2 = T\left(v_2^1\right)$, $v_1^3 = T^2\left(v_2^1\right), \ldots$; etc. Hence A_1 takes the form

$$\begin{bmatrix} 0 & 1 & 0 & & & 0 \\ 0 & 0 & 1 & & & 0 \\ \vdots & & 0 & 0 & \ddots & 0 \\ \vdots & \vdots & & 0 & & 0 \\ \vdots & \vdots & & \vdots & & 1 \\ 0 & 0 & & 0 & & 0 \end{bmatrix}_{t \times t}.$$

Notation This matrix is denoted by M_t. Thus

$$M_t \equiv \begin{bmatrix} 0 & 1 & 0 & & 0 \\ 0 & 0 & 1 & & 0 \\ \vdots & & 0 & 0 & \ddots & 0 \\ \vdots & \vdots & & 0 & & 0 \\ \vdots & \vdots & \vdots & & & 1 \\ 0 & 0 & 0 & & 0 \end{bmatrix}_{t \times t} .$$

Similarly, A_2 takes the form M_s for some positive integer s, etc.

3.2.24.6 Let V be any n-dimensional vector space over the field F. Let $T \in A(V)$. Suppose that T is nilpotent. Let n_1 be a positive integer. Let n_1 be the index of nilpotence of T.

By 3.2.24.3, there exist linear subspaces V_1, W of V such that

1. $V = V_1 \oplus W$,
2. V_1, W are invariant under T.

Now, by 3.2.24.4, there exists a basis $\{v_1^1, v_1^2, \ldots, v_1^{n_1}\}$ of V_1 such that for every basis $\mathcal{B}$ of W, the matrix of $T (\in A(V))$ relative to the basis $\{v_1^1, v_1^2, \ldots, v_1^{n_1}\} \cup \mathcal{B}$ has the canonical form

$$\begin{bmatrix} M_{n_1} & 0 \\ 0 & A_2 \end{bmatrix}_{n \times n},$$

where M_{n_1} is the $n_1 \times n_1$ matrix of the linear transformation $T|_{V_1}$ relative to the basis $\{v_1^1, v_1^2, \ldots, v_1^{n_1}\}$, and A_2 is the $(n - n_1) \times (n - n_1)$ matrix of the linear transformation $T|_W$ relative to the "arbitrary" basis $\mathcal{B}$.

Since $T^{n_1} = 0$ and W is invariant under T, we have $(T|_W)^{n_1} = 0$, and hence there exists a smallest integer n_2 such that $n_2 \le n_1$ and $(T|_W)^{n_2} = 0$. Thus $(T_2)^{n_2} = 0$, where $T_2 \equiv T|_W \in A(W)$. Also, n_2 is the index of nilpotence of T_2.

Again by 3.2.24.3, there exist linear subspaces V_2, X of W such that

1. $W = V_2 \oplus X$,
2. V_2, X are invariant under T_2.

Now, by 3.2.24.4, there exists a basis $\{v_2^1, v_2^2, \ldots, v_2^{n_2}\}$ of V_2 such that for every basis $\mathcal{C}$ of X, the matrix of $T_2 (\in A(W))$ relative to the basis $\{v_2^1, v_2^2, \ldots, v_2^{n_2}\} \cup \mathcal{C}$ has the canonical form

$$\begin{bmatrix} M_{n_2} & 0 \\ 0 & A_3 \end{bmatrix}_{(n-n_1) \times (n-n_1)},$$

where M_{n_2} is the $n_2 \times n_2$ matrix of the linear transformation $T|_{V_2}$ relative to the basis $\{v_2^1, v_2^2, \ldots, v_2^{n_2}\}$, and A_3 is the $((n - n_1) - n_2) \times ((n - n_1) - n_2)$ matrix of the linear transformation T_2 relative to the "arbitrary" basis $\mathcal{C}$.

Now, since A_2 is the $(n - n_1) \times (n - n_1)$ matrix of the linear transformation T_2 relative to the "arbitrary" basis $\mathcal{B}$, the matrix of $T (\in A(V))$ relative to the basis $\{v_1^1, v_1^2, \ldots, v_1^{n_1}\} \cup \{v_2^1, v_2^2, \ldots, v_2^{n_2}\} \cup \mathcal{C}$ has the canonical form

$$
\begin{bmatrix}
M_{n_1} & 0 & 0 \\
0 & M_{n_2} & 0 \\
0 & 0 & A_3
\end{bmatrix}_{n \times n}.
$$

We can repeat the above process finitely many times, obtaining finally the following result.

3.2.25 Conclusion Let V be any n-dimensional vector space over the field F. Let $T \in A(V)$. Suppose that T is nilpotent. Then there exist linear subspaces $V_1, V_2, \ldots, V_k$ of V such that

1. $V = V_1 \oplus V_2 \oplus \cdots \oplus V_k$,
2. each V_i is invariant under T.

Also there exist a basis $\{v_1^1, v_1^2, \ldots, v_1^{n_1}\}$ of V_1, a basis $\{v_2^1, v_2^2, \ldots, v_2^{n_2}\}$ of $V_2, \ldots$, a basis $\{v_k^1, v_k^2, \ldots, v_k^{n_k}\}$ of V_k such that the matrix of $T (\in A(V))$ relative to the basis $\{v_1^1, v_1^2, \ldots, v_1^{n_1}; v_2^1, v_2^2, \ldots, v_2^{n_2}; \ldots; v_k^1, v_k^2, \ldots, v_k^{n_k}\}$ has the canonical form

$$
\begin{bmatrix}
M_{n_1} & 0 & 0 \\
0 & \ddots & 0 \\
0 & 0 & M_{n_k}
\end{bmatrix}_{n \times n},
$$

where M_{n_1} is the matrix of the linear transformation $T|_{V_1}$ relative to $\{v_1^1, v_1^2, \ldots, v_1^{n_1}\}$, M_{n_2} is the matrix of the linear transformation $T|_{V_2}$ relative to $\{v_2^1, v_2^2, \ldots, v_2^{n_2}\}$, etc. Further,

$$
n = n_1 + \cdots + n_k
$$

and

$$
n_1 \geq n_2 \geq \cdots \geq n_k.
$$

3.3 The Cayley–Hamilton Theorem

3.3.1 Definition By the *transpose* of an $m \times n$ matrix $A \equiv [a_{ij}]$, we mean the $n \times m$ matrix whose (i,j)-entry is a_{ji}. The transpose of A is denoted by A^T. By the *conjugate* of an $m \times n$ matrix $A \equiv [a_{ij}]$ having complex numbers as entries, we mean the $m \times n$ matrix whose (i,j)-entry is $\overline{a_{ij}}$ where $\overline{a_{ij}}$ denotes the complex conjugate of the complex number a_{ij}. The conjugate of A is denoted by $\overline{A}$. By A^* we mean $\left(\overline{A}\right)^T$, and this matrix is called the *conjugate transpose of A*. The $n \times n$ matrix whose (i,j)-entry is

$$\left.\begin{array}{l} 1 \text{ if } i = j \\ 0 \text{ if } i \neq j \end{array}\right\}$$

is denoted by I_n, or simply I. By a *scalar matrix* we mean a scalar multiple of I. By a *zero matrix* we mean a matrix each entry of which is 0.

Definition Let $A \equiv [a_{ij}]$ be a square complex matrix. If $(i \neq j \Rightarrow a_{ij} = 0)$, then we say that A *is diagonal*. If $(i > j \Rightarrow a_{ij} = 0)$, then we say that A *is upper triangular*. If $A^T = A$, then we say that A *is symmetric*. If $A^T A = AA^T = I$, then we say that A *is orthogonal*. If $A^* A = AA^* = I$, then we say that A *is unitary*. If $A^* = A$, then we say that A *is Hermitian*. If $A^* A = AA^*$, then we say that A *is normal*.

Definition Let A, B be square complex matrices of the same size. If there exists an invertible matrix P such that $P^{-1} A P = B$, then we say that A *and B are similar*. Clearly, the relation of similarity is an equivalence relation.

Definition A *submatrix of a matrix* is obtained by suppressing some rows and/or suppressing some columns from the given matrix. Some nomenclatures are self-explanatory. For example, if

$$A \equiv \begin{bmatrix} 1 & 3 & -1 \\ 4 & 0 & \frac{3}{5} \\ 0 & 3 & -2 \end{bmatrix}$$

is the given matrix, then $\begin{bmatrix} 0 & \frac{3}{5} \end{bmatrix}$ is a submatrix of A. Using submatrices, we can form matrices like

$$\begin{bmatrix} \begin{bmatrix} 1 \\ 4 \\ 0 \end{bmatrix} & \begin{bmatrix} 3 & -1 \\ 0 & \frac{3}{5} \\ 3 & -2 \end{bmatrix} \end{bmatrix},$$

which is an example of a *partitioned form of A* into a 2×2 block matrix. The manipulation of matrices in partitioned form is a basic technique in linear algebra. For example:

$$
\begin{bmatrix} \begin{bmatrix} [1] \\ \begin{bmatrix} 4 \\ 0 \end{bmatrix} \end{bmatrix} & \begin{bmatrix} 3 & -1 \\ 0 & \frac{3}{5} \\ 3 & -2 \end{bmatrix} \end{bmatrix}_{2\times2} \begin{bmatrix} \begin{bmatrix} [1] \\ \begin{bmatrix} 4 \\ 0 \end{bmatrix} \end{bmatrix} & \begin{bmatrix} [3] \\ 0 \\ 3 \end{bmatrix} & \begin{bmatrix} [-1] \\ \frac{3}{5} \\ -2 \end{bmatrix} \end{bmatrix}_{2\times3}
$$

$$
= \begin{bmatrix} [1][1] + [3 \ \ -1]\begin{bmatrix} 4 \\ 0 \end{bmatrix} & [1][3] + [3 \ \ -1]\begin{bmatrix} 0 \\ 3 \end{bmatrix} & [1][-1] + [3 \ \ -1]\begin{bmatrix} \frac{3}{5} \\ -2 \end{bmatrix} \\ \begin{bmatrix} 4 \\ 0 \end{bmatrix}[1] + \begin{bmatrix} 0 & \frac{3}{5} \\ 3 & -2 \end{bmatrix}\begin{bmatrix} 4 \\ 0 \end{bmatrix} & \begin{bmatrix} 4 \\ 0 \end{bmatrix}[3] + \begin{bmatrix} 0 & \frac{3}{5} \\ 3 & -2 \end{bmatrix}\begin{bmatrix} 0 \\ 3 \end{bmatrix} & \begin{bmatrix} 4 \\ 0 \end{bmatrix}[-1] + \begin{bmatrix} 0 & \frac{3}{5} \\ 3 & -2 \end{bmatrix}\begin{bmatrix} \frac{3}{5} \\ -2 \end{bmatrix} \end{bmatrix}_{2\times3}
$$

$$
= \begin{bmatrix} [1] + [12] & [3] + [-3] & [-1] + [\frac{19}{5}] \\ \begin{bmatrix} 4 \\ 0 \end{bmatrix} + \begin{bmatrix} 0 \\ 12 \end{bmatrix} & \begin{bmatrix} 12 \\ 0 \end{bmatrix} + \begin{bmatrix} \frac{9}{5} \\ -6 \end{bmatrix} & \begin{bmatrix} -4 \\ 0 \end{bmatrix} + \begin{bmatrix} \frac{-6}{5} \\ \frac{29}{5} \end{bmatrix} \end{bmatrix}_{2\times3} = \begin{bmatrix} [13] & [0] & [\frac{14}{5}] \\ \begin{bmatrix} 4 \\ 12 \end{bmatrix} & \begin{bmatrix} \frac{69}{5} \\ -6 \end{bmatrix} & \begin{bmatrix} \frac{-26}{5} \\ \frac{29}{5} \end{bmatrix} \end{bmatrix}_{2\times3}.
$$

Further,

$$
\begin{bmatrix} 1 & 3 & -1 \\ 4 & 0 & \frac{3}{5} \\ 0 & 3 & -2 \end{bmatrix} \begin{bmatrix} 1 & 3 & -1 \\ 4 & 0 & \frac{3}{5} \\ 0 & 3 & -2 \end{bmatrix} = \begin{bmatrix} 13 & 0 & \frac{14}{5} \\ 4 & \frac{69}{5} & \frac{-26}{5} \\ 12 & -6 & \frac{29}{5} \end{bmatrix}.
$$

3.3.2 Note By an elementary *row operation* for matrices, we mean any one of the following:

1. interchange any two rows,
2. multiply a row by a nonzero constant,
3. add a multiple of a row to another row.

Definition An $n \times n$ (or *n*-square for short) matrix E is called an *elementary matrix* if there exists an elementary row operation R such that E is obtained by a single application of R on the unit matrix I_n.

For $n = 3$, examples of elementary matrices are

$$
\begin{bmatrix} 1 & 0 & 0 \\ 0 & 0 & 1 \\ 0 & 1 & 0 \end{bmatrix}, \begin{bmatrix} 1 & 0 & 0 \\ 0 & -3 & 0 \\ 0 & 0 & 1 \end{bmatrix}, \begin{bmatrix} 1 & 0 & 0 \\ 0 & 1 & 5 \\ 0 & 0 & 1 \end{bmatrix}.
$$

3.3.3 Example Observe that

$$
\begin{bmatrix} 1 & 2 & 3 \\ 4 & 5 & 6 \end{bmatrix} \xrightarrow{R_2 \to R_2 + (-4)R_1} \begin{bmatrix} 1 & 2 & 3 \\ 0 & -3 & -6 \end{bmatrix},
$$

$$
\begin{bmatrix} 1 & 0 \\ 0 & 1 \end{bmatrix} \xrightarrow{R_2 \to R_2 + (-4)R_1} \begin{bmatrix} 1 & 0 \\ -4 & 1 \end{bmatrix},
$$

and

$$\begin{bmatrix} 1 & 0 \\ -4 & 1 \end{bmatrix} \begin{bmatrix} 1 & 2 & 3 \\ 4 & 5 & 6 \end{bmatrix} = \begin{bmatrix} 1 & 2 & 3 \\ 0 & -3 & -6 \end{bmatrix}.$$

Again

$$\begin{bmatrix} 1 & 2 & 3 \\ 0 & -3 & -6 \end{bmatrix} \xrightarrow{R_2 \to \frac{1}{-3}R_2} \begin{bmatrix} 1 & 2 & 3 \\ 0 & 1 & 2 \end{bmatrix},$$

$$\begin{bmatrix} 1 & 0 \\ 0 & 1 \end{bmatrix} \xrightarrow{R_2 \to \frac{1}{-3}R_2} \begin{bmatrix} 1 & 0 \\ 0 & \frac{1}{-3} \end{bmatrix},$$

and

$$\begin{bmatrix} 1 & 0 \\ 0 & \frac{1}{-3} \end{bmatrix} \begin{bmatrix} 1 & 2 & 3 \\ 0 & -3 & -6 \end{bmatrix} = \begin{bmatrix} 1 & 2 & 3 \\ 0 & 1 & 2 \end{bmatrix}.$$

Next,

$$\begin{bmatrix} 1 & 2 & 3 \\ 0 & 1 & 2 \end{bmatrix} \xrightarrow{R_1 \to R_1 + (-2)R_2} \begin{bmatrix} 1 & 0 & -1 \\ 0 & 1 & 2 \end{bmatrix},$$

$$\begin{bmatrix} 1 & 0 \\ 0 & 1 \end{bmatrix} \xrightarrow{R_1 \to R_1 + (-2)R_2} \begin{bmatrix} 1 & -2 \\ 0 & 1 \end{bmatrix},$$

and

$$\begin{bmatrix} 1 & -2 \\ 0 & 1 \end{bmatrix} \begin{bmatrix} 1 & 2 & 3 \\ 0 & 1 & 2 \end{bmatrix} = \begin{bmatrix} 1 & 0 & -1 \\ 0 & 1 & 2 \end{bmatrix}.$$

Now we apply certain elementary column operations on $\begin{bmatrix} 1 & 0 & -1 \\ 0 & 1 & 2 \end{bmatrix}$ to make things simpler. The following are self-explanatory:

$$\begin{bmatrix} 1 & 0 & -1 \\ 0 & 1 & 2 \end{bmatrix} \xrightarrow{C_3 \to C_3 + (-2)C_2} \begin{bmatrix} 1 & 0 & -1 \\ 0 & 1 & 0 \end{bmatrix},$$

$$\begin{bmatrix} 1 & 0 & 0 \\ 0 & 1 & 0 \\ 0 & 0 & 1 \end{bmatrix} \xrightarrow{C_3 \to C_3 + (-2)C_2} \begin{bmatrix} 1 & 0 & 0 \\ 0 & 1 & -2 \\ 0 & 0 & 1 \end{bmatrix},$$

$$\begin{bmatrix} 1 & 0 & -1 \\ 0 & 1 & 2 \end{bmatrix} \begin{bmatrix} 1 & 0 & 0 \\ 0 & 1 & -2 \\ 0 & 0 & 1 \end{bmatrix} = \begin{bmatrix} 1 & 0 & -1 \\ 0 & 1 & 0 \end{bmatrix}.$$

Next,

$$\begin{bmatrix} 1 & 0 & -1 \\ 0 & 1 & 0 \end{bmatrix} \xrightarrow{C_3 \to C_3 + 1C_1} \begin{bmatrix} 1 & 0 & 0 \\ 0 & 1 & 0 \end{bmatrix},$$

$$\begin{bmatrix} 1 & 0 & 0 \\ 0 & 1 & 0 \\ 0 & 0 & 1 \end{bmatrix} \xrightarrow{C_3 \to C_3 + 1C_1} \begin{bmatrix} 1 & 0 & 1 \\ 0 & 1 & 0 \\ 0 & 0 & 1 \end{bmatrix},$$

$$\begin{bmatrix} 1 & 0 & -1 \\ 0 & 1 & 0 \end{bmatrix} \begin{bmatrix} 1 & 0 & 1 \\ 0 & 1 & 0 \\ 0 & 0 & 1 \end{bmatrix} = \begin{bmatrix} 1 & 0 & 0 \\ 0 & 1 & 0 \end{bmatrix}.$$

If we collect the above information, we get

$$\begin{bmatrix} 1 & 2 & 3 \\ 4 & 5 & 6 \end{bmatrix} \xrightarrow{R_2 \to R_2 + (-4)R_1} \begin{bmatrix} 1 & 2 & 3 \\ 0 & -3 & -6 \end{bmatrix} \xrightarrow{R_2 \to -\frac{1}{3}R_2} \begin{bmatrix} 1 & 2 & 3 \\ 0 & 1 & 2 \end{bmatrix}$$

$$\xrightarrow{R_1 \to R_1 + (-2)R_2} \begin{bmatrix} 1 & 0 & -1 \\ 0 & 1 & 2 \end{bmatrix} \xrightarrow{C_3 \to C_3 + (-2)C_2} \begin{bmatrix} 1 & 0 & -1 \\ 0 & 1 & 0 \end{bmatrix} \xrightarrow{C_3 \to C_3 + 1C_1} \begin{bmatrix} 1 & 0 & 0 \\ 0 & 1 & 0 \end{bmatrix}$$

and

$$\left(\begin{bmatrix} 1 & -2 \\ 0 & 1 \end{bmatrix} \begin{bmatrix} 1 & 0 \\ 0 & \frac{1}{-3} \end{bmatrix} \begin{bmatrix} 1 & 0 \\ -4 & 1 \end{bmatrix} \right) \begin{bmatrix} 1 & 2 & 3 \\ 4 & 5 & 6 \end{bmatrix} \left(\begin{bmatrix} 1 & 0 & 0 \\ 0 & 1 & -2 \\ 0 & 0 & 1 \end{bmatrix} \begin{bmatrix} 1 & 0 & 1 \\ 0 & 1 & 0 \\ 0 & 0 & 1 \end{bmatrix} \right)$$

$$= \begin{bmatrix} 1 & 0 & 0 \\ 0 & 1 & 0 \end{bmatrix}.$$

3.3.4 Conclusion Let A be a nonzero $m \times n$ matrix with complex numbers as entries. Then there exist an invertible $m \times m$ matrix P, an invertible $n \times n$ matrix Q, and a positive integer r such that

1. $PAQ = \begin{bmatrix} I_r & 0 \\ 0 & 0 \end{bmatrix}_{m \times n}$,
2. P is a product of elementary matrices of size $m \times m$,
3. Q is a product of elementary matrices of size $n \times n$.
4. r is the rank of A.

3.3.5 Observe that

$$Q^T (A^T) P^T = (PAQ)^T = \underbrace{\left(\begin{bmatrix} I_r & 0 \\ 0 & 0 \end{bmatrix}_{m \times n} \right)^T}_{} = \begin{bmatrix} I_r & 0 \\ 0 & 0 \end{bmatrix}_{n \times m},$$

so

$$R(A^T)S = \begin{bmatrix} I_r & 0 \\ 0 & 0 \end{bmatrix}_{n \times m},$$

where $R \equiv Q^T$ and $S \equiv P^T$. Since Q is invertible, $(R =) Q^T$ is invertible, and hence R is invertible. Similarly, S is invertible. It follows that $\text{rank}(A^T) = r \,(= \text{rank}(A))$.

Thus $\text{rank}(A^T) = \text{rank}(A)$. Similarly, $\text{rank}(\overline{A}) = \text{rank}(A)$ and $\text{rank}(A^*) = \text{rank}(A)$.

3.3.6 Note It is easy to see that for a partitioned matrix $\begin{bmatrix} A & B \\ 0 & C \end{bmatrix}$, we have

$$\det \begin{bmatrix} A & B \\ 0 & C \end{bmatrix} = \det(A) \cdot \det(C).$$

It is also written as

$$\begin{vmatrix} A & B \\ 0 & C \end{vmatrix} = \det(A) \cdot \det(C).$$

3.3.7 Example

$$\begin{vmatrix} \begin{bmatrix} a_{11} & a_{12} & a_{13} \\ a_{21} & a_{22} & a_{23} \\ a_{31} & a_{32} & a_{33} \end{bmatrix} & \begin{bmatrix} b_{11} & b_{12} \\ b_{21} & b_{22} \\ b_{31} & b_{32} \end{bmatrix} \\ \begin{bmatrix} 0 & 0 & 0 \\ 0 & 0 & 0 \end{bmatrix} & \begin{bmatrix} c_{11} & c_{12} \\ c_{21} & c_{22} \end{bmatrix} \end{vmatrix}$$

$$= -c_{21} \begin{vmatrix} a_{11} & a_{12} & a_{13} & b_{12} \\ a_{21} & a_{22} & a_{23} & b_{22} \\ a_{31} & a_{32} & a_{33} & b_{32} \\ 0 & 0 & 0 & c_{12} \end{vmatrix} + c_{22} \begin{vmatrix} a_{11} & a_{12} & a_{13} & b_{11} \\ a_{21} & a_{22} & a_{23} & b_{21} \\ a_{31} & a_{32} & a_{33} & b_{31} \\ 0 & 0 & 0 & c_{11} \end{vmatrix}$$

$$= -c_{21} \left(c_{12} \begin{vmatrix} a_{11} & a_{12} & a_{13} \\ a_{21} & a_{22} & a_{23} \\ a_{31} & a_{32} & a_{33} \end{vmatrix} \right) + c_{22} \left(c_{11} \begin{vmatrix} a_{11} & a_{12} & a_{13} \\ a_{21} & a_{22} & a_{23} \\ a_{31} & a_{32} & a_{33} \end{vmatrix} \right)$$

$$= \begin{vmatrix} a_{11} & a_{12} & a_{13} \\ a_{21} & a_{22} & a_{23} \\ a_{31} & a_{32} & a_{33} \end{vmatrix} (c_{11}c_{22} - c_{12}c_{21}) = \begin{vmatrix} a_{11} & a_{12} & a_{13} \\ a_{21} & a_{22} & a_{23} \\ a_{31} & a_{32} & a_{33} \end{vmatrix} \begin{vmatrix} c_{11} & c_{12} \\ c_{21} & c_{22} \end{vmatrix},$$

and hence

$$\begin{vmatrix} \begin{bmatrix} a_{11} & a_{12} & a_{13} \\ a_{21} & a_{22} & a_{23} \\ a_{31} & a_{32} & a_{33} \\ 0 & 0 & 0 \\ 0 & 0 & 0 \end{bmatrix} \begin{bmatrix} b_{11} & b_{12} \\ b_{21} & b_{22} \\ b_{31} & b_{32} \\ c_{11} & c_{12} \\ c_{21} & c_{22} \end{bmatrix} \end{vmatrix} = \begin{vmatrix} a_{11} & a_{12} & a_{13} \\ a_{21} & a_{22} & a_{23} \\ a_{31} & a_{32} & a_{33} \end{vmatrix} \begin{vmatrix} c_{11} & c_{12} \\ c_{21} & c_{22} \end{vmatrix}.$$

3.3.8 Problem Let V be an n-dimensional vector space over $\mathbb{C}$. Let $\mathcal{A} : V \to V$ be a linear transformation. Let λ_1 and λ_2 be distinct eigenvalues of $\mathcal{A}$. Let v_1 be an eigenvector corresponding to the eigenvalue λ_1. Let v_2 be an eigenvector corresponding to the eigenvalue λ_2. Then v_1, v_2 are linearly independent.

Proof Suppose to the contrary that $v_1 = kv_2$, where k is a complex number. We seek a contradiction.

Since v_2 is an eigenvector corresponding to the eigenvalue λ_2, we have $v_2 \neq 0$. Similarly, $v_1 \neq 0$. Now, since $v_1 = kv_2$, we have $k \neq 0$. Also, $\mathcal{A}(v_1) = \lambda_1 v_1$ and $\mathcal{A}(v_2) = \lambda_2 v_2$. Hence

$$k\mathcal{A}(v_2) = \mathcal{A}(kv_2) = \underbrace{\mathcal{A}(v_1) = \lambda_1 v_1} = \lambda_1 kv_2.$$

Since $k\mathcal{A}(v_2) = \lambda_1 kv_2$ and k is nonzero, we have $\lambda_2 v_2 = \underbrace{\mathcal{A}(v_2) = \lambda_1 v_2}$, and

hence $(\lambda_2 - \lambda_1)v_2 = 0$. Since λ_1 and λ_2 are distinct, $\lambda_2 - \lambda_1$ is nonzero. Now, since $(\lambda_2 - \lambda_1)v_2 = 0$, we have $v_2 = 0$. This is a contradiction. ∎

3.3.9 Conclusion The eigenvectors corresponding to distinct eigenvalues are linearly independent.

3.3.10 Note By $\mathbb{C}^n$ we shall mean the collection of all column matrices of size $n \times 1$ with complex entries. Such matrices are also called *column vectors*. We know that $\mathbb{C}^n$ is a vector space over $\mathbb{C}$. For every $x \equiv [x_1, \ldots, x_n]^T$ and $y \equiv [y_1, \ldots, y_n]^T$ in $\mathbb{C}^n$, we define

$$\langle x, y \rangle \equiv x_1 \overline{y_1} + \ldots + x_n \overline{y_n} (= x^T \bar{y} = y^* x).$$

Clearly, $\mathbb{C}^n$ is an inner product space. For every $x \equiv [x_1, \ldots, x_n]^T$ in $\mathbb{C}^n$, we define the length x of x as follows:

$$\|x\| \equiv \sqrt{|x_1|^2 + \cdots + |x_n|^2} \left(= \sqrt{\langle x, x \rangle} \right).$$

In $\mathbb{C}^n$, an m-tuple $(v_1, \ldots, v_m)$ of vectors in $\mathbb{C}^n$ can be thought of as a matrix $[v_{ij}]_{n \times m}$, where $v_1 \equiv [v_{11}, \ldots, v_{n1}]^T$, $v_1 \equiv [v_{12}, \ldots, v_{n2}]^T$, etc.

3.3.11 Note Let x be a nonzero vector of $\mathbb{C}^n$.

There exist distinct nonzero vectors $v_2, \ldots, v_n$ in $\mathbb{C}^n$ such that $\{x, v_2, \ldots, v_n\}$ is a basis of $\mathbb{C}^n$. Assume that for some complex number α, $\alpha x + v_2$ is orthogonal to x, that is,

$$(\alpha \langle x, x \rangle + \langle v_2, x \rangle =) \langle \alpha x + v_2, x \rangle = 0.$$

Since x is nonzero, $\langle x, x \rangle$ is nonzero. This shows that $\frac{-\langle v_2, x \rangle}{\langle x, x \rangle} x + v_2$ is orthogonal to x. Let us put

$$y_2 \equiv \frac{-\langle v_2, x \rangle}{\langle x, x \rangle} x + v_2.$$

Since $\{x, v_2, \ldots, v_n\}$ is a basis of $\mathbb{C}^n$, $\left\{ x, \frac{-\langle v_2, x \rangle}{\langle x, x \rangle} x + v_2, v_3, \ldots, v_n \right\}$ is a basis of $\mathbb{C}^n$, and hence $\{x, y_2, v_3, \ldots, v_n\}$ is a basis of $\mathbb{C}^n$. Since, $(y_2 =) \frac{-\langle v_2, x \rangle}{\langle x, x \rangle} x + v_2$ is orthogonal to x, it follows that y_2 is orthogonal to x.

Thus $\{x, y_2, v_3, \ldots, v_n\}$ is a basis of $\mathbb{C}^n$ such that y_2 is orthogonal to x.

Assume that for some complex numbers α, β, $\alpha x + \beta y_2 + v_3$ is orthogonal to x and y_2, that is,

$$(\alpha \langle x, x \rangle + \langle v_3, x \rangle = \alpha \langle x, x \rangle + \beta 0 + \langle v_3, x \rangle = \alpha \langle x, x \rangle + \beta \langle y_2, x \rangle + \langle v_3, x \rangle =)$$
$$\langle \alpha x + \beta y_2 + v_3, x \rangle = 0$$

and

$$(\beta \langle y_2, y_2 \rangle + \langle v_3, y_2 \rangle = \alpha 0 + \beta \langle y_2, y_2 \rangle + \langle v_3, y_2 \rangle = \alpha \langle x, y_2 \rangle + \beta \langle y_2, y_2 \rangle + \langle v_3, y_2 \rangle =)$$
$$\langle \alpha x + \beta y_2 + v_3, y_2 \rangle = 0.$$

It follows that $-\frac{\langle v_3, x \rangle}{\langle x, x \rangle} x - \frac{\langle v_3, y_2 \rangle}{\langle y_2, y_2 \rangle} y_2 + v_3$ is orthogonal to x and y_2. Let us put

$$y_3 \equiv -\frac{\langle v_3, x \rangle}{\langle x, x \rangle} x - \frac{\langle v_3, y_2 \rangle}{\langle y_2, y_2 \rangle} y_2 + v_3.$$

Since $\{x, y_2, v_3, \ldots, v_n\}$ is a basis of $\mathbb{C}^n$, $\left\{ x, y_2, -\frac{\langle v_3, x \rangle}{\langle x, x \rangle} x - \frac{\langle v_3, y_2 \rangle}{\langle y_2, y_2 \rangle} y_2 + v_3, v_4, \ldots, v_n \right\}$ is a basis of $\mathbb{C}^n$, and hence $\{x, y_2, y_3, v_4, \ldots, v_n\}$ is a basis of $\mathbb{C}^n$. Since $(y_3 =) - \frac{\langle v_3, x \rangle}{\langle x, x \rangle} x - \frac{\langle v_3, y_2 \rangle}{\langle y_2, y_2 \rangle} y_2 + v_3$ is orthogonal to x and y_2, it follows that y_3 is orthogonal to x and y_2.

Thus $\{x, y_2, y_3, v_4, \ldots, v_n\}$ is a basis of $\mathbb{C}^n$ such that the members of $\{x, y_2, y_3\}$ are mutually orthogonal, etc.

Finally, we get a basis $\{x, y_2, y_3, y_4, \ldots, y_n\}$ of $\mathbb{C}^n$ such that the members of $\{x, y_2, y_3, y_4, \ldots, y_n\}$ are mutually orthogonal.

It follows that $\left\{\frac{1}{\|x\|}x, \frac{1}{\|y_2\|}y_2, \frac{1}{\|y_3\|}y_3, \frac{1}{\|y_4\|}y_4, \ldots, \frac{1}{\|y_n\|}y_n\right\}$ is an orthonormal basis of $\mathbb{C}^n$.

3.3.12 Conclusion Let $u_1, \ldots, u_k$ be orthonormal vectors in $\mathbb{C}^n$. Then there exist $\underbrace{u_{k+1}, \ldots, u_n}_{n-k}$ in $\mathbb{C}^n$ such that $\{u_1, \ldots, u_k, u_{k+1}, \ldots, u_n\}$ is an orthonormal basis of $\mathbb{C}^n$. By the definition of unitary matrix $(A^*A = AA^* = I)$, the n-square matrix $[u_1, \ldots, u_k, u_{k+1}, \ldots, u_n]$ is unitary.

The construction used here is known as the **Gram–Schmidt orthogonalization process**.

3.3.13 Note Let $A \equiv [a_{ij}]$ be an n-square complex matrix.

For every $x \equiv [x_1, \ldots, x_n]^T$ in $\mathbb{C}^n$, the product Ax of matrices A and x is a matrix of size $n \times 1$, and hence Ax is a member of $\mathbb{C}^n$. Let us define a function $\mathcal{A} : \mathbb{C}^n \to \mathbb{C}^n$ as follows: for every $x \equiv [x_1, \ldots, x_n]^T$ in $\mathbb{C}^n$,

$$\mathcal{A}(x) \equiv Ax.$$

Clearly, $\mathcal{A} : \mathbb{C}^n \to \mathbb{C}^n$ is a linear transformation.

3.3.14 Note Let $\mathcal{A} : \mathbb{C}^n \to \mathbb{C}^n$ be a linear transformation. Put

$$e_1 \equiv [1, 0, \ldots, 0]^T (\in \mathbb{C}^n), e_2 \equiv [0, 1, 0, \ldots, 0]^T (\in \mathbb{C}^n), \text{ etc.}$$

Clearly, $\{e_1, \ldots, e_n\}$ is an orthonormal basis of the inner product space $\mathbb{C}^n$. This basis is called the *standard basis of* $\mathbb{C}^n$. Let $A \equiv [\alpha_{ij}]$ be the matrix of $\mathcal{A}$ relative to the basis $\{e_1, \ldots, e_n\}$. Hence

$$\mathcal{A}(e_1) = a_{11}e_1 + a_{21}e_2 + \cdots + a_{n1}e_n \left(= \sum_{i=1}^{n} a_{i1}e_i = \begin{bmatrix} a_{11} \\ \vdots \\ a_{n1} \end{bmatrix} = [a_{11}, \ldots, a_{n1}]^T \right),$$

$$\mathcal{A}(e_2) = a_{12}e_1 + a_{22}e_2 + \cdots + a_{n2}a_n \left(= \sum_{i=1}^{n} a_{i2}e_i = \begin{bmatrix} a_{12} \\ \vdots \\ a_{n2} \end{bmatrix} = [a_{12}, \ldots, a_{n2}]^T \right),$$

etc.

In short, $\mathcal{A}(e_j) = [a_{1j}, \ldots, a_{nj}]^T$. By 3.3.13,

$$\widehat{\mathcal{A}} : x \mapsto Ax$$

is a linear transformation from $\mathbb{C}^n$ to $\mathbb{C}^n$. It follows that

$$\widehat{\mathcal{A}}(e_1) = Ae_1 = \begin{bmatrix} a_{11} \\ \vdots \\ a_{n1} \end{bmatrix} = \mathcal{A}(e_1),$$

$$\widehat{\mathcal{A}}(e_2) = Ae_2 = \begin{bmatrix} a_{12} \\ \vdots \\ a_{n2} \end{bmatrix} = \mathcal{A}(e_2), \text{ etc.}$$

It follows that $\widehat{\mathcal{A}} = \mathcal{A}$.

3.3.15 Conclusion The n-square complex matrix $A \equiv [a_{ij}]$ can be represented by the linear transformation $x \mapsto Ax$ from $\mathbb{C}^n$ to $\mathbb{C}^n$.

3.3.16 Note Let a_1, a_2, a_3, a_4 be any complex numbers. Observe that

$$\begin{vmatrix} 1 & 1 \\ a_1 & a_2 \end{vmatrix} = (a_2 - a_1).$$

Next,

$$\begin{vmatrix} 1 & 1 & 1 \\ a_1 & a_2 & a_3 \\ (a_1)^2 & (a_2)^2 & (a_3)^2 \end{vmatrix}$$

$$= \begin{vmatrix} 1 & 1 & 1 \\ a_1 & a_2 & a_3 \\ 0 & a_2(a_2 - a_1) & a_3(a_3 - a_1) \end{vmatrix} \quad (R_3 \rightarrow R_3 - a_1 R_2)$$

$$= \begin{vmatrix} 1 & 1 & 1 \\ 0 & a_2 - a_1 & a_3 - a_1 \\ 0 & a_2(a_2 - a_1) & a_3(a_3 - a_1) \end{vmatrix} \quad (R_2 \rightarrow R_2 - a_1 R_1)$$

$$= (a_2 - a_1)(a_3 - a_1) \begin{vmatrix} 1 & 1 & 1 \\ 0 & 1 & 1 \\ 0 & a_2 & a_3 \end{vmatrix}$$

$$= (a_2 - a_1)(a_3 - a_1) \begin{vmatrix} 1 & 1 \\ a_2 & a_3 \end{vmatrix} = (a_2 - a_1)(a_3 - a_1)(a_3 - a_2),$$

so

$$\begin{vmatrix} 1 & 1 & 1 \\ a_1 & a_2 & a_3 \\ (a_1)^2 & (a_2)^2 & (a_3)^2 \end{vmatrix} = (a_2 - a_1)(a_3 - a_1)(a_3 - a_2).$$

Again,

$$\begin{vmatrix} 1 & 1 & 1 & 1 \\ a_1 & a_2 & a_3 & a_4 \\ (a_1)^2 & (a_2)^2 & (a_3)^2 & (a_4)^2 \\ (a_1)^3 & (a_2)^3 & (a_3)^3 & (a_4)^3 \end{vmatrix}$$

$$= \begin{vmatrix} 1 & 1 & 1 & 1 \\ 0 & (a_2 - a_1) & (a_3 - a_1) & (a_4 - a_1) \\ 0 & a_2(a_2 - a_1) & a_3(a_3 - a_1) & a_4(a_4 - a_1) \\ 0 & (a_2)^2(a_2 - a_1) & (a_3)^2(a_3 - a_1) & (a_4)^2(a_4 - a_1) \end{vmatrix} \begin{pmatrix} R_4 \to R_4 - a_1 R_3 \\ R_3 \to R_3 - a_1 R_2 \\ R_2 \to R_2 - a_1 R_1 \end{pmatrix}$$

$$= (a_2 - a_1)(a_3 - a_1)(a_4 - a_1) \begin{vmatrix} 1 & 1 & 1 & 1 \\ 0 & 1 & 1 & 1 \\ 0 & a_2 & a_3 & a_4 \\ 0 & (a_2)^2 & (a_3)^2 & (a_4)^2 \end{vmatrix}$$

$$= (a_2 - a_1)(a_3 - a_1)(a_4 - a_1) \begin{vmatrix} 1 & 1 & 1 \\ a_2 & a_3 & a_4 \\ (a_2)^2 & (a_3)^2 & (a_4)^2 \end{vmatrix}$$

$$= (a_2 - a_1)(a_3 - a_1)(a_4 - a_1) \cdot (a_3 - a_2)(a_4 - a_2)(a_4 - a_3),$$

so

$$\begin{vmatrix} 1 & 1 & 1 & 1 \\ a_1 & a_2 & a_3 & a_4 \\ (a_1)^2 & (a_2)^2 & (a_3)^2 & (a_4)^2 \\ (a_1)^3 & (a_2)^3 & (a_3)^3 & (a_4)^3 \end{vmatrix}$$
$$= (a_2 - a_1)(a_3 - a_1)(a_3 - a_2)(a_4 - a_1)(a_4 - a_2)(a_4 - a_3), \text{ etc.}$$

3.3.17 Conclusion If each a_i is a complex number, then

$$\begin{vmatrix} 1 & 1 & & 1 \\ a_1 & a_2 & \cdots & a_n \\ \vdots & \vdots & & \vdots \\ (a_1)^{n-1} & (a_2)^{n-1} & & (a_n)^{n-1} \end{vmatrix} = \prod_{n \geq j > i \geq 1} (a_j - a_i).$$

3.3.18 Note Let $A \equiv [a_{ij}]$ be an n-square complex matrix. Let $\lambda_1, \ldots, \lambda_n$ be the eigenvalues of A. Suppose that all the eigenvalues of A are distinct. Let $B \equiv [b_{ij}]$ be an n-square complex matrix such that $AB = BA$, that is, B commutes with A.

Since λ_1 is an eigenvalue of A, there exists a nonzero $u_1 \in \mathbb{C}^n$ such that $Au_1 = \lambda_1 u_1$. Similarly, there exists a nonzero $u_2 \in \mathbb{C}^n$ such that $Au_2 = \lambda_2 u_2$, etc. Since all the eigenvalues of A are distinct, by 3.3.9, $\{u_1, \ldots, u_n\}$ is linearly independent, and hence $\{u_1, \ldots, u_n\}$ is a basis of $\mathbb{C}^n$. It follows that the $n \times n$ matrix $(u_1, \ldots, u_n)$ in invertible. Next,

$$\underbrace{\begin{aligned} (u_1, \ldots, u_n)^{-1} A(u_1, \ldots, u_n) &= (u_1, \ldots, u_n)^{-1} (A(u_1, \ldots, u_n)) \\ &= (u_1, \ldots, u_n)^{-1} (Au_1, \ldots, Au_n) = (u_1, \ldots, u_n)^{-1} (\lambda_1 u_1, \ldots, \lambda_n u_n) \end{aligned}}$$

$$= \left((u_1, \ldots, u_n)^{-1} (\lambda_1 u_1), \ldots, (u_1, \ldots, u_n)^{-1} (\lambda_n u_n) \right)$$

$$= \left(\lambda_1 (u_1, \ldots, u_n)^{-1} u_1, \ldots, \lambda_n (u_1, \ldots, u_n)^{-1} u_n \right)$$

$$= (\lambda_1 e_1, \ldots, \lambda_n e_n) = \text{diag}(\lambda_1, \ldots, \lambda_n),$$

so

$$(u_1, \ldots, u_n)^{-1} A(u_1, \ldots, u_n) = \text{diag}(\lambda_1, \ldots, \lambda_n).$$

Next,

$$\begin{aligned} (\text{diag}(\lambda_1, \ldots, \lambda_n)) & \left((u_1, \ldots, u_n)^{-1} B(u_1, \ldots, u_n) \right) \\ &= \left((u_1, \ldots, u_n)^{-1} A(u_1, \ldots, u_n) \right) \left((u_1, \ldots, u_n)^{-1} B(u_1, \ldots, u_n) \right) \\ &= (u_1, \ldots, u_n)^{-1} A \left((u_1, \ldots, u_n)(u_1, \ldots, u_n)^{-1} \right) B(u_1, \ldots, u_n) \\ &= (u_1, \ldots, u_n)^{-1} AIB(u_1, \ldots, u_n) = (u_1, \ldots, u_n)^{-1} AB(u_1, \ldots, u_n) \\ &= (u_1, \ldots, u_n)^{-1} BA(u_1, \ldots, u_n) = (u_1, \ldots, u_n)^{-1} BIA(u_1, \ldots, u_n) \\ &= (u_1, \ldots, u_n)^{-1} B \left((u_1, \ldots, u_n)(u_1, \ldots, u_n)^{-1} \right) A(u_1, \ldots, u_n) \\ &= \left((u_1, \ldots, u_n)^{-1} B(u_1, \ldots, u_n) \right) \left((u_1, \ldots, u_n)^{-1} A(u_1, \ldots, u_n) \right) \\ &= \left((u_1, \ldots, u_n)^{-1} B(u_1, \ldots, u_n) \right) (\text{diag}(\lambda_1, \ldots, \lambda_n)), \end{aligned}$$

so $(\text{diag}(\lambda_1, \ldots, \lambda_n))D = D(\text{diag}(\lambda_1, \ldots, \lambda_n))$, where $D \equiv (u_1, \ldots, u_n)^{-1} B(u_1, \ldots, u_n)$.

Suppose that $D \equiv (v_1, \ldots, v_n)$, where each $v_i \equiv [v_{1i}, \ldots, v_{ni}]^T$ is in $\mathbb{C}^n$. It follows that

$$
\begin{aligned}
(\mathrm{diag}(\lambda_1,\ldots,\lambda_n))D &= (\mathrm{diag}(\lambda_1,\ldots,\lambda_n))(v_1,\ldots,v_n) \\
&= ((\mathrm{diag}(\lambda_1,\ldots,\lambda_n))v_1,\ldots,(\mathrm{diag}(\lambda_1,\ldots,\lambda_n))v_n) \\
&= ((\mathrm{diag}(\lambda_1,\ldots,\lambda_n))[v_{11},\ldots,v_{n1}]^T,\ldots, \\
&\quad\ (\mathrm{diag}(\lambda_1,\ldots,\lambda_n))[v_{1n},\ldots,v_{nn}]^T) \\
&= ([\lambda_1 v_{11},\ldots,\lambda_n v_{n1}]^T,\ldots,[\lambda_1 v_{1n},\ldots,\lambda_n v_{nn}]^T) \\
&=
\begin{bmatrix}
\lambda_1 v_{11} & & \lambda_1 v_{1n} \\
\vdots & \ddots & \vdots \\
\lambda_n v_{n1} & & \lambda_n v_{nn}
\end{bmatrix},
\end{aligned}
$$

and

$$
\begin{aligned}
D(\mathrm{diag}(\lambda_1,\ldots,\lambda_n)) &= (v_1,\ldots,v_n)(\mathrm{diag}(\lambda_1,\ldots,\lambda_n)) \\
&= (v_1,\ldots,v_n)(\lambda_1 e_1,\ldots,\lambda_n e_n) \\
&= ((v_1,\ldots,v_n)(\lambda_1 e_1),\ldots,(v_1,\ldots,v_n)(\lambda_n e_n)) \\
&= (\lambda_1(v_1,\ldots,v_n)e_1,\ldots,\lambda_n(v_1,\ldots,v_n)e_n) \\
&= (\lambda_1 v_1,\ldots,\lambda_n v_n) = \lambda_1 v_1 + \cdots + \lambda_n v_n \\
&=
\begin{bmatrix}
\lambda_1 v_{11} & & \lambda_n v_{1n} \\
\vdots & \ddots & \vdots \\
\lambda_1 v_{n1} & & \lambda_n v_{nn}
\end{bmatrix}.
\end{aligned}
$$

Next, since

$$
\begin{bmatrix}
\lambda_1 v_{11} & & \lambda_1 v_{1n} \\
\vdots & \ddots & \vdots \\
\lambda_n v_{n1} & & \lambda_n v_{nn}
\end{bmatrix}
= \underbrace{(\mathrm{diag}(\lambda_1,\ldots,\lambda_n))D = D(\mathrm{diag}(\lambda_1,\ldots,\lambda_n))}
$$

$$
=
\begin{bmatrix}
\lambda_1 v_{11} & & \lambda_n v_{1n} \\
\vdots & \ddots & \vdots \\
\lambda_1 v_{n1} & & \lambda_n v_{nn}
\end{bmatrix},
$$

we have

$$
\begin{bmatrix}
\lambda_1 v_{11} & & \lambda_1 v_{1n} \\
\vdots & \ddots & \vdots \\
\lambda_n v_{n1} & & \lambda_n v_{nn}
\end{bmatrix}
=
\begin{bmatrix}
\lambda_1 v_{11} & & \lambda_n v_{1n} \\
\vdots & \ddots & \vdots \\
\lambda_1 v_{n1} & & \lambda_n v_{nn}
\end{bmatrix}.
$$

Now, since all the eigenvalues $\lambda_1,\ldots,\lambda_n$ of A are distinct, $i \neq j \Rightarrow v_{ij} = 0$. Next, since $D = (v_1,\ldots,v_n)$, and each v_i equals $[v_{1i},\ldots,v_{ni}]^T$, it follows that

$\left((u_1,\ldots,u_n)^{-1}B(u_1,\ldots,u_n)=\right)D$ is a diagonal matrix, and hence $P^{-1}BP$ is a diagonal matrix, where $P\equiv(u_1,\ldots,u_n)$.

Since $P^{-1}BP$ is a diagonal matrix, we can suppose that

$$P^{-1}BP\equiv\operatorname{diag}(\mu_1,\ldots,\mu_n),$$

where each μ_i is a complex number.

Let us consider the following system of linear equations in n variables $x_0,x_1,\ldots,x_{n-1}$:

$$\left.\begin{array}{l}1x_0+\lambda_1 x_1+(\lambda_1)^2 x_2+\cdots+(\lambda_1)^{n-1}x_{n-1}=\mu_1\\1x_0+\lambda_2 x_1+(\lambda_2)^2 x_2+\cdots+(\lambda_2)^{n-1}x_{n-1}=\mu_2\\\quad\vdots\\1x_0+\lambda_n x_1+(\lambda_n)^2 x_2+\cdots+(\lambda_n)^{n-1}x_{n-1}=\mu_n\end{array}\right\}.$$

Since $\lambda_1,\ldots,\lambda_n$ are distinct, by 3.3.17, we have

$$\begin{vmatrix}1&\lambda_1&(\lambda_1)^2&&(\lambda_1)^{n-1}\\1&\lambda_2&(\lambda_2)^2&&(\lambda_2)^{n-1}\\\vdots&\vdots&\vdots&\cdots&\vdots\\1&\lambda_n&(\lambda_n)^2&&(\lambda_n)^{n-1}\end{vmatrix}=\prod_{n\geq j>i\geq 1}(\lambda_j-\lambda_i)\neq 0,$$

and the above system of linear equations has a unique solution $(x_0,x_1,\ldots,x_{n-1})=(a_0,a_1,\ldots,a_{n-1})$.

Hence

$$\left.\begin{array}{l}1a_0+\lambda_1 a_1+(\lambda_1)^2 a_2+\cdots+(\lambda_1)^{n-1}a_{n-1}=\mu_1\\1a_0+\lambda_2 a_1+(\lambda_2)^2 a_2+\cdots+(\lambda_2)^{n-1}a_{n-1}=\mu_2\\\quad\vdots\\1a_0+\lambda_n a_1+(\lambda_n)^2 a_2+\cdots+(\lambda_n)^{n-1}a_{n-1}=\mu_n\end{array}\right\},$$

that is,

$$\left.\begin{array}{l}1a_0+a_1\lambda_1+a_2(\lambda_1)^2+\cdots+a_{n-1}(\lambda_1)^{n-1}=\mu_1\\1a_0+a_1\lambda_2+a_2(\lambda_2)^2+\cdots+a_{n-1}(\lambda_2)^{n-1}=\mu_2\\\quad\vdots\\1a_0+a_1\lambda_n+a_2(\lambda_n)^2+\cdots+a_{n-1}(\lambda_n)^{n-1}=\mu_n\end{array}\right\}.$$

Let us denote the polynomial

$$1a_0 + a_1 x + a_2 x^2 + \cdots + a_{n-1} x^{n-1}$$

by $p(x)$. Observe that $\deg(p(x)) \geq (n-1)$. Also,

$$\left.\begin{array}{c} p(\lambda_1) = \mu_1 \\ p(\lambda_2) = \mu_2 \\ \vdots \\ p(\lambda_n) = \mu_n \end{array}\right\}.$$

Next,

$$p(A) = a_0 I + a_1 A + a_2 A^2 + \cdots + a_{n-1} A^{n-1}$$

and

$$\begin{aligned} p(P^{-1}AP) &= a_0 I + a_1 (P^{-1}AP) + a_2 (P^{-1}AP)(P^{-1}AP) + \cdots + a_{n-1} (P^{-1}AP)^{n-1} \\ &= a_0 (P^{-1}IP) + a_1 (P^{-1}AP) + a_2 (P^{-1}A^2 P) + \cdots + a_{n-1} (P^{-1}A^{n-1}P) \\ &= P^{-1}(a_0 I + a_1 A + a_2 A^2 + \cdots + a_{n-1} A^{n-1})P = P^{-1}p(A)P, \end{aligned}$$

so $p(P^{-1}AP) = P^{-1}p(A)P$. Since

$$\left(P^{-1}AP = \right)(u_1, \ldots, u_n)^{-1} A(u_1, \ldots, u_n) = \operatorname{diag}(\lambda_1, \ldots, \lambda_n),$$

we have

$$P^{-1}p(A)P = \underbrace{p\left(P^{-1}AP\right) = p(\operatorname{diag}(\lambda_1, \ldots, \lambda_n))}$$

$$= a_0 I + a_1 \operatorname{diag}(\lambda_1, \ldots, \lambda_n) + a_2 (\operatorname{diag}(\lambda_1, \ldots, \lambda_n))^2$$
$$+ \cdots + a_{n-1}(\operatorname{diag}(\lambda_1, \ldots, \lambda_n))^{n-1}$$
$$= a_0 \operatorname{diag}(1, \ldots, 1) + a_1 \operatorname{diag}(\lambda_1, \ldots, \lambda_n) + a_2 \operatorname{diag}\left((\lambda_1)^2, \ldots, (\lambda_n)^2\right)$$
$$+ \cdots + a_{n-1}\operatorname{diag}\left((\lambda_1)^{n-1}, \ldots, (\lambda_n)^{n-1}\right)$$
$$= \operatorname{diag}(a_0, \ldots, a_0) + \operatorname{diag}(a_1 \lambda_1, \ldots, a_1 \lambda_n) + \operatorname{diag}\left(a_2(\lambda_1)^2, \ldots, a_2(\lambda_n)^2\right)$$
$$+ \cdots + \operatorname{diag}\left(a_{n-1}(\lambda_1)^{n-1}, \ldots, a_{n-1}(\lambda_n)^{n-1}\right)$$
$$= \operatorname{diag}\left(\begin{array}{c} a_0 + a_1 \lambda_1 + a_2(\lambda_1)^2 + \cdots + a_{n-1}(\lambda_1)^{n-1}, \ldots, a_0 + a_1 \lambda_n + a_2(\lambda_n)^2 \\ + \cdots + a_{n-1}(\lambda_n)^{n-1} \end{array}\right)$$
$$= \operatorname{diag}(p(\lambda_1), \ldots, p(\lambda_n)) = \operatorname{diag}(\mu_1, \ldots, \mu_n) = P^{-1}BP,$$

and hence

$$P^{-1}p(A)P = P^{-1}BP.$$

It follows that $p(A) = B$.

3.3.19 Conclusion Let $A \equiv [a_{ij}]$ be an n-square complex matrix. Suppose that all the eigenvalues of A are distinct. Let $B \equiv [b_{ij}]$ be an n-square complex matrix such that B commutes with A. Then there exists a polynomial $p(x)$ such that

1. $\deg(p(x)) \leq n - 1$,
2. $p(A) = B$.

3.3.20 Problem Let $A \equiv [a_{ij}]$ and $B \equiv [b_{ij}]$ be any n-square complex matrices. Suppose that A commutes with B, that is, $AB = BA$. Then there exists a unitary matrix U such that U^*AU and U^*BU are both upper triangular matrices.

Proof (Induction on n) The assertion is trivially true for $n = 1$. Next, suppose that the assertion is true for $n - 1$. We have to show that the assertion is true for n.

Let us take any eigenvalue μ of B.

Clearly, $\{v : v \in \mathbb{C}^n \text{ and } Bv = \mu v\}$ is a linear subspace of $\mathbb{C}^n$ such that its dimension is ≥ 1. Let $\mathcal{A} : v \mapsto Av$ be a mapping from $\mathbb{C}^n$ to $\mathbb{C}^n$. Clearly, $\mathcal{A} : \mathbb{C}^n \to \mathbb{C}^n$ is a linear transformation. Observe that for every $v \in \mathbb{C}^n$ satisfying $Bv = \mu v$, we have

$$B(\mathcal{A}(v)) = B(Av) = (BA)v = (AB)v = A(Bv) = A(\mu v) = \mu(Av) = \mu(\mathcal{A}(v)),$$

and hence $B(\mathcal{A}(v)) = \mu(\mathcal{A}(v))$. This shows that the subspace $\{v : v \in \mathbb{C}^n \text{ and } Bv = \mu v\}$ is invariant under $\mathcal{A} : \mathbb{C}^n \to \mathbb{C}^n$. Hence the restriction

$$\mathcal{A}|_{V_\mu} : V_\mu \to V_\mu,$$

where $V_\mu \equiv \{v : v \in \mathbb{C}^n \text{ and } Bv = \mu v\}$, is a linear transformation. Also, $\dim(V_\mu) \geq 1$.

Let λ be an eigenvalue of $\mathcal{A}|_{V_\mu}$. Then there exists a nonzero vector v_1 in V_μ such that

$$Av_1 = \mathcal{A}(v_1) = \underbrace{\left(\mathcal{A}|_{V_\mu}\right)(v_1)}_{} = \lambda v_1,$$

and hence $Av_1 = \lambda v_1$. Next, since v_1 is in $V_\mu(= \{v : v \in \mathbb{C}^n \text{ and } Bv = \mu v\})$, we have $Bv_1 = \mu v_1$. Since $v_1 \neq 0$, $w_1 \equiv \frac{1}{v_1}v_1$ is a unit vector. Also, $Aw_1 = \lambda w_1$ and $Bw_1 = \mu w_1$. By 3.3.12, there exist $\underbrace{w_2, \ldots, w_n}_{n-1}$ in $\mathbb{C}^n$ such that $\{w_1, w_2, \ldots, w_n\}$ is an orthonormal basis of $\mathbb{C}^n$. By the definition of unitary matrix $(A^*A = AA^* = I)$, the n-square matrix $[w_1, w_2, \ldots, w_n]$ is unitary. Observe that

$$
\begin{aligned}
[w_1, w_2, \ldots, w_n]^* A[w_1, w_2, \ldots, w_n] &= [w_1, w_2, \ldots, w_n]^* (A[w_1, w_2, \ldots, w_n]) \\
&= [w_1, w_2, \ldots, w_n]^* [Aw_1, Aw_2, \ldots, Aw_n] \\
&= [w_1, w_2, \ldots, w_n]^* [\lambda w_1, Aw_2, \ldots, Aw_n] \\
&= [\overline{w_1}, \overline{w_2}, \ldots, \overline{w_n}]^T [\lambda w_1, Aw_2, \ldots, Aw_n] \\
&= \big[[\overline{w_1}, \overline{w_2}, \ldots, \overline{w_n}]^T (\lambda w_1), [\overline{w_1}, \overline{w_2}, \ldots, \overline{w_n}]^T \\
&\qquad (Aw_2), \ldots, [\overline{w_1}, \overline{w_2}, \ldots, \overline{w_n}]^T (Aw_n) \big] \\
&= \big[\lambda \big([\overline{w_1}, \overline{w_2}, \ldots, \overline{w_n}]^T w_1 \big), [\overline{w_1}, \overline{w_2}, \ldots, \overline{w_n}]^T \\
&\qquad (Aw_2), \ldots, [\overline{w_1}, \overline{w_2}, \ldots, \overline{w_n}]^T (Aw_n) \big] \\
&= \big[\lambda [\langle w_1, w_1 \rangle, \langle w_1, w_2 \rangle, \ldots, \langle w_1, w_n \rangle]^T, \\
&\qquad [\langle Aw_2, w_1 \rangle, \langle Aw_2, w_2 \rangle, \ldots, \langle Aw_2, w_n \rangle]^T, \ldots, \\
&\qquad [\langle Aw_n, w_1 \rangle, \langle Aw_n, w_2 \rangle, \ldots, \langle Aw_n, w_n \rangle]^T \big] \\
&= \big[\lambda [1, 0, \ldots, 0]^T, [\langle Aw_2, w_1 \rangle, \langle Aw_2, w_2 \rangle, \ldots, \langle Aw_2, w_n \rangle]^T, \ldots, \\
&\qquad [\langle Aw_n, w_1 \rangle, \langle Aw_n, w_2 \rangle, \ldots, \langle Aw_n, w_n \rangle]^T \big] \\
&= \big[[\lambda, 0, \ldots, 0]^T, [\langle Aw_2, w_1 \rangle, \langle Aw_2, w_2 \rangle, \ldots, \langle Aw_2, w_n \rangle]^T, \ldots, \\
&\qquad [\langle Aw_n, w_1 \rangle, \langle Aw_n, w_2 \rangle, \ldots, \langle Aw_n, w_n \rangle]^T \big],
\end{aligned}
$$

and

$$
\big[[\lambda, 0, \ldots, 0]^T, [\langle Aw_2, w_1 \rangle, \langle Aw_2, w_2 \rangle, \ldots, \langle Aw_2, w_n \rangle]^T, \ldots, \\
[\langle Aw_n, w_1 \rangle, \langle Aw_n, w_2 \rangle, \ldots, \langle Aw_n, w_n \rangle]^T \big]
$$

is of the form

$$
\begin{bmatrix}
\lambda & * & & * \\
0 & * & & * \\
\vdots & \vdots & \cdots & \vdots \\
0 & * & & *
\end{bmatrix},
$$

so $[w_1, w_2, \ldots, w_n]^* A[w_1, w_2, \ldots, w_n]$ is of the partitioned form

$$
\begin{bmatrix}
\lambda & \alpha \\
0 & C
\end{bmatrix},
$$

where C is a matrix of size $(n-1) \times (n-1)$, and α is a matrix of size $1 \times (n-1)$.

Next,

$$[w_1, w_2, \ldots, w_n]^* B[w_1, w_2, \ldots, w_n]$$
$$= [w_1, w_2, \ldots, w_n]^* (B[w_1, w_2, \ldots, w_n])$$
$$= [w_1, w_2, \ldots, w_n]^* [Bw_1, Bw_2, \ldots, Bw_n]$$
$$= [w_1, w_2, \ldots, w_n]^* [\mu w_1, Bw_2, \ldots, Bw_n]$$
$$= [\overline{w_1}, \overline{w_2}, \ldots, \overline{w_n}]^T [\mu w_1, Bw_2, \ldots, Bw_n]$$
$$= \left[[\overline{w_1}, \overline{w_2}, \ldots, \overline{w_n}]^T (\mu w_1), [\overline{w_1}, \overline{w_2}, \ldots, \overline{w_n}]^T (Bw_2), \ldots, \right.$$
$$\left. [\overline{w_1}, \overline{w_2}, \ldots, \overline{w_n}]^T (Bw_n) \right]$$
$$= \left[\mu \big([\overline{w_1}, \overline{w_2}, \ldots, \overline{w_n}]^T w_1 \big), [\overline{w_1}, \overline{w_2}, \ldots, \overline{w_n}]^T (Bw_2), \ldots, \right.$$
$$\left. [\overline{w_1}, \overline{w_2}, \ldots, \overline{w_n}]^T (Bw_n) \right]$$
$$= \left[\mu [\langle w_1, w_1 \rangle, \langle w_1, w_2 \rangle, \ldots, \langle w_1, w_n \rangle]^T, \right.$$
$$[\langle Bw_2, w_1 \rangle, \langle Bw_2, w_2 \rangle, \ldots, \langle Bw_2, w_n \rangle]^T, \ldots,$$
$$\left. [\langle Bw_n, w_1 \rangle, \langle Bw_n, w_2 \rangle, \ldots, \langle Bw_n, w_n \rangle]^T \right]$$
$$= \left[\mu [1, 0, \ldots, 0]^T, [\langle Bw_2, w_1 \rangle, \langle Bw_2, w_2 \rangle, \ldots, \langle Bw_2, w_n \rangle]^T, \ldots, \right.$$
$$\left. [\langle Bw_n, w_1 \rangle, \langle Bw_n, w_2 \rangle, \ldots, \langle Bw_n, w_n \rangle]^T \right]$$
$$= \left[[\mu, 0, \ldots, 0]^T, [\langle Bw_2, w_1 \rangle, \langle Bw_2, w_2 \rangle, \ldots, \langle Bw_2, w_n \rangle]^T, \ldots, \right.$$
$$\left. [\langle Bw_n, w_1 \rangle, \langle Bw_n, w_2 \rangle, \ldots, \langle Bw_n, w_n \rangle]^T \right]$$

and

$$\left[[\mu, 0, \ldots, 0]^T, [\langle Bw_2, w_1 \rangle, \langle Bw_2, w_2 \rangle, \ldots, \langle Bw_2, w_n \rangle]^T, \ldots, \right.$$
$$\left. [\langle Bw_n, w_1 \rangle, \langle Bw_n, w_2 \rangle, \ldots, \langle Bw_n, w_n \rangle]^T \right]$$

is of the form

$$\begin{bmatrix} \mu & * & & * \\ 0 & * & & * \\ \vdots & \vdots & \cdots & \vdots \\ 0 & * & & * \end{bmatrix},$$

so $[w_1, w_2, \ldots, w_n]^* B[w_1, w_2, \ldots, w_n]$ is of the partitioned form

$$\begin{bmatrix} \mu & \beta \\ 0 & D \end{bmatrix},$$

where D is a matrix of size $(n-1) \times (n-1)$, and β is a matrix of size $1 \times (n-1)$. It follows that

$$\begin{bmatrix} \mu\lambda & \mu\alpha + \beta C \\ 0 & DC \end{bmatrix} = \begin{bmatrix} \mu & \beta \\ 0 & D \end{bmatrix}\begin{bmatrix} \lambda & \alpha \\ 0 & C \end{bmatrix}$$

$$= ([w_1, w_2, \ldots, w_n]^* B[w_1, w_2, \ldots, w_n])([w_1, w_2, \ldots, w_n]^* A[w_1, w_2, \ldots, w_n])$$

$$= [w_1, w_2, \ldots, w_n]^* B([w_1, w_2, \ldots, w_n][w_1, w_2, \ldots, w_n]^*)A[w_1, w_2, \ldots, w_n]$$

$$= [w_1, w_2, \ldots, w_n]^* BIA[w_1, w_2, \ldots, w_n] = w_1, w_2, \ldots, w_n * BA[w_1, w_2, \ldots, w_n]$$

$$= [w_1, w_2, \ldots, w_n]^* AB[w_1, w_2, \ldots, w_n] = [w_1, w_2, \ldots, w_n]^* AIB[w_1, w_2, \ldots, w_n]$$

$$= [w_1, w_2, \ldots, w_n]^* A([w_1, w_2, \ldots, w_n][w_1, w_2, \ldots, w_n]^*)B[w_1, w_2, \ldots, w_n]$$

$$([w_1, w_2, \ldots, w_n]^* A[w_1, w_2, \ldots, w_n])([w_1, w_2, \ldots, w_n]^* B[w_1, w_2, \ldots, w_n])$$

$$= \underbrace{= \begin{bmatrix} \lambda & \alpha \\ 0 & C \end{bmatrix}\begin{bmatrix} \mu & \beta \\ 0 & D \end{bmatrix}}$$

$$= \begin{bmatrix} \lambda\mu & \lambda\beta + \alpha D \\ 0 & CD \end{bmatrix},$$

and hence

$$\begin{bmatrix} \mu\lambda & \mu\alpha + \beta C \\ 0 & DC \end{bmatrix} = \begin{bmatrix} \lambda\mu & \lambda\beta + \alpha D \\ 0 & CD \end{bmatrix}.$$

This shows that $CD = DC$. Further, C and D are $(n-1)$-square complex matrices. By the induction hypothesis, there exists a unitary matrix V of size $(n-1) \times (n-1)$ such that V^*CV and V^*DV are both upper triangular matrices of size $(n-1) \times (n-1)$. It follows that

$$\begin{bmatrix} 1 & 0 \\ 0 & V \end{bmatrix}$$

is a partitioned form of an $n \times n$ matrix, and hence

$$[w_1, w_2, \ldots, w_n]\begin{bmatrix} 1 & 0 \\ 0 & V \end{bmatrix}$$

is an $n \times n$ matrix. Next,

$$\left(\left([w_1, w_2, \ldots, w_n] \begin{bmatrix} 1 & 0 \\ 0 & V \end{bmatrix} \right)^* A \left([w_1, w_2, \ldots, w_n] \begin{bmatrix} 1 & 0 \\ 0 & V \end{bmatrix} \right) \right)$$

$$= \left(\begin{bmatrix} 1 & 0 \\ 0 & V \end{bmatrix}^* [w_1, w_2, \ldots, w_n]^* \right) A \left([w_1, w_2, \ldots, w_n] \begin{bmatrix} 1 & 0 \\ 0 & V \end{bmatrix} \right)$$

$$= \left(\begin{bmatrix} 1 & 0 \\ 0 & V^* \end{bmatrix} [w_1, w_2, \ldots, w_n]^* \right) A \left([w_1, w_2, \ldots, w_n] \begin{bmatrix} 1 & 0 \\ 0 & V \end{bmatrix} \right)$$

$$= \begin{bmatrix} 1 & 0 \\ 0 & V^* \end{bmatrix} \left([w_1, w_2, \ldots, w_n]^* A [w_1, w_2, \ldots, w_n] \right) \begin{bmatrix} 1 & 0 \\ 0 & V \end{bmatrix}$$

$$= \begin{bmatrix} 1 & 0 \\ 0 & V^* \end{bmatrix} \begin{bmatrix} \lambda & \alpha \\ 0 & C \end{bmatrix} \begin{bmatrix} 1 & 0 \\ 0 & V \end{bmatrix}$$

$$= \begin{bmatrix} 1 & 0 \\ 0 & V^* \end{bmatrix} \left(\begin{bmatrix} \lambda & \alpha \\ 0 & C \end{bmatrix} \begin{bmatrix} 1 & 0 \\ 0 & V \end{bmatrix} \right)$$

$$= \begin{bmatrix} 1 & 0 \\ 0 & V^* \end{bmatrix} \begin{bmatrix} \lambda & \alpha V \\ 0 & CV \end{bmatrix} = \begin{bmatrix} \lambda & \alpha V \\ 0 & V^* CV \end{bmatrix},$$

so

$$\left(\left([w_1, w_2, \ldots, w_n] \begin{bmatrix} 1 & 0 \\ 0 & V \end{bmatrix} \right)^* A \left([w_1, w_2, \ldots, w_n] \begin{bmatrix} 1 & 0 \\ 0 & V \end{bmatrix} \right) \right) = \begin{bmatrix} \lambda & \alpha V \\ 0 & V^* CV \end{bmatrix}.$$

Now, since $V^* CV$ is an upper triangular matrix of size $(n - 1) \times (n - 1)$, $\begin{bmatrix} \lambda & \alpha V \\ 0 & V^* CV \end{bmatrix}$ is an upper triangular matrix of size $n \times n$, and hence $U^* AU$ is an upper triangular matrix of size $n \times n$, where

$$U = [w_1, w_2, \ldots, w_n] \begin{bmatrix} 1 & 0 \\ 0 & V \end{bmatrix}.$$

Similarly, $U^* BU$ is an upper triangular matrix of size $n \times n$. It remains to show that U is unitary, that is, $U^* U = UU^* = I$.

Observe that

$$U^*U = \left([w_1, w_2, \ldots, w_n]\begin{bmatrix} 1 & 0 \\ 0 & V \end{bmatrix}\right)^* \left([w_1, w_2, \ldots, w_n]\begin{bmatrix} 1 & 0 \\ 0 & V \end{bmatrix}\right)$$

$$= \left(\begin{bmatrix} 1 & 0 \\ 0 & V \end{bmatrix}^* [w_1, w_2, \ldots, w_n]^*\right)\left([w_1, w_2, \ldots, w_n]\begin{bmatrix} 1 & 0 \\ 0 & V \end{bmatrix}\right)$$

$$= \left(\begin{bmatrix} 1 & 0 \\ 0 & V^* \end{bmatrix}[w_1, w_2, \ldots, w_n]^*\right)\left([w_1, w_2, \ldots, w_n]\begin{bmatrix} 1 & 0 \\ 0 & V \end{bmatrix}\right)$$

$$= \begin{bmatrix} 1 & 0 \\ 0 & V^* \end{bmatrix}\left([w_1, w_2, \ldots, w_n]^*[w_1, w_2, \ldots, w_n]\right)\begin{bmatrix} 1 & 0 \\ 0 & V \end{bmatrix}$$

$$= \begin{bmatrix} 1 & 0 \\ 0 & V^* \end{bmatrix}I\begin{bmatrix} 1 & 0 \\ 0 & V \end{bmatrix} = \begin{bmatrix} 1 & 0 \\ 0 & V^* \end{bmatrix}\begin{bmatrix} 1 & 0 \\ 0 & V \end{bmatrix} = \begin{bmatrix} 1 & 0 \\ 0 & V^*V \end{bmatrix} = \begin{bmatrix} 1 & 0 \\ 0 & I_{n-1} \end{bmatrix} = I,$$

so $U^*U = I$. Similarly, $UU^* = I$. ∎

3.3.21 Problem Let $A \equiv [a_{ij}]$ be an n-square complex matrix. Then there exists a unitary matrix U such that U^*AU is an upper triangular matrix.

Proof Since A commutes with A, by 3.3.20, there exists a unitary matrix U such that U^*AU is an upper triangular matrix. ∎

3.3.22 Theorem Let $A \equiv [a_{ij}]$ be an n-square complex matrix. Then there exists a unitary matrix U such that

1. U^*AU is an upper triangular matrix,
2. the eigenvalues of A are the diagonal entries of U^*AU.

This result is due to **Issai Schur** (1875–1941).

Proof By 3.3.21, there exists a unitary matrix U such that U^*AU is an upper triangular matrix, say C. Since U is unitary, $U^*U = UU^* = I$, we have $U^{-1} = U^*$. Thus $C = U^{-1}AU$, and hence $A = UCU^{-1}$. Now,

$$A - \lambda I = UCU^{-1} - \lambda I = UCU^{-1} - \lambda UU^{-1} = (UC - \lambda U)U^{-1}$$
$$= (UC - \lambda(UI))U^{-1}$$
$$= (UC - U(\lambda I))U^{-1}$$
$$= U(C - \lambda I)U^{-1},$$

so $(A - \lambda I) = U(C - \lambda I)U^{-1}$, and hence

$$\underbrace{\det(A - \lambda I)} = \det\left(U(C - \lambda I)U^{-1}\right) = \det(U) \cdot \det(C - \lambda I) \cdot \det\left(U^{-1}\right)$$

$$= \det(U) \cdot \det(C - \lambda I) \cdot \frac{1}{\det(U)}$$
$$= \det(C - \lambda I).$$

Thus $\det(A - \lambda I) = \det(C - \lambda I)$. Since C is an upper triangular matrix, we have

$$\det(A - \lambda I) = \underbrace{\det(C - \lambda I) = (c_1 - \lambda)(c_2 - \lambda) \cdots (c_n - \lambda)},$$

where the diagonal entries of $(U^*AU =)C$ are $c_1, c_2, \ldots, c_n$. Also,

$$\det(A - \lambda I) = (c_1 - \lambda)(c_2 - \lambda) \cdots (c_n - \lambda).$$

Hence the roots of the polynomial $\det(A - \lambda I)$ in λ are $c_1, c_2, \ldots, c_n$. This shows that $c_1, c_2, \ldots, c_n$ are the eigenvalues of A.   ■

3.3.23 Problem If C is a normal upper triangular matrix of size $n \times n$, then C is diagonal.

Proof (Induction on n) The assertion is trivially true for $n = 1$. Next suppose that the assertion is true for $n - 1$. We have to show that the assertion is true for n.

Let C be any normal upper triangular matrix of size $n \times n$. We have to show that C is diagonal.

Since C is an upper triangular matrix, C is of the form

$$\begin{bmatrix} \lambda & \alpha \\ 0 & D \end{bmatrix},$$

where D is an upper triangular matrix of size $(n - 1) \times (n - 1)$, α is a matrix of size $1 \times (n - 1)$, and λ is a complex number. It suffices to show that $\alpha = 0$, and D is a diagonal matrix of size $(n - 1) \times (n - 1)$. Here,

$$C^* = \begin{bmatrix} \lambda & \alpha \\ 0 & D \end{bmatrix}^* = \begin{bmatrix} \bar{\lambda} & \bar{\alpha} \\ 0 & \bar{D} \end{bmatrix}^T = \begin{bmatrix} \bar{\lambda} & 0 \\ \alpha^* & D^* \end{bmatrix},$$

so

$$C^* = \begin{bmatrix} \bar{\lambda} & 0 \\ \alpha^* & D^* \end{bmatrix}.$$

Now, since C is normal, we have

$$\begin{bmatrix} \bar{\lambda}\lambda & \bar{\lambda}\alpha \\ \alpha^*\lambda & \alpha^*\alpha + D^*D \end{bmatrix} = \begin{bmatrix} \bar{\lambda} & 0 \\ \alpha^* & D^* \end{bmatrix}\begin{bmatrix} \lambda & \alpha \\ 0 & D \end{bmatrix} = \underbrace{C^*C = CC^*} = \begin{bmatrix} \lambda & \alpha \\ 0 & D \end{bmatrix}\begin{bmatrix} \bar{\lambda} & 0 \\ \alpha^* & D^* \end{bmatrix}$$

$$= \begin{bmatrix} \lambda\bar{\lambda} + \alpha\alpha^* & \alpha D^* \\ D\alpha^* & DD^* \end{bmatrix},$$

and hence

$$\begin{bmatrix} \bar{\lambda}\lambda & \bar{\lambda}\alpha \\ \alpha^*\lambda & \alpha^*\alpha + D^*D \end{bmatrix} = \begin{bmatrix} \lambda\bar{\lambda} + \alpha\alpha^* & \alpha D^* \\ D\alpha^* & DD^* \end{bmatrix}.$$

It follows that

$$\left. \begin{array}{l} \bar{\lambda}\lambda = \lambda\bar{\lambda} + \alpha\alpha^* \\ \alpha^*\alpha + D^*D = DD^* \end{array} \right\},$$

that is,

$$\left. \begin{array}{l} \alpha\alpha^* = 0 \\ \alpha^*\alpha + D^*D = DD^* \end{array} \right\}.$$

Hence

$$\left. \begin{array}{l} \alpha = 0 \\ \alpha^*\alpha + D^*D = DD^* \end{array} \right\},$$

that is,

$$\left. \begin{array}{l} \alpha = 0 \\ D^*D = DD^* \end{array} \right\},$$

that is, $\alpha = 0$, and D is normal. Since D is a normal upper triangular matrix of size $(n-1) \times (n-1)$, it follows by the induction hypothesis that D is diagonal. ■

3.3.24 Problem Let $A \equiv [a_{ij}]$ be an n-square complex matrix. Suppose that A is a normal matrix. Then there exists a unitary matrix U such that

1. U^*AU is a diagonal matrix,
2. the eigenvalues of A are the diagonal entries of U^*AU.

Proof By 3.3.22, there exists a unitary matrix U such that

1. U^*AU is an upper triangular matrix,
2. the eigenvalues of A are the diagonal entries of U^*AU.

It suffices to show that U^*AU is a diagonal matrix.

Let us denote the upper triangular matrix U^*AU by C. Since U is unitary, we have $U^*U = UU^* = I$. It follows that $U^{-1} = U^*$. Now, since $C = U^*AU$, we have

$$\underbrace{UCU^* = U(U^*AU)U^*}_{} = (UU^*)A(UU^*) = IAI = A,$$

and hence $A = UCU^*$. It follows that

$$\underbrace{A^* = (UCU^*)^* = (U^*)^* C^* U^* = UC^* U^*},$$

and hence $A^* = UC^* U^*$. Since A is a normal matrix, we have

$$U(C^*C)U^{-1} = U(C^*C)U^* = UC^* ICU^* = UC^*(U^*U)CU^*$$
$$= (UC^*U^*)(UCU^*) = \underbrace{A^*A = AA^*} = A(UC^*U^*)$$

$$= (UCU^*)(UC^*U^*) = UC(U^*U)C^*U^*$$
$$= UCIC^*U^* = U(CC^*)U^* = U(CC^*)U^{-1},$$

and hence $U(C^*C)U^{-1} = U(CC^*)U^{-1}$. It follows that $C^*C = CC^*$, and hence C is normal. Since C is a normal upper triangular matrix, by 3.3.23, $(U^*AU =)C$ is diagonal, and hence U^*AU is diagonal. ■

Note 3.3.25 Problem Let $A \equiv [a_{ij}]$ be an n-square complex matrix. Let U be a unitary matrix such that U^*AU is a diagonal matrix. Then A is a normal matrix.

Proof We have to show that $A^*A = AA^*$.

Let us denote the diagonal matrix U^*AU by $\mathrm{diag}(\lambda_1, \ldots, \lambda_n)$. Since U is unitary, we have $U^*U = UU^* = I$. Now, since

$$U(\mathrm{diag}(\lambda_1, \ldots, \lambda_n))U^* = U(U^*AU)U^* = \underbrace{(UU^*)A(UU^*) = IAI} = A,$$

we have $A = U(\mathrm{diag}(\lambda_1, \ldots, \lambda_n))U^*$. It follows that

$$A^* = (U(\mathrm{diag}(\lambda_1, \ldots, \lambda_n))U^*)^*$$
$$= (U^*)^*(\mathrm{diag}(\lambda_1, \ldots, \lambda_n))^* U^*$$
$$= U(\mathrm{diag}(\lambda_1, \ldots, \lambda_n))^* U^*$$
$$= U(\mathrm{diag}(\overline{\lambda_1}, \ldots, \overline{\lambda_n}))U^*,$$

so $A^* = U(\mathrm{diag}(\overline{\lambda_1}, \ldots, \overline{\lambda_n}))U^*$. Here

$$A^*A = \left(U\left(\mathrm{diag}\left(\overline{\lambda_1}, \ldots, \overline{\lambda_n}\right)\right)U^*\right)A$$
$$= \left(U\left(\mathrm{diag}\left(\overline{\lambda_1}, \ldots, \overline{\lambda_n}\right)\right)U^*\right)\left(U(\mathrm{diag}(\lambda_1, \ldots, \lambda_n))U^*\right)$$
$$= U\left(\mathrm{diag}\left(\overline{\lambda_1}, \ldots, \overline{\lambda_n}\right)\right)(U^*U)(\mathrm{diag}(\lambda_1, \ldots, \lambda_n))U^*$$
$$= U\left(\mathrm{diag}\left(\overline{\lambda_1}, \ldots, \overline{\lambda_n}\right)\right)I(\mathrm{diag}(\lambda_1, \ldots, \lambda_n))U$$
$$= U\left(\left(\mathrm{diag}\left(\overline{\lambda_1}, \ldots, \overline{\lambda_n}\right)\right)(\mathrm{diag}(\lambda_1, \ldots, \lambda_n))\right)U^*$$
$$= U\left(\mathrm{diag}\left(|\lambda_1|^2, \ldots, |\lambda_n|^2\right)\right)U^*,$$

and

$$AA^* = A\left(U\left(\mathrm{diag}\left(\overline{\lambda_1}, \ldots, \overline{\lambda_n}\right)\right)U^*\right)$$
$$= \left(U(\mathrm{diag}(\lambda_1, \ldots, \lambda_n))U^*\right)\left(U\left(\mathrm{diag}\left(\overline{\lambda_1}, \ldots, \overline{\lambda_n}\right)\right)U^*\right)$$
$$= U(\mathrm{diag}(\lambda_1, \ldots, \lambda_n))(U^*U)\left(\mathrm{diag}\left(\overline{\lambda_1}, \ldots, \overline{\lambda_n}\right)\right)U^*$$
$$= U(\mathrm{diag}(\lambda_1, \ldots, \lambda_n))I\left(\mathrm{diag}\left(\overline{\lambda_1}, \ldots, \overline{\lambda_n}\right)\right)U^*$$
$$= U\left((\mathrm{diag}(\lambda_1, \ldots, \lambda_n))\left(\mathrm{diag}\left(\overline{\lambda_1}, \ldots, \overline{\lambda_n}\right)\right)\right)U^*$$
$$= U\left(\mathrm{diag}\left(|\lambda_1|^2, \ldots, |\lambda_n|^2\right)\right)U^*,$$

so $A^*A = AA^*$.   ∎

3.3.26 Theorem Let $A \equiv [a_{ij}]$ be an n-square complex matrix. Let A be a Hermitian matrix, that is, $A^* = A$. Then the eigenvalues of A are real numbers.

Proof   Let $\lambda_1, \ldots, \lambda_n$ be the eigenvalues of A. It suffices to show that $\mathrm{diag}\left(\overline{\lambda_1}, \ldots, \overline{\lambda_n}\right) = \mathrm{diag}(\lambda_1, \ldots, \lambda_n)$, that is,

$$(\mathrm{diag}(\lambda_1, \ldots, \lambda_n))^* = \mathrm{diag}(\lambda_1, \ldots, \lambda_n).$$

Since $A^* = A$, we have $A^*A = AA = AA^*$, and hence $A^*A = AA^*$. Thus A is normal. Now, by 3.3.24, there exists a unitary matrix U such that

$$U^*AU = \mathrm{diag}(\lambda_1, \ldots, \lambda_n).$$

Hence

$$\underbrace{(\mathrm{diag}(\lambda_1, \ldots, \lambda_n))^* = (U^*AU)^*} = U^*A^*(U^*)^* = U^*A^*U = U^*AU$$

$$= \mathrm{diag}(\lambda_1, \ldots, \lambda_n).$$

Thus $(\mathrm{diag}(\lambda_1, \ldots, \lambda_n))^* = \mathrm{diag}(\lambda_1, \ldots, \lambda_n)$.   ∎

3.3.27 Theorem Let $A \equiv \begin{bmatrix} a_{ij} \end{bmatrix}$ be an n-square complex matrix. Let A be a normal matrix. Suppose that all the eigenvalues of A are real numbers. Then A is a Hermitian matrix, that is, $A^* = A$.

Proof   By 3.3.22, there exists a unitary matrix U such that

1. U^*AU is an upper triangular matrix,
2. the eigenvalues of A are the diagonal entries of U^*AU.

Let us denote U^*AU by C. Thus $C = U^*AU$. Since U is unitary, we have $U^*U = UU^* = I$. It follows that $U^{-1} = U^*$ and

$$UCU^* = U(U^*AU)U^* = (UU^*)A(UU^*) = IAI = A,$$

and hence $A = UCU^*$. Next,

$$A^* = (UCU^*)^* = (U^*)^*C^*U^* = UC^*U^*,$$

so $A^* = UC^*U^*$. Thus it suffices to show that $C^* = C$.

Since A is normal, we have

$$U(C^*C)U^{-1} = UC^*CU^* = UC^*ICU^* = UC^*(U^*U)CU^*$$
$$= (UC^*U^*)(UCU^*) = \underbrace{A^*A = AA^*} = (UCU^*)(UC^*U^*)$$

$$= UC(U^*U)C^*U^* = UCIC^*U^* = UCC^*U^* = U(CC^*)U^{-1},$$

and hence $U(C^*C)U^{-1} = U(CC^*)U^{-1}$. It follows that $C^*C = CC^*$, and hence C is normal. Now, since C is an upper triangular matrix, by 3.3.23, C is diagonal. Since C is diagonal, and the diagonal entries of C are real numbers, we have $C^* = C$. ∎

Definition Let $A \equiv \begin{bmatrix} a_{ij} \end{bmatrix}$ be an n-square complex matrix. If for every $x \in \mathbb{C}^n$, the 1×1 matrix x^*Ax has a nonnegative real number as its sole entry, then we say that *A is a positive semidefinite matrix* or *A is a nonnegative definite matrix*. This is expressed as $A \geq 0$.

In short, A is a positive semidefinite matrix if for every $x \in \mathbb{C}^n$, $x^*Ax \geq 0$.

3.3.28 Problem Let $A \equiv \begin{bmatrix} a_{ij} \end{bmatrix}$ be an n-square complex matrix. Suppose that A is positive semidefinite. Then A is a Hermitian matrix, that is, $A^* = A$. Also, the diagonal entries of A are nonnegative real numbers.

Proof Since A is a positive semidefinite matrix and $[1, 0, \ldots, 0]^T \in \mathbb{C}^n$, we have

$$a_{11} = [1, 0, \ldots, 0] \left(A[1, 0, \ldots, 0]^T \right) = \underbrace{\left([1, 0, \ldots, 0]^T \right)^* A[1, 0, \ldots, 0]^T \geq 0},$$

and hence a_{11} is a nonnegative real number. Similarly, a_{22} is a nonnegative real number, a_{33} is a nonnegative real number, etc.

Also,

$$
\begin{aligned}
(a_{11}+a_{22})+(a_{12}+a_{21}) &= 1(a_{11}1+a_{12}1)+1(a_{21}1+a_{22}1) \\
&= [1,1,0,\ldots,0]\left(A[1,1,0,\ldots,0]^T\right) \\
&= \underbrace{\left([1,1,0,\ldots,0]^T\right)^* A[1,1,0,\ldots,0]^T \geq 0,}
\end{aligned}
$$

so

$$
\left.\begin{array}{r}
(a_{11}+a_{22})+Re(a_{12})+Re(a_{21}) \geq 0 \\
Im(a_{12}) = -Im(a_{21})
\end{array}\right\}.
$$

Next,

$$
\begin{aligned}
(a_{11}+a_{22})+(ia_{12}-ia_{21}) &= 1(a_{11}1+a_{12}i)+(-i)(a_{21}1+a_{22}i) \\
&= [1,-i,0,\ldots,0]\left(A[1,i,0,\ldots,0]^T\right) \\
&= \underbrace{\left([1,i,0,\ldots,0]^T\right)^* A[1,i,0,\ldots,0]^T \geq 0,}
\end{aligned}
$$

so

$$
\left.\begin{array}{r}
(a_{11}+a_{22})-Im(a_{12})+Im(a_{21}) \geq 0 \\
Re(a_{12}) = Re(a_{21})
\end{array}\right\}.
$$

Since

$$
\left.\begin{array}{r}
Re(a_{12}) = Re(a_{21}) \\
Im(a_{12}) = -Im(a_{21})
\end{array}\right\},
$$

we have $a_{12} = \overline{a_{21}}$. Similarly, $a_{13} = \overline{a_{31}}, a_{23} = \overline{a_{32}}$, etc. This shows that $[a_{ij}] = [\overline{a_{ij}}]^T$, that is, $A = A^*$. ∎

3.3.29 Problem Let $A \equiv [a_{ij}]$ be an n-square complex matrix. Suppose that A is positive semidefinite. Then the eigenvalues of A are nonnegative real numbers.

Proof Let λ be an eigenvalue of A. We have to show that λ is a nonnegative real number.

Since λ is an eigenvalue of A, there exists a nonzero $x \in \mathbb{C}^n$ such that $Ax = \lambda x$. Since A is positive semidefinite, $(\lambda\langle x,x\rangle = \lambda(x^*x) = x^*(\lambda x) = x^*(Ax) =)x^*Ax$ is a nonnegative real number, and hence $\lambda\langle x,x\rangle$ is a nonnegative real number. Since

$x \neq 0$, $\langle x, x \rangle$ is a positive real number. Since $\lambda \langle x, x \rangle$ is a nonnegative real number, λ is a nonnegative real number. ∎

3.3.30 Problem Let $A \equiv [a_{ij}]$ be an n-square complex matrix. Let A be normal. Suppose that the eigenvalues of A are nonnegative real numbers. Then A is positive semidefinite.

Proof Let x be a member of $\mathbb{C}^n$. We have to show that $x^* A x$ is a nonnegative real number.

By 3.3.24, there exists a unitary matrix U such that

1. $U^* A U$ is a diagonal matrix,
2. the eigenvalues of A are the diagonal entries of $U^* A U$.

So we can write $U^* A U = \mathrm{diag}(t_1, \ldots, t_n)$, where $t_1, \ldots, t_n$ are the eigenvalues of A. By assumption, $t_1, \ldots, t_n$ are nonnegative real numbers. Since U is unitary, we have $U^* U = U U^* = I$. It follows that $U^{-1} = U^*$ and

$$U(\mathrm{diag}(t_1, \ldots, t_n))U^* = U(U^* A U)U^* = (U U^*)A(U U^*) = IAI = A,$$

and hence $A = U(\mathrm{diag}(t_1, \ldots, t_n))U^*$. Next,

$$x^* A x = x^*(U(\mathrm{diag}(t_1, \ldots, t_n))U^*)x = (U^* x)^*(\mathrm{diag}(t_1, \ldots, t_n))(U^* x),$$

so

$$x^* A x = ([y_1, \ldots, y_n]^T)^*(\mathrm{diag}(t_1, \ldots, t_n))[y_1, \ldots, y_n]^T,$$

where $[y_1, \ldots, y_n]^T \equiv U^* x$. We now have

$$\underbrace{x^* A x = ([y_1, \ldots, y_n]^T)^*(\mathrm{diag}(t_1, \ldots, t_n))[y_1, \ldots, y_n]^T}$$

$$= [\overline{y_1}, \ldots, \overline{y_n}]((\mathrm{diag}(t_1, \ldots, t_n))[y_1, \ldots, y_n]^T)$$
$$= [\overline{y_1}, \ldots, \overline{y_n}][t_1 y_1, \ldots, t_n y_n]^T = \overline{y_1}(t_1 y_1) + \cdots + \overline{y_n}(t_n y_n)$$
$$= t_1 |y_1|^2 + \cdots + t_n |y_n|^2 \geq 0,$$

and conclude that $x^* A x \geq 0$. ∎

3.3.31 Note Let $A \equiv [a_{ij}]$ be an n-square complex matrix.

Let $\mathbb{M}_n$ be the collection of all n-square complex matrices. We know that $\mathbb{M}_n$ is a vector space over $\mathbb{C}$. The collection of all C in $\mathbb{M}_n$ such that C has exactly one entry 1 and 0 entries elsewhere constitutes a basis of $\mathbb{M}_n$. This basis has n^2 members, so $\dim(\mathbb{M}_n) = n^2$. Since $I, A, A^2, \ldots, A^{n^2}$ is a collection of size $(n^2 + 1)$ ($> \dim(\mathbb{M}_n)$) and $I, A, A^2, \ldots, A^{n^2}$ are in $\mathbb{M}_n$, it follows that $I, A, A^2, \ldots, A^{n^2}$ are linearly dependent in $\mathbb{M}_n$. Thus there exist complex numbers $a_0, a_1, a_2, \ldots, a_{n^2}$, not all zero, such that

$$a_0I + a_1A + a_2A^2 + \cdots + a_{n^2}A^n = 0.$$

Thus $p(A) = 0$, where $p(x)$ denotes the polynomial $a_0 + a_1x + a_2x^2 + \cdots + a_{n^2}x^{n^2}$. Clearly, $\deg(p(x)) \leq n^2$.

3.3.32 Conclusion Let $A \equiv [a_{ij}]$ be an n-square complex matrix. Then there exists a polynomial $p(x)$ with complex coefficients such that $p(A) = 0$. In short, there exists an "annihilating polynomial" for A.

3.3.33 Problem Let $A \equiv [a_{ij}]$ be an n-square complex matrix. Let A be an upper triangular matrix. Let $\lambda_1, \lambda_2, \ldots, \lambda_n$ be the diagonal entries of A. Then $(\lambda_1I - A)(\lambda_2I - A) \cdots (\lambda_nI - A) = 0$.

Proof Observe that the first column of $(\lambda_1I - A)$ is 0. So we can suppose that $(\lambda_1I - A) = [0, \alpha_1, \ldots, \alpha_{n-1}]$, where each α_i is in $\mathbb{C}^n$. Similarly, we can suppose that $(\lambda_2I - A) = [\beta_1, 0, \beta_2, \ldots, \beta_{n-1}]$, where each β_i is in $\mathbb{C}^n$, etc. Hence

$$
\begin{aligned}
&(\lambda_1I - A)(\lambda_2I - A) \\
&= [0, \alpha_1, \ldots, \alpha_{n-1}][\beta_1, 0, \beta_2, \ldots, \beta_{n-1}] \\
&= [[0, \alpha_1, \ldots, \alpha_{n-1}]\beta_1, [0, \alpha_1, \ldots, \alpha_{n-1}]0, [0, \alpha_1, \ldots, \alpha_{n-1}]\beta_2, \ldots, \\
&\quad [0, \alpha_1, \ldots, \alpha_{n-1}]\beta_{n-1}] \\
&= [[0, \alpha_1, \ldots, \alpha_{n-1}]\beta_1, 0, [0, \alpha_1, \ldots, \alpha_{n-1}]\beta_2, \ldots, [0, \alpha_1, \ldots, \alpha_{n-1}]\beta_{n-1}].
\end{aligned}
$$

Thus

$$
\begin{aligned}
&(\lambda_1I - A)(\lambda_2I - A) \\
&= [[0, \alpha_1, \ldots, \alpha_{n-1}]\beta_1, 0, [0, \alpha_1, \ldots, \alpha_{n-1}]\beta_2, \ldots, [0, \alpha_1, \ldots, \alpha_{n-1}]\beta_{n-1}], \quad (*)
\end{aligned}
$$

and hence the second column of $(\lambda_1I - A)(\lambda_2I - A)$ is 0. Since $(\lambda_2I - A) = [\beta_1, 0, \beta_2, \ldots, \beta_{n-1}]$ and A is an upper triangular matrix, we have $\beta_1 = [\lambda_2 - \lambda_1, 0, \ldots, 0]^T \in \mathbb{C}^n$, and hence

$$[0, \alpha_1, \ldots, \alpha_{n-1}]\beta_1 = [0, \alpha_1, \ldots, \alpha_{n-1}][\lambda_2 - \lambda_1, 0, \ldots, 0]^T = [0, 0, \ldots, 0]^T \in \mathbb{C}^n.$$

Thus $[0, \alpha_1, \ldots, \alpha_{n-1}]\beta_1 = 0$. Now from $(*)$,

$$(\lambda_1I - A)(\lambda_2I - A) = [0, 0, [0, \alpha_1, \ldots, \alpha_{n-1}]\beta_2, \ldots, [0, \alpha_1, \ldots, \alpha_{n-1}]\beta_{n-1}].$$

Thus the first and second columns of $(\lambda_1 I - A)(\lambda_2 I - A)$ are 0. Similarly, the first three columns of $(\lambda_1 I - A)(\lambda_2 I - A)(\lambda_3 I - A)$ are 0, etc. Finally, all the n columns of $(\lambda_1 I - A)(\lambda_2 I - A) \cdots (\lambda_n I - A)$ are 0. Thus,

$$(\lambda_1 I - A)(\lambda_2 I - A) \cdots (\lambda_n I - A) = 0.$$

∎

3.3.34 Note Let $A \equiv [a_{ij}]$ be an n-square complex matrix. Let $\lambda_1, \lambda_2, \ldots, \lambda_n$ be the eigenvalues of A.

By 3.3.22, there exists a unitary matrix U such that

1. $U^* A U$ is an upper triangular matrix,
2. $\lambda_1, \lambda_2, \ldots, \lambda_n$ are the diagonal entries of $U^* A U$.

By 3.3.12,

$$\begin{aligned}
U^*&((\lambda_1 I - A)(\lambda_2 I - A) \cdots (\lambda_n I - A))U \\
&= (U^*(\lambda_1 I - A)U)(U^*(\lambda_2 I - A)U) \cdots (U^*(\lambda_n I - A)U) \\
&= \underbrace{(\lambda_1 I - U^* A U)(\lambda_2 I - U^* A U) \cdots (\lambda_n I - U^* A U)}_{} = 0,
\end{aligned}$$

and hence

$$U^*((\lambda_1 I - A)(\lambda_2 I - A) \cdots (\lambda_n I - A))U = 0.$$

Since U is unitary, we have

$$(\lambda_1 I - A)(\lambda_2 I - A) \cdots (\lambda_n I - A) = 0.$$

It follows that $p(A) = 0$, where $p(x)$ denotes the polynomial $(x - \lambda_1)$ $(x - \lambda_2) \cdots (x - \lambda_n)$. Observe that $\lambda_1, \lambda_2, \ldots, \lambda_n$ are the roots of the monic polynomial $(x - \lambda_1)(x - \lambda_2) \cdots (x - \lambda_n) (= p(x))$. Since $\lambda_1, \lambda_2, \ldots, \lambda_n$ are the eigenvalues of A, $\lambda_1, \lambda_2, \ldots, \lambda_n$ are the roots of the monic polynomial $\det(\lambda I - A)$, and since $\lambda_1, \lambda_2, \ldots, \lambda_n$ are the roots of the monic polynomial $p(x)$, we have

$$p(\lambda) = \det(\lambda I - A).$$

3.3.35 Conclusion Let $A \equiv [a_{ij}]$ be an n-square complex matrix. Let $\lambda_1, \lambda_2, \ldots, \lambda_n$ be the eigenvalues of A. Then $p(A) = 0$, where $p(\lambda) = \det(\lambda I - A)$.

This result is known as the **Cayley–Hamilton theorem**.

Here the polynomial $\det(\lambda I - A)$ is called the *characteristic polynomial of A*. Thus the characteristic polynomial of A is an annihilating polynomial of A. It follows that the minimal polynomial of A divides the characteristic polynomial of A.

Exercises

1. Let V be an n-dimensional inner product space. Let $T : V \to V$ be a linear transformation. Let $v, w_1, w_2 \in V$. Suppose that

$$u \in V \Rightarrow \langle u, w_1 \rangle = \langle T(u), v \rangle = \langle u, w_2 \rangle.$$

Show that $w_1 = w_2$.

2. Let V be an n-dimensional inner product space. Let $S_1 : V \to V$, $S_2 : V \to V$, and $S_3 : V \to V$ be linear transformations. Let λ, μ be any complex numbers. Show that $((\lambda S_1 + \mu S_2)S_3)^* = \bar{\lambda}(S_3)^*(S_1)^* + \bar{\mu}(S_3)^*(S_2)^*$.

3. Let V be any n-dimensional vector space. Let $S, T \in A(V)$ be such that $ST = 0$ and $TS \neq 0$. Show that T is not invertible.

4. Let V be an n-dimensional inner product space. Let $T : V \to V$ be a normal linear transformation. Let $v \in V$. Suppose that

$$T^3(v) = 0.$$

Show that v is a member of the null space of T.

5. Let $T \in A(\mathbb{C}^3)$. Suppose that $\mathbb{C}$ is invariant under T, and $\mathbb{C}^2$ is invariant under T. Let $p_1(x)$ be a minimal polynomial of $T|_{\mathbb{C}}$, and $p_2(x)$ a minimal polynomial of $T|_{\mathbb{C}^2}$. Show that the least common multiple of $p_1(x)$ and $p_2(x)$ is a minimal polynomial of T.

6. Let $T \in A(\mathbb{C}^n)$. Suppose that T is nilpotent. Show that there exist linear subspaces $V_1, V_2, \ldots, V_k$ of $\mathbb{C}^n$ such that
   1. $\mathbb{C}^n = V_1 \oplus V_2 \oplus \cdots \oplus V_k$,
   2. each V_i is invariant under T.

   Also there exist a basis $\{v_1^1, v_1^2, \ldots, v_1^{n_1}\}$ of V_1, a basis $\{v_2^1, v_2^2, \ldots, v_2^{n_2}\}$ of $V_2, \ldots$, a basis $\{v_k^1, v_k^2, \ldots, v_k^{n_k}\}$ of V_k such that the matrix of T relative to the basis $\{v_1^1, v_1^2, \ldots, v_1^{n_1}; v_2^1, v_2^2, \ldots, v_2^{n_2}; \ldots; v_k^1, v_k^2, \ldots, v_k^{n_k}\}$ has the canonical form

$$\begin{bmatrix} M_{n_1} & 0 & 0 \\ 0 & \ddots & 0 \\ 0 & 0 & M_{n_k} \end{bmatrix}_{n \times n}.$$

7. Let A be a nonzero 3×5 matrix with complex numbers as entries. Suppose that $\text{rank}(A) = 2$. Show that there exist an invertible 3×3 matrix P and an invertible 5×5 matrix Q such that

$$PAQ = \begin{bmatrix} I_2 & 0 \\ 0 & 0 \end{bmatrix}_{3 \times 5}.$$

8. Let A and B be any n-square complex matrices. Suppose that $AB = BA$. Show that there exists a unitary matrix U such that U^*AU and U^*BU are both upper triangular matrices.

9. Let A be an n-square complex matrix. Suppose that A is a normal matrix. Show that there exists a unitary matrix U such that the diagonal entries of U^*AU are the eigenvalues of A.

10. Find the characteristic polynomial $p(x)$ of the matrix

$$A \equiv \begin{bmatrix} 1 & 2 & -3 \\ 5 & -2 & -3 \\ 1 & -1 & 4 \end{bmatrix}.$$

Verify that $p(A) = 0$.

Chapter 4
Sylvester's Law of Inertia

Sylvester's law characterizes an equivalence relation called *congruence*. This remarkable result introduces a new concept of a matrix, called its *signature*. It is similar to the rank of a matrix. Finally, a beautiful method of obtaining the signature of a real quadratic form is introduced.

4.1 Positive Definite Matrices

4.1.1 Theorem Let V be any n-dimensional vector space over the field F. Let $T : V \to V$ be a linear transformation. Then there exists a positive integer k such that $\mathcal{N}(T^k) = \mathcal{N}(T^{k+1}) = \mathcal{N}(T^{k+2}) = \cdots$, and $\mathcal{N}(T^{k-1})$ is a proper subset of $\mathcal{N}(T^k)$.

Proof It is clear that

$$(\{v : v \in V \text{ and } T(v) = 0\} =)\mathcal{N}(T) \subset \mathcal{N}(T \circ T)(= \mathcal{N}(T^2)),$$

so $\mathcal{N}(T) \subset \mathcal{N}(T^2)$. Similarly, $\mathcal{N}(T^2) \subset \mathcal{N}(T^3)$, etc. Since

$$\{0\} \subset \mathcal{N}(T) \subset \mathcal{N}(T^2) \subset \mathcal{N}(T^3) \subset \cdots \subset V,$$

each "null space" $\mathcal{N}(T^k)$ is a linear subspace of V, and V is a finite-dimensional vector space, the chain $\mathcal{N}(T) \subset \mathcal{N}(T^2) \subset \mathcal{N}(T^3) \subset \cdots$ cannot continue to increase indefinitely. Hence there exists a positive integer k such that $\mathcal{N}(T^k) = \mathcal{N}(T^{k+1}) = \mathcal{N}(T^{k+2}) = \cdots$, and $\mathcal{N}(T^{k-1})$ is a proper subset of $\mathcal{N}(T^k)$. ∎

© Springer Nature Singapore Pte Ltd. 2020
R. Sinha, *Galois Theory and Advanced Linear Algebra*,
https://doi.org/10.1007/978-981-13-9849-0_4

4.1.2 Theorem $\mathcal{N}(T^k)$ is invariant under T.

Proof To show this, let us take an arbitrary $v \in \mathcal{N}(T^k)$, that is, $T^k(v) = 0$. We have to show that $T(v) \in \mathcal{N}(T^k)$, that is,

$$T(0) = T(T^k(v)) = T^{k+1}(v) = \underbrace{T^k(T(v)) = 0},$$

that is, $T(0) = 0$. This is known to be true, because $T : V \to V$ is a linear transformation. ∎

4.1.3 Theorem The restriction $T|_{\mathcal{N}(T^k)} : \mathcal{N}(T^k) \to \mathcal{N}(T^k)$ is a nilpotent transformation.

Proof To show this, let us take an arbitrary $v \in \mathcal{N}(T^k)$. It suffices to show that $\left(\left(T|_{\mathcal{N}(T^k)} \right)^k \right)(v) = 0$.

Since $v \in \mathcal{N}(T^k)$, we have $(T^k)(v) = 0$. Since $v \in \mathcal{N}(T^k)$, we have $\left(T|_{\mathcal{N}(T^k)} \right)(v) = 0$. Now,

$$\text{LHS} = \left(\left(T|_{\mathcal{N}(T^k)} \right)^k \right)(v) = \left(\left(T|_{\mathcal{N}(T^k)} \right)^{k-1} \right) \left(\left(T|_{\mathcal{N}(T^k)} \right)(v) \right)$$
$$= \left(\left(T|_{\mathcal{N}(T^k)} \right)^{k-1} \right)(0) = 0 = \text{RHS}.$$

∎

4.1.4 Theorem $\mathcal{N}(T^k) \cap \text{ran}(T^k) = \{0\}$.

Proof Suppose to the contrary that there exists a nonzero v in $\mathcal{N}(T^k) \cap \text{ran}(T^k)$, that is, $v \neq 0$, $T^k(v) = 0$, and for some nonzero $w \in V$, $T^k(w) = v$. We seek a contradiction.

Since

$$T^{2k}(w) = T^k(T^k(w)) = \underbrace{T^k(v) = 0},$$

we have $T^{2k}(w) = 0$, and hence $w \in \mathcal{N}(T^{2k})$. Now, since $\mathcal{N}(T^k) = \mathcal{N}(T^{k+1}) = \mathcal{N}(T^{k+2}) = \cdots$, we have $\mathcal{N}(T^k) = \mathcal{N}(T^{2k})(\ni w)$, and hence $w \in \mathcal{N}(T^k)$. It follows that $v = \underbrace{T^k(w) = 0}$, and hence $v = 0$. This is a contradiction. ∎

4.1.5 Theorem $V = \mathcal{N}(T^k) \oplus \operatorname{ran}(T^k)$.

Proof From 4.1.4, it remains to show that $V = \mathcal{N}(T^k) + \operatorname{ran}(T^k)$. Since $\mathcal{N}(T^k), \operatorname{ran}(T^k)$ are subspaces of V, $\mathcal{N}(T^k) + \operatorname{ran}(T^k)$ is a subspace of V, and hence it suffices to show that

$$\dim\left(\text{domain of } T^k\right) = \underbrace{\dim(V) = \dim\left(\mathcal{N}(T^k) + \operatorname{ran}(T^k)\right)}$$

$$= \dim\left(\mathcal{N}(T^k)\right) + \dim\left(\operatorname{ran}(T^k)\right) - \dim\left(\mathcal{N}(T^k) \cap \operatorname{ran}(T^k)\right)$$
$$= \dim\left(\mathcal{N}(T^k)\right) + \dim\left(\operatorname{ran}(T^k)\right) - \dim(\{0\}) = \dim\left(\mathcal{N}(T^k)\right) + \dim\left(\operatorname{ran}(T^k)\right) - 0$$
$$= \dim\left(\mathcal{N}(T^k)\right) + \dim\left(\operatorname{ran}(T^k)\right),$$

that is, $\dim\left(\text{domain of } T^k\right) = \dim\left(\mathcal{N}(T^k)\right) + \dim\left(\operatorname{ran}(T^k)\right)$.

Since $T^k : V \to V$ is a linear transformation, we obtain from the well-known result (nullity + rank = dimension of domain) that

$$\dim\left(\mathcal{N}(T^k)\right) + \dim\left(\operatorname{ran}(T^k)\right) = \dim\left(\text{domain of } T^k\right).$$

Thus $V = \mathcal{N}(T^k) \oplus \operatorname{ran}(T^k)$.   ∎

4.1.6 Theorem $T|_{\operatorname{ran}(T^k)} : \operatorname{ran}(T^k) \to \operatorname{ran}(T^k)$ is a mapping, that is, $\operatorname{ran}(T^k)$ is invariant under T.

Proof To show this, let us take an arbitrary $v \in V$. We have to show that $\left(T|_{\operatorname{ran}(T^k)}\right)\left(T^k(v)\right) \in \operatorname{ran}(T^k)$, that is, $T\left(T^k(v)\right) \in \operatorname{ran}(T^k)$, that is, $T^{k+1}(v) \in \operatorname{ran}(T^k)$, that is, $T^k(T(v)) \in \operatorname{ran}(T^k)$. This is clearly true.   ∎

4.1.7 Theorem The restriction $T|_{\operatorname{ran}(T^k)} : \operatorname{ran}(T^k) \to \operatorname{ran}(T^k)$ is invertible.

Proof To show this, let us take an arbitrary $v \in V$ such that $\left(T|_{\operatorname{ran}(T^k)}\right)\left(T^k(v)\right) = 0$, that is, $T\left(T^k(v)\right) = 0$, that is, $T^{k+1}(v) = 0$, that is, $v \in \mathcal{N}(T^{k+1})$. It suffices to show that $T^k(v) = 0$, that is, $v \in \mathcal{N}(T^k)$.

Since $v \in \mathcal{N}(T^{k+1})$, and $\mathcal{N}(T^k) = \mathcal{N}(T^{k+1})$, we have $V \in \mathcal{N}(T^k)$.   ∎

Thus we have shown that the linear transformation $T|_{\operatorname{ran}(T^k)} : \operatorname{ran}(T^k) \to \operatorname{ran}(T^k)$ is invertible.

4.1.8 Conclusion Let V be any n-dimensional vector space over the field F. Let $T : V \to V$ be a linear transformation. Then there exists a positive integer k such that

1. $\mathcal{N}(T) \subset \mathcal{N}(T^2) \subset \cdots \subset \mathcal{N}(T^k) = \mathcal{N}(T^{k+1}) = \mathcal{N}(T^{k+2}) = \cdots$,
2. $V = \mathcal{N}(T^k) \oplus \operatorname{ran}(T^k)$,
3. $T|_{\mathcal{N}(T^k)} : \mathcal{N}(T^k) \to \mathcal{N}(T^k)$ is a nilpotent transformation,
4. $T|_{\operatorname{ran}(T^k)} : \operatorname{ran}(T^k) \to \operatorname{ran}(T^k)$ is invertible.

4.1.9 Theorem Let V be any n-dimensional vector space over the field F. Let $T : V \to V$ be a linear transformation. Then there exist unique linear subspaces H and K of V such that

1. $V = H \oplus K$,
2. $T|_H : H \to H$ is a nilpotent transformation,
3. $T|_K : K \to K$ is invertible.

Also, there exists a positive integer k such that $H = \mathcal{N}(T^k)$, and $K = \text{ran}(T^k)$.

Proof of existence: By 4.1.1, there exists a positive integer k such that

1. $\mathcal{N}(T) \subset \mathcal{N}(T^2) \subset \cdots \subset \mathcal{N}(T^k) = \mathcal{N}(T^{k+1}) = \mathcal{N}(T^{k+2}) = \cdots$,
2. $V = \mathcal{N}(T^k) \oplus \text{ran}(T^k)$,
3. $T|_{\mathcal{N}(T^k)} : \mathcal{N}(T^k) \to \mathcal{N}(T^k)$ is a nilpotent transformation,
4. $T|_{\text{ran}(T^k)} : \text{ran}(T^k) \to \text{ran}(T^k)$ is invertible.

Let us put $H \equiv \mathcal{N}(T^k)$, and $K \equiv \text{ran}(T^k)$. We get

1. $V = H \oplus K$,
2. $T|_H : H \to H$ is a nilpotent transformation,
3. $T|_K : K \to K$ is invertible.

Proof of uniqueness: Suppose that H_1 and K_1 are subspaces of V such that

1. $V = H_1 \oplus K_1$,
2. $T|_{H_1} : H_1 \to H_1$ is a nilpotent transformation,
3. $T|_{K_1} : K_1 \to K_1$ is invertible.

Suppose that H_2 and K_2 are subspaces of V such that

1. $V = H_2 \oplus K_2$,
2. $T|_{H_2} : H_2 \to H_2$ is a nilpotent transformation,
3. $T|_{K_2} : K_2 \to K_2$ is invertible.

We have to show that $H_1 = H_2$, and $K_1 = K_2$.

By 4.1.1, there exists a positive integer k such that

1. $\mathcal{N}(T) \subset \mathcal{N}(T^2) \subset \cdots \subset \mathcal{N}(T^k) = \mathcal{N}(T^{k+1}) = \mathcal{N}(T^{k+2}) = \cdots$,
2. $V = \mathcal{N}(T^k) \oplus \text{ran}(T^k)$,
3. $T|_{\mathcal{N}(T^k)} : \mathcal{N}(T^k) \to \mathcal{N}(T^k)$ is a nilpotent transformation,
4. $T|_{\text{ran}(T^k)} : \text{ran}(T^k) \to \text{ran}(T^k)$ is invertible.

Since $V = \mathcal{N}(T^k) \oplus \text{ran}(T^k)$, we have $\dim(V) = \dim(\mathcal{N}(T^k)) + \dim(\text{ran}(T^k))$.

Clearly, $H_1 \subset \mathcal{N}(T^k)$.

Proof To show this, let us take an arbitrary $v \in H_1$. We have to show that $v \in \mathcal{N}(T^k)$, that is, $T^k(v) = 0$. Since $T|_{H_1} : H_1 \to H_1$ is a nilpotent transformation, there exists a positive integer l such that $(T|_{H_1})^l = 0$. It follows that

$$(T^l)(v) = \cdots = (T|_{H_1})^{l-1}(T(v)) = (T|_{H_1})^{l-1}((T|_{H_1})(v)) = \underbrace{(T|_{H_1})^l(v) = 0},$$

and hence $(T^l)(v) = 0$. Thus $v \in \mathcal{N}(T^l)$. Since

$$\mathcal{N}(T) \subset \mathcal{N}(T^2) \subset \cdots \subset \mathcal{N}(T^k) = \mathcal{N}(T^{k+1}) = \mathcal{N}(T^{k+2}) = \cdots,$$

we have

$$v \in \underbrace{\mathcal{N}(T^l) \subset \mathcal{N}(T^k)},$$

and hence $v \in \mathcal{N}(T^k)$.   ∎

It follows that $\dim(H_1) \le \dim(\mathcal{N}(T^k))$.

We claim that $\dim(H_1) = \dim(\mathcal{N}(T^k))$.

Suppose to the contrary that $\dim(H_1) < \dim(\mathcal{N}(T^k))$. We seek a contradiction.

   Clearly, $K_1 \subset \mathrm{ran}(T)$.

Proof To show this, let us take an arbitrary $v \in K_1$. We have to show that $v \in \mathrm{ran}(T)$. Since $v \in K_1$, and $T|_{K_1} : K_1 \to K_1$ is invertible, there exists $w \in K_1$ such that $T(w) = \underbrace{(T|_{K_1})(w) = v}$, and hence $(\mathrm{ran}(T) \ni)T(w) = v$. Thus $v \in \mathrm{ran}(T)$.   ∎

   Clearly, $K_1 \subset \mathrm{ran}(T^2)$.

Proof To show this, let us take an arbitrary $v \in K_1$. We have to show that $v \in \mathrm{ran}(T^2)$. Since $T|_{K_1} : K_1 \to K_1$ is invertible, $(T|_{K_1})^2 : K_1 \to K_1$ is invertible. Now, since $v \in K_1$, there exists $w \in K_1$ such that

$$(T|_{K_1})(T(w)) = (T|_{K_1})((T|_{K_1})(w)) = \underbrace{((T|_{K_1})^2)(w) = v},$$

and hence $(T|_{K_1})(T(w)) = v$. Since $w \in K_1$ and $T|_{K_1} : K_1 \to K_1$, we have $T(w) = \underbrace{(T|_{K_1})(w) \in K_1}$, and hence $T(w) \in K_1$. It follows that

$$v = \underbrace{\left(T|_{K_1}\right)(T(w)) = T(T(w))}_{} = \left(T^2\right)(w) \in \operatorname{ran}\left(T^2\right),$$

and hence $v \in \operatorname{ran}\left(T^2\right)$.   ∎

Similarly, $K_1 \subset \operatorname{ran}\left(T^3\right)$. etc. Thus $K_1 \subset \operatorname{ran}\left(T^k\right)$.

It follows that $\dim(K_1) \leq \dim\left(\operatorname{ran}\left(T^k\right)\right)$. Since $V = H_1 \oplus K_1$, we have $\dim(V) = \dim(H_1) + \dim(K_1)$. Similarly, $\dim(V) = \dim(H_2) + \dim(K_2)$. Since

$$\dim(V) - \dim(H_1) = \underbrace{\dim(K_1) \leq \dim\left(\operatorname{ran}\left(T^k\right)\right)}_{} = \dim(V) - \dim\left(\mathcal{N}\left(T^k\right)\right),$$

we have

$$\dim(V) - \dim(H_1) \leq \dim(V) - \dim\left(\mathcal{N}\left(T^k\right)\right),$$

and hence $\dim\left(\mathcal{N}\left(T^k\right)\right) \leq \dim(H_1)$. This is a contradiction.

So our claim is substantiated, that is, $\dim(H_1) = \dim\left(\mathcal{N}\left(T^k\right)\right)$.

Now, since $H_1 \subset \mathcal{N}\left(T^k\right)$, we have $H_1 = \mathcal{N}\left(T^k\right)$. Similarly, $H_2 = \mathcal{N}\left(T^k\right)$. It follows that $H_1 = H_2$. It remains to show that $K_1 = K_2$.

Since $V = H_1 \oplus K_1$, we have $\dim(V) = \dim(H_1) + \dim(K_1)$. Similarly, $\dim(V) = \dim\left(\mathcal{N}\left(T^k\right)\right) + \dim\left(\operatorname{ran}\left(T^k\right)\right)$. Since

$$\dim(V) - \dim(K_1) = \underbrace{\dim(H_1) = \dim\left(\mathcal{N}\left(T^k\right)\right)}_{} = \dim(V) - \dim\left(\operatorname{ran}\left(T^k\right)\right),$$

we have $\dim(V) - \dim(K_1) = \dim(V) - \dim\left(\operatorname{ran}\left(T^k\right)\right)$, and hence $\dim(K_1) = \dim\left(\operatorname{ran}\left(T^k\right)\right)$. Now, since $K_1 \subset \operatorname{ran}\left(T^k\right)$, we have $K_1 = \operatorname{ran}\left(T^k\right)$. Similarly, $K_2 = \operatorname{ran}\left(T^k\right)$. Hence $K_1 = K_2$.   ∎

4.1.10 Note Let V be any n-dimensional vector space. Let $T : V \rightarrow V$ be a linear transformation.

Let $v_1, \ldots, v_n$ be any basis of V. Let $A \equiv \left[a_{ij}\right]_{n \times n}$ be the matrix of T relative to the basis $v_1, \ldots, v_n$. By 3.3.22, there exists a unitary matrix U such that

1. U^*AU is an upper triangular matrix,
2. the eigenvalues of A are the diagonal entries of U^*AU.

Since U is a unitary matrix, we have $U^*U = UU^* = I$, and hence $U^{-1} = U^*$. Thus U is invertible. Also

1. $U^{-1}AU$ is an upper triangular matrix,
2. the eigenvalues of T are the diagonal entries of $U^{-1}AU$.

Since U is invertible, by 3.1.35(b), there exists a basis $w_1, \ldots, w_n$ of V such that $U^{-1}AU$ is the matrix of T relative to the basis $w_1, \ldots, w_n$. Thus

1. the matrix of T relative to the basis $w_1, \ldots, w_n$ is upper triangular,
2. the eigenvalues of T are the diagonal entries of the matrix of T relative to the basis $w_1, \ldots, w_n$.

4.1.11 Conclusion Let V be any n-dimensional vector space. Let $T : V \to V$ be a linear transformation. Then there exists a basis $w_1, \ldots, w_n$ of V such that if B is the matrix of T relative to the basis $w_1, \ldots, w_n$, then

1. B is an upper triangular matrix,
2. the eigenvalues of T are the diagonal entries of B.

4.1.12 Note Let V be any n-dimensional vector space over the field $\mathbb{C}$. Let $T : V \to V$ be a linear transformation.

Suppose that $\lambda_1, \lambda_2, \ldots, \lambda_p$ are all the distinct eigenvalues of T. Suppose that the eigenvalue λ_1 has multiplicity m_1, the eigenvalue λ_2 has multiplicity m_2, etc. In other words, the list of all eigenvalues of T is $\underbrace{\lambda_1, \lambda_1, \ldots, \lambda_1}_{m_1 \text{ in number}}, \ldots, \underbrace{\lambda_p, \lambda_p, \ldots, \lambda_p}_{m_p \text{ in number}}$,

where

$$n = m_1 + \cdots + m_p.$$

Thus the characteristic polynomial of T is $(\lambda - \lambda_1)^{m_1} \cdots (\lambda - \lambda_p)^{m_p}$. Now, since the minimal polynomial of T divides the characteristic polynomial of T, we can suppose that the minimal polynomial of T is of the form

$$(\lambda - \lambda_1)^{l_1} \cdots (\lambda - \lambda_p)^{l_p},$$

where $l_i \leq m_i (i = 1, \ldots, p)$. Put

$$V_1 \equiv \left\{ v : v \in V \text{ and} \left((T - \lambda_1 I)^{l_1} \right)(v) = 0 \right\},$$
$$V_2 \equiv \left\{ v : v \in V \text{ and} \left((T - \lambda_2 I)^{l_2} \right)(v) = 0 \right\}, \text{etc.}$$

By 3.2.3,

1. each V_i is a nontrivial linear subspace of V,
2. each V_i is invariant under T,
3. $V = V_1 \oplus V_2 \oplus \cdots \oplus V_p$,
4. for each $i = 1, 2, \ldots, p$, the minimal polynomial of $T|_{V_i}$ is $(\lambda - \lambda_i)^{l_i}$.

From 4.1.5, $\left(((T - \lambda_1 I)|_{V_1})^{l_1} = \right)(T|_{V_1} - \lambda_1 I)^{l_1} = 0$, so $(T - \lambda_1 I)|_{V_1} : V_1 \to V_1$ is a nilpotent transformation, and hence by 3.2.22, 0 is the only eigenvalue of

$(T - \lambda_1 I)|_{V_1}$. It follows that λ_1 is the only eigenvalue of $T|_{V_1}$. Similarly, λ_2 is the only eigenvalue of $T|_{V_2}$, etc.

4.1.13 Conclusion 4.1.13 Let V be any n-dimensional vector space over the field $\mathbb{C}$. Let $T : V \rightarrow V$ be a linear transformation. Suppose that $\lambda_1, \lambda_2, \ldots, \lambda_p$ are all the distinct eigenvalues of T. Suppose that the minimal polynomial of T is of the form

$$(\lambda - \lambda_1)^{l_1} \cdots (\lambda - \lambda_p)^{l_p}.$$

Put

$$V_1 \equiv \left\{ v : v \in V \text{ and } \left((T - \lambda_1 I)^{l_1} \right)(v) = 0 \right\},$$
$$V_2 \equiv \left\{ v : v \in V \text{ and } \left((T - \lambda_2 I)^{l_2} \right)(v) = 0 \right\}, \text{etc.}$$

Then

1. each V_i is a nontrivial linear subspace of V,
2. each V_i is invariant under T,
3. $V = V_1 \oplus V_2 \oplus \cdots \oplus V_p$,
4. for each $i = 1, 2, \ldots, p$, λ_i is the only eigenvalue of $T|_{V_i}$.
5. there exists a basis B of V such that the matrix of T relative to B is of the block form

$$\begin{bmatrix} A_1 & & & \\ & A_2 & & \\ & & \ddots & \\ & & & A_p \end{bmatrix}_{n \times n},$$

where A_1 is a $(\dim V_1) \times (\dim V_1)$ matrix of $T|_{V_1}$, A_2 is a $(\dim V_2) \times (\dim V_2)$ matrix of $T|_{V_2}$, etc., and all other entries are 0.

Definition A square matrix of the form

$$\begin{bmatrix} \lambda & 1 & 0 & & & 0 \\ 0 & \lambda & 1 & & & 0 \\ \vdots & 0 & \lambda & \ddots & & 0 \\ \vdots & \vdots & 0 & & & 0 \\ \vdots & \vdots & \vdots & & & 1 \\ 0 & 0 & 0 & & & \lambda \end{bmatrix}$$

is called a *basic Jordan block belonging to* λ. The basic Jordan block belonging to λ'' of size $t \times t$ can also be written as

$$\lambda I + \begin{bmatrix} 0 & 1 & 0 & & & 0 \\ 0 & 0 & 1 & & & 0 \\ \vdots & & 0 & 0 & & 0 \\ \vdots & \vdots & & 0 & \ddots & 0 \\ \vdots & \vdots & \vdots & & & 1 \\ 0 & 0 & 0 & & & 0 \end{bmatrix}_{t \times t}$$

or $\lambda I + M_t$, where

$$M_t \equiv \begin{bmatrix} 0 & 1 & 0 & & & 0 \\ 0 & 0 & 1 & & & 0 \\ \vdots & & 0 & 0 & \ddots & 0 \\ \vdots & \vdots & & 0 & & 0 \\ \vdots & \vdots & \vdots & & & 1 \\ 0 & 0 & 0 & & & 0 \end{bmatrix}_{t \times t}.$$

4.1.14 Theorem Let V be any n-dimensional vector space over the field $\mathbb{C}$. Let $T : V \to V$ be a linear transformation. Suppose that $\lambda_1, \lambda_2, \ldots, \lambda_p$ are the distinct eigenvalues of T. Then there exists a basis B^* of V such that the matrix of T relative to B^* is of the form

$$\begin{bmatrix} J_1 & & & \\ & J_2 & & \\ & & \ddots & \\ & & & J_p \end{bmatrix}$$

such that for every $i \in \{1, 2, \ldots, p\}$, J_i is of the form

$$\begin{bmatrix} B_{i1} & & & \\ & B_{i2} & & \\ & & \ddots & \end{bmatrix},$$

where each B_{ij} is a basic Jordan block belonging to λ_i.

Here a matrix of the type

$$
\left[
\begin{array}{cccc}
\left[\begin{array}{ccc} B_{11} & & \\ & B_{12} & \\ & & \ddots \end{array}\right] & & & \\
& \left[\begin{array}{ccc} B_{21} & & \\ & B_{22} & \\ & & \ddots \end{array}\right] & & \\
& & & \ddots
\end{array}
\right],
$$

where each B_{ij} is a basic Jordan block belonging to λ_i, is called a *Jordan canonical form*.

Thus for a given square matrix A, there exists an invertible matrix C such that $C^{-1}AC$ is a Jordan canonical form.

Proof By 4.1.13, there exist linear subspaces $V_1, \ldots, V_p$ such that
1. each V_i is invariant under T,
2. $V = V_1 \oplus V_2 \oplus \cdots \oplus V_p$,
3. for each $i = 1, 2, \ldots, p$, λ_i is the only eigenvalue of $T|_{V_i}$.
4. there exists a basis B of V such that the matrix of T relative to B is of the block form

$$
\left[
\begin{array}{cccc}
A_1 & & & \\
& A_2 & & \\
& & \ddots & \\
& & & A_p
\end{array}
\right]_{n \times n},
$$

where A_1 is a $(\dim V_1) \times (\dim V_1)$ matrix of $T|_{V_1}$, A_2 is a $(\dim V_2) \times (\dim V_2)$ matrix of $T|_{V_2}$, etc., and all other entries are 0.

Now, by 3.3.22, there exists a basis C_1 of V_1 such that the matrix $K_1\left(= [C_1]^{-1}A_1[C_1]\right)$ of $T|_{V_1}$ relative to C_1 is upper triangular, and each diagonal entry of K_1 is λ_1. Similarly, there exists a basis C_2 of V_2 such that the matrix $K_2\left(= [C_2]^{-1}A_2[C_2]\right)$ of $T|_{V_2}$ relative to C_2 is upper triangular, and each diagonal entry of K_2 is λ_2, etc. Now, since $V = V_1 \oplus V_2 \oplus \cdots \oplus V_p$, there exists a basis D of V such that the matrix of T relative to D is of the block form

$$
\left[
\begin{array}{cccc}
K_1 & & & \\
& K_2 & & \\
& & \ddots & \\
& & & K_p
\end{array}
\right]_{n \times n}.
$$

Since λ_1 is the only eigenvalue of $T|_{V_1}$, 0 is the only eigenvalue of $(T|_{V_1} - \lambda_1 I)$, and hence the characteristic polynomial of $(T|_{V_1} - \lambda_1 I)$ is $\underbrace{(\lambda - 0) \cdots (\lambda - 0)}_{\dim(T|_{V_1}) \text{ in number}} \left(= \lambda^{\dim(T|_{V_1})}\right)$, and hence by the Cayley–Hamilton theorem, $(T|_{V_1} - \lambda_1 I)^{\dim(T|_{V_1})} = 0$. This shows that $(T|_{V_1} - \lambda_1 I)$ is nilpotent. Now, by 3.2.25 , there exists a basis D_1 of V_1 such that the matrix

$$[D_1]^{-1}(K_1 - \lambda_1 I)[D_1]\left(= [D_1]^{-1}(K_1 - \lambda_1 I)[D_1] = [D_1]^{-1}K_1[D_1] - \lambda_1 I\right)$$

of $(T|_{V_1} - \lambda_1 I)$ relative to the basis D_1 has the canonical form

$$\begin{bmatrix} M_{n_1} & 0 & 0 \\ 0 & M_{n_2} & 0 \\ 0 & 0 & \ddots \end{bmatrix}.$$

It follows that the matrix $[D_1]^{-1}K_1[D_1]\left(= [D_1]^{-1}\left([C_1]^{-1}A_1[C_1]\right)[D_1] = ([C_1][D_1])^{-1}A_1([C_1][D_1])\right)$ of $T|_{V_1}$ relative to the basis D_1 has the canonical form

$$\lambda_1 I + \begin{bmatrix} M_{n_1} & 0 & 0 \\ 0 & M_{n_2} & 0 \\ 0 & 0 & \ddots \end{bmatrix} = \begin{bmatrix} \lambda_1 I + M_{n_1} & 0 & 0 \\ 0 & \lambda_1 I + M_{n_2} & 0 \\ 0 & 0 & \ddots \end{bmatrix} = \begin{bmatrix} B_{11} & 0 & 0 \\ 0 & B_{12} & 0 \\ 0 & 0 & \ddots \end{bmatrix},$$

where $B_{11} \equiv \lambda_1 I + M_{n_1}, B_{12} \equiv \lambda_1 I + M_{n_2}$, etc. Clearly, each of $B_{11}, B_{12}, \ldots$ is a basic Jordan block belonging to λ_1. Thus the matrix of $T|_{V_1}$ relative to the basis D_1 has the canonical form

$$\begin{bmatrix} B_{11} & 0 & 0 \\ 0 & B_{12} & 0 \\ 0 & 0 & \ddots \end{bmatrix}.$$

Similarly, there exists a basis D_2 of V_2 such that the matrix of $T|_{V_2}$ relative to the basis D_2 has the canonical form

$$\begin{bmatrix} B_{21} & 0 & 0 \\ 0 & B_{22} & 0 \\ 0 & 0 & \ddots \end{bmatrix},$$

where each $B_{21}, B_{22}, \ldots$ is a basic Jordan block belonging to λ_2, etc.

Now, since $V = V_1 \oplus V_2 \oplus \cdots \oplus V_p$, there exists a basis B^* of V such that the matrix of T relative to B^* is of the form

$$\begin{bmatrix} J_1 & & & \\ & J_2 & & \\ & & \ddots & \\ & & & J_p \end{bmatrix}$$

such that for every $i \in \{1, 2, \ldots, p\}$, J_i is of the form

$$\begin{bmatrix} B_{i1} & & & \\ & B_{i2} & & \\ & & \ddots & \end{bmatrix},$$

where each B_{ij} is a basic Jordan block belonging to λ_i.   ∎

4.1.15 Note Let V be any n-dimensional inner product space over the field $\mathbb{C}$. Let $T : V \to V$ be a linear transformation. Let T be normal.

Suppose that $\lambda_1, \lambda_2, \ldots, \lambda_p$ are the distinct eigenvalues of T. Suppose that the minimal polynomial of T is of the form

$$(\lambda - \lambda_1)^{l_1} \cdots (\lambda - \lambda_p)^{l_p}.$$

Put

$$V_1 \equiv \left\{ v : v \in V \text{ and } \left((T - \lambda_1 I)^{l_1} \right)(v) = 0 \right\},$$
$$V_2 \equiv \left\{ v : v \in V \text{ and } \left((T - \lambda_2 I)^{l_2} \right)(v) = 0 \right\}, \text{etc.}$$

By 4.1.13,
1. each V_i is a nontrivial linear subspace of V,
2. each V_i is invariant under T,
3. $V = V_1 \oplus V_2 \oplus \cdots \oplus V_p$,
4. for each $i = 1, 2, \ldots, p$, λ_i is the only eigenvalue of $T|_{V_i}$.

Observe that

$$V_1 = E_1 \cup \{0\},$$

where E_1 is the set of all eigenvectors belonging to the eigenvalue λ_1 of T.

Proof To show this, let us take an arbitrary nonzero $v \in V_1(= \{v : v \in V \text{ and } \left((T - \lambda_1 I)^{l_1} \right)(v) = 0\})$. It follows that $\left((T - \lambda_1 I)^{l_1} \right)(v) = 0$. Since T is normal, by 3.1.24, we have $T(v) = \lambda v$. Next, since $v \neq 0$, v is an eigenvector belonging to the eigenvalue λ_1, and hence $v \in E_1$. Thus $V_1 \subset E_1 \cup \{0\}$. It suffices to show that $E_1 \subset V_1$.

To show this, let us take an arbitrary $v \in E_1$, that is, v is an eigenvector belonging to the eigenvalue λ_1 of T. Hence $T(v) = \lambda_1 v$. It follows that $(T - \lambda_1 I)(v) = 0$, and hence

$$\left((T - \lambda_1 I)^{l_1}\right)(v) = \underbrace{\left((T - \lambda_1 I)^{l_1 - 1}\right)((T - \lambda_1 I)(v))}_{} = \left((T - \lambda_1 I)^{l_1 - 1}\right)(0) = 0.$$

Thus $\left((T - \lambda_1 I)^{l_1}\right)(v) = 0$. It follows that $v \in V_1$.

We have shown that $V_1 = E_1 \cup \{0\}$.   ∎

Similarly, $V_2 = E_2 \cup \{0\}$, where E_2 is the set of all eigenvectors belonging to the eigenvalue λ_2 of T, etc.

Since V_1 is a nontrivial linear subspace of V, by 3.3.12, there exists an orthonormal basis B_1 of $V_1(= E_1 \cup \{0\})$. Since a basis does not contain the zero vector, we have $B_1 \subset E_1$, and hence each member of B_1 is an eigenvector belonging to the eigenvalue λ_1 of T. Thus for every $v \in B_1$, $T(v) = \lambda_1 v$.

Similarly, there exists an orthonormal basis B_2 of V_2 such that $B_2 \subset E_2$, and for every $w \in B_2$, $T(w) = \lambda_2 w$, etc.

Clearly, $B_1 \cup B_2 \cup \cdots \cup B_p$ is an orthonormal basis of V.

Proof Since each B_i is an orthonormal basis of V_i, and $V = V_1 \oplus V_2 \oplus \cdots \oplus V_p$, it suffices to show that for distinct $i, j \in \{1, \ldots, p\}$, $(v \in B_i, w \in B_j \Rightarrow \langle v, w \rangle = 0)$.

To show this, let us take arbitrary $i, j \in \{1, \ldots, p\}$ such that $i \neq j$. Next, let us take arbitrary $v \in B_i$, and $w \in B_j$. We have to show that $\langle v, w \rangle = 0$.

Since $v \in B_i$, we have $T(v) = \lambda_i v$. Since $v \in B_i$, and B_i is a basis, v is nonzero. Since $i \neq j$, and $\lambda_1, \lambda_2, \ldots, \lambda_p$ are distinct, we have $\lambda_i \neq \lambda_j$. Since $w \in B_j$, we have $T(w) = \lambda_j w$. Now, since T is normal, by 3.1.25, $\langle v, w \rangle = 0$.

We have shown that B is an orthonormal basis of V, where $B \equiv B_1 \cup B_2 \cup \cdots \cup B_p$.   ∎

Since for every $i \in \{1, \ldots, p\}, B_i \subset E_i$, we have

$$B = \underbrace{(B_1 \cup B_2 \cup \cdots \cup B_p) \subset (E_1 \cup E_2 \cup \cdots \cup E_p)}_{}$$

$$= (\text{the collection of all eigenvectors of } T),$$

and hence each member of B is an eigenvector of T. Suppose that

$$B = \{e_1, e_2, \ldots, e_n\} (\subset V).$$

Since $\{e_1, e_2, \ldots, e_n\}$ is an orthonormal basis B of V we have

$$T(e_1) = \langle T(e_1), e_1 \rangle e_1 = \langle T(e_1), e_1 \rangle e_1 + 0e_2 + \cdots + 0e_p,$$

and hence

$$T(e_1) = \langle T(e_1), e_1 \rangle e_1 + 0e_2 + \cdots + 0e_p.$$

Similarly,

$$T(e_2) = 0e_1 + \langle T(e_2), e_2 \rangle e_2 + 0e_3 + \cdots + 0e_p,$$

etc. Thus the matrix of T relative to the basis $\{e_1, e_2, \ldots, e_n\}$ is the diagonal matrix

$$\begin{bmatrix} \langle T(e_1), e_1 \rangle & & & \\ & \langle T(e_2), e_2 \rangle & & \\ & & \ddots & \end{bmatrix}_{n \times n}.$$

4.1.16 Conclusion Let V be any n-dimensional inner product space over the field $\mathbb{C}$. Let $T : V \to V$ be a linear transformation. Let T be normal. Then there exists an orthonormal basis B of V such that the matrix of T relative to the basis B is a diagonal matrix.

Since every Hermitian linear transformation is normal, and every unitary linear transformation is normal, the above conclusion is also valid when either T is Hermitian or T is unitary.

4.1.17 Theorem Let V be any n-dimensional inner product space over the field $\mathbb{C}$. Let $T : V \to V$ be a normal linear transformation. Then T is Hermitian if and only if all the eigenvalues of T are real.

Proof In view of 3.1.13, it remains to show that if all the eigenvalues of T are real, then T is Hermitian. So we suppose that all the eigenvalues of T are real. We have to show that T is Hermitian, that is, $T^* = T$.

Since T is normal, by 4.1.16, there exists an orthonormal basis $\{e_1, \ldots, e_n\}$ of V such that the matrix of T relative to the basis $\{e_1, \ldots, e_n\}$ is a diagonal matrix, say $\mathrm{diag}(\alpha_1, \ldots, \alpha_n)$. It follows that

$$T(e_1) = \alpha_1 e_1 + 0e_2 + \cdots + 0e_n = \alpha_1 e_1,$$

and hence $T(e_1) = \alpha_1 e_1$. Since $\{e_1, \ldots, e_n\}$ is a basis, we have $e_1 \neq 0$. Now, since $T(e_1) = \alpha_1 e_1$, α_1 is an eigenvalue of T. Next, by assumption, α_1 is a real number. Similarly, α_2 is a real number, etc. It follows that

$$\left(\mathrm{diag}(\alpha_1,\ldots,\alpha_n)\right)^* = \left(\left(\mathrm{diag}(\alpha_1,\ldots,\alpha_n)\right)^T\right)^- = \left(\mathrm{diag}(\alpha_1,\ldots,\alpha_n)\right)^-$$
$$= \underbrace{\mathrm{diag}(\overline{\alpha_1},\ldots,\overline{\alpha_n}) = \mathrm{diag}(\alpha_1,\ldots,\alpha_n)},$$

and hence

$$\left(\mathrm{diag}(\alpha_1,\ldots,\alpha_n)\right)^* = \mathrm{diag}(\alpha_1,\ldots,\alpha_n).$$

Since the matrix of T relative to the basis $\{e_1,\ldots,e_n\}$ is $\mathrm{diag}(\alpha_1,\ldots,\alpha_n)$, by 3.1.11, the matrix of T^* relative to the basis $\{e_1,\ldots,e_n\}$ is $\left(\mathrm{diag}(\alpha_1,\ldots,\alpha_n)\right)^*(=\mathrm{diag}(\alpha_1,\ldots,\alpha_n))$, and hence the matrix of T^* relative to the basis $\{e_1,\ldots,e_n\}$ is $\mathrm{diag}(\alpha_1,\ldots,\alpha_n)$. Now, since the matrix of T relative to the basis $\{e_1,\ldots,e_n\}$ is $\mathrm{diag}(\alpha_1,\ldots,\alpha_n)$, the matrices of T and T^* relative to the basis $\{e_1,\ldots,e_n\}$ are equal. It follows that $T(e_i)=T^*(e_i)(i=1,\ldots,n)$. Now, since T and T^* are linear, for every $v\in V$, $T(v)=T^*(v)$, and hence $T^*=T$. ∎

4.1.18 Theorem Let V be any n-dimensional inner product space over the field $\mathbb{C}$. Let $T:V\to V$ be a normal linear transformation. Then T is unitary if and only if the absolute value of each eigenvalue of T is 1.

Proof In view of 3.1.22, it remains to show that if the absolute value of each eigenvalue of T is 1, then T is unitary. So we suppose that the absolute value of each eigenvalue of T is 1 and show that T is unitary. In view of 3.1.10, it suffices to show that $T^*T=I$.

Since T is normal, by 4.1.16, there exists an orthonormal basis $\{e_1,\ldots,e_n\}$ of V such that the matrix of T relative to the basis $\{e_1,\ldots,e_n\}$ is a diagonal matrix, say $\mathrm{diag}(\alpha_1,\ldots,\alpha_n)$.

Since the matrix of T relative to the basis $\{e_1,\ldots,e_n\}$ is $\mathrm{diag}(\alpha_1,\ldots,\alpha_n)$, by 3.1.11, the matrix of T^* relative to the basis $\{e_1,\ldots,e_n\}$ is

$$\left(\mathrm{diag}(\alpha_1,\ldots,\alpha_n)\right)^*\left(=\left(\left(\mathrm{diag}(\alpha_1,\ldots,\alpha_n)\right)^T\right)^-\right.$$
$$=\left(\mathrm{diag}(\alpha_1,\ldots,\alpha_n)\right)^-=\mathrm{diag}(\overline{\alpha_1},\ldots,\overline{\alpha_n})\Big),$$

and hence the matrix of T^* relative to the basis $\{e_1,\ldots,e_n\}$ is $\mathrm{diag}(\overline{\alpha_1},\ldots,\overline{\alpha_n})$. Now, since the matrix of T relative to the basis $\{e_1,\ldots,e_n\}$ is $\mathrm{diag}(\alpha_1,\ldots,\alpha_n)$, by 3.1.33, the matrix of T^*T relative to the basis $\{e_1,\ldots,e_n\}$ is

$$\mathrm{diag}(\overline{\alpha_1},\ldots,\overline{\alpha_n})\cdot\mathrm{diag}(\alpha_1,\ldots,\alpha_n)$$
$$=\left(\mathrm{diag}(\overline{\alpha_1}\alpha_1,\ldots,\overline{\alpha_n}\alpha_n)=\mathrm{diag}\left(|\alpha_1|^2,\ldots,|\alpha_n|^2\right)\right),$$

and hence the matrix of T^*T relative to the basis $\{e_1,\ldots,e_n\}$ is $\mathrm{diag}\left(|\alpha_1|^2,\ldots,|\alpha_n|^2\right)$.

Since the matrix of T relative to the basis $\{e_1,\ldots,e_n\}$ is $\mathrm{diag}(\alpha_1,\ldots,\alpha_n)$, we have

$$T(e_1) = \underbrace{\alpha_1 e_1 + 0e_2 + \cdots + 0e_n} = \alpha_1 e_1,$$

and hence $T(e_1) = \alpha_1 e_1$. Since $\{e_1, \ldots, e_n\}$ is a basis, we have $e_1 \neq 0$. Now, since $T(e_1) = \alpha_1 e_1$, α_1 is an eigenvalue of T. Next, by assumption, $|\alpha_1| = 1$. Similarly, $|\alpha_2| = 1$, etc. Now, since the matrix of T^*T relative to the basis $\{e_1, \ldots, e_n\}$ is

$$\mathrm{diag}\left(|\alpha_1|^2, \ldots, |\alpha_n|^2\right) \left(= \mathrm{diag}\left(1^2, \ldots, 1^2\right) = \mathrm{diag}(1, \ldots, 1) = \left[\delta_{ij}\right]\right),$$

the matrix of T^*T relative to the basis $\{e_1, \ldots, e_n\}$ is $\left[\delta_{ij}\right]$. Also, the matrix of I relative to the basis $\{e_1, \ldots, e_n\}$ is $\left[\delta_{ij}\right]$. So, the matrices of T^*T and I relative to the basis $\{e_1, \ldots, e_n\}$ are equal. It follows that $(T^*T)(e_i) = I(e_i)(i = 1, \ldots, n)$. Now, since T^*T and I are linear, for every $v \in V$, $(T^*T)(v) = I(v)$, and hence $T^*T = I$.   ■

4.1.19 Theorem Let V be any n-dimensional inner product space over the field $\mathbb{C}$. Let $N : V \to V$ be a normal linear transformation. Let $T : V \to V$ be a linear transformation. Suppose that $TN = NT$. Then $TN^* = N^*T$.

Proof Let us put $X \equiv TN^* - N^*T$. We have to show that $X = 0$.

Since N is normal, we have $NN^* = N^*N$, and hence N commutes with N^*. Since $TN = NT$, N commutes with T. Since N commutes with T and N^*, N commutes with $(TN^* - N^*T)(= X)$, and hence N commutes with X

By 3.1.7, 3.1.8, and 3.1.6, we have

$$X^* = (TN^* - N^*T)^* = (TN^*)^* - (N^*T)^* = (N^*)^*T^* - T^*(N^*)^* = NT^* - T^*N,$$

and hence

$$\underbrace{XX^* = (TN^* - N^*T)(NT^* - T^*N)} = (TN^* - N^*T)NT^* - (TN^* - N^*T)T^*N$$

$$= ((TN^* - N^*T)N)T^* - (TN^* - N^*T)T^*N$$
$$= (N(TN^* - N^*T))T^* - (TN^* - N^*T)T^*N$$
$$= N((TN^* - N^*T)T^*) - (TN^* - N^*T)T^*N$$
$$= N((TN^* - N^*T)T^*) - ((TN^* - N^*T)T^*)N = NB - BN,$$

where $B \equiv (TN^* - N^*T)T^*$. Thus

$$XX^* = NB - BN.$$

Since N is normal, by 4.1.16, there exists an orthonormal basis $\{e_1, \ldots, e_n\}$ of V such that the matrix of T relative to the basis $\{e_1, \ldots, e_n\}$ is a diagonal matrix, say $\mathrm{diag}(\alpha_1, \ldots, \alpha_n)$.

Let $\left[b_{ij}\right]$ be the matrix of B relative to the basis $\{e_1, \ldots, e_n\}$.

By 3.1.33, the matrix of BN relative to the basis $\{e_1, \ldots, e_n\}$ is $[b_{ij}](\text{diag}(\alpha_1, \ldots, \alpha_n))$. Clearly, the diagonal entries of $[b_{ij}](\text{diag}(\alpha_1, \ldots, \alpha_n))$ are $b_{11}\alpha_1, b_{22}\alpha_2, \ldots, b_{nn}\alpha_n$. Thus the diagonal entries of the matrix of BN relative to the basis $\{e_1, \ldots, e_n\}$ are $b_{11}\alpha_1, b_{22}\alpha_2, \ldots, b_{nn}\alpha_n$.

By 3.1.33, the matrix of NB relative to the basis $\{e_1, \ldots, e_n\}$ is $(\text{diag}(\alpha_1, \ldots, \alpha_n))[b_{ij}]$. Clearly, the diagonal entries of $(\text{diag}(\alpha_1, \ldots, \alpha_n))[b_{ij}]$ are $\alpha_1 b_{11}, \alpha_2 b_{22}, \ldots, \alpha_n b_{nn}$. Thus the diagonal entries of the matrix of NB relative to the basis $\{e_1, \ldots, e_n\}$ are $\alpha_1 b_{11}, \alpha_2 b_{22}, \ldots, \alpha_n b_{nn}$.

It follows that the matrix of $NB - BN$ relative to the basis $\{e_1, \ldots, e_n\}$ is

$$(\text{diag}(\alpha_1, \ldots, \alpha_n))[b_{ij}] - [b_{ij}](\text{diag}(\alpha_1, \ldots, \alpha_n)),$$

and hence the diagonal entries of the matrix of $NB - BN (= XX^*)$ relative to the basis $\{e_1, \ldots, e_n\}$ are all 0.

Thus, the diagonal entries of the matrix of XX^* relative to the basis $\{e_1, \ldots, e_n\}$ are all 0.

Let $[x_{ij}]$ be the matrix of X relative to the basis $\{e_1, \ldots, e_n\}$. By 3.1.11, the matrix of X^* relative to the basis $\{e_1, \ldots, e_n\}$ is

$$([x_{ij}])^* \left(= \left([x_{ij}]^T\right)^- = [\overline{x_{ij}}]^T\right),$$

and hence by 3.1.33, the matrix of XX^* relative to the basis $\{e_1, \ldots, e_n\}$ is $[x_{ij}][\overline{x_{ij}}]^T$. Here the diagonal entries of $[x_{ij}][\overline{x_{ij}}]^T$ are

$$\sum_{j=1}^{n} x_{1j}\overline{x_{1J}}, \sum_{j=1}^{n} x_{2j}\overline{x_{2J}}, \ldots, \sum_{j=1}^{n} x_{nj}\overline{x_{nJ}},$$

that is, the diagonal entries of $[x_{ij}][\overline{x_{iJ}}]^T$ are

$$\sum_{j=1}^{n} |x_{1j}|^2, \sum_{j=1}^{n} |x_{2j}|^2, \ldots, \sum_{j=1}^{n} |x_{nj}|^2.$$

Hence the diagonal entries of the matrix of XX^* relative to the basis $\{e_1, \ldots, e_n\}$ are

$$\sum_{j=1}^{n} |x_{1j}|^2, \sum_{j=1}^{n} |x_{2j}|^2, \ldots, \sum_{j=1}^{n} |x_{nj}|^2.$$

Now, since the diagonal entries of the matrix of XX^* relative to the basis $\{e_1, \ldots, e_n\}$ are all 0, we have

$$\sum_{j=1}^{n}|x_{1j}|^{2}=0, \sum_{j=1}^{n}|x_{2j}|^{2}=0, \text{etc.}$$

Since $\sum_{j=1}^{n}|x_{1j}|^{2}=0,$ we have $x_{1j}=0(j=1,\ldots,n)$. Similarly, $x_{2j}=0$ $(j=1,\ldots,n)$, etc. Thus each x_{ij} is 0, and hence the matrix $[x_{ij}]$ of X relative to the basis $\{e_1,\ldots,e_n\}$ is the zero matrix. This shows that $X=0$.   ∎

4.1.20 Theorem Let V be any n-dimensional inner product space over the field $\mathbb{C}$. Let $T : V \to V$ be a linear transformation. Then T is Hermitian if and only if for every $v \in V$, $\langle T(v), v \rangle$ is a real number.

Proof Let T be Hermitian, that is, $T^* = T$. We have to show that for every $v \in V$, $\langle T(v), v \rangle$ is a real number.

To do so, let us take an arbitrary $v \in V$. We have to show that $\langle T(v), v \rangle$ is a real number, that is, $\overline{\langle T(v), v \rangle} = \langle T(v), v \rangle$:

$$\text{LHS} = \overline{\langle T(v), v \rangle} = \overline{\langle v, T^*(v) \rangle} = \overline{\langle v, T(v) \rangle} = \overline{\overline{\langle T(v), v \rangle}} = \text{RHS}.$$

Conversely, suppose that for every $v \in V$, $\langle T(v), v \rangle$ is a real number. We have to show that T is Hermitian, that is, $T^* = T$, that is, $X = 0$, where $X \equiv T^* - T$. By 3.1.1, it suffices to show that for every $v \in V$, $\langle X(v), v \rangle = 0$:

$$\begin{aligned}
\text{LHS} &= \langle X(v), v \rangle = \langle ((T^* - T)(v), v \rangle = \langle T^*(v) - T(v), v \rangle \\
&= \langle T^*(v), v \rangle - \langle T(v), v \rangle = \overline{\langle v, T^*(v) \rangle} - \langle T(v), v \rangle \\
&= \overline{\langle T(v), v \rangle} - \langle T(v), v \rangle = \langle T(v), v \rangle - \langle T(v), v \rangle = 0 = \text{RHS}.
\end{aligned}$$

Definition Let V be any n-dimensional inner product space over the field $\mathbb{C}$. Let $T : V \to V$ be a linear transformation. If for every $v \in V$, $\langle T(v), v \rangle$ is a nonnegative real number, then we write $T \geq 0$, and we say that T *is nonnegative (definite)*.

By 4.1.20, if $T \geq 0$, then T is Hermitian.

Theorem 4.1.21 Let V be any n-dimensional inner product space over the field $\mathbb{C}$. Let $T : V \to V$ be a linear transformation. Suppose that T is nonnegative. Then all the eigenvalues of T are nonnegative.

Proof To show this, let us take an arbitrary eigenvalue λ of T. We have to show that λ is a nonnegative real number.

Since λ is an eigenvalue of T, there exists a nonzero $v \in V$ such that $T(v) = \lambda v$. Since T is nonnegative, $(\lambda\langle v, v \rangle = \langle \lambda v, v \rangle =)\langle T(v), v \rangle$ is a nonnegative real number, and hence $\lambda\langle v, v \rangle$ is a nonnegative real number. Since v is nonzero, $\langle v, v \rangle$ is a positive real number. Now, since $\lambda\langle v, v \rangle$ is a nonnegative real number, λ is a nonnegative real number.   ∎

Theorem 4.1.22 Let V be any n-dimensional inner product space over the field $\mathbb{C}$. Let $T : V \to V$ be a Hermitian linear transformation. Suppose that all the eigenvalues of T are nonnegative. Then T is nonnegative.

Proof To show this, let us take an arbitrary nonzero $v \in V$. We have to show that $\langle T(v), v \rangle$ is a nonnegative real number.

Since T is Hermitian, T is normal, and hence by 4.1.16, there exists an orthonormal basis $\{e_1, \ldots, e_n\}$ of V such that the matrix of T relative to the basis $\{e_1, \ldots, e_n\}$ is a diagonal matrix, say $\operatorname{diag}(t_1, \ldots, t_n)$. It follows that

$$T(e_1) = t_1 e_1 + 0 e_2 + \cdots + 0 e_n (= t_1 e_1),$$

and hence $T(e_1) = t_1 e_1$. Since $\{e_1, \ldots, e_n\}$ is a basis, e_1 is nonzero. Now, since $T(e_1) = t_1 e_1$, t_1 is an eigenvalue of T. Here by assumption, t_1 is a nonnegative real number. Similarly, $T(e_2) = t_2 e_2$, and t_2 is a nonnegative real number, etc.

Since $v \in V$, and $\{e_1, \ldots, e_n\}$ is an orthonormal basis of V, we have

$$v = \langle v, e_1 \rangle e_1 + \cdots + \langle v, e_n \rangle e_n,$$

and hence

$$
\langle T(v), v \rangle = \left\langle T\left(\sum_{i=1}^{n} \langle v, e_i \rangle e_i \right), \sum_{i=1}^{n} \langle v, e_i \rangle e_i \right\rangle = \left\langle \sum_{i=1}^{n} \langle v, e_i \rangle T(e_i), \sum_{i=1}^{n} \langle v, e_i \rangle e_i \right\rangle
$$

$$
= \left\langle \sum_{i=1}^{n} \langle v, e_i \rangle t_i e_i, \sum_{i=1}^{n} \langle v, e_i \rangle e_i \right\rangle = \sum_{i=1}^{n} \langle v, e_i \rangle t_i \left\langle e_i, \sum_{j=1}^{n} \langle v, e_j \rangle e_j \right\rangle
$$

$$
= \sum_{i=1}^{n} \langle v, e_i \rangle t_i \left(\sum_{j=1}^{n} \langle \overline{v, e_j} \rangle \langle e_i, e_j \rangle \right) = \sum_{i=1}^{n} \langle v, e_i \rangle t_i \left(\sum_{j=1}^{n} \langle \overline{v, e_j} \rangle \delta_{ij} \right)
$$

$$
= \sum_{i=1}^{n} \langle v, e_i \rangle t_i \langle \overline{v, e_i} \rangle = \sum_{i=1}^{n} |\langle v, e_i \rangle|^2 t_i.
$$

Thus $\langle T(v), v \rangle$ is a nonnegative real number. ∎

Definition Let V be any n-dimensional inner product space over the field $\mathbb{C}$. Let $T : V \to V$ be a linear transformation. If for every nonzero $v \in V$, $\langle T(v), v \rangle$ is a positive real number, then we write $T > 0$, and we say that T is positive (definite).

By 4.1.20, if $T > 0$, then T is Hermitian.

4.1.23 Theorem Let V be any n-dimensional inner product space over the field $\mathbb{C}$. Let $T : V \to V$ be a Hermitian linear transformation. Suppose that T is positive. Then all the eigenvalues of T are positive.

Proof To show this, let us take an arbitrary eigenvalue λ of T We have to show that λ is a positive real number.

Since λ is an eigenvalue of T there exists a nonzero $v \in V$ such that $T(v) = \lambda v$. Since T is positive, $(\lambda \langle v, v \rangle = \langle \lambda v, v \rangle =) \langle T(v), v \rangle$ is a positive real number, and hence $\lambda \langle v, v \rangle$ is a positive real number. Since v is nonzero, $\langle v, v \rangle$ is a positive real number. Now, since $\lambda \langle v, v \rangle$ is a positive real number, λ is a positive real number. $\blacksquare$

Theorem 4.1.24 Let V be any n-dimensional inner product space over the field $\mathbb{C}$. Let $T : V \to V$ be a Hermitian linear transformation. Suppose that all the eigenvalues of T are positive. Then T is positive.

Proof To show this, let us take an arbitrary nonzero $v \in V$. We have to show that $\langle T(v), v \rangle$ is a positive real number.

Since T is Hermitian, T is normal, and hence by 4.1.16, there exists an orthonormal basis $\{e_1, \ldots, e_n\}$ of V such that the matrix of T relative to the basis $\{e_1, \ldots, e_n\}$ is a diagonal matrix, say $\mathrm{diag}(t_1, \ldots, t_n)$. It follows that

$$T(e_1) = t_1 e_1 + 0 e_2 + \cdots + 0 e_n (= t_1 e_1),$$

and hence $T(e_1) = t_1 e_1$. Since $\{e_1, \ldots, e_n\}$ is a basis, e_1 is nonzero. Now, since $T(e_1) = t_1 e_1$, t_1 is an eigenvalue of T. Here by assumption, t_1 is a positive real number. Similarly, $T(e_2) = t_2 e_2$, and t_2 is a positive real number, etc.

Since $v \in V$, and $\{e_1, \ldots, e_n\}$ is an orthonormal basis of V, we have

$$v = \langle v, e_1 \rangle e_1 + \cdots + \langle v, e_n \rangle e_n,$$

and hence

$$
\langle T(v), v \rangle = \left\langle T\left(\sum_{i=1}^{n} \langle v, e_i \rangle e_i \right), \sum_{i=1}^{n} \langle v, e_i \rangle e_i \right\rangle = \left\langle \sum_{i=1}^{n} \langle v, e_i \rangle T(e_i), \sum_{i=1}^{n} \langle v, e_i \rangle e_i \right\rangle
$$

$$
= \left\langle \sum_{i=1}^{n} \langle v, e_i \rangle t_i e_i, \sum_{i=1}^{n} \langle v, e_i \rangle e_i \right\rangle = \sum_{i=1}^{n} \langle v, e_i \rangle t_i \left\langle e_i, \sum_{j=1}^{n} \langle v, e_j \rangle e_j \right\rangle
$$

$$
= \sum_{i=1}^{n} \langle v, e_i \rangle t_i \left(\sum_{j=1}^{n} \overline{\langle v, e_j \rangle} \langle e_i, e_j \rangle \right) = \sum_{i=1}^{n} \langle v, e_i \rangle t_i \left(\sum_{j=1}^{n} \overline{\langle v, e_j \rangle} \delta_{ij} \right)
$$

$$
= \sum_{i=1}^{n} \langle v, e_i \rangle t_i \overline{\langle v, e_i \rangle} = \sum_{i=1}^{n} |\langle v, e_i \rangle|^2 t_i.
$$

Since $v = \langle v, e_1 \rangle e_1 + \cdots + \langle v, e_n \rangle e_n$, and v is nonzero, there exists $j \in \{1, \ldots, n\}$ such that $\langle v, e_j \rangle \neq 0$, and hence $|\langle v, e_j \rangle|^2 > 0$. Also $t_j > 0$, so

$$
\langle T(v), v \rangle = \sum_{i=1}^{n} |\langle v, e_i \rangle|^2 t_i \geq \underbrace{|\langle v, e_j \rangle|^2 t_j > 0}.
$$

Thus $\langle T(v), v \rangle$ is a positive real number. $\blacksquare$

4.1.25 Theorem Let V be any n-dimensional inner product space over the field $\mathbb{C}$. Let $T : V \to V$ be a Hermitian linear transformation. Let $\{e_1, \ldots, e_n\}$ be an orthonormal basis of T. Let $A \equiv [a_{ij}]$ be the matrix of T relative to $\{e_1, \ldots, e_n\}$. Then A is a Hermitian matrix.

Proof We have to show that $A^* = A$. By 3.1.11, it suffices to show that $\overline{a_{ji}} = a_{ij}$. Since $A \equiv [a_{ij}]$ is the matrix of T relative to $\{e_1, \ldots, e_n\}$, we have

$$T(e_1) = a_{11}e_1 + a_{21}e_2 + \cdots + a_{n1}e_n,$$

and hence

$$\underbrace{\langle T(e_1), e_i \rangle = \langle a_{11}e_1 + a_{21}e_2 + \cdots + a_{n1}e_n, e_i \rangle}_{} = \sum_{j=1}^{n} \langle a_{j1}e_j, e_i \rangle$$

$$= \sum_{j=1}^{n} a_{j1} \langle e_j, e_i \rangle = \sum_{j=1}^{n} a_{j1} \delta_{ji} = a_{i1},$$

etc. Thus $a_{ij} = \langle T(e_j), e_i \rangle$. Hence $[\langle T(e_j), e_i \rangle]$ is the matrix of T relative to $\{e_1, \ldots, e_n\}$. Similarly, $[\langle T^*(e_j), e_i \rangle]$ is the matrix of T^* relative to $\{e_1, \ldots, e_n\}$. Since T is Hermitian, we have $T^* = T$. Since $A = [a_{ij}]$, we have $A^T = [b_{ij}]$, where $b_{ij} \equiv a_{ji} (= \langle T(e_i), e_j \rangle)$. It follows that $A^* = \underbrace{(A^T)^{-} = [b_{ij}]^{-}}_{} = [\overline{b_{ij}}]$. Also,

$$\overline{a_{ji}} = \overline{b_{ij}} = \overline{\langle T(e_i), e_j \rangle} = \langle e_j, T(e_i) \rangle = \langle T^*(e_j), e_i \rangle = \langle T(e_j), e_i \rangle = a_{ij},$$

so $\overline{a_{ji}} = a_{ij}$. ∎

4.1.26 Note Let V be any n-dimensional inner product space over the field $\mathbb{C}$. Let $T : V \to V$ be a linear transformation. Let T be nonnegative.

Let $\{e_1, \ldots, e_n\}$ be an orthonormal basis of T. Let $A \equiv [a_{ij}]$ be the matrix of T relative to $\{e_1, \ldots, e_n\}$.

It follows that

$$T(e_1) = a_{11}e_1 + a_{21}e_2 + \cdots + a_{n1}e_n,$$

and hence

$$\langle T(e_1), e_i \rangle = \langle a_{11}e_1 + a_{21}e_2 + \cdots + a_{n1}e_n, e_i \rangle = a_{i1},$$

etc. Thus $a_{ij} = \langle T(e_j), e_i \rangle$. Hence $[\langle T(e_j), e_i \rangle]$ is the matrix of T relative to $\{e_1, \ldots, e_n\}$.

Since $T \geq 0$, by 4.1.20, T is a Hermitian linear transformation, and hence by 4.1.25, A is a Hermitian matrix. Since T is a Hermitian linear transformation, we have $T^* = T$.

Suppose that $t_1, \ldots, t_n$ are the eigenvalues of the linear transformation T.

Since $T \geq 0$, by 4.1.21, all the eigenvalues of the linear transformation T are nonnegative, that is, each t_i is a nonnegative real number. Hence each $\sqrt{t_i}$ is a nonnegative real number. Since A is a Hermitian matrix, $A \equiv [a_{ij}]$ is a normal matrix. Now, by 3.3.24, there exists a unitary matrix $U \equiv [u_{ij}]$ such that

1. U^*AU is a diagonal matrix,
2. the eigenvalues of the matrix A (that is, the eigenvalues of the linear transformation T) are the diagonal entries of U^*AU.

Thus

$$U^*AU = \operatorname{diag}(t_1, \ldots, t_n).$$

Put

$$u_j \equiv \sum_{i=1}^{n} u_{ij}e_i (j = 1, \ldots, n).$$

Clearly, $\{u_1, \ldots, u_n\}$ is an orthonormal basis of V.

Proof It suffices to show that $\langle u_j, u_k \rangle = \delta_{ij}$. Since

$$\langle u_j, u_k \rangle = \left\langle \sum_{i=1}^{n} u_{ij}e_i, \sum_{l=1}^{n} u_{lk}e_l \right\rangle = \sum_{i=1}^{n} u_{ij} \left\langle e_i, \sum_{l=1}^{n} u_{lk}e_l \right\rangle = \sum_{i=1}^{n} u_{ij} \left(\sum_{l=1}^{n} \overline{u_{lk}} \langle e_i, e_l \rangle \right)$$
$$= \sum_{i=1}^{n} u_{ij} \left(\sum_{l=1}^{n} \overline{u_{lk}} \delta_{il} \right) = \sum_{i=1}^{n} u_{ij} \overline{u_{ik}} = \sum_{i=1}^{n} \overline{u_{ik}} u_{ij},$$

we have $\langle u_j, u_k \rangle = \sum_{i=1}^{n} \overline{u_{ik}} u_{ij}$. Since $[u_{ij}]$ is a unitary matrix, we have

$$[\langle u_j, u_i \rangle] = \left[\sum_{k=1}^{n} \overline{u_{kl}} u_{kj} \right] = [\overline{u_{iJ}}]^T [u_{ij}] = \underbrace{[u_{ij}]^* [u_{ij}] = [\delta_{ij}]},$$

and hence $[\langle u_j, u_i \rangle] = [\delta_{ij}]$. This shows that $\langle u_i, u_j \rangle = \delta_{ij}$. ∎

Since U is a unitary matrix, U is invertible, and $U^{-1} = U^*$.

Let us define a linear transformation $W : V \rightarrow V$ as follows: for every $i \in \{1, \ldots, n\}$, $W(e_i) \equiv u_i \left(= \sum_{k=1}^{n} u_{ki}e_k \right)$. It follows that the matrix of W relative to $\{e_1, \ldots, e_n\}$ is $[u_{ij}]$.

Now, since $A \equiv [a_{ij}]$ is the matrix of T relative to $\{e_1, \ldots, e_n\}$, by 3.1.35(a), the matrix of T relative to $\{u_1, \ldots, u_n\}$ is $[u_{ij}]^{-1} A [u_{ij}] (= U^{-1}AU = U^*AU = \operatorname{diag}(t_1, \ldots, t_n))$, and hence the matrix of T relative to $\{u_1, \ldots, u_n\}$ is $\operatorname{diag}(t_1, \ldots, t_n)$.

Let us define a linear transformation $S : V \to V$ as follows:

$$S(u_i) \equiv \sqrt{t_i}u_i(i = 1, \ldots, n).$$

This shows that $\sqrt{t_1}, \ldots, \sqrt{t_1}$ are the eigenvalues of the linear transformation S. Also, the eigenvalues of the linear transformation S are nonnegative real numbers. Further, the matrix of the linear transformation S relative to the basis $\{u_1, \ldots, u_n\}$ is $\mathrm{diag}(\sqrt{t_1}, \ldots, \sqrt{t_n})$. Next, by 3.1.11, the matrix of the linear transformation S^* relative to the basis $\{u_1, \ldots, u_n\}$ is

$$(\mathrm{diag}(\sqrt{t_1}, \ldots, \sqrt{t_n}))^* (= \mathrm{diag}(\sqrt{t_1}, \ldots, \sqrt{t_n})).$$

It follows, by 3.1.33, that the matrix of the linear transformation SS^* relative to the basis $\{u_1, \ldots, u_n\}$ is

$$\mathrm{diag}(\sqrt{t_1}, \ldots, \sqrt{t_n})(\mathrm{diag}(\sqrt{t_1}, \ldots, \sqrt{t_n}))^*$$
$$= \mathrm{diag}(\sqrt{t_1}, \ldots, \sqrt{t_n})\mathrm{diag}(\sqrt{t_1}, \ldots, \sqrt{t_n}) = \mathrm{diag}(t_1, \ldots, t_n).$$

Thus the matrix of the linear transformation SS^* relative to the basis $\{u_1, \ldots, u_n\}$ is $\mathrm{diag}(t_1, \ldots, t_n)$. Now, since the matrix of T relative to $\{u_1, \ldots, u_n\}$ is $\mathrm{diag}(t_1, \ldots, t_n)$, we have $SS^* = T$.

Clearly, S is a Hermitian linear transformation, that is, $S^* = S$.

Proof By 4.1.20, it suffices to show that for every $v \in V$, $\langle S(v), v \rangle$ is a real number.

To this end, let us take an arbitrary $v \equiv a_1 u_1 + \cdots + a_n u_n$ in V. Since

$$\langle S(v), v \rangle = \langle S(a_1 u_1 + \cdots + a_n u_n), a_1 u_1 + \cdots + a_n u_n \rangle$$
$$= \langle a_1 S(u_1) + \cdots + a_n S(u_n), a_1 u_1 + \cdots + a_n u_n \rangle$$
$$= \langle a_1 \sqrt{t_1} u_1 + \cdots + a_n \sqrt{t_n} u_n, a_1 u_1 + \cdots + a_n u_n \rangle$$
$$= a_1 \sqrt{t_1} \overline{a_1} + \cdots + a_n \sqrt{t_n} \overline{a_n} = |a_1|^2 \sqrt{t_1} + \cdots + |a_n|^2 \sqrt{t_n},$$

$\langle S(v), v \rangle$ is a real number. ∎

We have shown that S is a Hermitian linear transformation. Next, since the eigenvalues of the linear transformation S are nonnegative real numbers, by 4.1.22, S is nonnegative.

4.1.27 Conclusion Let V be any n-dimensional inner product space over the field $\mathbb{C}$. Let $T : V \to V$ be a linear transformation. Let T be nonnegative. Then there exists a linear transformation $S : V \to V$ such that

1. $S \geq 0$,
2. $SS^* = T$,
3. $S^2 = T$.

Definition Let $A \equiv [a_{ij}]$ be an n-square complex matrix. Observe that for every $x \in \mathbb{C}^n$, x^*Ax is a 1×1 matrix. By $x^*Ax > 0$, we mean that the entry of the 1×1 matrix x^*Ax is a positive real number.

If for every nonzero $x \in \mathbb{C}^n$, $x^*Ax > 0$, then we say that A *is a positive definite matrix*, and we write $A > 0$.

4.1.28 Problem Let V be any n-dimensional inner product space over the field $\mathbb{C}$. Let $T : V \to V$ be a nonnegative linear transformation. Let $\{v_1, \ldots, v_n\}$ be an orthonormal basis of V. Let $A \equiv [a_{ij}]$ be the matrix of T relative to the basis $\{v_1, \ldots, v_n\}$. Then A is a nonnegative definite matrix.

Proof To show this, let us take an arbitrary $x \equiv [\alpha_1, \ldots, \alpha_n]^T \in \mathbb{C}^n$. We have to show that $x^*Ax \geq 0$.

Since $[a_{ij}]$ is the matrix of T relative to the basis $\{v_1, \ldots, v_n\}$, it follows that $T(v_j) = \sum_{i=1}^{n} a_{ij} v_i$. Now,

$$
\begin{aligned}
x^*Ax &= \left([\alpha_1, \ldots, \alpha_n]^T\right)^* A [\alpha_1, \ldots, \alpha_n]^T = \left(\left([\alpha_1, \ldots, \alpha_n]^T\right)^T\right)^{-} A [\alpha_1, \ldots, \alpha_n]^T \\
&= [\overline{\alpha_1}, \ldots, \overline{\alpha_n}] A [\alpha_1, \ldots, \alpha_n]^T = ([\overline{\alpha_1}, \ldots, \overline{\alpha_n}] A)[\alpha_1, \ldots, \alpha_n]^T \\
&= \left[\sum_{i=1}^{n} \overline{\alpha_i} a_{i1}, \ldots, \sum_{i=1}^{n} \overline{\alpha_i} a_{in}\right] [\alpha_1, \ldots, \alpha_n]^T \\
&= \left(\sum_{i=1}^{n} \overline{\alpha_i} a_{i1}\right) \alpha_1 + \cdots + \left(\sum_{i=1}^{n} \overline{\alpha_i} a_{in}\right) \alpha_n \\
&= \sum_{j=1}^{n}\left(\sum_{i=1}^{n} \overline{\alpha_i} a_{ij}\right) \alpha_j = \sum_{j=1}^{n}\left(\sum_{i=1}^{n} \overline{\alpha_i} a_{ij} \alpha_j\right),
\end{aligned}
$$

so

$$
x^*Ax = \sum_{j=1}^{n}\left(\sum_{i=1}^{n} \overline{\alpha_i} a_{ij} \alpha_j\right).
$$

We have to show that $\sum_{j=1}^{n}\left(\sum_{i=1}^{n} \overline{\alpha_i} a_{ij} \alpha_j\right) \geq 0$.

Since $T : V \to V$ is a nonnegative linear transformation, we have

$$
\langle T(\alpha_1 v_1 + \cdots + \alpha_n v_n), (\alpha_1 v_1 + \cdots + \alpha_n v_n)\rangle \geq 0.
$$

It suffices to show that

$$\langle T(\alpha_1 v_1 + \cdots + \alpha_n v_n), (\alpha_1 v_1 + \cdots + \alpha_n v_n)\rangle = \sum_{j=1}^{n}\left(\sum_{i=1}^{n}\overline{\alpha_i}a_{ij}\alpha_j\right):$$

$$\text{LHS} = \langle T(\alpha_1 v_1 + \cdots + \alpha_n v_n), (\alpha_1 v_1 + \cdots + \alpha_n v_n)\rangle$$

$$= \langle \alpha_1 T(v_1) + \cdots + \alpha_n T(v_n), \alpha_1 v_1 + \cdots + \alpha_n v_n\rangle = \left\langle \sum_{i=1}^{n}\alpha_i T(v_i), \sum_{j=1}^{n}\alpha_j v_j\right\rangle$$

$$= \sum_{i=1}^{n}\alpha_i\left\langle T(v_i), \sum_{j=1}^{n}\alpha_j v_j\right\rangle = \sum_{i=1}^{n}\alpha_i\left\langle \sum_{k=1}^{n}a_{ki}v_k, \sum_{j=1}^{n}\alpha_j v_j\right\rangle$$

$$= \sum_{i=1}^{n}\alpha_i\left(\sum_{k=1}^{n}a_{ki}\left\langle v_k, \sum_{j=1}^{n}\alpha_j v_j\right\rangle\right) = \sum_{i=1}^{n}\left(\sum_{k=1}^{n}\alpha_i a_{ki}\left\langle v_k, \sum_{j=1}^{n}\alpha_j v_j\right\rangle\right)$$

$$= \sum_{i=1}^{n}\left(\sum_{k=1}^{n}\alpha_i a_{ki}\left(\sum_{j=1}^{n}\overline{\alpha_j}\langle v_k, v_j\rangle\right)\right) = \sum_{i=1}^{n}\left(\sum_{k=1}^{n}\alpha_i a_{ki}\left(\sum_{j=1}^{n}\overline{\alpha_j}\delta_{kj}\right)\right)$$

$$= \sum_{i=1}^{n}\left(\sum_{k=1}^{n}\alpha_i a_{ki}\overline{\alpha_k}\right) = \sum_{k=1}^{n}\left(\sum_{i=1}^{n}\alpha_k a_{ik}\overline{\alpha_i}\right)$$

$$= \sum_{j=1}^{n}\left(\sum_{i=1}^{n}\alpha_j a_{ij}\overline{\alpha_i}\right) = \sum_{j=1}^{n}\left(\sum_{i=1}^{n}\overline{\alpha_i}a_{ij}\alpha_j\right) = \text{RHS}.$$

∎

4.1.29 Problem Let $A \equiv [a_{ij}]$ be an n-square complex matrix. Suppose that A is a nonnegative definite matrix. Let $T : x \mapsto Ax$ be the linear transformation from the inner product space $\mathbb{C}^n$ to $\mathbb{C}^n$. Then T is a nonnegative linear transformation.

Proof To show this, let us take an arbitrary $x \equiv [x_1, \ldots, x_n]^T \in \mathbb{C}^n$. We have to show that $\langle T(x), x\rangle \geq 0$, that is, $\langle Ax, x\rangle \geq 0$.

Since A is a nonnegative definite matrix, and $x \in \mathbb{C}^n$, we have $x^*Ax \geq 0$. It suffices to show that $\langle Ax, x\rangle = x^*Ax$. By the definition of inner product of $\mathbb{C}^n$, $\langle Ax, x\rangle = x^*(Ax) = x^*Ax$, we have $\langle Ax, x\rangle = x^*Ax$. ∎

4.1.30 Problem Let $A \equiv [a_{ij}]$ be an n-square complex matrix. Suppose that A is a nonnegative definite matrix. Let $X \equiv [x_{ij}]_{n\times m} = [x_1, \ldots, x_m]$ be any complex matrix of size $n \times m$. (It is clear that X^*AX is an $m \times m$ matrix.) Then $X^*AX \geq 0$.

Proof To show this, let us take an arbitrary $x \in \mathbb{C}^m$. We have to show that $x^*(X^*AX)x \geq 0$. Since

$$x^*(X^*AX)x = (x^*X^*)A(Xx) = (Xx)^*A(Xx),$$

we have to show that $y^*Ay \geq 0$, where $y \equiv Xx$. Since X is a complex matrix of size $n \times m$, and $x \in \mathbb{C}^m$, we have $(y =)Xx \in \mathbb{C}^n$, and hence $y \in \mathbb{C}^n$. Now, since A is a nonnegative definite matrix of size $n \times n$, we have $y^*Ay \geq 0$. ∎

4.1.31 Problem Let $A \equiv [a_{ij}]$ be an n-square complex matrix. Suppose that for every $n \times m$ complex matrix $X \equiv [x_{ij}]_{n \times m} = [x_1, \ldots, x_m]$, $X^*AX \geq 0$. Then $A \geq 0$.

Proof To show this, let us take an arbitrary $x \in \mathbb{C}^n$. We have to show that $x^*Ax \geq 0$. Let us take $x_1 = x$, $x_2 = 0$, ..., $x_m = 0$. By assumption, $X^*AX \geq 0$. Observe that

$$
\begin{aligned}
X^*AX &= [x_1, \ldots, x_m]^* A[x_1, \ldots, x_m] = [(x_1)^*, \ldots, (x_m)^*]^T A[x_1, \ldots, x_m] \\
&= [(x_1)^*, \ldots, (x_m)^*]^T (A[x_1, \ldots, x_m]) = [(x_1)^*, \ldots, (x_m)^*]^T [Ax_1, \ldots, Ax_m] \\
&= \begin{bmatrix} (x_1)^*Ax_1 & (x_1)^*Ax_2 & \cdots \\ (x_2)^*Ax_1 & (x_2)^*Ax_2 & \cdots \\ \vdots & \vdots & \ddots \end{bmatrix} = \begin{bmatrix} (x_1)^*Ax_1 & (x_1)^*A0 & \cdots \\ 0^*Ax_1 & 0^*A0 & \cdots \\ \vdots & \vdots & \ddots \end{bmatrix} \\
&= \begin{bmatrix} (x_1)^*Ax_1 & 0 & \cdots \\ 0 & 0 & \cdots \\ \vdots & \vdots & \ddots \end{bmatrix},
\end{aligned}
$$

so

$$X^*AX = \begin{bmatrix} (x_1)^*Ax_1 & 0 & \cdots \\ 0 & 0 & \cdots \\ \vdots & \vdots & \ddots \end{bmatrix}.$$

Now, since $X^*AX \geq 0$, and $[1, 0, \ldots, 0]^T \in \mathbb{C}^n$, we have

$$\underbrace{0 \leq ([1, 0, \ldots, 0]^T)^*(X^*AX)([1, \quad 0, \quad \cdots \quad 0]^T)}$$

$$= [1 \quad 0 \quad \cdots \quad 0] \begin{bmatrix} (x_1)^*Ax_1 & 0 & \cdots \\ 0 & 0 & \cdots \\ \vdots & \vdots & \ddots \end{bmatrix} \begin{bmatrix} 1 \\ 0 \\ \vdots \\ 0 \end{bmatrix}$$

$$= [1 \quad 0 \quad \cdots \quad 0] \begin{bmatrix} (x_1)^*Ax_1 \\ 0 \\ \vdots \\ 0 \end{bmatrix} = (x_1)^*Ax_1 = x^*Ax,$$

and hence, $0 \leq x^*Ax$. ∎

4.1.32 Note Let $A \equiv [a_{ij}]$ be an n-square complex matrix. Suppose that A is a nonnegative definite matrix.

By 3.3.28, A is a Hermitian matrix, and hence $A \equiv [a_{ij}]$ is a normal matrix. Now, by 3.3.24, there exists a unitary matrix $U \equiv [u_{ij}]$ such that

1. U^*AU is a diagonal matrix, say $\text{diag}(\lambda_1, \ldots, \lambda_n)$,
2. the eigenvalues of the matrix A are $\lambda_1, \ldots, \lambda_n$.

Since U is unitary, we have $U^*U = UU^* = I$, and hence $U^{-1} = U^*$. Now, since $U^*AU = \text{diag}(\lambda_1, \ldots, \lambda_n)$, we have $A = U(\text{diag}(\lambda_1, \ldots, \lambda_n))U^*$.

Let $T : x \mapsto Ax$ be the linear transformation from the inner product space $\mathbb{C}^n$ to $\mathbb{C}^n$. By 4.1.29, T is a nonnegative linear transformation, and by 4.1.21, all the eigenvalues of the linear transformation T are nonnegative. Here $\{e_1, \ldots, e_n\}$ is an orthonormal basis of V, where $e_1 \equiv [1, 0, \ldots, 0]^T$, $e_2 \equiv [0, 1, 0, \ldots, 0]^T$, etc. Since

$$T(e_1) = Ae_1 = [a_{ij}][1, 0, \ldots, 0]^T = [a_{11}, a_{21}, \ldots, a_{n1}]^T$$
$$= a_{11}e_1 + a_{21}e_2 + \cdots + a_{n1}e_n,$$

we have

$$T(e_1) = a_{11}e_1 + a_{21}e_2 + \cdots + a_{n1}e_n.$$

Similarly,

$$T(e_2) = a_{12}e_1 + a_{22}e_2 + \cdots + a_{n2}e_n,$$

etc. Thus $T(e_j) = \sum_{j=1}^{n} a_{ij}e_j$. It follows that the matrix of T relative to $\{e_1, \ldots, e_n\}$ is $[a_{ij}](= A)$. Now, since the eigenvalues of the matrix A are $\lambda_1, \ldots, \lambda_n$, the eigenvalues of the linear transformation T are also $\lambda_1, \ldots, \lambda_n$. Next, since the eigenvalues of the linear transformation T are nonnegative real numbers, each λ_i is a nonnegative real number.

Since

$$A = U(\text{diag}(\lambda_1, \ldots, \lambda_n))U^* (= U(\text{diag}(\lambda_1, \ldots, \lambda_n))U^{-1}),$$

we have

$$\det(A) = \det(U(\text{diag}(\lambda_1, \ldots, \lambda_n))U^{-1}) = \det(U) \cdot \det(\text{diag}(\lambda_1, \ldots, \lambda_n)) \cdot \det(U^{-1})$$

$$= \det(U) \cdot \det(\text{diag}(\lambda_1, \ldots, \lambda_n)) \cdot \tfrac{1}{\det(U)} = \det(\text{diag}(\lambda_1, \ldots, \lambda_n)) = \lambda_1\lambda_2\cdots\lambda_n,$$

and hence

$$\det(A) = \lambda_1\lambda_2\ldots\lambda_n.$$

Since each λ_i is a nonnegative real number, $\det(A)$ is a nonnegative real number.

4.1.33 Conclusion Let $A \equiv [a_{ij}]$ be an n-square complex matrix. Suppose that A is a nonnegative definite matrix. Then there exists a unitary matrix U such that

1. $A = U(\text{diag}(\lambda_1, \ldots, \lambda_n))U^*$,
2. $\lambda_1, \ldots, \lambda_n$ are the eigenvalues of the matrix A,
3. each λ_i is a nonnegative real number,
4. $\det(A)$ is a nonnegative real number.

4.1.34 Note Let $A \equiv [a_{ij}]$ be an n-square complex matrix. Suppose that A is a nonnegative definite matrix.

By 4.1.33, there exists a unitary matrix U such that

1. $A = U(\text{diag}(\lambda_1, \ldots, \lambda_n))U^*$,
2. $\lambda_1, \ldots, \lambda_n$ are the eigenvalues of the matrix A,
3. each λ_i is a nonnegative real number,
4. $\det(A)$ is a nonnegative real number.

Since each λ_i is a nonnegative real number, each $\sqrt{\lambda_i}$ is a nonnegative real number. Observe that

$$
\begin{aligned}
&\left(U\left(\text{diag}\left(\sqrt{\lambda_1}, \ldots, \sqrt{\lambda_n}\right)\right)U^*\right)^2 \\
&= \left(U\left(\text{diag}\left(\sqrt{\lambda_1}, \ldots, \sqrt{\lambda_n}\right)\right)U^*\right)\left(U\left(\text{diag}\left(\sqrt{\lambda_1}, \ldots, \sqrt{\lambda_n}\right)\right)U^*\right) \\
&= U\left(\text{diag}\left(\sqrt{\lambda_1}, \ldots, \sqrt{\lambda_n}\right)\right)(U^*U)\left(\text{diag}\left(\sqrt{\lambda_1}, \ldots, \sqrt{\lambda_n}\right)\right)U^* \\
&= U\left(\text{diag}\left(\sqrt{\lambda_1}, \ldots, \sqrt{\lambda_n}\right)\right)I\left(\text{diag}\left(\sqrt{\lambda_1}, \ldots, \sqrt{\lambda_n}\right)\right)U^* \\
&= U\left(\left(\text{diag}\left(\sqrt{\lambda_1}, \ldots, \sqrt{\lambda_n}\right)\right)\left(\text{diag}\left(\sqrt{\lambda_1}, \ldots, \sqrt{\lambda_n}\right)\right)\right)U^* \\
&= U(\text{diag}(\lambda_1, \ldots, \lambda_n))U^* = A,
\end{aligned}
$$

so

$$
\left(U\left(\text{diag}\left(\sqrt{\lambda_1}, \ldots, \sqrt{\lambda_n}\right)\right)U^*\right)^2 = A.
$$

Thus

$$
B^2 = A,
$$

where $B \equiv U\left(\text{diag}\left(\sqrt{\lambda_1}, \ldots, \sqrt{\lambda_n}\right)\right)U^*$. Since for every $x \equiv [x_1, \ldots, x_n]^T \in \mathbb{C}^n$,

$$
\begin{aligned}
&x^*\left(\text{diag}\left(\sqrt{\lambda_1}, \ldots, \sqrt{\lambda_n}\right)\right)x \\
&= \left([x_1, \ldots, x_n]^T\right)^*\left(\text{diag}\left(\sqrt{\lambda_1}, \ldots, \sqrt{\lambda_n}\right)\right)[x_1, \ldots, x_n]^T \\
&= [\overline{x_1}, \ldots, \overline{x_n}]\left(\text{diag}\left(\sqrt{\lambda_1}, \ldots, \sqrt{\lambda_n}\right)\right)[x_1, \ldots, x_n]^T \\
&= \left([\overline{x_1}, \ldots, \overline{x_n}]\left(\text{diag}\left(\sqrt{\lambda_1}, \ldots, \sqrt{\lambda_n}\right)\right)\right)[x_1, \ldots, x_n]^T \\
&= [\overline{x_1}\sqrt{\lambda_1}, \ldots, \overline{x_n}\sqrt{\lambda_n}][x_1, \ldots, x_n]^T = \overline{x_1}\sqrt{\lambda_1}x_1 + \ldots + \overline{x_n}\sqrt{\lambda_n}x_n \\
&= \sqrt{\lambda_1}|x_1|^2 + \cdots + \sqrt{\lambda_n}|x_n|^2 \geq 0,
\end{aligned}
$$

.

we have

$$x^* \left(\mathrm{diag} \left(\sqrt{\lambda_1}, \ldots, \sqrt{\lambda_n} \right) \right) x \geq 0.$$

This shows that $\mathrm{diag} \left(\sqrt{\lambda_1}, \ldots, \sqrt{\lambda_n} \right)$ is a nonnegative definite matrix, and hence by 4.1.30, $(U^*)^* \left(\mathrm{diag} \left(\sqrt{\lambda_1}, \ldots, \sqrt{\lambda_n} \right) \right) (U^*) \left(= U \left(\mathrm{diag} \left(\sqrt{\lambda_1}, \ldots, \sqrt{\lambda_n} \right) \right) U^* = B \right)$ is a nonnegative definite matrix. Thus B is a nonnegative matrix.

Since

$$
\begin{aligned}
\lambda I - B &= \lambda I - U \left(\mathrm{diag} \left(\sqrt{\lambda_1}, \ldots, \sqrt{\lambda_n} \right) \right) U^* \\
&= \lambda U U^* - U \left(\mathrm{diag} \left(\sqrt{\lambda_1}, \ldots, \sqrt{\lambda_n} \right) \right) U^* \\
&= U(\lambda I) U^* - U \left(\mathrm{diag} \left(\sqrt{\lambda_1}, \ldots, \sqrt{\lambda_n} \right) \right) U^* \\
&= U \left(\lambda I - \mathrm{diag} \left(\sqrt{\lambda_1}, \ldots, \sqrt{\lambda_n} \right) \right) U^* \\
&= U \left(\mathrm{diag}(\lambda, \ldots, \lambda) - \mathrm{diag} \left(\sqrt{\lambda_1}, \ldots, \sqrt{\lambda_n} \right) \right) U^* \\
&= U \left(\mathrm{diag} \left(\lambda - \sqrt{\lambda_1}, \ldots, \lambda - \sqrt{\lambda_n} \right) \right) U^*,
\end{aligned}
$$

we have

$$\lambda I - B = U \left(\mathrm{diag} \left(\lambda - \sqrt{\lambda_1}, \ldots, \lambda - \sqrt{\lambda_n} \right) \right) U^*,$$

and hence

$$\underline{\det(\lambda I - B) = \det \left(U \left(\mathrm{diag} \left(\lambda - \sqrt{\lambda_1}, \ldots, \lambda - \sqrt{\lambda_n} \right) \right) U^* \right)}$$

$$
\begin{aligned}
&= \det(U) \cdot \det \left(\mathrm{diag} \left(\lambda - \sqrt{\lambda_1}, \ldots, \lambda - \sqrt{\lambda_n} \right) \right) \cdot \det(U^*) \\
&= \det(U) \cdot \det \left(\mathrm{diag} \left(\lambda - \sqrt{\lambda_1}, \ldots, \lambda - \sqrt{\lambda_n} \right) \right) \cdot \det(U^{-1}) \\
&= \det(U) \cdot \det \left(\mathrm{diag} \left(\lambda - \sqrt{\lambda_1}, \ldots, \lambda - \sqrt{\lambda_n} \right) \right) \cdot \frac{1}{\det(U)} \\
&= \det \left(\mathrm{diag} \left(\lambda - \sqrt{\lambda_1}, \ldots, \lambda - \sqrt{\lambda_n} \right) \right) = \left(\lambda - \sqrt{\lambda_1} \right) \left(\lambda - \sqrt{\lambda_2} \right) \cdots \left(\lambda - \sqrt{\lambda_n} \right).
\end{aligned}
$$

Thus

$$\det(\lambda I - B) = \left(\lambda - \sqrt{\lambda_1} \right) \left(\lambda - \sqrt{\lambda_2} \right) \cdots \left(\lambda - \sqrt{\lambda_n} \right).$$

Hence the characteristic polynomial of the matrix B is $\left(\lambda - \sqrt{\lambda_1}\right)$ $\left(\lambda - \sqrt{\lambda_2}\right) \cdots \left(\lambda - \sqrt{\lambda_n}\right)$. Its roots are $\sqrt{\lambda_1}, \ldots, \sqrt{\lambda_n}$.

So the eigenvalues of the matrix B are $\sqrt{\lambda_1}, \ldots, \sqrt{\lambda_n}$. Also, $\lambda_1, \ldots, \lambda_n$ are the eigenvalues of the matrix A. Thus the eigenvalues of the matrix B are the square roots of the eigenvalues of the matrix A.

4.1.35 Conclusion Let $A \equiv \begin{bmatrix} a_{ij} \end{bmatrix}$ be an n-square complex matrix. Suppose that A is a nonnegative definite matrix. Let $\lambda_1, \ldots, \lambda_n$ be the eigenvalues of the matrix A. Then there exists a matrix B such that

1. $B^2 = A$,
2. B is a nonnegative definite matrix,
3. $\sqrt{\lambda_1}, \ldots, \sqrt{\lambda_n}$ are the eigenvalues of the matrix B,
4. there exists a unitary matrix U such that $U\left(\text{diag}\left(\sqrt{\lambda_1}, \ldots, \sqrt{\lambda_n}\right)\right)U^* = B$.

4.1.36 Problem Let $A \equiv \begin{bmatrix} a_{ij} \end{bmatrix}$ be an n-square complex matrix. Suppose that A is a nonnegative definite matrix. Let $\lambda_1, \ldots, \lambda_n$ be the eigenvalues of the matrix A. Then there exists a unique matrix B such that

1. $B^2 = A$,
2. there exists a unitary matrix U such that $U\left(\text{diag}\left(\sqrt{\lambda_1}, \ldots, \sqrt{\lambda_n}\right)\right)U^* = B$.

Proof In view of 4.1.25, it remains to prove the uniqueness part.

Uniqueness: Suppose that B is a matrix such that

1. $B^2 = A$,
2. there exists a unitary matrix U such that $U\left(\text{diag}\left(\sqrt{\lambda_1}, \ldots, \sqrt{\lambda_n}\right)\right)U^* = B$.

Suppose that C is a matrix such that

1. $C^2 = A$,
2. there exists a unitary matrix V such that $V\left(\text{diag}\left(\sqrt{\lambda_1}, \ldots, \sqrt{\lambda_n}\right)\right)V^* = C$.

We have to show that $B = C$, that is, $U\left(\text{diag}\left(\sqrt{\lambda_1}, \ldots, \sqrt{\lambda_n}\right)\right)U^* = V\left(\text{diag}\left(\sqrt{\lambda_1}, \ldots, \sqrt{\lambda_n}\right)\right)V^*$, that is,

$$U\left(\text{diag}\left(\sqrt{\lambda_1}, \ldots, \sqrt{\lambda_n}\right)\right) = V\left(\text{diag}\left(\sqrt{\lambda_1}, \ldots, \sqrt{\lambda_n}\right)\right)V^*U,$$

that is,

$$V^*U\left(\text{diag}\left(\sqrt{\lambda_1}, \ldots, \sqrt{\lambda_n}\right)\right) = \left(\text{diag}\left(\sqrt{\lambda_1}, \ldots, \sqrt{\lambda_n}\right)\right)V^*U,$$

that is,

$$W\left(\text{diag}\left(\sqrt{\lambda_1},\ldots,\sqrt{\lambda_n}\right)\right) = \left(\text{diag}\left(\sqrt{\lambda_1},\ldots,\sqrt{\lambda_n}\right)\right)W,$$

where $W \equiv V^*U$.

Suppose that $W \equiv [w_{ij}]$. Observe that $\text{diag}\left(\sqrt{\lambda_1},\ldots,\sqrt{\lambda_n}\right) = [\tau_{ij}]$, where $\tau_{ij} \equiv \sqrt{\lambda_i}\delta_{ij}$. So we have to show that

$$[w_{ij}][\tau_{ij}] = [\tau_{ij}][w_{ij}],$$

that is,

$$\left(\sqrt{\lambda_j}w_{ij} = w_{ij}\sqrt{\lambda_j} = \sum_{k=1}^{n} w_{ik}\left(\sqrt{\lambda_k}\delta_{kj}\right) = \right)\underbrace{\sum_{k=1}^{n} w_{ik}\tau_{kj} = \sum_{k=1}^{n} \tau_{ik}w_{kj}}$$

that is,

$$\sqrt{\lambda_j}w_{ij} = \sqrt{\lambda_i}w_{ij},$$

that is,

$$\left(\sqrt{\lambda_i} - \sqrt{\lambda_j}\right)w_{ij} = 0.$$

Thus it suffices to show that

$$\boxed{\left(\sqrt{\lambda_i} - \sqrt{\lambda_j}\right)w_{ij} = 0.}$$

Since

$$A = B^2 = BB = \left(U\left(\text{diag}\left(\sqrt{\lambda_1},\ldots,\sqrt{\lambda_n}\right)\right)U^*\right)\left(U\left(\text{diag}\left(\sqrt{\lambda_1},\ldots,\sqrt{\lambda_n}\right)\right)U^*\right)$$
$$= U\left(\text{diag}\left(\sqrt{\lambda_1},\ldots,\sqrt{\lambda_n}\right)\right)(U^*U)\left(\text{diag}\left(\sqrt{\lambda_1},\ldots,\sqrt{\lambda_n}\right)\right)U^*$$
$$= U\left(\text{diag}\left(\sqrt{\lambda_1},\ldots,\sqrt{\lambda_n}\right)\right)I\left(\text{diag}\left(\sqrt{\lambda_1},\ldots,\sqrt{\lambda_n}\right)\right)U^*$$
$$= U\left(\left(\text{diag}\left(\sqrt{\lambda_1},\ldots,\sqrt{\lambda_n}\right)\right)\left(\text{diag}\left(\sqrt{\lambda_1},\ldots,\sqrt{\lambda_n}\right)\right)\right)U^*$$
$$= U(\text{diag}(\lambda_1,\ldots,\lambda_n))U^* = U(\text{diag}(\lambda_1,\ldots,\lambda_n))U^*,$$

we have

$$A = U(\text{diag}(\lambda_1,\ldots,\lambda_n))U^*.$$

Similarly,

$$A = V(\text{diag}(\lambda_1, \ldots, \lambda_n))V^*.$$

It follows that

$$U(\text{diag}(\lambda_1, \ldots, \lambda_n))U^* = V(\text{diag}(\lambda_1, \ldots, \lambda_n))V^*,$$

that is,

$$V^*U(\text{diag}(\lambda_1, \ldots, \lambda_n))U^* = (\text{diag}(\lambda_1, \ldots, \lambda_n))V^*,$$

that is,

$$(V^*U)(\text{diag}(\lambda_1, \ldots, \lambda_n)) = (\text{diag}(\lambda_1, \ldots, \lambda_n))(V^*U),$$

that is,

$$W(\text{diag}(\lambda_1, \ldots, \lambda_n)) = (\text{diag}(\lambda_1, \ldots, \lambda_n))W,$$

that is,

$$[w_{ij}](\text{diag}(\lambda_1, \ldots, \lambda_n)) = (\text{diag}(\lambda_1, \ldots, \lambda_n))[w_{ij}].$$

Observe that $\text{diag}(\lambda_1, \ldots, \lambda_n) = [\kappa_{ij}]$, where $\kappa_{ij} \equiv \lambda_i \delta_{ij}$. It follows that

$$[w_{ij}][\kappa_{ij}] = [\kappa_{ij}][w_{ij}],$$

and hence

$$\sum_{k=1}^{n} w_{ik}\kappa_{kj} = \sum_{k=1}^{n} \kappa_{ik}w_{kj}.$$

Now, since

$$\sum_{k=1}^{n} w_{ik}\kappa_{kj} = \sum_{k=1}^{n} w_{ik}(\lambda_k \delta_{kj}) = w_{ij}\lambda_j,$$

and

$$\sum_{k=1}^{n} \kappa_{ik}w_{kj} = \sum_{k=1}^{n} (\lambda_i \delta_{ik})w_{kj} = \lambda_i w_{ij} = w_{ij}\lambda_i,$$

we have $w_{ij}\lambda_j = w_{ij}\lambda_i$, and hence $(\lambda_i - \lambda_j)w_{ij} = 0$. It follows that for distinct λ_i and λ_j, $w_{ij} = 0$. Hence $(\sqrt{\lambda_i} - \sqrt{\lambda_j})w_{ij} = 0$.   ∎

4.1.37 Theorem Let $A \equiv [a_{ij}]$ be an n-square complex matrix. Suppose that A is a nonnegative definite matrix. Then there exists a unique matrix B such that

1. $B^2 = A$,
2. B is a nonnegative definite matrix.

Here the unique matrix B is denoted by $\sqrt{A}$ and is called the *square root of the nonnegative definite matrix A*.

Proof In view of 4.1.35, it remains to prove the uniqueness part.
   Uniqueness: Let $\lambda_1, \ldots, \lambda_n$ be the eigenvalues of the matrix A.
   Suppose that B is a matrix such that

1. $B^2 = A$,
2. B is a nonnegative definite matrix.

   Suppose that C is a matrix such that

1. $C^2 = A$,
2. C is a nonnegative definite matrix.

   We have to show that $B = C$.
   Since B is a nonnegative definite matrix, by 3.3.28, B is a Hermitian matrix, and hence by 3.3.24, there exists a unitary matrix U such that

1. U^*BU is a diagonal matrix,
2. the eigenvalues of B are the diagonal entries of U^*BU.

   Hence $U^*BU = \text{diag}(\mu_1, \ldots, \mu_n)$, where $\mu_1, \ldots, \mu_n$ are the eigenvalues of B. It follows that

$$B = U(\text{diag}(\mu_1, \ldots, \mu_n))U^*.$$

   Now, since

$$A = \underbrace{B^2 = (U(\text{diag}(\mu_1, \ldots, \mu_n))U^*)^2}$$

$$= (U(\text{diag}(\mu_1, \ldots, \mu_n))U^*)(U(\text{diag}(\mu_1, \ldots, \mu_n))U^*)$$
$$= U(\text{diag}(\mu_1, \ldots, \mu_n))(\text{diag}(\mu_1, \ldots, \mu_n)))U^*$$
$$= U\left(\text{diag}\left((\mu_1)^2, \ldots, (\mu_n)^2\right)\right)U^*,$$

we have

$$A = U\left(\text{diag}\left((\mu_1)^2, \ldots, (\mu_n)^2\right)\right)U^*.$$

Since

$$
\begin{aligned}
\lambda I - A &= \lambda I - U\Big(\mathrm{diag}\big((\mu_1)^2, \ldots, (\mu_n)^2\big)\Big)U^* \\
&= \lambda UU^* - U\Big(\mathrm{diag}\big((\mu_1)^2, \ldots, (\mu_n)^2\big)\Big)U^* \\
&= U(\lambda I)U^* - U\Big(\mathrm{diag}\big((\mu_1)^2, \ldots, (\mu_n)^2\big)\Big)U^* \\
&= U\Big(\lambda I - \mathrm{diag}\big((\mu_1)^2, \ldots, (\mu_n)^2\big)\Big)U^* \\
&= U\Big(\mathrm{diag}(\lambda, \ldots, \lambda) - \mathrm{diag}\big((\mu_1)^2, \ldots, (\mu_n)^2\big)\Big)U^* \\
&= U\Big(\mathrm{diag}\big(\lambda - (\mu_1)^2, \ldots, \lambda - (\mu_n)^2\big)\Big)U^*,
\end{aligned}
$$

we have

$$
\lambda I - A = U\Big(\mathrm{diag}\big(\lambda - (\mu_1)^2, \ldots, \lambda - (\mu_n)^2\big)\Big)U^*,
$$

and hence

$$
\det(\lambda I - A) = \underbrace{\det\Big(U\Big(\mathrm{diag}\big(\lambda - (\mu_1)^2, \ldots, \lambda - (\mu_n)^2\big)\Big)U^*\Big)}
$$

$$
\begin{aligned}
&= \det(U) \cdot \det\Big(\mathrm{diag}\big(\lambda - (\mu_1)^2, \ldots, \lambda - (\mu_n)^2\big)\Big) \cdot \det(U^*) \\
&= \det(U) \cdot \det\Big(\mathrm{diag}\big(\lambda - (\mu_1)^2, \ldots, \lambda - (\mu_n)^2\big)\Big) \cdot \det(U^{-1}) \\
&= \det(U) \cdot \det\Big(\mathrm{diag}\big(\lambda - (\mu_1)^2, \ldots, \lambda - (\mu_n)^2\big)\Big) \cdot \frac{1}{\det(U)} \\
&= \det\Big(\mathrm{diag}\big(\lambda - (\mu_1)^2, \ldots, \lambda - (\mu_n)^2\big)\Big) \\
&= \Big(\lambda - (\mu_1)^2\Big)\Big(\lambda - (\mu_2)^2\Big) \cdots \Big(\lambda - (\mu_n)^2\Big).
\end{aligned}
$$

Thus

$$
\det(\lambda I - A) = \Big(\lambda - (\mu_1)^2\Big)\Big(\lambda - (\mu_2)^2\Big) \cdots \Big(\lambda - (\mu_n)^2\Big).
$$

Hence the characteristic polynomial of the matrix A is $\Big(\lambda - (\mu_1)^2\Big)$ $\Big(\lambda - (\mu_2)^2\Big) \cdots \Big(\lambda - (\mu_n)^2\Big)$. Its roots are $(\mu_1)^2, \ldots, (\mu_n)^2$. So the eigenvalues of the matrix A are $(\mu_1)^2, \ldots, (\mu_n)^2$.

Now, since the eigenvalues of the matrix A are $\lambda_1, \ldots, \lambda_n$, we can suppose that $(\mu_1)^2 = \lambda_1$. Since μ_1 is an eigenvalue of B, and B is a nonnegative definite matrix, by 3.3.30, μ_1 is a nonnegative real number. It follows that $\sqrt{\lambda_1} = \underbrace{\sqrt{(\mu_1)^2} = \mu_1}$,

and hence $\mu_1 = \sqrt{\lambda_1}$. Similarly, $\mu_2 = \sqrt{\lambda_2}$, etc.

Next, since

$$U^*BU = \mathrm{diag}(\mu_1, \ldots, \mu_n) \left(= \mathrm{diag}\left(\sqrt{\lambda_1}, \ldots, \sqrt{\lambda_2} \right) \right),$$

we have

$$U^*BU = \mathrm{diag}\left(\sqrt{\lambda_1}, \ldots, \sqrt{\lambda_2} \right),$$

and hence

$$B = U\left(\mathrm{diag}\left(\sqrt{\lambda_1}, \ldots, \sqrt{\lambda_2} \right) \right) U^*.$$

Similarly,

$$C = V\left(\mathrm{diag}\left(\sqrt{\lambda_1}, \ldots, \sqrt{\lambda_2} \right) \right) V^*.$$

Thus

1. $B^2 = A$, 2; $U\left(\mathrm{diag}\left(\sqrt{\lambda_1}, \ldots, \sqrt{\lambda_n}\right)\right)U^* = B$, where U is a unitary matrix; 3. $C^2 = A$; 4. $V\left(\mathrm{diag}\left(\sqrt{\lambda_1}, \ldots, \sqrt{\lambda_n}\right)\right)V^* = C$, where V is a unitary matrix. Now, by 4.1.36, $B = C$. ∎

4.1.38 Theorem Let V be any n-dimensional inner product space over the field $\mathbb{C}$. Let $T : V \rightarrow V$ be a nonnegative linear transformation. Then there exists a unique nonnegative linear transformation $S : V \rightarrow V$ such that $S^2 = T$.

Here the unique linear transformation S is denoted by $\sqrt{T}$, and is called the *square root of the nonnegative linear transformation T.*

Proof In view of 4.1.27, it remains to prove the uniqueness part.

Uniqueness: Let $R : V \rightarrow V$ be a nonnegative linear transformation such that $R^2 = T$. Let $S : V \rightarrow V$ be a nonnegative linear transformation such that $S^2 = T$. We have to show that $R = S$.

Let us take an orthonormal basis $\{e_1, \ldots, e_n\}$ of V. Let $A \equiv \left[a_{ij}\right]$ be the matrix of T relative to the basis $\{e_1, \ldots, e_n\}$. Let $B \equiv \left[b_{ij}\right]$ be the matrix of R relative to the basis $\{e_1, \ldots, e_n\}$. Let $C \equiv \left[c_{ij}\right]$ be the matrix of S relative to the basis $\{e_1, \ldots, e_n\}$. Thus

$$\left. \begin{array}{l} T(e_j) = \sum\limits_{i=1}^{n} a_{ij}e_i \\ R(e_j) = \sum\limits_{i=1}^{n} b_{ij}e_i \\ S(e_j) = \sum\limits_{i=1}^{n} c_{ij}e_i \end{array} \right\}.$$

290 4 Sylvester's Law of Inertia

It suffices to show that $b_{ij} = c_{ij} (i, j \in \{1, \ldots, n\})$, that is, $B = C$.

Since T is a nonnegative linear transformation, by 4.1.28, A is a nonnegative definite matrix. Similarly, B is a nonnegative definite matrix, and C is a nonnegative definite matrix. By 3.1.33, the matrix of $R^2 (= R \circ R = T)$ relative to the basis $\{e_1, \ldots, e_n\}$ is $BB (= B^2)$. Now, since the matrix of T relative to the basis $\{e_1, \ldots, e_n\}$ is A, We have $B^2 = A$. Similarly, $C^2 = A$. Now, by 4.1.37, $B = C$. ∎

4.2 Sylvester's Law

4.2.1 Note Let $A \equiv [a_{ij}]$ be an n-square real matrix.

Since $\overline{a_{ij}} = a_{ij}$, A is Hermitian if and only if A is symmetric (that is, $A^T = A$). Here A is unitary if and only if A is orthogonal (that is, $AA^T = A^T A = I$).

Let $A \equiv [a_{ij}]$ be a real symmetric matrix.

It follows that A is a Hermitian matrix, and hence A is a normal matrix. Now, by 3.3.24, there exists a unitary matrix U such that

1. $U^* A U$ is a diagonal matrix,
2. the eigenvalues of A are the diagonal entries of $U^* A U$.

Hence $U^* A U = \text{diag}(\lambda_1, \ldots, \lambda_n)$, where $\lambda_1, \ldots, \lambda_n$ are the eigenvalues of the matrix A. It follows that

$$\text{diag}(\lambda_1, \ldots, \lambda_n) = U^* A U = U^* A^* U = U^* A^* (U^*)^*$$
$$= \underbrace{(U^* A U)^* = (\text{diag}(\lambda_1, \ldots, \lambda_n))^*} = \text{diag}(\overline{\lambda_1}, \ldots, \overline{\lambda_n}),$$

and hence $\text{diag}(\lambda_1, \ldots, \lambda_n) = \text{diag}(\overline{\lambda_1}, \ldots, \overline{\lambda_n})$. It follows that $\overline{\lambda_i} = \lambda_i (i = 1, \ldots, n)$, and hence each λ_i is a real number.

4.2.2 Conclusion Let $A \equiv [a_{ij}]$ be a real symmetric matrix. Then there exists a unitary matrix U such that $U^* A U = \text{diag}(\lambda_1, \ldots, \lambda_n)$, where $\lambda_1, \ldots, \lambda_n$ are real numbers.

4.2.3 Theorem Let $A \equiv [a_{ij}]$ be an n-square real matrix. Let $B \equiv [b_{ij}]$ be an n-square real matrix. Let P be an invertible n-square complex matrix such that $B = P^{-1} A P$. Then there exists an invertible n-square real matrix Q such that $B = Q^{-1} A Q$.

Proof Since P is an n-square complex matrix, we can write $P = P_1 + i P_2$, where P_1, P_2 are n-square real matrices.

Case I: $P_2 = 0$. In this case, $P = P_1$. Now, since $B = P^{-1} A P$, we have $B = (P_1)^{-1} A P_1$, where P_1 is an n-square real matrix.

Case II: $P_2 \neq 0$. Since $B = P^{-1}AP$, we have

$$P_1B + iP_2B = (P_1 + iP_2)B = \underbrace{PB = AP} = A(P_1 + iP_2) = AP_1 + iAP_2,$$

and hence

$$(P_1B) + i(P_2B) = (AP_1) + i(AP_2).. \tag{*}$$

Since P_1, B are real matrices, P_1B is a real matrix. Similarly, P_2B, AP_1, AP_2 are real matrices. Now, from (*)

$$\left.\begin{array}{l} P_1B = AP_1 \\ P_2B = AP_2 \end{array}\right\}. \tag{**}$$

Since $P_2 \neq 0$, $\det(P_1 + xP_2)$ is a polynomial in x. Suppose that $\{\alpha_1, \ldots, \alpha_k\}$ is the collection of all the roots of the polynomial $\det(P_1 + xP_2)$. We can find a real number $t_0 \notin \{\alpha_1, \ldots, \alpha_k\}$. It follows that

$$\det(P_1 + t_0P_2) \neq 0.$$

Hence $P_1 + t_0P_2$ is an invertible n-square matrix. Since P_1, P_2 are real matrices and t_0 is a real number, $P_1 + t_0P_2$ is an n-square real matrix. Thus Q is an invertible n-square real matrix, where $Q \equiv P_1 + t_0P_2$. It remains to show that $Q^{-1}AQ = B$, that is, $AQ = QB$, that is, $A(P_1 + t_0P_2) = (P_1 + t_0P_2)B$, that is, $AP_1 + t_0(AP_2) = P_1B + t_0(P_2B)$. This is clearly true from (**) ∎

4.2.4 Note Let λ and μ be distinct complex numbers. Observe that

$$\begin{bmatrix} 1 & 0 & 0 \\ 0 & 0 & 1 \\ 0 & 1 & 0 \end{bmatrix} (\mathrm{diag}(\lambda, \mu, \lambda)) \begin{bmatrix} 1 & 0 & 0 \\ 0 & 0 & 1 \\ 0 & 1 & 0 \end{bmatrix}$$

$$= \left(\begin{bmatrix} 1 & 0 & 0 \\ 0 & 0 & 1 \\ 0 & 1 & 0 \end{bmatrix} (\mathrm{diag}(\lambda, \mu, \lambda)) \right) \begin{bmatrix} 1 & 0 & 0 \\ 0 & 0 & 1 \\ 0 & 1 & 0 \end{bmatrix}$$

$$= \left(\begin{bmatrix} 1 & 0 & 0 \\ 0 & 0 & 1 \\ 0 & 1 & 0 \end{bmatrix} \begin{bmatrix} \lambda & 0 & 0 \\ 0 & \mu & 0 \\ 0 & 0 & \lambda \end{bmatrix} \right) \begin{bmatrix} 1 & 0 & 0 \\ 0 & 0 & 1 \\ 0 & 1 & 0 \end{bmatrix}$$

$$= \begin{bmatrix} \lambda & 0 & 0 \\ 0 & 0 & \lambda \\ 0 & \mu & 0 \end{bmatrix} \begin{bmatrix} 1 & 0 & 0 \\ 0 & 0 & 1 \\ 0 & 1 & 0 \end{bmatrix} = \begin{bmatrix} \lambda & 0 & 0 \\ 0 & \lambda & 0 \\ 0 & 0 & \mu \end{bmatrix} = \mathrm{diag}\left(\underbrace{\lambda, \lambda}_{2}, \underbrace{\mu}_{1} \right),$$

and hence

$$\begin{bmatrix} 1 & 0 & 0 \\ 0 & 0 & 1 \\ 0 & 1 & 0 \end{bmatrix} (\operatorname{diag}(\lambda, \mu, \lambda)) \begin{bmatrix} 1 & 0 & 0 \\ 0 & 0 & 1 \\ 0 & 1 & 0 \end{bmatrix} = \operatorname{diag}\left(\underbrace{\lambda, \lambda}_{2}, \underbrace{\mu}_{1}\right).$$

Notation $\operatorname{diag}\left(\underbrace{\lambda, \lambda}_{2}, \underbrace{\mu}_{1}\right)$ is denoted by $\lambda I_2 \oplus \mu I_1$.

Thus

$$\begin{bmatrix} 1 & 0 & 0 \\ 0 & 0 & 1 \\ 0 & 1 & 0 \end{bmatrix} (\operatorname{diag}(\lambda, \mu, \lambda)) \begin{bmatrix} 1 & 0 & 0 \\ 0 & 0 & 1 \\ 0 & 1 & 0 \end{bmatrix} = \lambda I_2 \oplus \mu I_1. \tag{$*$}$$

Observe that

$$\begin{bmatrix} 1 & 0 & 0 \\ 0 & 0 & 1 \\ 0 & 1 & 0 \end{bmatrix} \begin{bmatrix} 1 & 0 & 0 \\ 0 & 0 & 1 \\ 0 & 1 & 0 \end{bmatrix} = \begin{bmatrix} 1 & 0 & 0 \\ 0 & 1 & 0 \\ 0 & 0 & 1 \end{bmatrix} = I_3,$$

so

$$\begin{bmatrix} 1 & 0 & 0 \\ 0 & 0 & 1 \\ 0 & 1 & 0 \end{bmatrix}^{-1} = \begin{bmatrix} 1 & 0 & 0 \\ 0 & 0 & 1 \\ 0 & 1 & 0 \end{bmatrix}.$$

Thus $\begin{bmatrix} 1 & 0 & 0 \\ 0 & 0 & 1 \\ 0 & 1 & 0 \end{bmatrix}$ is invertible. It is clear that $\begin{bmatrix} 1 & 0 & 0 \\ 0 & 0 & 1 \\ 0 & 1 & 0 \end{bmatrix}$ is symmetric and

Hermitian. Thus $\begin{bmatrix} 1 & 0 & 0 \\ 0 & 0 & 1 \\ 0 & 1 & 0 \end{bmatrix}$ is unitary.

Also, from (*)

$$\operatorname{diag}(\lambda, \mu, \lambda) = \begin{bmatrix} 1 & 0 & 0 \\ 0 & 0 & 1 \\ 0 & 1 & 0 \end{bmatrix} (\lambda I_2 \oplus \mu I_1) \begin{bmatrix} 1 & 0 & 0 \\ 0 & 0 & 1 \\ 0 & 1 & 0 \end{bmatrix}.$$

Hence

$$\operatorname{diag}(\lambda, \mu, \lambda) = \begin{bmatrix} 1 & 0 & 0 \\ 0 & 0 & 1 \\ 0 & 1 & 0 \end{bmatrix}^{*} (\lambda I_2 \oplus \mu I_1) \begin{bmatrix} 1 & 0 & 0 \\ 0 & 0 & 1 \\ 0 & 1 & 0 \end{bmatrix}.$$

4.2.5 Conclusion Let D be a diagonal matrix. Then D can be expressed as $Q^*(\lambda I_r \oplus \cdots \oplus \mu I_s)Q$, where $\lambda, \ldots, \mu$ are the distinct members of the diagonal entries of D, and Q is a unitary matrix.

4.2.6 Note Let $A \equiv [a_{ij}]$ be an n-square complex matrix. Let A be unitary (that is, $A^*A = AA^* = I$, that is, $A^* = A^{-1}$). Let λ be an eigenvalue of the matrix A. Then $|\lambda| = 1$.

Proof Let $T : x \mapsto Ax$ be the linear transformation from the inner product space $\mathbb{C}^n$ to $\mathbb{C}^n$. Since λ is an eigenvalue of the matrix A, we have $\det(\lambda I - A) = 0$, and hence there exists a nonzero $x \in \mathbb{C}^n$ such that

$$\lambda x - T(x) = \lambda x - Ax = \lambda Ix - Ax = \underbrace{(\lambda I - A)x = 0}.$$

Thus $\lambda x - T(x) = 0$, and hence $T(x) = \lambda x$, where $x \neq 0$. This shows that λ is an eigenvalue of the linear transformation T.

By 3.1.22, it suffices to show that T is a unitary transformation. To this end, let us take an arbitrary $x \in \mathbb{C}^n$. By 3.1.2, it suffices to show that $\langle T(x), T(x) \rangle = \langle x, x \rangle$:

$$\begin{aligned} \text{LHS} &= \langle T(x), T(x) \rangle = \langle Ax, Ax \rangle = (Ax)^*(Ax) \\ &= (x^*A^*)(Ax) = x^*(A^*A)x = x^*Ix = x^*x = \langle x, x \rangle = \text{RHS}. \end{aligned}$$

4.2.7 Note Let $A \equiv [a_{ij}]$ be an n-square complex matrix. Let A be symmetric (that is, $A^T = A$). Let A be unitary (that is, $A^*A = AA^* = I$, that is, $A^* = A^{-1}$).

Since A is unitary, A is a normal matrix, and hence by 3.3.24, there exists a unitary matrix U such that

1. U^*AU is a diagonal matrix,
2. the eigenvalues of A are the diagonal entries of U^*AU.

Hence $U^*AU = \text{diag}(\lambda_1, \ldots, \lambda_n)$, where $\lambda_1, \ldots, \lambda_n$ are the eigenvalues of the matrix A.

Now, since A is a unitary matrix, by 4.2.6, $|\lambda_1| = 1$, $|\lambda_2| = 1$, etc. Since $U^*AU = \text{diag}(\lambda_1, \ldots, \lambda_n)$, and U is a unitary matrix, we have $A = U(\text{diag}(\lambda_1, \ldots, \lambda_n))U^*$.

Suppose that $\mu_1, \ldots, \mu_k$ are the distinct members of $\lambda_1, \ldots, \lambda_n$.

By 4.2.5, $\text{diag}(\lambda_1, \ldots, \lambda_n)$ can be expressed as $Q^*(\mu_1 I_{r_1} \oplus \cdots \oplus \mu_k I_{r_k})Q$, where Q is a unitary matrix. Since

$$A = U(\text{diag}(\lambda_1, \ldots, \lambda_n))U^*,$$

we have

$$\underbrace{A = U(Q^*(\mu_1 I_{r_1} \oplus \cdots \oplus \mu_k I_{r_k})Q)U^*}= (UQ^*)(\mu_1 I_{r_1} \oplus \cdots \oplus \mu_k I_{r_k})(UQ^*)^*$$

$$= V(\mu_1 I_{r_1} \oplus \cdots \oplus \mu_k I_{r_k})V^*,$$

where $V \equiv UQ^*$. Thus

$$A = V(\mu_1 I_{r_1} \oplus \cdots \oplus \mu_k I_{r_k})V^*.$$

Since $\mu_1, \ldots, \mu_k$ are the distinct members of $\lambda_1, \ldots, \lambda_n$, and each $|\lambda_i| = 1$, we have $|\mu_1| = 1$, $|\mu_2| = 1$, etc. Hence we can suppose that $\mu_1 \equiv e^{i\theta_1}$, $\mu_2 \equiv e^{i\theta_2}$, etc., where $\theta_1, \theta_2, \ldots$ are real numbers. Thus

$$A = V\left(e^{i\theta_1} I_{r_1} \oplus \cdots \oplus e^{i\theta_k} I_{r_k}\right)V^*.$$

Since Q is unitary, Q^* is unitary. Since U is unitary, $(V =)UQ^*$ is unitary, and hence V is unitary. Next, since

$$A = V\left(e^{i\theta_1} I_{r_1} \oplus \cdots \oplus e^{i\theta_k} I_{r_k}\right)V^*,$$

we have

$$\left(e^{i\theta_1} I_{r_1} \oplus \cdots \oplus e^{i\theta_k} I_{r_k}\right) = V^*AV.$$

Put

$$S \equiv V\left(e^{i\frac{\theta_1}{2}} I_{r_1} \oplus \cdots \oplus e^{i\frac{\theta_k}{2}} I_{r_k}\right)V^*.$$

It follows that

$$\left(e^{i\frac{\theta_1}{2}} I_{r_1} \oplus \cdots \oplus e^{i\frac{\theta_k}{2}} I_{r_k}\right) = V^*SV.$$

Here

$$S^2 = \left(V\left(e^{i\frac{\theta_1}{2}} I_{r_1} \oplus \cdots \oplus e^{i\frac{\theta_k}{2}} I_{r_k}\right)V^*\right)\left(V\left(e^{i\frac{\theta_1}{2}} I_{r_1} \oplus \cdots \oplus e^{i\frac{\theta_k}{2}} I_{r_k}\right)V^*\right)$$

$$= V\left(e^{i\frac{\theta_1}{2}} I_{r_1} \oplus \cdots \oplus e^{i\frac{\theta_k}{2}} I_{r_k}\right)(V^*V)\left(e^{i\frac{\theta_1}{2}} I_{r_1} \oplus \cdots \oplus e^{i\frac{\theta_k}{2}} I_{r_k}\right)V^*$$

$$= V\left(\left(e^{i\frac{\theta_1}{2}} I_{r_1} \oplus \cdots \oplus e^{i\frac{\theta_k}{2}} I_{r_k}\right)\left(e^{i\frac{\theta_1}{2}} I_{r_1} \oplus \cdots \oplus e^{i\frac{\theta_k}{2}} I_{r_k}\right)\right)V^*$$

$$= V\left(\left(e^{i\frac{\theta_1}{2}}\right)^2 I_{r_1} \oplus \cdots \oplus \left(e^{i\frac{\theta_k}{2}}\right)^2 I_{r_k}\right)V^* = V\left(e^{i\theta_1} I_{r_1} \oplus \cdots \oplus e^{i\theta_k} I_{r_k}\right)V^* = A,$$

so $\boxed{S^2 = A.}$

Clearly, S is unitary, that is, $S^*S = SS^* = I$.

Proof Here,

$$
\begin{aligned}
S^* &= \left(V\left(e^{i\frac{\theta_1}{2}}I_{r_1} \oplus \cdots \oplus e^{i\frac{\theta_k}{2}}I_{r_k}\right)V^*\right)^* = (V^*)^*\left(e^{i\frac{\theta_1}{2}}I_{r_1} \oplus \cdots \oplus e^{i\frac{\theta_k}{2}}I_{r_k}\right)^* V^* \\
&= V\left(e^{i\frac{\theta_1}{2}}I_{r_1} \oplus \cdots \oplus e^{i\frac{\theta_k}{2}}I_{r_k}\right)^* V^* = V\left(\left(e^{i\frac{\theta_1}{2}}I_{r_1} \oplus \cdots \oplus e^{i\frac{\theta_k}{2}}I_{r_k}\right)^-\right)^T V^* \\
&= V\left(\overline{\left(e^{i\frac{\theta_1}{2}}\right)}I_{r_1} \oplus \cdots \oplus \overline{\left(e^{i\frac{\theta_k}{2}}\right)}I_{r_k}\right)^T V^* \\
&= V\left(e^{-i\frac{\theta_1}{2}}I_{r_1} \oplus \cdots \oplus e^{-i\frac{\theta_k}{2}}I_{r_k}\right)^T V^* = V\left(e^{-i\frac{\theta_1}{2}}I_{r_1} \oplus \cdots \oplus e^{-i\frac{\theta_k}{2}}I_{r_k}\right)V^*,
\end{aligned}
$$

so

$$
S^* = V\left(e^{-i\frac{\theta_1}{2}}I_{r_1} \oplus \cdots \oplus e^{-i\frac{\theta_k}{2}}I_{r_k}\right)V^*.
$$

Now,

$$
\begin{aligned}
S^*S &= \left(V\left(e^{-i\frac{\theta_1}{2}}I_{r_1} \oplus \cdots \oplus e^{-i\frac{\theta_k}{2}}I_{r_k}\right)V^*\right)\left(V\left(e^{i\frac{\theta_1}{2}}I_{r_1} \oplus \cdots \oplus e^{i\frac{\theta_k}{2}}I_{r_k}\right)V^*\right) \\
&= V\left(\left(e^{-i\frac{\theta_1}{2}}I_{r_1} \oplus \cdots \oplus e^{-i\frac{\theta_k}{2}}I_{r_k}\right)\left(e^{i\frac{\theta_1}{2}}I_{r_1} \oplus \cdots \oplus e^{i\frac{\theta_k}{2}}I_{r_k}\right)\right)V^* \\
&= V\left(e^{-i\frac{\theta_1}{2}}e^{i\frac{\theta_1}{2}}I_{r_1} \oplus \cdots \oplus e^{-i\frac{\theta_k}{2}}e^{i\frac{\theta_k}{2}}I_{r_k}\right)V^* = V(1I_{r_1} \oplus \cdots \oplus 1I_{r_k})V^* \\
&= VIV^* = VV^* = I,
\end{aligned}
$$

so $S^*S = I$. Similarly, $SS^* = I$.

Thus we have shown that $\boxed{S \text{ is unitary.}}$. ∎

Suppose that B is any n-square complex matrix. Suppose that B commutes with A, that is, $AB = BA$.

Now clearly, V^*BV commutes with $\left(e^{i\theta_1}I_{r_1} \oplus \cdots \oplus e^{i\theta_k}I_{r_k}\right)$, that is,

$$
\left(e^{i\theta_1}I_{r_1} \oplus \cdots \oplus e^{i\theta_k}I_{r_k}\right)(V^*BV) = (V^*BV)\left(e^{i\theta_1}I_{r_1} \oplus \cdots \oplus e^{i\theta_k}I_{r_k}\right).
$$

Proof Here,

$$
\begin{aligned}
\text{LHS} &= \left(e^{i\theta_1}I_{r_1} \oplus \cdots \oplus e^{i\theta_k}I_{r_k}\right)(V^*BV) = (V^*AV)(V^*BV) = V^*A(BV) \\
&= V^*(AB)V,
\end{aligned}
$$

and

$$
\begin{aligned}
\text{RHS} &= (V^*BV)\left(e^{i\theta_1}I_{r_1} \oplus \cdots \oplus e^{i\theta_k}I_{r_k}\right) = (V^*BV)(V^*AV) = V^*B(AV) \\
&= V^*(BA)V = V^*(AB)V,
\end{aligned}
$$

so LHS = RHS. ∎

Again, it is clear that $S\left(= V\left(e^{i\frac{\theta_1}{2}}I_{r_1} \oplus \cdots \oplus e^{i\frac{\theta_k}{2}}I_{r_k}\right)V^*\right)$ commutes with B, that is, $SB = BS$, that is, $B = S^*BS$.

Proof Here,

$$
\begin{aligned}
\text{RHS} = S^*BS &= \left(V\left(e^{-i\frac{\theta_1}{2}}I_{r_1} \oplus \cdots \oplus e^{-i\frac{\theta_k}{2}}I_{r_k}\right)V^*\right)B\left(V\left(e^{i\frac{\theta_1}{2}}I_{r_1} \oplus \cdots \oplus e^{i\frac{\theta_k}{2}}I_{r_k}\right)V^*\right) \\
&= V\left(e^{-i\frac{\theta_1}{2}}I_{r_1} \oplus \cdots \oplus e^{-i\frac{\theta_k}{2}}I_{r_k}\right)\left((V^*BV)\left(e^{i\frac{\theta_1}{2}}I_{r_1} \oplus \cdots \oplus e^{i\frac{\theta_k}{2}}I_{r_k}\right)\right)V^* \\
&= V\left(e^{-i\frac{\theta_1}{2}}I_{r_1} \oplus \cdots \oplus e^{-i\frac{\theta_k}{2}}I_{r_k}\right)\left(\left(e^{i\frac{\theta_1}{2}}I_{r_1} \oplus \cdots \oplus e^{i\frac{\theta_k}{2}}I_{r_k}\right)(V^*BV)\right)V^* \\
&= V\left(\left(e^{-i\frac{\theta_1}{2}}I_{r_1} \oplus \cdots \oplus e^{-i\frac{\theta_k}{2}}I_{r_k}\right)\left(e^{i\frac{\theta_1}{2}}I_{r_1} \oplus \cdots \oplus e^{i\frac{\theta_k}{2}}I_{r_k}\right)\right)V^*B \\
&= V\left(e^{-i\frac{\theta_1}{2}}e^{i\frac{\theta_1}{2}}I_{r_1} \oplus \cdots \oplus e^{-i\frac{\theta_k}{2}}e^{i\frac{\theta_k}{2}}I_{r_k}\right)V^*B = V(1I_{r_1} \oplus \cdots \oplus 1I_{r_k})V^*B \\
&= VIV^*B = B = \text{LHS}.
\end{aligned}
$$

Thus we have shown that $\boxed{\text{if } B \text{ commutes with } A, \text{ then } B \text{ commutes with } S.}$ $\quad(*)$
Since A is unitary, we have $A^*A = AA^* = I$, and hence the inverse of the matrix A is A^*. Since A is unitary, we have $A^*A = I$, and hence

$$
A^T\bar{A} = A^T(A^*)^T = \underbrace{(A^*A)^T = I^T} = I.
$$

Thus $A^T\bar{A} = I$. Similarly, $\bar{A}A^T = I$. Thus the inverse of the matrix A^T is $\bar{A}$. Similarly, since V is unitary, the inverse of the matrix V^T is $\bar{V}$, and the inverse of the matrix V is V^*.

Since

$$
\begin{aligned}
\bar{V}\left(e^{i\theta_1}I_{r_1} \oplus \cdots \oplus e^{i\theta_k}I_{r_k}\right)V^T &= \bar{V}\left(e^{i\theta_1}I_{r_1} \oplus \cdots \oplus e^{i\theta_k}I_{r_k}\right)^T V^T \\
&= (V^*)^T\left(e^{i\theta_1}I_{r_1} \oplus \cdots \oplus e^{i\theta_k}I_{r_k}\right)^T V^T \\
&= \left(V\left(e^{i\theta_1}I_{r_1} \oplus \cdots \oplus e^{i\theta_k}I_{r_k}\right)V^*\right)^T = \underbrace{A^T = A} = V\left(e^{i\theta_1}I_{r_1} \oplus \cdots \oplus e^{i\theta_k}I_{r_k}\right)V^*,
\end{aligned}
$$

we have

$$
\bar{V}\left(e^{i\theta_1}I_{r_1} \oplus \cdots \oplus e^{i\theta_k}I_{r_k}\right)V^T = V\left(e^{i\theta_1}I_{r_1} \oplus \cdots \oplus e^{i\theta_k}I_{r_k}\right)V^*.
$$

Hence

$$
\bar{V}\left(e^{i\theta_1}I_{r_1} \oplus \cdots \oplus e^{i\theta_k}I_{r_k}\right)V^TV = V\left(e^{i\theta_1}I_{r_1} \oplus \cdots \oplus e^{i\theta_k}I_{r_k}\right).
$$

It follows that

$$\left(e^{i\theta_1}I_{r_1} \oplus \cdots \oplus e^{i\theta_k}I_{r_k}\right)V^TV = V^TV\left(e^{i\theta_1}I_{r_1} \oplus \cdots \oplus e^{i\theta_k}I_{r_k}\right).$$

Thus we have shown that V^TV commutes with $\left(e^{i\theta_1}I_{r_1} \oplus \cdots \oplus e^{i\theta_k}I_{r_k}\right)$.
   Clearly, VV^T commutes with A.

Proof We have to show that

$$(VV^T)\left(V\left(e^{i\theta_1}I_{r_1} \oplus \cdots \oplus e^{i\theta_k}I_{r_k}\right)V^*\right) = \underbrace{(VV^T)A = A(VV^T)}$$

$$= \left(V\left(e^{i\theta_1}I_{r_1} \oplus \cdots \oplus e^{i\theta_k}I_{r_k}\right)V^*\right)(VV^T),$$

that is,

$$\left((V^TV)\left(e^{i\theta_1}I_{r_1} \oplus \cdots \oplus e^{i\theta_k}I_{r_k}\right)V^*\right) = V\left(\left(e^{i\theta_1}I_{r_1} \oplus \cdots \oplus e^{i\theta_k}I_{r_k}\right)(V^*V)V^T\right),$$

that is,

$$\left(V^TV\right)\left(e^{i\theta_1}I_{r_1} \oplus \cdots \oplus e^{i\theta_k}I_{r_k}\right)V^* = \left(e^{i\theta_1}I_{r_1} \oplus \cdots \oplus e^{i\theta_k}I_{r_k}\right)(V^*V)V^T,$$

that is,

$$\left(V^TV\right)\left(e^{i\theta_1}I_{r_1} \oplus \cdots \oplus e^{i\theta_k}I_{r_k}\right)V^* = \left(e^{i\theta_1}I_{r_1} \oplus \cdots \oplus e^{i\theta_k}I_{r_k}\right)IV^T,$$

that is,

$$\left(V^TV\right)\left(e^{i\theta_1}I_{r_1} \oplus \cdots \oplus e^{i\theta_k}I_{r_k}\right) = \left(e^{i\theta_1}I_{r_1} \oplus \cdots \oplus e^{i\theta_k}I_{r_k}\right)\left(V^TV\right).$$

   This is known to be true.   ∎
   It follows, from $(*)$, that VV^T commutes with S, that is,

$$(VV^T)\left(V\left(e^{i\frac{\theta_1}{2}}I_{r_1} \oplus \cdots \oplus e^{i\frac{\theta_k}{2}}I_{r_k}\right)V^*\right) = \underbrace{(VV^T)S = S(VV^T)}$$

$$= \left(V\left(e^{i\frac{\theta_1}{2}}I_{r_1} \oplus \cdots \oplus e^{i\frac{\theta_k}{2}}I_{r_k}\right)V^*\right)(VV^T) = V\left(e^{i\frac{\theta_1}{2}}I_{r_1} \oplus \cdots \oplus e^{i\frac{\theta_k}{2}}I_{r_k}\right)V^T,$$

that is,

$$V\left(V^TV\left(e^{i\frac{\theta_1}{2}}I_{r_1} \oplus \cdots \oplus e^{i\frac{\theta_k}{2}}I_{r_k}\right)V^*\right) = V\left(\left(e^{i\frac{\theta_1}{2}}I_{r_1} \oplus \cdots \oplus e^{i\frac{\theta_k}{2}}I_{r_k}\right)V^T\right),$$

that is,

$$V^T V \left(e^{i\frac{\theta_1}{2}} I_{r_1} \oplus \cdots \oplus e^{i\frac{\theta_k}{2}} I_{r_k} \right) V^* = \left(e^{i\frac{\theta_1}{2}} I_{r_1} \oplus \cdots \oplus e^{i\frac{\theta_k}{2}} I_{r_k} \right) V^T ,$$

that is,

$$\left(V^T V \right) \left(e^{i\frac{\theta_1}{2}} I_{r_1} \oplus \cdots \oplus e^{i\frac{\theta_k}{2}} I_{r_k} \right) = \left(e^{i\frac{\theta_1}{2}} I_{r_1} \oplus \cdots \oplus e^{i\frac{\theta_k}{2}} I_{r_k} \right) \left(V^T V \right) .$$

Thus $V^T V$ commutes with $\left(e^{i\frac{\theta_1}{2}} I_{r_1} \oplus \cdots \oplus e^{i\frac{\theta_k}{2}} I_{r_k} \right)$.

Clearly, S is symmetric.

Proof We have to show that

$$\bar{V} \left(e^{i\frac{\theta_1}{2}} I_{r_1} \oplus \cdots \oplus e^{i\frac{\theta_k}{2}} I_{r_k} \right) V^T = (V^*)^T \left(e^{i\frac{\theta_1}{2}} I_{r_1} \oplus \cdots \oplus e^{i\frac{\theta_k}{2}} I_{r_k} \right)^T V^T$$

$$= \left(V \left(e^{i\frac{\theta_1}{2}} I_{r_1} \oplus \cdots \oplus e^{i\frac{\theta_k}{2}} I_{r_k} \right) V^* \right)^T = \underbrace{S^T = S} = V \left(e^{i\frac{\theta_1}{2}} I_{r_1} \oplus \cdots \oplus e^{i\frac{\theta_k}{2}} I_{r_k} \right) V^* ,$$

that is,

$$\bar{V} \left(e^{i\frac{\theta_1}{2}} I_{r_1} \oplus \cdots \oplus e^{i\frac{\theta_k}{2}} I_{r_k} \right) V^T = V \left(e^{i\frac{\theta_1}{2}} I_{r_1} \oplus \cdots \oplus e^{i\frac{\theta_k}{2}} I_{r_k} \right) V^* ,$$

that is,

$$\bar{V} \left(e^{i\frac{\theta_1}{2}} I_{r_1} \oplus \cdots \oplus e^{i\frac{\theta_k}{2}} I_{r_k} \right) V^T V = V \left(e^{i\frac{\theta_1}{2}} I_{r_1} \oplus \cdots \oplus e^{i\frac{\theta_k}{2}} I_{r_k} \right) ,$$

that is,

$$\left(e^{i\frac{\theta_1}{2}} I_{r_1} \oplus \cdots \oplus e^{i\frac{\theta_k}{2}} I_{r_k} \right) \left(V^T V \right) = \left(V^T V \right) \left(e^{i\frac{\theta_1}{2}} I_{r_1} \oplus \cdots \oplus e^{i\frac{\theta_k}{2}} I_{r_k} \right) .$$

This is known to be true.
Thus $\boxed{S \text{ is symmetric}}$. ∎

4.2.8 Conclusion Let $A \equiv \left[a_{ij} \right]$ be an n-square complex matrix. Let A be symmetric. Let A be unitary. Then there exists a complex matrix S such that

1. $S^2 = A$,
2. S is unitary,
3. if B commutes with A, then B commutes with S,
4. S is symmetric.

4.2.9 Note Let $A \equiv \left[a_{ij} \right]$ be an n-square real matrix. Let $B \equiv \left[a_{ij} \right]$ be an n-square real matrix. Let U be a unitary complex matrix such that $A = UBU^*$ (that is, $U^* A U = B$).

Since U is unitary, we have $U^*U = UU^* = I$, and hence the inverse of the matrix U is U^*. Since U is unitary, we have $U^*U = I$, and hence

$$U^T \bar{U} = U^T (U^*)^T = \underbrace{(U^*U)^T = I^T} = I.$$

Thus $U^T \bar{U} = I$. Similarly, $\bar{U} U^T = I$. Thus the inverse of the matrix U^T is $\bar{U}$. Clearly, $U^T U$ is symmetric.

Proof Since

$$\left(U^T U\right)^T = U^T \left(U^T\right)^T = U^T U,$$

we have $\left(U^T U\right)^T = U^T U$, so $U^T U$ is symmetric. ∎

Clearly, $U^T U$ is unitary.

Proof Since U is unitary, we have $U^*U = I$, and hence

$$\left(U^T\right)\left(U^T\right)^* = U^T \bar{U} = U^T (U^*)^T = \underbrace{(U^*U)^T = I^T} = I.$$

Thus $(U^T)(U^T)^* = I$. Similarly, $(U^T)^*(U^T) = I$. It follows that U^T is unitary, and hence $\left(U^T\right)^{-1} = \left(U^T\right)^*$. Since U is unitary, we have $U^{-1} = U^*$. We have to show that $\left(U^T U\right)^{-1} = \left(U^T U\right)^*$:

$$\text{RHS} = \left(U^T U\right)^* = U^* \left(U^T\right)^* = U^* \left(U^T\right)^{-1} = U^{-1} \left(U^T\right)^{-1} = \left(U^T U\right)^{-1} = \text{LHS}.$$

∎

Since $U^T U$ is unitary and symmetric, by 4.2.8, there exists a complex matrix S such that

1. $S^2 = U^T U$,
2. S is unitary,
3. if C commutes with $U^T U$, then C commutes with S,
4. S is symmetric.

Clearly, $U^T U$ commutes with B, that is, $(U^T U)B = B(U^T U)$.

Proof Since A is a real matrix, we have

$$\bar{U} B U^T = \bar{U} \bar{B} U^T = \bar{U} \bar{B} \bar{U}^* = \overline{(UBU^*)} = \bar{A} = A = UBU^*,$$

and hence $\bar{U} B U^T = UBU^*$. This shows that $BU^T = U^T(UBU^*)$, and hence $(BU^T)U = U^T(UB)$. Thus

$$B(U^T U) = (U^T U)B.$$

∎

Thus we have shown that $U^T U$ commutes with B. Now, by (3), B commutes with S. (∗)

Let us put $Q \equiv US^{-1}$. Clearly, Q is unitary.

Proof We have to show that $(US^{-1})^{-1} = (US^{-1})^*$. Since S is unitary, we have $S^{-1} = S^*$. Now we have to show that

$$\underbrace{(US^*)^{-1} = (US^{-1})^*} = (US^*)^* = (S^*)^* U^* = SU^* = SU^{-1},$$

that is, $(US^*)^{-1} = SU^{-1}$:

$$\text{LHS} = (US^*)^{-1} = (US^{-1})^{-1} = SU^{-1} = \text{RHS}.$$

∎

Clearly, Q is orthogonal.

Proof We have to show that $Q^T Q = I$:

$$\text{LHS} = (US^{-1})^T (US^{-1}) = (US^*)^T (US^{-1}) = ((S^*)^T U^T)(US^{-1}) = (\bar{S}(U^T))(US^{-1})$$
$$= (\overline{(S^T)}(U^T))(US^{-1}) = (S^*(U^T))(US^{-1}) = S^*(U^T U)S^{-1}$$
$$= S^{-1}(U^T U)S^{-1} = S^{-1}(S^2)S^{-1} = I = \text{RHS}.$$

∎

Since Q is orthogonal, we have $Q^T Q = I$, and hence $(\bar{Q})^T = Q^* = \underbrace{Q^{-1} = Q^T}$.

Thus $(\bar{Q})^T = Q^T$, and therefore $\bar{Q} = Q$. Hence Q is a real matrix.

Clearly, $A = QBQ^T$.

Proof Here

$$QBQ^T = QB\bar{Q}^T = QBQ^* = QB(US^{-1})^* = QB(US^*)^* = QB(SU^*)$$
$$= (QBS)U^* = ((US^{-1})BS)U^* = U(S^{-1}BS)U^*,$$

so

$$QBQ^T = U(S^{-1}BS)U^*.$$

Now, since $A = UBU^*$, it suffices to show that $B = (S^{-1}BS)$, that is, $SB = BS$. From $(*)$, this is true. ∎

4.2.10 Conclusion Let A and B be n-square real matrices. Let U be a unitary complex matrix such that $A = UBU^*$. Then there exists a real orthogonal matrix Q such that $A = QBQ^T$.

4.2.11 Note Let $A \equiv [a_{ij}]$ be an n-square real matrix (that is, $\bar{A} = A$). Let A be symmetric (that is, $A^T = A$).

It follows that $A^* = (\bar{A})^T = A^T = A$, and hence $A^* = A$. This shows that A is Hermitian, and hence A is a normal matrix. Now, by 3.3.24, there exists a unitary matrix U such that

1. U^*AU is a diagonal matrix,
2. the eigenvalues of A are the diagonal entries of U^*AU.

Hence $U^*AU = \operatorname{diag}(\lambda_1, \ldots, \lambda_n)$, where $\lambda_1, \ldots, \lambda_n$ are the eigenvalues of the matrix A. Since A is Hermitian, by 3.3.26, $\lambda_1, \ldots, \lambda_n$ are real numbers. Since $U^*AU = \operatorname{diag}(\lambda_1, \ldots, \lambda_n)$, and U is a unitary matrix, we have $A = U(\operatorname{diag}(\lambda_1, \ldots, \lambda_n))U^*$. Also A and $\operatorname{diag}(\lambda_1, \ldots, \lambda_n)$ are real matrices of the same size. Now, by 4.2.10, there exists a real orthogonal matrix Q such that

$$A = Q(\operatorname{diag}(\lambda_1, \ldots, \lambda_n))Q^T. \quad (*).$$

Since Q is orthogonal, we have $Q^{-1} = Q^T$. Now from $(*)$,

$$A = Q(\operatorname{diag}(\lambda_1, \ldots, \lambda_n))Q^{-1}.$$

It follows that

$$\underbrace{\operatorname{diag}(\lambda_1, \ldots, \lambda_n)}_{} = Q^{-1}AQ = Q^TAQ = (Q^T)A(Q^T)^T = PAP^T,$$

where $P \equiv Q^T$. Since Q is a real matrix, $(P =)Q^T$ is a real matrix, and hence P is a real matrix. Since Q is orthogonal, we have $QQ^T = I$, and hence $P^{-1} = \underbrace{(Q^T)^{-1} = Q}_{} = P^T$. Thus $P^{-1} = P^T$, and hence P is orthogonal. Also,

$$PAP^T = \operatorname{diag}(\lambda_1, \ldots, \lambda_n).$$

4.2.12 Conclusion Let A be an n-square real symmetric matrix. Then there exists a real orthogonal matrix P such that $PAP^T = \operatorname{diag}(\lambda_1, \ldots, \lambda_n)$, where $\lambda_1, \ldots, \lambda_n$ are the eigenvalues of A.

In short, a real symmetric matrix can be brought to diagonal form by a real orthogonal matrix.

Definition Let A be an n-square real symmetic matrix. Let B be an n-square real symmetic matrix. If there exists a real invertible matrix S such that $A = SBS^T$, then we say that A *and* B *are congruent.*

4.2.13 Note Suppose that S is an invertible matrix. It follows that S^{-1} exists, and $SS^{-1} = I$. Hence

$$\left(S^{-1}\right)^T S^T = \underbrace{\left(SS^{-1}\right)^T = I^T} = I.$$

Thus $\left(S^{-1}\right)^T S^T = I$. It follows that $\left(S^T\right)^{-1}$ exists, and $\left(S^T\right)^{-1} = \left(S^{-1}\right)^T$.

4.2.14 Conclusion If S is a real invertible matrix, then S^T is a real invertible matrix, and $\left(S^T\right)^{-1} = \left(S^{-1}\right)^T$.

4.2.15 Problem Congruence is an equivalence relation.

Proof

(i)   Let us take an arbitrary real symmetric n-square matrix A. Since $A = IAI^T$, and I is a real invertible matrix, A and A are congruent.

(ii) Let us take arbitrary real symmetric n-square matrices A and B that are congruent. We have to show that B and A are congruent.

Since A and B are congruent, there exists a real invertible matrix S such that $A = SBS^T$. It follows that $\left(S^T\right)^{-1} = \left(S^{-1}\right)^T$, and $B = \left(S^{-1}\right)A\left(S^{-1}\right)^T$, and hence $B = RAR^T$, where $R \equiv S^{-1}$. Since S is a real invertible matrix, $(R =)S^{-1}$ is a real invertible matrix, and hence R is a real invertible matrix. Thus B and A are congruent.

(iii)   Let us take any real symmetric n-square matrices A, B, C. Suppose that A and B are congruent. Suppose that B and C are congruent. We have to show that A and C are congruent.

Since A and B are congruent, there exists a real invertible matrix S such that $A = SBS^T$. Since B and C are congruent, there exists a real invertible matrix R such that $B = RCR^T$.

It follows that $A = S(RCR^T)S^T \left(= (SR)C(SR)^T\right)$, and hence $A = (SR)C(SR)^T$. Since S,R are real invertible matrices, SR is also a real invertible matrix. Thus A and C are congruent.

Hence, congruence is an equivalence relation.   ∎

4.2.16 Note Let A be an n-square real symmetric matrix.

Since $A^* = (\bar{A})^T = A^T = A$, we have $A^* = A$. This shows that A is Hermitian, and hence by 3.3.26, its eigenvalues are real numbers. Let $\underbrace{\mu_1, \ldots, \mu_r}_{r}$ be the positive

distinct eigenvalues of A, and let $\underbrace{-\mu_{r+1}, \ldots, -\mu_{r+s}}_{s}$ be the negative distinct

eigenvalues of A.

By 4.2.12, there exists a real orthogonal matrix P such that

$$PAP^T = \mu_1 I_{k_1} \oplus \cdots \oplus \mu_r I_{k_r} \oplus (-\mu_{r+1})I_{k_{r+1}} \oplus \cdots \oplus (-\mu_{r+s})I_{k_{r+s}}$$
$$\oplus 0I_{n-(k_1+\cdots+k_{r+s})}.$$

Put

$$D \equiv \frac{1}{\sqrt{\mu_1}}I_{k_1} \oplus \cdots \oplus \frac{1}{\sqrt{\mu_r}}I_{k_r} \oplus \frac{1}{\sqrt{\mu_{r+1}}}I_{k_{r+1}} \oplus \cdots \oplus \frac{1}{\sqrt{\mu_{r+s}}}I_{k_{r+s}}$$
$$\oplus 1I_{n-(k_1+\cdots+k_{r+s})}.$$

It follows that

$$D^T = \left(\frac{1}{\sqrt{\mu_1}}I_{k_1} \oplus \cdots \oplus \frac{1}{\sqrt{\mu_r}}I_{k_r} \oplus \frac{1}{\sqrt{\mu_{r+1}}}I_{k_{r+1}} \oplus \cdots \oplus \frac{1}{\sqrt{\mu_{r+s}}}I_{k_{r+s}} \oplus 1I_{n-(k_1+\cdots+k_{r+s})}\right)^T$$
$$= \frac{1}{\sqrt{\mu_1}}I_{k_1} \oplus \cdots \oplus \frac{1}{\sqrt{\mu_r}}I_{k_r} \oplus \frac{1}{\sqrt{\mu_{r+1}}}I_{k_{r+1}} \oplus \cdots \oplus \frac{1}{\sqrt{\mu_{r+s}}}I_{k_{r+s}} \oplus 1I_{n-(k_1+\cdots+k_{r+s})},$$

and hence

$$(DP)A(DP)^T = D(PAP^T)D^T$$

$$= \left(\underbrace{\frac{1}{\sqrt{\mu_1}}\mu_1\frac{1}{\sqrt{\mu_1}}I_{k_1} \oplus \cdots}_{r} \oplus \underbrace{\frac{1}{\sqrt{\mu_{r+1}}}(-\mu_{r+1})\frac{1}{\sqrt{\mu_{r+1}}}I_{k_{r+1}} \oplus \cdots}_{s}\right.$$

$$\left. \oplus 1 \cdot 0 \cdot 1I_{n-(k_1+\cdots+k_{r+s})}\right)$$

$$= \left(1I_{k_1} \oplus \cdots \oplus 1I_{k_r} \oplus (-1)I_{k_{r+1}} \oplus \cdots \oplus (-1)I_{k_{r+s}} \oplus 0I_{n-(k_1+\cdots+k_{r+s})}\right)$$

$$= 1I_l \oplus (-1)I_m \oplus 0I_{n-(l+m)}$$

where $l \equiv k_1 + \cdots + k_r$, $m \equiv k_{r+1} + \cdots + k_{r+s}$.

Thus

$$RAR^T = \left(1I_l \oplus (-1)I_m \oplus 0I_{n-(l+m)}\right),$$

where $R \equiv DP$. Since

$$D = \frac{1}{\sqrt{\mu_1}}I_{k_1} \oplus \cdots \oplus \frac{1}{\sqrt{\mu_r}}I_{k_r} \oplus \frac{1}{\sqrt{\mu_{r+1}}}I_{k_{r+1}} \oplus \cdots \oplus \frac{1}{\sqrt{\mu_{r+s}}}I_{k_{r+s}}$$
$$\oplus 1I_{n-(k_1 + \cdots + k_{r+s})},$$

D is a real invertible matrix. Since P is a real orthogonal matrix, P is a real invertible matrix, and $P^{-1} = P^T$. Since D, P are real invertible matrices, $(R =)DP$ is a real invertible matrix, and hence R is a real invertible matrix. Since $(RAR^T =)$ $\left(1I_l \oplus (-1)I_m \oplus 0I_{n-(l+m)}\right)$ is a real symmetric matrix, RAR^T is a real symmetric matrix. Now, since R is a real invertible matrix, A and $RAR^T (= (1I_l \oplus (-1) I_m \oplus 0I_{n-(l+m)}))$ are congruent, and hence A and $\left(1I_l \oplus (-1)I_m \oplus 0I_{n-(l+m)}\right)$ are congruent. Next, by 4.2.15, $1I_l \oplus (-1)I_m \oplus 0I_{n-(l+m)}$ is a member of the congruence class of A. It is clear that $l+m$ is the rank of A.

Now we want to show that l and m are unique.

To this end, suppose that

$$1I_r \oplus (-1)I_s \oplus 0I_{n-(r+s)} \text{ and } 1I_{r'} \oplus (-1)I_{s'} \oplus 0I_{n-(r'+s')}$$

are congruent. We have to show that $r = r'$.

Suppose to the contrary that $r < r'$. We seek a contradiction.
Since

$$\left(1I_{r'} \oplus (-1)I_{s'} \oplus 0I_{n-(r'+s')}\right) \text{ and } \left(1I_r \oplus (-1)I_s \oplus 0I_{n-(r+s)}\right)$$

are congruent, there exists a real invertible matrix S such that

$$\left(1I_{r'} \oplus (-1)I_{s'} \oplus 0I_{n-(r'+s')}\right) = S\left(1I_r \oplus (-1)I_s \oplus 0I_{n-(r+s)}\right)S^T.$$

Since S is invertible, by 4.2.14, S^T is invertible, and hence

$$r' + s' = \text{rank}\left(1I_{r'} \oplus (-1)I_{s'} \oplus 0I_{n-(r'+s')}\right)$$
$$= \underbrace{\text{rank}\left(S\left(1I_r \oplus (-1)I_s \oplus 0I_{n-(r+s)}\right)S^T\right) = \text{rank}\left(1I_r \oplus (-1)I_s \oplus 0I_{n-(r+s)}\right)}$$

$$= r + s.$$

Thus $r' + s' = r + s$. Now, since $r < r'$, we have $s' < s$. (∗)
Observe that the set

$$U \equiv \left\{ \left[\underbrace{0, \ldots, 0}_{r}, \underbrace{y_1, \ldots, y_s}_{s}, \underbrace{0, \ldots, 0}_{n-(r+s)} \right]^T : y_1, \ldots, y_s \in \mathbb{R} \right\}$$

is an s-dimensional subspace of the real inner product space $\mathbb{R}^n$, where $\mathbb{R}^n$ denotes the collection of all $n \times 1$ column matrices with real entries. Also

$$
W \equiv \left\{ \left[\underbrace{x_1, \ldots, x_{r'}}_{r'}, \underbrace{0, \ldots, 0}_{s'}, \underbrace{z_1, \ldots, z_{n-(r'+s')}}_{n-(r'+s')} \right]^T : x_1, \ldots, x_{r'}, z_1, \ldots, z_{n-(r'+s')} \in \mathbb{R} \right\}
$$

is an $(n - s')$-dimensional subspace of the real inner product space $\mathbb{R}^n$.

Observe that, for every nonzero $\left[\underbrace{0, \ldots, 0}_{r}, \underbrace{y_1, \ldots, y_s}_{s}, \underbrace{0, \ldots, 0}_{n-(r+s)} \right]^T \in U$,

$$
\left\langle \left(1I_r \oplus (-1)I_s \oplus 0I_{n-(r+s)} \right) \left[\underbrace{0, \ldots, 0}_{r}, \underbrace{y_1, \ldots, y_s}_{s}, \underbrace{0, \ldots, 0}_{n-(r+s)} \right]^T , \right.
$$

$$
\left. \left[\underbrace{0, \ldots, 0}_{r}, \underbrace{y_1, \ldots, y_s}_{s}, \underbrace{0, \ldots, 0}_{n-(r+s)} \right]^T \right\rangle
$$

$$
= \left(\left[\underbrace{0, \ldots, 0}_{r}, \underbrace{y_1, \ldots, y_s}_{s}, \underbrace{0, \ldots, 0}_{n-(r+s)} \right]^T \right)^T
$$

$$
\left(\left(1I_r \oplus (-1)I_s \oplus 0I_{n-(r+s)} \right) \left[\underbrace{0, \ldots, 0}_{r}, \underbrace{y_1, \ldots, y_s}_{s}, \underbrace{0, \ldots, 0}_{n-(r+s)} \right]^T \right)
$$

$$
= \left[\underbrace{0, \ldots, 0}_{r}, \underbrace{y_1, \ldots, y_s}_{s}, \underbrace{0, \ldots, 0}_{n-(r+s)} \right]
$$

$$
\left(\left(1I_r \oplus (-1)I_s \oplus 0I_{n-(r+s)} \right) \left[\underbrace{0, \ldots, 0}_{r}, \underbrace{y_1, \ldots, y_s}_{s}, \underbrace{0, \ldots, 0}_{n-(r+s)} \right]^T \right)
$$

$$
= \left[\underbrace{0, \ldots, 0}_{r}, \underbrace{y_1, \ldots, y_s}_{s}, \underbrace{0, \ldots, 0}_{n-(r+s)} \right] \left[\underbrace{0, \ldots, 0}_{r}, \underbrace{(-1)y_1, \ldots, (-1)y_s}_{s}, \underbrace{0, \ldots, 0}_{n-(r+s)} \right]^T
$$

$$= \underbrace{0 + \cdots + 0}_{r} + \underbrace{\left(-(y_1)^2\right) + \cdots + \left(-(y_s)^2\right)}_{s} + \underbrace{0 + \cdots + 0}_{n-(r+s)} =$$

$$-\left((y_1)^2 + \cdots + (y_s)^2\right) < 0,$$

so for every nonzero $u \in U$, $\left\langle \left(1I_r \oplus (-1)I_s \oplus 0I_{n-(r+s)}\right)u, u\right\rangle$ is negative.

Observe that, for every $\left[\underbrace{x_1, \ldots, x_{r'}}_{r'}, \underbrace{0, \ldots, 0}_{s'}, \underbrace{z_1, \ldots, z_{n-(r'+s')}}_{n-(r'+s')}\right]^T \in W,$

$$\left\langle \left(1I_{r'} \oplus (-1)I_{s'} \oplus 0I_{n-(r'+s')}\right)\left[\underbrace{x_1, \cdots,}_{r'} \underbrace{0, \cdots,}_{s'} \underbrace{z_1, \cdots}_{n-(r'+s')}\right]^T,\right.$$

$$\left.\left[\underbrace{x_1, \cdots,}_{r'} \underbrace{0, \cdots, 0,}_{s'} \underbrace{z_1, \cdots}_{n-(r'+s')}\right]^T\right\rangle$$

$$= \left(\left[\underbrace{x_1, \cdots,}_{r'} \underbrace{0, \cdots,}_{s'} \underbrace{z_1, \cdots}_{n-(r'+s')}\right]^T\right)^T$$

$$\left(\left(1I_{r'} \oplus (-1)I_{s'} \oplus 0I_{n-(r'+s')}\right)\left[\underbrace{x_1, \cdots,}_{r'} \underbrace{0, \cdots}_{s'} \underbrace{z_1, \cdots}_{n-(r'+s')}\right]^T\right)$$

$$= \left[\underbrace{x_1, \cdots,}_{r'} \underbrace{0, \cdots,}_{s'} \underbrace{z_1, \cdots}_{n-(r'+s')}\right]$$

$$\left(\left(1I_{r'} \oplus (-1)I_{s'} \oplus 0I_{n-(r'+s')}\right)\left[\underbrace{x_1, \cdots,}_{r'} \underbrace{0, \cdots,}_{s'} \underbrace{z_1, \cdots}_{n-(r'+s')}\right]^T\right)$$

$$= \left[\underbrace{x_1, \cdots, x_{r'}}_{r'}, \underbrace{0, \cdots, 0,}_{s'} \underbrace{z_1, \cdots, z_{n-(r'+s')}}_{n-(r'+s')}\right]\left[\underbrace{x_1, \cdots, x_{r'}}_{r'}, \underbrace{0, \cdots, 0,}_{s'} \underbrace{0, \cdots, 0}_{n-(r'+s')}\right]^T$$

$$= \underbrace{(x_1)^2 + \cdots + (x_{r'})^2}_{r'} + \underbrace{0 + \cdots + 0}_{s'} + \underbrace{0 + \cdots + 0}_{n-(r'+s')}$$

$$= (x_1)^2 + \cdots + (x_{r'})^2 \geq 0,$$

so for every $w \in W,$

$$\left\langle \left(1I_r \oplus (-1)I_s \oplus 0I_{n-(r+s)} \right) \left(S^T w \right), \left(S^T w \right) \right\rangle$$
$$= \left(S^T w \right)^T \left(\left(1I_r \oplus (-1)I_s \oplus 0I_{n-(r+s)} \right) \left(S^T w \right) \right)$$
$$= w^T \left(\left(S \left(1I_r \oplus (-1)I_s \oplus 0I_{n-(r+s)} \right) S^T \right) w \right)$$
$$= \left\langle \left(S \left(1I_r \oplus (-1)I_s \oplus 0I_{n-(r+s)} \right) S^T \right) w, w \right\rangle$$
$$= \underbrace{\left\langle \left(1I_{r'} \oplus (-1)I_{s'} \oplus 0I_{n-(r'+s')} \right) w, w \right\rangle \geq 0}.$$

Thus for every $v \in \{ S^T w : w \in W \}$,

$$\left\langle \left(1I_r \oplus (-1)I_s \oplus 0I_{n-(r+s)} \right) v, v \right\rangle \geq 0.$$

Further, we have seen that for every nonzero $u \in U$,

$$\left\langle \left(1I_r \oplus (-1)I_s \oplus 0I_{n-(r+s)} \right) u, u \right\rangle \ngeq 0.$$

It follows that $\{ S^T w : w \in W \} \cap U = \{ 0 \}$.

Clearly, $\{ S^T w : w \in W \}$ is an $(n - s')$-dimensional real vector space.

Proof Since S is invertible, S^T is invertible. The map $T : x \mapsto S^T x$ from the real vector space $\mathbb{R}^n$ to $\mathbb{R}^n$ is a linear transformation. Since S^T is invertible, T is one-to-one, and hence

$$\underbrace{\dim \left(\{ S^T w : w \in W \} \right) = \dim(W)}_{} = n - s'.$$

Thus we have shown that $\{ S^T w : w \in W \}$ is an $(n - s')$-dimensional real vector space. ∎

Next, since $\dim(U) = s$, we have

$$n = \dim(\mathbb{R}^n) \geq \dim \left(\{ S^T w : w \in W \} \cup U \right)$$
$$= \dim \left(\{ S^T w : w \in W \} \right) + \dim(U) - \dim \left(\{ S^T w : w \in W \} \cap U \right)$$
$$= \dim \left(\{ S^T w : w \in W \} \right) + \dim(U) - \dim(\{ 0 \}) = \dim \left(\{ S^T w : w \in W \} \right) + \dim(U) - 0$$
$$= \dim \left(\{ S^T w : w \in W \} \right) + \dim(U) = (n - s') + \dim(U) = (n - s') + s = n + (s - s'),$$

and hence $n \geq n + (s - s')$. It follows that $s \leq s'$. This contradicts $(*)$.
Thus we have shown that $r = r'$.
Finally, we have to show that $s = s'$.

Since

$$1I_r \oplus (-1)I_s \oplus 0I_{n-(r+s)} \text{ and } 1I_{r'} \oplus (-1)I_{s'} \oplus 0I_{n-(r'+s')}$$

are congruent, there exists a real invertible matrix S such that

$$\left(1I_{r'} \oplus (-1)I_{s'} \oplus 0I_{n-(r'+s')}\right) = S\left(1I_r \oplus (-1)I_s \oplus 0I_{n-(r+s)}\right)S^T.$$

Since S is invertible, by 4.2.14, S^T is invertible, and hence

$$r' + s' = \text{rank}\left(1I_{r'} \oplus (-1)I_{s'} \oplus 0I_{n-(r'+s')}\right)$$
$$= \underbrace{\text{rank}\left(S\left(1I_r \oplus (-1)I_s \oplus 0I_{n-(r+s)}\right)S^T\right) = \text{rank}\left(1I_r \oplus (-1)I_s \oplus 0I_{n-(r+s)}\right)}_{} = r + s.$$

Thus $r' + s' = r + s$. Now, since $r = r'$, we have $s = s'$.   ∎

4.2.17 Conclusion Let A be an n-square real symmetric matrix. There exist a real invertible matrix R, and two nonnegative integers r and s such that

$$RAR^T = \left(1I_r \oplus (-1)I_s \oplus 0I_{n-(r+s)}\right).$$

Also, r and s are unique. Further, $\text{rank}(A) = r + s$.
The integer $(r - s)$ is called the *signature of* A, and is denoted by $\text{sg}(A)$.
Thus there exists a real invertible matrix R such that

$$RAR^T = \left(1I_{\frac{\text{rank}(A) + \text{sg}(A)}{2}} \oplus (-1)I_{\frac{\text{rank}(A) - \text{sg}(A)}{2}} \oplus 0I_{n - \text{rank}(A)}\right).$$

This result is known as **Sylvester's law.**

4.2.18 Theorem Let V be an n-dimensional real inner product space. Let $S : V \to V$ be a linear transformation. Let $v \in V$. Then there exists a unique $w \in V$ such that

$$u \in V \Rightarrow u, w = \langle S(u), v \rangle.$$

We denote w by $S^T(v)$. Thus $S^T : V \to V$, and for every $u, v \in V$, $\langle u, S^T(v) \rangle = \langle S(u), v \rangle$. Also, $S^T : V \to V$ is linear.

Proof Existence: Since V is an n-dimensional real inner product space, there exists an orthonormal basis $\{u_1, \ldots, u_n\}$ of V. Put

$$w \equiv \langle S(u_1), v \rangle u_1 + \cdots + \langle S(u_n), v \rangle u_n.$$

Let us fix an arbitrary $u \equiv \sum_{i=1}^{n} \alpha_i u_i$. We have to show that

$$\left\langle \sum_{i=1}^{n} \alpha_i u_i, \sum_{j=1}^{n} \langle S(u_j), v\rangle u_j \right\rangle = \left\langle S\left(\sum_{i=1}^{n} \alpha_i u_i \right), v \right\rangle :$$

$$\text{LHS} = \left\langle \sum_{i=1}^{n} \alpha_i u_i, \sum_{j=1}^{n} \langle S(u_j), v\rangle u_j \right\rangle = \sum_{i,j} \alpha_i \langle S(u_j), v\rangle \langle u_i, u_j \rangle = \sum_{i,j} \alpha_i \langle S(u_j), v\rangle \delta^{ij}$$

$$= \sum_{i=1}^{n} \alpha_i \langle S(u_i), v\rangle = \left\langle \sum_{i=1}^{n} \alpha_i S(u_i), v \right\rangle = \left\langle S\left(\sum_{i=1}^{n} \alpha_i u_i \right), v \right\rangle = \text{RHS.}$$

Uniqueness: Suppose that there exist $w_1, w_2 \in V$ such that

$$u \in V \Rightarrow \langle u, w_1 \rangle = \langle S(u), v\rangle, \text{ and } \langle u, w_2 \rangle = \langle S(u), v\rangle.$$

We have to show that $w_1 = w_2$, that is, $\langle w_1 - w_2, w_1 - w_2 \rangle = 0$. Here

$$u \in V \Rightarrow \langle u, w_1 \rangle = \langle u, w_2 \rangle,$$

so for every $u \in V$, $\langle u, w_1 - w_2 \rangle = 0$. It follows that $\langle w_1 - w_2, w_1 - w_2 \rangle = 0$.

Linearity: Let us take arbitrary $v_1, v_2 \in V$. Let α, β be arbitrary real numbers. We have to show that

$$S^T(\alpha v_1 + \beta v_2) = \alpha S^T(v_1) + \beta S^T(v_2).$$

It suffices to show that for every $u \in V$,

$$\langle u, S^T(\alpha v_1 + \beta v_2) \rangle = \langle u, \alpha S^T(v_1) + \beta S^T(v_2) \rangle.$$

To this end, let us fix an arbitrary $u \in V$. We have to show that

$$\langle u, S^T(\alpha v_1 + \beta v_2) \rangle = \langle u, \alpha S^T(v_1) + \beta S^T(v_2) \rangle :$$
$$\text{LHS} = \langle u, S^T(\alpha v_1 + \beta v_2) \rangle = \langle S(u), \alpha v_1 + \beta v_2 \rangle = \alpha \langle S(u), v_1 \rangle + \beta \langle S(u), v_2 \rangle$$
$$= \alpha \langle u, S^T(v_1) \rangle + \beta \langle u, S^T(v_2) \rangle = \langle u, \alpha S^T(v_1) + \beta S^T(v_2) \rangle = \text{RHS.}$$

∎

Definition Let V be an n-dimensional real inner product space. Let $S : V \to V$ be a linear transformation. By 4.2.18, $S^T : V \to V$ is a linear transformation such that for all $u, v \in V$, $\langle u, S^T(v) \rangle = \langle S(u), v\rangle$. Here S^T is called the *transpose of S*.

4.2.19 Theorem Let V be an n-dimensional real inner product space. Let $S : V \to V$ be a linear transformation. Then $(S^T)^T = S$.

Proof Let us take an arbitrary $v \in V$. We have to show that

$$\left(S^T\right)^T(v) = S(v).$$

To this end, let us take an arbitrary $u \in V$. It suffices to show that

$$\left\langle u, \left(S^T\right)^T(v) \right\rangle = \langle u, S(v) \rangle :$$

$$\text{LHS} = \left\langle u, \left(S^T\right)^T(v) \right\rangle = \langle S^T(u), v \rangle = \langle v, S^T(u) \rangle = \langle S(v), u \rangle = \langle u, S(v) \rangle = \text{RHS}.$$

∎

4.2.20 Theorem Let V be an n-dimensional real inner product space. Let $R : V \to V$ and $S : V \to V$ be linear transformations. Let λ, μ be any real numbers. Then $(\lambda R + \mu S)^T = \lambda R^T + \mu S^T$.

Proof Let us take an arbitrary $v \in V$. We have to show that

$$(\lambda R + \mu S)^T(v) = \left(\lambda R^T + \mu S^T\right)(v),$$

that is,

$$(\lambda R + \mu S)^T(v) = \lambda R^T(v) + \mu S^T(v).$$

To this end, let us take an arbitrary $u \in V$. It suffices to show that

$$\langle u, (\lambda R + \mu S)^T(v) \rangle = \langle u, \lambda R^T(v) + \mu S^T(v) \rangle :$$
$$\text{LHS} = \left\langle u, (\lambda R + \mu S)^T(v) \right\rangle = \langle (\lambda R + \mu S)(u), v \rangle$$
$$= \langle \lambda R(u) + \mu S(u), v \rangle = \lambda \langle R(u), v + \mu S(u), v \rangle$$
$$= \lambda \langle u, R^T(v) \rangle + \mu \langle u, S^T(v) \rangle = \langle u, \lambda R^T(v) + \mu S^T(v) \rangle = \text{RHS}.$$

∎

4.2.21 Theorem Let V be an n-dimensional real inner product space. Let $R : V \to V$ and $S : V \to V$ be linear transformations. Then $(RS)^T = S^T R^T$.

Proof Let us take an arbitrary $v \in V$. We have to show that

$$(RS)^T(v) = \left(S^T R^T\right)(v),$$

that is,

$$(RS)^T(v) = S^T\left(R^T(v)\right).$$

To this end, let us take an arbitrary $u \in V$. It suffices to show that

$$\langle u, (RS)^T(v) \rangle = \langle u, S^T(R^T(v)) \rangle :$$
$$\text{LHS} = \langle u, (RS)^T(v) \rangle = \langle (RS)(u), v = R(S(u)), v \rangle$$
$$= \langle S(u), R^T(v) \rangle = \langle u, S^T(R^T(v)) \rangle = \text{RHS}.$$

∎

4.2.22 Theorem Let V be an n-dimensional real inner product space. Let $S : V \to V$ be a linear transformation. Let $\{v_1, \ldots, v_n\}$ be an orthonormal basis of V. Let $[\alpha_{ij}]$ be the matrix of S relative to the basis $\{v_1, \ldots, v_n\}$, in the sense that

$$S(v_1) = a_{11}v_1 + a_{21}v_2 + \cdots + a_{n1}v_n \left(= \sum_{i=1}^{n} a_{i1}v_i \right),$$
$$S(v_2) = a_{12}v_1 + a_{22}v_2 + \cdots + a_{n2}v_n,$$
$$\vdots$$
$$S(v_n) = a_{1n}v_1 + a_{2n}v_2 + \cdots + a_{nn}v_n.$$

In short, $S(v_j) = \sum_{i=1}^{n} a_{ij}v_i$.

Then the matrix of S^T relative to the basis $\{v_1, \ldots, v_n\}$ is $[\beta_{ij}]$, where, $\beta_{ij} = \alpha_{ji}$. In short, $S^T(v_j) = \sum_{i=1}^{n} \beta_{ij}v_i$.

Proof By the proof of 4.2.18,

$$S^T(v_1) = \langle S(v_1), v_1 \rangle v_1 + \cdots + \langle S(v_n), v_1 \rangle v_n,$$
$$S^T(v_2) = \langle S(v_1), v_2 \rangle v_1 + \cdots + \langle S(v_n), v_2 \rangle v_n,$$
$$\vdots$$
$$S^T(v_n) = \langle S(v_1), v_n \rangle v_1 + \cdots + \langle S(v_n), v_n \rangle v_n.$$

Since

$$S^T(v_1) = \sum_{i=1}^{n} \langle S(v_i), v_1 \rangle v_i = \sum_{i=1}^{n} \langle a_{1i}v_1 + a_{2i}v_2 + \cdots + a_{ni}v_n, v_1 \rangle v_i$$
$$= \sum_{i=1}^{n} (a_{1i}\langle v_1, v_1 \rangle + a_{2i}\langle v_2, v_1 \rangle + \cdots + a_{ni}\langle v_n, v_1 \rangle)v_i$$
$$= \sum_{i=1}^{n} (a_{1i}1 + a_{2i}0 + \cdots + a_{ni}0)v_i = \sum_{i=1}^{n} a_{1i}v_i$$
$$= a_{11}v_1 + a_{12}v_2 + \cdots + a_{1n}v_n,$$

we have

$$S^T(v_1) = a_{11}v_1 + a_{12}v_2 + \cdots + a_{1n}v_n \left(= \sum_{i=1}^{n} a_{1i}v_i \right).$$

Similarly,

$$S^T(v_2) = a_{21}v_1 + a_{22}v_2 + \cdots + a_{2n}v_n,$$

etc. In short, $S^T(v_j) = \sum_{i=1}^{n} a_{ji}v_i$. If the matrix of S^T relative to the basis $\{v_1, \ldots, v_n\}$ is $[\beta_{ij}]$, then $\beta_{ij} = \alpha_{ji}$.   ∎

Definition Let V be an n-dimensional real inner product space. Let $S : V \to V$ be a linear transformation. If $S^T = S$, then we say that S *is symmetric*.

4.2.23 Theorem Let V be an n-dimensional real inner product space. Let $S : V \to V$ be a linear transformation. Let S be symmetric. Let $\{e_1, \ldots, e_n\}$ be any orthonormal basis of V. Let $[a_{ij}]$ be the matrix of S relative to $\{e_1, \ldots, e_n\}$, that is, $S(e_j) = \sum_{i=1}^{n} a_{ij}e_i$. Let $Q : v \mapsto \langle S(v), v \rangle$ be a function from V to $\mathbb{R}$. Let $v \equiv \sum_{i=1}^{n} x_i e_i$. Then

$$Q\left(\sum_{i=1}^{n} x_i e_i\right) = a_{11}(x_1)^2 + \cdots + a_{nn}(x_n)^2 + 2\sum_{i<j}^{n} a_{ij}x_i x_j.$$

Here $a_{11}(x_1)^2 + \cdots + a_{nn}(x_n)^2 + 2\sum_{i<j}^{n} a_{ij}x_i x_j$ is called the *real quadratic form of S*.

Proof Since S is symmetric and $[a_{ij}]$ is the matrix of S relative to $\{e_1, \ldots, e_n\}$, by 4.2.22, $\alpha_{ji} = \alpha_{ij}$. It follows that

$$Q\left(\sum_{i=1}^{n} x_i e_i\right) = Q(v) = \langle S(v), v \rangle = \left\langle S\left(\sum_{i=1}^{n} x_i e_i\right), \sum_{i=1}^{n} x_i e_i \right\rangle$$

$$= \left\langle \sum_{i=1}^{n} x_i S(e_i), \sum_{j=1}^{n} x_j e_j \right\rangle = \sum_{i=1}^{n} x_i \left\langle S(e_i), \sum_{j=1}^{n} x_j e_j \right\rangle$$

$$= \sum_{i=1}^{n} x_i \left(\sum_{j=1}^{n} x_j \langle S(e_i), e_j \rangle\right) = \sum_{i=1}^{n} \left(\sum_{j=1}^{n} x_i x_j \langle S(e_i), e_j \rangle\right)$$

$$= \sum_{i=1}^{n} \left(\sum_{j=1}^{n} x_i x_j \left\langle \sum_{k=1}^{n} a_{ki} e_k, e_j \right\rangle\right) = \sum_{i=1}^{n} \left(\sum_{j=1}^{n} x_i x_j \left(\sum_{k=1}^{n} a_{ki} \langle e_k, e_j \rangle\right)\right)$$

$$= \sum_{i=1}^{n} \left(\sum_{j=1}^{n} x_i x_j \left(\sum_{k=1}^{n} a_{ki} \delta_{kj}\right)\right) = \sum_{i=1}^{n} \left(\sum_{j=1}^{n} x_i x_j a_{ji}\right)$$

$$= \sum_{i=1}^{n} \left(\sum_{j=1}^{n} a_{ji} x_i x_j\right) = \sum_{i=1}^{n} \left(\sum_{j=1}^{n} a_{ij} x_i x_j\right)$$

$$= a_{11}(x_1)^2 + \cdots + a_{nn}(x_n)^2 + 2\sum_{i<j}^{n} a_{ij}x_i x_j,$$

and hence

$$Q\left(\sum_{i=1}^{n} x_i e_i\right) = a_{11}(x_1)^2 + \cdots + a_{nn}(x_n)^2 + 2\sum_{i<j}^{n} a_{ij} x_i x_j.$$

∎

4.2.24 Example Let us consider the following real quadratic form:

$$(x_1)^2 + 2(x_2)^2 + 2(x_3)^2 + x_1 x_2 + 2x_1 x_3 + 4x_2 x_3.$$

Here

$$(x_1)^2 + 2(x_2)^2 + 2(x_3)^2 + x_1 x_2 + 2x_1 x_3 + 4x_2 x_3$$
$$= \left((x_1)^2 + \tfrac{1}{2}x_1 x_2 + x_1 x_3\right) + \left(\tfrac{1}{2}x_1 x_2 + 2(x_2)^2 + 2x_2 x_3\right) + \left(x_1 x_3 + 2x_2 x_3 + 2(x_3)^2\right)$$
$$= (x_1 + \tfrac{1}{2}x_2 + x_3)x_1 + (\tfrac{1}{2}x_1 + x_2 + 2x_3)x_2 + (x_1 + 2x_2 + 2x_3)x_3$$
$$= [x_1 x_2 x_3]\begin{bmatrix} 1 & \tfrac{1}{2} & 1 \\ \tfrac{1}{2} & 2 & 2 \\ 1 & 2 & 2 \end{bmatrix}\begin{bmatrix} x_1 \\ x_2 \\ x_3 \end{bmatrix} = [x_1 x_2 x_3] A [x_1 x_2 x_3]^T,$$

where

$$A \equiv \begin{bmatrix} a_{11} & a_{12} & a_{13} \\ a_{21} & a_{22} & a_{23} \\ a_{31} & a_{32} & a_{33} \end{bmatrix} = \begin{bmatrix} 1 & \tfrac{1}{2} & 1 \\ \tfrac{1}{2} & 2 & 2 \\ 1 & 2 & 2 \end{bmatrix}.$$

Observe that

$$\begin{bmatrix} 1 & \tfrac{1}{2} & 1 \\ \tfrac{1}{2} & 2 & 2 \\ 1 & 2 & 2 \end{bmatrix} \xrightarrow{R_2 \to R_2 - \tfrac{1}{2}R_1} \begin{bmatrix} 1 & \tfrac{1}{2} & 1 \\ 0 & \tfrac{7}{4} & \tfrac{3}{2} \\ 1 & 2 & 2 \end{bmatrix} \xrightarrow{C_2 \to C_2 - \tfrac{1}{2}C_1} \begin{bmatrix} 1 & 0 & 1 \\ 0 & \tfrac{7}{4} & \tfrac{3}{2} \\ 1 & \tfrac{3}{2} & 2 \end{bmatrix}.$$

Thus

$$\begin{bmatrix} 1 & 0 & 0 \\ -\tfrac{1}{2} & 1 & 0 \\ 0 & 0 & 1 \end{bmatrix}\begin{bmatrix} 1 & \tfrac{1}{2} & 1 \\ \tfrac{1}{2} & 2 & 2 \\ 1 & 2 & 2 \end{bmatrix}\begin{bmatrix} 1 & -\tfrac{1}{2} & 0 \\ 0 & 1 & 0 \\ 0 & 0 & 1 \end{bmatrix} = \begin{bmatrix} 1 & 0 & 1 \\ 0 & \tfrac{7}{4} & \tfrac{3}{2} \\ 1 & \tfrac{3}{2} & 2 \end{bmatrix},$$

or

$$\begin{bmatrix} 1 & 0 & 0 \\ -\tfrac{1}{2} & 1 & 0 \\ 0 & 0 & 1 \end{bmatrix}\begin{bmatrix} 1 & \tfrac{1}{2} & 1 \\ \tfrac{1}{2} & 2 & 2 \\ 1 & 2 & 2 \end{bmatrix}\begin{bmatrix} 1 & 0 & 0 \\ -\tfrac{1}{2} & 1 & 0 \\ 0 & 0 & 1 \end{bmatrix}^T = \begin{bmatrix} 1 & 0 & 1 \\ 0 & \tfrac{7}{4} & \tfrac{3}{2} \\ 1 & \tfrac{3}{2} & 2 \end{bmatrix}.$$

Since

$$\begin{vmatrix} 1 & 0 & 0 \\ -\frac{1}{2} & 1 & 0 \\ 0 & 0 & 1 \end{vmatrix} = \begin{vmatrix} 1 & 0 & 0 \\ 0 & 1 & 0 \\ 0 & 0 & 1 \end{vmatrix} = 1 \neq 0,$$

it follows that

$$\begin{vmatrix} 1 & 0 & 0 \\ -\frac{1}{2} & 1 & 0 \\ 0 & 0 & 1 \end{vmatrix}$$

is invertible.

Observe that

$$\begin{bmatrix} 1 & 0 & 1 \\ 0 & \frac{7}{4} & \frac{3}{2} \\ 1 & \frac{3}{2} & 2 \end{bmatrix} \xrightarrow{R_3 \to R_3 - R_1} \begin{bmatrix} 1 & 0 & 1 \\ 0 & \frac{7}{4} & \frac{3}{2} \\ 0 & \frac{3}{2} & 1 \end{bmatrix} \xrightarrow{C_3 \to C_3 - C_1} \begin{bmatrix} 1 & 0 & 0 \\ 0 & \frac{7}{4} & \frac{3}{2} \\ 0 & \frac{3}{2} & 1 \end{bmatrix}.$$

Thus

$$\begin{bmatrix} 1 & 0 & 0 \\ 0 & 1 & 0 \\ -1 & 0 & 1 \end{bmatrix} \begin{bmatrix} 1 & 0 & 1 \\ 0 & \frac{7}{4} & \frac{3}{2} \\ 1 & \frac{3}{2} & 2 \end{bmatrix} \begin{bmatrix} 1 & 0 & -1 \\ 0 & 1 & 0 \\ 0 & 0 & 1 \end{bmatrix} = \begin{bmatrix} 1 & 0 & 0 \\ 0 & \frac{7}{4} & \frac{3}{2} \\ 0 & \frac{3}{2} & 1 \end{bmatrix},$$

or

$$\begin{bmatrix} 1 & 0 & 0 \\ 0 & 1 & 0 \\ -1 & 0 & 1 \end{bmatrix} \begin{bmatrix} 1 & 0 & 1 \\ 0 & \frac{7}{4} & \frac{3}{2} \\ 1 & \frac{3}{2} & 2 \end{bmatrix} \begin{bmatrix} 1 & 0 & 0 \\ 0 & 1 & 0 \\ -1 & 0 & 1 \end{bmatrix}^T = \begin{bmatrix} 1 & 0 & 0 \\ 0 & \frac{7}{4} & \frac{3}{2} \\ 0 & \frac{3}{2} & 1 \end{bmatrix}.$$

Since

$$\begin{vmatrix} 1 & 0 & 0 \\ 0 & 1 & 0 \\ -1 & 0 & 1 \end{vmatrix} = \begin{vmatrix} 1 & 0 & 0 \\ 0 & 1 & 0 \\ 0 & 0 & 1 \end{vmatrix} = 1 \neq 0,$$

it follows that

$$\begin{vmatrix} 1 & 0 & 0 \\ 0 & 1 & 0 \\ -1 & 0 & 1 \end{vmatrix}$$

is invertible.

Observe that

$$
\begin{bmatrix} 1 & 0 & 0 \\ 0 & \frac{7}{4} & \frac{3}{2} \\ 0 & \frac{3}{2} & 1 \end{bmatrix}
\xrightarrow{R_2 \to 2R_2}
\begin{bmatrix} 1 & 0 & 0 \\ 0 & \frac{7}{2} & 3 \\ 0 & \frac{3}{2} & 1 \end{bmatrix}
\xrightarrow{C_2 \to 2C_2}
\begin{bmatrix} 1 & 0 & 0 \\ 0 & 7 & 3 \\ 0 & 3 & 1 \end{bmatrix}.
$$

Thus

$$
\begin{bmatrix} 1 & 0 & 0 \\ 0 & 2 & 0 \\ 0 & 0 & 1 \end{bmatrix}
\begin{bmatrix} 1 & 0 & 0 \\ 0 & \frac{7}{4} & \frac{3}{2} \\ 0 & \frac{3}{2} & 1 \end{bmatrix}
\begin{bmatrix} 1 & 0 & 0 \\ 0 & 2 & 0 \\ 0 & 0 & 1 \end{bmatrix}
=
\begin{bmatrix} 1 & 0 & 0 \\ 0 & 7 & 3 \\ 0 & 3 & 1 \end{bmatrix},
$$

or

$$
\begin{bmatrix} 1 & 0 & 0 \\ 0 & 2 & 0 \\ 0 & 0 & 1 \end{bmatrix}
\begin{bmatrix} 1 & 0 & 0 \\ 0 & \frac{7}{4} & \frac{3}{2} \\ 0 & \frac{3}{2} & 1 \end{bmatrix}
\begin{bmatrix} 1 & 0 & 0 \\ 0 & 2 & 0 \\ 0 & 0 & 1 \end{bmatrix}^{T}
=
\begin{bmatrix} 1 & 0 & 0 \\ 0 & 7 & 3 \\ 0 & 3 & 1 \end{bmatrix}.
$$

Since

$$
\begin{vmatrix} 1 & 0 & 0 \\ 0 & 2 & 0 \\ 0 & 0 & 1 \end{vmatrix}
= 2
\begin{vmatrix} 1 & 0 & 0 \\ 0 & 1 & 0 \\ 0 & 0 & 1 \end{vmatrix}
= 2 \neq 0,
$$

it follows that

$$
\begin{vmatrix} 1 & 0 & 0 \\ 0 & 2 & 0 \\ 0 & 0 & 1 \end{vmatrix}
$$

is invertible.

Observe that

$$
\begin{bmatrix} 1 & 0 & 0 \\ 0 & 7 & 3 \\ 0 & 3 & 1 \end{bmatrix}
\xrightarrow{R_2 \to R_2 - 2R_3}
\begin{bmatrix} 1 & 0 & 0 \\ 0 & 1 & 1 \\ 0 & 3 & 1 \end{bmatrix}
\xrightarrow{C_2 \to C_2 - 2C_3}
\begin{bmatrix} 1 & 0 & 0 \\ 0 & -1 & 1 \\ 0 & 1 & 1 \end{bmatrix}.
$$

Thus

$$
\begin{bmatrix} 1 & 0 & 0 \\ 0 & 1 & -2 \\ 0 & 0 & 1 \end{bmatrix}
\begin{bmatrix} 1 & 0 & 0 \\ 0 & 7 & 3 \\ 0 & 3 & 1 \end{bmatrix}
\begin{bmatrix} 1 & 0 & 0 \\ 0 & 1 & 0 \\ 0 & -2 & 1 \end{bmatrix}
=
\begin{bmatrix} 1 & 0 & 0 \\ 0 & -1 & 1 \\ 0 & 1 & 1 \end{bmatrix},
$$

or

$$\begin{bmatrix} 1 & 0 & 0 \\ 0 & 1 & -2 \\ 0 & 0 & 1 \end{bmatrix} \begin{bmatrix} 1 & 0 & 0 \\ 0 & 7 & 3 \\ 0 & 3 & 1 \end{bmatrix} \begin{bmatrix} 1 & 0 & 0 \\ 0 & 1 & -2 \\ 0 & 0 & 1 \end{bmatrix}^T = \begin{bmatrix} 1 & 0 & 0 \\ 0 & -1 & 1 \\ 0 & 1 & 1 \end{bmatrix}.$$

Since

$$\begin{vmatrix} 1 & 0 & 0 \\ 0 & 1 & -2 \\ 0 & 0 & 1 \end{vmatrix} = \begin{vmatrix} 1 & 0 & 0 \\ 0 & 1 & 0 \\ 0 & 0 & 1 \end{vmatrix} = 1 \neq 0,$$

it follows that

$$\begin{vmatrix} 1 & 0 & 0 \\ 0 & 1 & -2 \\ 0 & 0 & 1 \end{vmatrix}$$

is invertible.

Observe that

$$\begin{bmatrix} 1 & 0 & 0 \\ 0 & -1 & 1 \\ 0 & 1 & 1 \end{bmatrix} \xrightarrow{R_3 \to R_3 + R_2} \begin{bmatrix} 1 & 0 & 0 \\ 0 & -1 & 1 \\ 0 & 0 & 2 \end{bmatrix} \xrightarrow{C_3 \to C_3 + C_2} \begin{bmatrix} 1 & 0 & 0 \\ 0 & -1 & 0 \\ 0 & 0 & 2 \end{bmatrix}.$$

Thus

$$\begin{bmatrix} 1 & 0 & 0 \\ 0 & 1 & 0 \\ 0 & 1 & 1 \end{bmatrix} \begin{bmatrix} 1 & 0 & 0 \\ 0 & -1 & 1 \\ 0 & 1 & 1 \end{bmatrix} \begin{bmatrix} 1 & 0 & 0 \\ 0 & 1 & 1 \\ 0 & 0 & 1 \end{bmatrix} = \begin{bmatrix} 1 & 0 & 0 \\ 0 & -1 & 0 \\ 0 & 0 & 2 \end{bmatrix},$$

or

$$\begin{bmatrix} 1 & 0 & 0 \\ 0 & 1 & 0 \\ 0 & 1 & 1 \end{bmatrix} \begin{bmatrix} 1 & 0 & 0 \\ 0 & -1 & 1 \\ 0 & 1 & 1 \end{bmatrix} \begin{bmatrix} 1 & 0 & 0 \\ 0 & 1 & 0 \\ 0 & 1 & 1 \end{bmatrix}^T = \begin{bmatrix} 1 & 0 & 0 \\ 0 & -1 & 0 \\ 0 & 0 & 2 \end{bmatrix}.$$

Since

$$\begin{vmatrix} 1 & 0 & 0 \\ 0 & 1 & 0 \\ 0 & 1 & 1 \end{vmatrix} = \begin{vmatrix} 1 & 0 & 0 \\ 0 & 1 & 0 \\ 0 & 0 & 1 \end{vmatrix} = 1 \neq 0,$$

it follows that

$$\begin{vmatrix} 1 & 0 & 0 \\ 0 & 1 & 0 \\ 0 & 1 & 1 \end{vmatrix}$$

is invertible.

Observe that

$$\begin{bmatrix} 1 & 0 & 0 \\ 0 & -1 & 0 \\ 0 & 0 & 2 \end{bmatrix} \xrightarrow{R_3 \to \frac{1}{\sqrt{2}} R_3} \begin{bmatrix} 1 & 0 & 0 \\ 0 & -1 & 0 \\ 0 & 0 & \sqrt{2} \end{bmatrix} \xrightarrow{C_3 \to \frac{1}{\sqrt{2}} C_3} \begin{bmatrix} 1 & 0 & 0 \\ 0 & -1 & 0 \\ 0 & 0 & 1 \end{bmatrix}.$$

Thus

$$\begin{bmatrix} 1 & 0 & 0 \\ 0 & 1 & 0 \\ 0 & 0 & \frac{1}{\sqrt{2}} \end{bmatrix} \begin{bmatrix} 1 & 0 & 0 \\ 0 & -1 & 0 \\ 0 & 0 & 2 \end{bmatrix} \begin{bmatrix} 1 & 0 & 0 \\ 0 & 1 & 0 \\ 0 & 0 & \frac{1}{\sqrt{2}} \end{bmatrix} = \begin{bmatrix} 1 & 0 & 0 \\ 0 & -1 & 0 \\ 0 & 0 & 1 \end{bmatrix},$$

or

$$\begin{bmatrix} 1 & 0 & 0 \\ 0 & 1 & 0 \\ 0 & 0 & \frac{1}{\sqrt{2}} \end{bmatrix} \begin{bmatrix} 1 & 0 & 0 \\ 0 & -1 & 0 \\ 0 & 0 & 2 \end{bmatrix} \begin{bmatrix} 1 & 0 & 0 \\ 0 & 1 & 0 \\ 0 & 0 & \frac{1}{\sqrt{2}} \end{bmatrix}^T = \begin{bmatrix} 1 & 0 & 0 \\ 0 & -1 & 0 \\ 0 & 0 & 1 \end{bmatrix}.$$

Since

$$\begin{vmatrix} 1 & 0 & 0 \\ 0 & 1 & 0 \\ 0 & 0 & \frac{1}{\sqrt{2}} \end{vmatrix} = \frac{1}{\sqrt{2}} \begin{vmatrix} 1 & 0 & 0 \\ 0 & 1 & 0 \\ 0 & 0 & 1 \end{vmatrix} = \frac{1}{\sqrt{2}} \neq 0,$$

it follows that

$$\begin{vmatrix} 1 & 0 & 0 \\ 0 & 1 & 0 \\ 0 & 0 & \frac{1}{\sqrt{2}} \end{vmatrix}$$

is invertible.

Observe that

$$\begin{bmatrix} 1 & 0 & 0 \\ 0 & -1 & 0 \\ 0 & 0 & 1 \end{bmatrix} \xrightarrow{R_{23}} \begin{bmatrix} 1 & 0 & 0 \\ 0 & 0 & 1 \\ 0 & -1 & 0 \end{bmatrix} \xrightarrow{C_{23}} \begin{bmatrix} 1 & 0 & 0 \\ 0 & 1 & 0 \\ 0 & 0 & -1 \end{bmatrix}.$$

Thus

$$\begin{bmatrix} 1 & 0 & 0 \\ 0 & 0 & 1 \\ 0 & 1 & 0 \end{bmatrix} \begin{bmatrix} 1 & 0 & 0 \\ 0 & -1 & 0 \\ 0 & 0 & 1 \end{bmatrix} \begin{bmatrix} 1 & 0 & 0 \\ 0 & 0 & 1 \\ 0 & 1 & 0 \end{bmatrix} = \begin{bmatrix} 1 & 0 & 0 \\ 0 & 1 & 0 \\ 0 & 0 & -1 \end{bmatrix},$$

or

$$\begin{bmatrix} 1 & 0 & 0 \\ 0 & 0 & 1 \\ 0 & 1 & 0 \end{bmatrix} \begin{bmatrix} 1 & 0 & 0 \\ 0 & -1 & 0 \\ 0 & 0 & 1 \end{bmatrix} \begin{bmatrix} 1 & 0 & 0 \\ 0 & 0 & 1 \\ 0 & 1 & 0 \end{bmatrix}^T = \begin{bmatrix} 1 & 0 & 0 \\ 0 & 1 & 0 \\ 0 & 0 & -1 \end{bmatrix}.$$

Since

$$\begin{vmatrix} 1 & 0 & 0 \\ 0 & 0 & 1 \\ 0 & 1 & 0 \end{vmatrix} = - \begin{vmatrix} 1 & 0 & 0 \\ 0 & 1 & 0 \\ 0 & 0 & 1 \end{vmatrix} = -1 \neq 0,$$

it follows that

$$\begin{vmatrix} 1 & 0 & 0 \\ 0 & 0 & 1 \\ 0 & 1 & 0 \end{vmatrix}$$

is invertible.
   From

$$\begin{bmatrix} 1 & 0 & 0 \\ 0 & 1 & 0 \\ 0 & 0 & -1 \end{bmatrix},$$

we find that $r = 2$ and $s = 1$. Hence the signature of the real quadratic form is $r - s (= 2 - 1 = 1)$.
   If we collect the above results, we get

$$RAR^T = \begin{bmatrix} 1 & 0 & 0 \\ 0 & 1 & 0 \\ 0 & 0 & -1 \end{bmatrix},$$

where

  R stands for

$$\begin{bmatrix} 1 & 0 & 0 \\ 0 & 0 & 1 \\ 0 & 1 & 0 \end{bmatrix} \begin{bmatrix} 1 & 0 & 0 \\ 0 & 1 & 0 \\ 0 & 0 & \frac{1}{\sqrt{2}} \end{bmatrix} \begin{bmatrix} 1 & 0 & 0 \\ 0 & 1 & 0 \\ 0 & 1 & 1 \end{bmatrix} \begin{bmatrix} 1 & 0 & 0 \\ 0 & 1 & -2 \\ 0 & 0 & 1 \end{bmatrix} \begin{bmatrix} 1 & 0 & 0 \\ 0 & 2 & 0 \\ 0 & 0 & 1 \end{bmatrix} \begin{bmatrix} 1 & 0 & 0 \\ 0 & 1 & 0 \\ -1 & 0 & 1 \end{bmatrix} \begin{bmatrix} 1 & 0 & 0 \\ -\frac{1}{2} & 1 & 0 \\ 0 & 0 & 1 \end{bmatrix}$$

$$= \begin{bmatrix} 1 & 0 & 0 \\ 0 & 0 & \frac{1}{\sqrt{2}} \\ 0 & 1 & 0 \end{bmatrix} \begin{bmatrix} 1 & 0 & 0 \\ 0 & 1 & 0 \\ 0 & 1 & 1 \end{bmatrix} \begin{bmatrix} 1 & 0 & 0 \\ 0 & 1 & -2 \\ 0 & 0 & 1 \end{bmatrix} \begin{bmatrix} 1 & 0 & 0 \\ 0 & 2 & 0 \\ 0 & 0 & 1 \end{bmatrix} \begin{bmatrix} 1 & 0 & 0 \\ 0 & 1 & 0 \\ -1 & 0 & 1 \end{bmatrix} \begin{bmatrix} 1 & 0 & 0 \\ -\frac{1}{2} & 1 & 0 \\ 0 & 0 & 1 \end{bmatrix}$$

$$= \begin{bmatrix} 1 & 0 & 0 \\ 0 & \frac{1}{\sqrt{2}} & \frac{1}{\sqrt{2}} \\ 0 & 1 & 0 \end{bmatrix} \begin{bmatrix} 1 & 0 & 0 \\ 0 & 1 & -2 \\ 0 & 0 & 1 \end{bmatrix} \begin{bmatrix} 1 & 0 & 0 \\ 0 & 2 & 0 \\ 0 & 0 & 1 \end{bmatrix} \begin{bmatrix} 1 & 0 & 0 \\ 0 & 1 & 0 \\ -1 & 0 & 1 \end{bmatrix} \begin{bmatrix} 1 & 0 & 0 \\ -\frac{1}{2} & 1 & 0 \\ 0 & 0 & 1 \end{bmatrix}$$

$$= \begin{bmatrix} 1 & 0 & 0 \\ 0 & \frac{1}{\sqrt{2}} & \frac{-1}{\sqrt{2}} \\ 0 & 1 & -2 \end{bmatrix} \begin{bmatrix} 1 & 0 & 0 \\ 0 & 2 & 0 \\ 0 & 0 & 1 \end{bmatrix} \begin{bmatrix} 1 & 0 & 0 \\ 0 & 1 & 0 \\ -1 & 0 & 1 \end{bmatrix} \begin{bmatrix} 1 & 0 & 0 \\ -\frac{1}{2} & 1 & 0 \\ 0 & 0 & 1 \end{bmatrix}$$

$$= \begin{bmatrix} 1 & 0 & 0 \\ 0 & \sqrt{2} & \frac{-1}{\sqrt{2}} \\ 0 & 2 & -2 \end{bmatrix} \begin{bmatrix} 1 & 0 & 0 \\ 0 & 1 & 0 \\ -1 & 0 & 1 \end{bmatrix} \begin{bmatrix} 1 & 0 & 0 \\ -\frac{1}{2} & 1 & 0 \\ 0 & 0 & 1 \end{bmatrix}$$

$$= \begin{bmatrix} 1 & 0 & 0 \\ \frac{1}{\sqrt{2}} & \sqrt{2} & \frac{-1}{\sqrt{2}} \\ 2 & 2 & -2 \end{bmatrix} \begin{bmatrix} 1 & 0 & 0 \\ -\frac{1}{2} & 1 & 0 \\ 0 & 0 & 1 \end{bmatrix} = \begin{bmatrix} 1 & 0 & 0 \\ 0 & \sqrt{2} & \frac{-1}{\sqrt{2}} \\ 1 & 2 & -2 \end{bmatrix}.$$

Thus

$$\begin{bmatrix} 1 & 0 & 0 \\ 0 & \sqrt{2} & \frac{-1}{\sqrt{2}} \\ 1 & 2 & -2 \end{bmatrix} \begin{bmatrix} 1 & \frac{1}{2} & 1 \\ \frac{1}{2} & 2 & 2 \\ 1 & 2 & 2 \end{bmatrix} \begin{bmatrix} 1 & 0 & 0 \\ 0 & \sqrt{2} & \frac{-1}{\sqrt{2}} \\ 1 & 2 & -2 \end{bmatrix}^T = \begin{bmatrix} 1 & 0 & 0 \\ 0 & 1 & 0 \\ 0 & 0 & -1 \end{bmatrix}.$$

Clearly, $\begin{bmatrix} 1 & 0 & 0 \\ 0 & \sqrt{2} & \frac{-1}{\sqrt{2}} \\ 1 & 2 & -2 \end{bmatrix}$ $(=R)$ is invertible. It follows that

$$\begin{bmatrix} 1 & \frac{1}{2} & 1 \\ \frac{1}{2} & 2 & 2 \\ 1 & 2 & 2 \end{bmatrix} = \begin{bmatrix} 1 & 0 & 0 \\ 0 & \sqrt{2} & \frac{-1}{\sqrt{2}} \\ 1 & 2 & -2 \end{bmatrix}^{-1} \begin{bmatrix} 1 & 0 & 0 \\ 0 & 1 & 0 \\ 0 & 0 & -1 \end{bmatrix} \left(\begin{bmatrix} 1 & 0 & 0 \\ 0 & \sqrt{2} & \frac{-1}{\sqrt{2}} \\ 1 & 2 & -2 \end{bmatrix}^T \right)^{-1}$$

$$= \begin{bmatrix} 1 & 0 & 0 \\ 0 & \sqrt{2} & \frac{-1}{\sqrt{2}} \\ 1 & 2 & -2 \end{bmatrix}^{-1} \begin{bmatrix} 1 & 0 & 0 \\ 0 & 1 & 0 \\ 0 & 0 & -1 \end{bmatrix} \left(\begin{bmatrix} 1 & 0 & 0 \\ 0 & \sqrt{2} & \frac{-1}{\sqrt{2}} \\ 1 & 2 & -2 \end{bmatrix}^{-1} \right)^T,$$

and hence

$$(x_1)^2 + 2(x_2)^2 + 2(x_3)^2 + x_1x_2 + 2x_1x_3 + 4x_2x_3 = [x_1x_2x_3]\begin{bmatrix} 1 & \frac{1}{2} & 1 \\ \frac{1}{2} & 2 & 2 \\ 1 & 2 & 2 \end{bmatrix}[x_1x_2x_3]^T$$

$$= [x_1x_2x_3]\begin{bmatrix} 1 & 0 & 0 \\ 0 & \sqrt{2} & \frac{-1}{\sqrt{2}} \\ 1 & 2 & -2 \end{bmatrix}^{-1}\begin{bmatrix} 1 & 0 & 0 \\ 0 & 1 & 0 \\ 0 & 0 & -1 \end{bmatrix}\left(\begin{bmatrix} 1 & 0 & 0 \\ 0 & \sqrt{2} & \frac{-1}{\sqrt{2}} \\ 1 & 2 & -2 \end{bmatrix}^{-1}\right)^T [x_1x_2x_3]^T$$

$$= \left([x_1x_2x_3]\begin{bmatrix} 1 & 0 & 0 \\ 0 & \sqrt{2} & \frac{-1}{\sqrt{2}} \\ 1 & 2 & -2 \end{bmatrix}^{-1}\right)\begin{bmatrix} 1 & 0 & 0 \\ 0 & 1 & 0 \\ 0 & 0 & -1 \end{bmatrix}\left([x_1x_2x_3]\begin{bmatrix} 1 & 0 & 0 \\ 0 & \sqrt{2} & \frac{-1}{\sqrt{2}} \\ 1 & 2 & -2 \end{bmatrix}^{-1}\right)^T$$

$$= [y_1y_2y_3]\begin{bmatrix} 1 & 0 & 0 \\ 0 & 1 & 0 \\ 0 & 0 & -1 \end{bmatrix}[y_1y_2y_3]^T,$$

where

$$[y_1y_2y_3] \equiv [x_1x_2x_3]\begin{bmatrix} 1 & 0 & 0 \\ 0 & \sqrt{2} & \frac{-1}{\sqrt{2}} \\ 1 & 2 & -2 \end{bmatrix}^{-1}.$$

It follows that

$$[x_1x_2x_3] = [y_1y_2y_3]\begin{bmatrix} 1 & 0 & 0 \\ 0 & \sqrt{2} & \frac{-1}{\sqrt{2}} \\ 1 & 2 & -2 \end{bmatrix} = \left[y_1 + y_3 \sqrt{2}y_2 + 2y_3 \frac{-1}{\sqrt{2}}y_2 - 2y_3\right],$$

or

$$\left.\begin{array}{l} x_1 = y_1 + y_3 \\ x_2 = \sqrt{2}y_2 + 2y_3 \\ x_3 = \frac{-1}{\sqrt{2}}y_2 - 2y_3 \end{array}\right\}.$$

Also,

$$(x_1)^2 + 2(x_2)^2 + 2(x_3)^2 + x_1x_2 + 2x_1x_3 + 4x_2x_3 = [y_1y_2y_3]\begin{bmatrix} 1 & 0 & 0 \\ 0 & 1 & 0 \\ 0 & 0 & -1 \end{bmatrix}[y_1y_2y_3]^T$$

$$= (y_1)^2 + (y_2)^2 - (y_3)^2,$$

that is,

$$(x_1)^2 + 2(x_2)^2 + 2(x_3)^2 + x_1x_2 + 2x_1x_3 + 4x_2x_3 = (y_1)^2 + (y_2)^2 - (y_3)^2,$$

where

$$\left. \begin{array}{l} x_1 = y_1 + y_3 \\ x_2 = \sqrt{2}y_2 + 2y_3 \\ x_3 = \frac{-1}{\sqrt{2}}y_2 - 2y_3 \end{array} \right\}.$$

Verification: Here

$$\text{LHS} = (x_1)^2 + 2(x_2)^2 + 2(x_3)^2 + x_1x_2 + 2x_1x_3 + 4x_2x_3$$

$$= (y_1 + y_3)^2 + 2\left(\sqrt{2}y_2 + 2y_3\right)^2 + 2\left(\frac{-1}{\sqrt{2}}y_2 - 2y_3\right)^2 + (y_1 + y_3)\left(\sqrt{2}y_2 + 2y_3\right)$$

$$+ 2(y_1 + y_3)\left(\frac{-1}{\sqrt{2}}y_2 - 2y_3\right) + 4\left(\sqrt{2}y_2 + 2y_3\right)\left(\frac{-1}{\sqrt{2}}y_2 - 2y_3\right)$$

$$= (y_1)^2 + (y_2)^2(4 + 1 - 4) + (y_3)^2(1 + 8 + 8 + 2 - 4 - 16)$$

$$+ y_1y_2\left(\sqrt{2} - \sqrt{2}\right) + y_1y_3(2 + 2 - 4) + y_2y_3\left(8\sqrt{2} + 4\sqrt{2} + \sqrt{2} - \sqrt{2} - 12\sqrt{2}\right)$$

$$= (y_1)^2 + (y_2)^2 - (y_3)^2 = \text{RHS}.$$

Verified.

4.2.25 Conclusion Let $A \equiv [a_{ij}]$ be an n-square real symmetric matrix. Let

$$\phi(x_1, \ldots, x_n) \equiv \sum_{i,j} a_{ij}x_ix_j \left(= [x_1, \ldots, x_n]A[x_1, \ldots, x_n]^T\right)$$

be a real quadratic form. Then there exists a real invertible matrix $C \equiv [c_{ij}]$ such that the transformation

$$\left. \begin{array}{l} x_1 = c_{11}y_1 + \cdots + c_{1n}y_n \\ \vdots \\ x_n = c_{n1}y_1 + \cdots + c_{nn}y_n \end{array} \right\}$$

reduces the form $\sum_{i,j} a_{ij}x_ix_j$ to the "normal form"

$$(y_1)^2 + \cdots + (y_r)^2 - (y_{r+1})^2 - \cdots - (y_{r+s})^2.$$

Definition If $r = 0$, then the normal form becomes

$$-(y_1)^2 - \cdots - (y_s)^2,$$

and $\phi(x_1, \ldots, x_n) \leq 0$ for every real $x_i (i = 1, \ldots, n)$. In this case, we say that ϕ is *negative definite*.

If $s = 0$, then the normal form becomes

$$(y_1)^2 + \cdots + (y_r)^2,$$

and $\phi(x_1, \ldots, x_n) \geq 0$ for every real $x_i (i = 1, \ldots, n)$. In this case, we say that ϕ is *positive definite*.

By ϕ is *definite* we mean that either ϕ is negative definite or ϕ is positive definite. If ϕ is not definite, then we say that ϕ is *indefinite*.

In the above example, the quadratic form is indefinite.

Definition Let

$$\left. \begin{array}{l} \phi \equiv \sum a_{ij} x_i x_j \\ \psi \equiv \sum b_{ij} x_i x_j \end{array} \right\}$$

be a pair of real quadratic forms. For a parameter $\lambda \in \mathbb{C}$, the quadratic form $\sum (a_{ij} - \lambda b_{ij}) x_i x_j$ is denoted by $\phi - \lambda \psi$. By the *discriminant of* ϕ we mean $\det[a_{ij}]$. Similarly, the discriminant of ψ is $\det[b_{ij}]$, and the discriminant of $\phi - \lambda \psi$ is

$$\det[a_{ij} - \lambda b_{ij}].$$

Clearly, $\det[a_{ij} - \lambda b_{ij}]$ is a polynomial in λ. The polynomial equation

$$\det[a_{ij} - \lambda b_{ij}] = 0$$

is called the λ-*equation of the pair of quadratic forms* $\sum a_{ij} x_i x_j$ and $\sum b x_i x_j$.

4.2.26 Theorem Let

$$\left. \begin{array}{l} \phi \equiv \sum a_{ij} x_i x_j \\ \psi \equiv \sum b_{ij} x_i x_j \end{array} \right\}$$

be a pair of real quadratic forms. Let $[b_{ij}]$ be invertible. Then all the roots of the λ-equation of ϕ and ψ are real.

Proof Let us denote $[a_{ij}]$ by A, and $[b_{ij}]$ by B. Now, $[a_{ij} - \lambda b_{ij}] = A - \lambda B$, and the λ-equation of ϕ and ψ becomes

$$\det(A - \lambda B) = 0.$$

Since $[b_{ij}]$ is invertible, B^{-1} exists, and $\det(B) \neq 0$. Now,

$$A - \lambda B = (AB^{-1} - \lambda I)B,$$

and hence

$$\det(A - \lambda B) = \det\big((AB^{-1} - \lambda I)B\big) = \det\big(AB^{-1} - \lambda I\big)\det(B).$$

Thus

$$\det(A - \lambda B) = \det\big(AB^{-1} - \lambda I\big)\det(B).$$

Since $\det(B) \neq 0$, every root of the λ-equation of ϕ and ψ is an eigenvalue of AB^{-1}. Since B is real and symmetric, B^{-1} is real and symmetric. Since A is real and symmetric, the product AB^{-1} is real and symmetric, and hence $(AB^{-1})^* = (AB^{-1})$. This shows that (AB^{-1}) is Hermitian, and hence by 3.3.26, all the eigenvalues of (AB^{-1}) are real. Now, since every root of the λ-equation of ϕ and ψ is an eigenvalue of AB^{-1}, every root of the λ-equation of ϕ and ψ is real. ∎

4.3 Application to Riemannian Geometry

4.3.1 Note Let $A \equiv [a_{ij}]$ be a symmetric n-square real matrix. Let

$$\phi(x_1, \ldots, x_n) \equiv \sum a_{ij}x_i x_j = [x_1, \ldots, x_n]A[x_1, \ldots, x_n]^T$$

be a real quadratic form. Suppose that $a_{11} \neq 0$.

Clearly, $\phi(x_1, \ldots, x_n) - \frac{1}{a_{11}}\left(\sum_{i=1}^{n} a_{1i}x_i\right)^2$ is a real quadratic form independent of x_1.

Proof Observe that

$$
\begin{aligned}
\phi(x_1, \ldots, x_n) &- \frac{1}{a_{11}}\left(\sum_{i=1}^{n} a_{1i}x_i\right)^2 \\
&= \phi(x_1, \ldots, x_n) - \frac{1}{a_{11}}\left(\sum_{i=1}^{n} a_{1i}x_i\right)\left(\sum_{j=1}^{n} a_{1j}x_j\right) \\
&= \phi(x_1, \ldots, x_n) - \frac{1}{a_{11}}\sum_{i=1}^{n}\left((a_{1i}x_i)\sum_{j=1}^{n} a_{1j}x_j\right) \\
&= \sum_{i=1}^{n}\left(\sum_{j=1}^{n} a_{ij}x_i x_j\right) - \frac{1}{a_{11}}\sum_{i=1}^{n}\left(\sum_{j=1}^{n} a_{1i}x_i \cdot a_{1j}x_j\right) \\
&= \sum_{i=1}^{n}\left(\sum_{j=1}^{n}\left(\frac{a_{11}a_{ij} - a_{1i}a_{1j}}{a_{11}}\right)x_i x_j\right),
\end{aligned}
$$

so

$$\phi(x_1, \ldots, x_n) - \frac{1}{a_{11}} \left(\sum_{i=1}^{n} a_{1i} x_i \right)^2 = \sum c_{ij} x_i x_j,$$

where $c_{ij} \equiv \frac{a_{11} a_{ij} - a_{1i} a_{1j}}{a_{11}}$. Here

$$c_{ji} = \frac{a_{11} a_{ji} - a_{1j} a_{1i}}{a_{11}} = \frac{a_{11} a_{ji} - a_{1i} a_{1j}}{a_{11}} = \frac{a_{11} a_{ij} - a_{1i} a_{1j}}{a_{11}} = c_{ij},$$

so $c_{ji} = c_{ij}$. Thus $[c_{ij}]$ is a symmetric n-square real matrix, and hence

$$\phi(x_1, \ldots, x_n) - \frac{1}{a_{11}} \left(\sum_{i=1}^{n} a_{1i} x_i \right)^2$$

is a real quadratic form. It suffices to show that $c_{1j} = 0$. Here

$$\text{LHS} = c_{1j} = \frac{a_{11} a_{1j} - a_{11} a_{1j}}{a_{11}} = 0 = \text{RHS}.$$

Thus we have shown that $\phi(x_1, \ldots, x_n) - \frac{1}{a_{11}} \left(\sum_{i=1}^{n} a_{1i} x_i \right)^2$ is a real quadratic form independent of x_1. We can denote it by $\phi_1(x_2, \ldots, x_n)$. Thus

$$\phi(x_1, \ldots, x_n) \equiv \frac{1}{a_{11}} \left(\sum_{i=1}^{n} a_{1i} x_i \right)^2 + \phi_1(x_2, \ldots, x_n).$$

Put

$$[y_1, \ldots, y_n]^T \equiv \begin{bmatrix} a_{11} & a_{12} & \cdots & a_{1n} \\ 0 & 1 & \cdots & 0 \\ \vdots & \vdots & \ddots & \vdots \\ 0 & 0 & \cdots & 1 \end{bmatrix} [x_1, \ldots, x_n]^T.$$

Since

$$\det \begin{bmatrix} a_{11} & a_{12} & \cdots & a_{1n} \\ 0 & 1 & \cdots & 0 \\ \vdots & \vdots & \ddots & \vdots \\ 0 & 0 & \cdots & 1 \end{bmatrix} = a_{11} \neq 0,$$

$$Q \equiv \begin{bmatrix} a_{11} & a_{12} & \cdots & a_{1n} \\ 0 & 1 & \cdots & 0 \\ \vdots & \vdots & \ddots & \vdots \\ 0 & 0 & \cdots & 1 \end{bmatrix} \text{ is invertible, and hence}$$

$$[x_1, \ldots, x_n]^T \mapsto Q[x_1, \ldots, x_n]^T$$

from $\mathbb{R}^3$ to $\mathbb{R}^3$ is a one-to-one linear transformation. Since

$$\phi(x_1, \ldots, x_n) \equiv \frac{1}{a_{11}} \left(\sum_{i=1}^{n} a_{1i} x_i \right)^2 + \phi_1(x_2, \ldots, x_n),$$

we have

$$\phi(x_1, \ldots, x_n) \equiv \frac{1}{a_{11}} (y_1)^2 + \phi_1(y_2, \ldots, y_n).$$

4.3.2 Conclusion Let $A \equiv [a_{ij}]$ be a symmetric n-square real matrix. Let

$$\phi(x_1, \ldots, x_n) \equiv \sum a_{ij} x_i x_j = [x_1, \ldots, x_n] A [x_1, \ldots, x_n]^T$$

be a real quadratic form. Suppose that $a_{11} \neq 0$. Then the one-to-one linear transformation

$$[y_1, \ldots, y_n]^T = \begin{bmatrix} a_{11} & a_{12} & \cdots & a_{1n} \\ 0 & 1 & \cdots & 0 \\ \vdots & \vdots & \ddots & \vdots \\ 0 & 0 & \cdots & 1 \end{bmatrix} [x_1, \ldots, x_n]^T$$

reduces $\phi(x_1, \ldots, x_n)$ to $\frac{1}{a_{11}} (y_1)^2 + \phi_1(y_2, \ldots, y_n)$, where $\phi_1(y_2, \ldots, y_n)$ is a quadratic form.

This result is known as **Lagrangian reduction**.

4.3.3 Note Let $A \equiv [a_{ij}]$ be a symmetric n-square real matrix. Let

$$\phi(x_1, \ldots, x_n) \equiv \sum a_{ij} x_i x_j = [x_1, \ldots, x_n] A [x_1, \ldots, x_n]^T$$

be a real quadratic form. Suppose that $a_{11} = 0$, $a_{22} = 0$, and $a_{12} \neq 0$.
    Observe that

$$\phi(x_1, \ldots, x_n) \equiv \sum a_{ij} x_i x_j$$

$$= \left(a_{11}(x_1)^2 + a_{22}(x_2)^2 + a_{12} x_1 x_2 + a_{21} x_2 x_1 \right)$$

$$+ (2a_{13} x_1 x_3 + 2a_{14} x_1 x_4 + \cdots + 2a_{1n} x_1 x_n)$$

$$+ (2a_{23} x_2 x_3 + 2a_{24} x_2 x_4 + \cdots + 2a_{2n} x_2 x_n) + \sum_{i=3}^{n} \left(\sum_{j=3}^{n} a_{ij} x_i x_j \right)$$

$$= 2a_{12} x_1 x_2 + 2x_1 (a_{13} x_3 + a_{14} x_4 + \cdots + a_{1n} x_n)$$

$$+ 2x_2 (a_{23} x_3 + a_{24} x_4 + \cdots + a_{2n} x_n) + \sum_{i=3}^{n} \left(\sum_{j=3}^{n} a_{ij} x_i x_j \right)$$

$$= 2\left(a_{12} x_1 x_2 + x_1 \sum_{i=3}^{n} a_{1i} x_i + x_2 \sum_{i=3}^{n} a_{2i} x_i \right) + \sum_{i=3}^{n} \left(\sum_{j=3}^{n} a_{ij} x_i x_j \right)$$

$$= \frac{2}{a_{12}} \left((a_{12} x_1)(a_{12} x_2) + (a_{12} x_1) \sum_{i=3}^{n} a_{1i} x_i \right.$$

$$\left. + (a_{12} x_2) \sum_{i=3}^{n} a_{2i} x_i \right) + \sum_{i=3}^{n} \left(\sum_{j=3}^{n} a_{ij} x_i x_j \right)$$

$$= \frac{2}{a_{12}} \left(\left(a_{12} x_2 + \sum_{i=3}^{n} a_{1i} x_i \right) \left(a_{12} x_1 + \sum_{i=3}^{n} a_{2i} x_i \right) \right.$$

$$\left. - \left(\sum_{i=3}^{n} a_{1i} x_i \right) \left(\sum_{i=3}^{n} a_{2i} x_i \right) \right) + \sum_{i=3}^{n} \left(\sum_{j=3}^{n} a_{ij} x_i x_j \right)$$

$$= \frac{2}{a_{12}} \left(\left(a_{12} x_2 + \sum_{i=3}^{n} a_{1i} x_i \right) \left(a_{21} x_1 + \sum_{i=3}^{n} a_{2i} x_i \right) \right.$$

$$\left. - \left(\sum_{i=3}^{n} a_{1i} x_i \right) \left(\sum_{i=3}^{n} a_{2i} x_i \right) \right) + \sum_{i=3}^{n} \left(\sum_{j=3}^{n} a_{ij} x_i x_j \right)$$

$$= \frac{2}{a_{12}} \left(a_{12} x_2 + \sum_{i=3}^{n} a_{1i} x_i \right) \left(a_{21} x_1 + \sum_{i=3}^{n} a_{2i} x_i \right)$$

$$+ \frac{-2}{a_{12}} \left(\sum_{i=3}^{n} a_{1i} x_i \right) \left(\sum_{j=3}^{n} a_{2j} x_j \right) + \sum_{i=3}^{n} \left(\sum_{j=3}^{n} a_{ij} x_i x_j \right)$$

$$= \frac{2}{a_{12}} \left(a_{12} x_2 + \sum_{i=3}^{n} a_{1i} x_i \right) \left(a_{21} x_1 + \sum_{i=3}^{n} a_{2i} x_i \right)$$

$$+ \frac{-2}{a_{12}} \left(\sum_{i=3}^{n} \left(a_{1i} x_i \sum_{j=3}^{n} a_{2j} x_j \right) \right) + \sum_{i=3}^{n} \left(\sum_{j=3}^{n} a_{ij} x_i x_j \right)$$

$$= \frac{2}{a_{12}} \left(a_{12}x_2 + \sum_{i=3}^{n} a_{1i}x_i \right) \left(a_{21}x_1 + \sum_{i=3}^{n} a_{2i}x_i \right)$$

$$+ \frac{-2}{a_{12}} \left(\sum_{i=3}^{n} \left(\sum_{j=3}^{n} (a_{1i}x_i \cdot a_{2j}x_j) \right) \right) + \sum_{i=3}^{n} \left(\sum_{j=3}^{n} a_{ij}x_ix_j \right)$$

$$= \frac{2}{a_{12}} \left(a_{12}x_2 + \sum_{i=3}^{n} a_{1i}x_i \right) \left(a_{21}x_1 + \sum_{i=3}^{n} a_{2i}x_i \right)$$

$$+ \sum_{i=3}^{n} \left(\sum_{j=3}^{n} \left(\frac{-2a_{1i}a_{2j}}{a_{12}} + a_{ij} \right) x_ix_j \right),$$

so

$$\phi(x_1, \ldots, x_n) \equiv \frac{2}{a_{12}} \left(a_{12}x_2 + \sum_{i=3}^{n} a_{1i}x_i \right) \left(a_{21}x_1 + \sum_{i=3}^{n} a_{2i}x_i \right) + \sum_{i=3}^{n} \left(\sum_{j=3}^{n} c_{ij}x_ix_j \right),$$

where $c_{ij} \equiv \frac{-2a_{1i}a_{2j}}{a_{12}} + a_{ij}$. Now, since

$$c_{ji} = \frac{-2a_{1j}a_{2i}}{a_{12}} + a_{ji} = \frac{-2a_{2i}a_{1j}}{a_{12}} + a_{ji} = \frac{-2a_{2i}a_{1j}}{a_{12}} + a_{ij} = c_{ij},$$

we have $c_{ji} = c_{ij}$, and hence $\sum_{i=3}^{n} \left(\sum_{j=3}^{n} c_{ij}x_ix_j \right)$ is a real quadratic form independent of x_1 and x_2. So we can denote the real quadratic form $\sum_{i=3}^{n} \left(\sum_{j=3}^{n} c_{ij}x_ix_j \right)$ by $\phi_1(x_3, \ldots, x_n)$. Thus

$$\phi(x_1, \ldots, x_n) \equiv \frac{2}{a_{12}} \left(a_{12}x_2 + \sum_{i=3}^{n} a_{1i}x_i \right) \left(a_{21}x_1 + \sum_{i=3}^{n} a_{2i}x_i \right) + \phi_1(x_3, \ldots, x_n).$$

Put

$$[y_1, \ldots, y_n]^T \equiv \begin{bmatrix} 0 & a_{12} & a_{13} & \cdots & a_{1n} \\ a_{21} & 0 & a_{23} & \cdots & a_{2n} \\ 0 & 0 & 1 & \cdots & 0 \\ \vdots & \vdots & \vdots & \ddots & \vdots \\ 0 & 0 & 0 & \cdots & 1 \end{bmatrix} [x_1, \ldots, x_n]^T.$$

Since

$$
\det
\begin{bmatrix}
0 & a_{12} & a_{13} & \cdots & a_{1n} \\
a_{21} & 0 & a_{23} & \cdots & a_{2n} \\
0 & 0 & 1 & \cdots & 0 \\
\vdots & \vdots & \vdots & \ddots & \vdots \\
0 & 0 & 0 & \cdots & 1
\end{bmatrix}
= -a_{21}
\begin{bmatrix}
a_{12} & a_{13} & \cdots & a_{1n} \\
0 & 1 & \cdots & 0 \\
\vdots & \vdots & \ddots & \vdots \\
0 & 0 & \cdots & 1
\end{bmatrix}
= -(a_{12})^2 \neq 0,
$$

$$
Q \equiv
\begin{bmatrix}
0 & a_{12} & a_{13} & \cdots & a_{1n} \\
a_{21} & 0 & a_{23} & \cdots & a_{2n} \\
0 & 0 & 1 & \cdots & 0 \\
\vdots & \vdots & \vdots & \ddots & \vdots \\
0 & 0 & 0 & \cdots & 1
\end{bmatrix}
$$

is invertible, and hence

$$
[x_1, \ldots, x_n]^T \mapsto Q[x_1, \ldots, x_n]^T
$$

from $\mathbb{R}^3$ to $\mathbb{R}^3$ is a one-to-one linear transformation. Since

$$
\phi(x_1, \ldots, x_n) \equiv \frac{2}{a_{12}} \left(a_{12}x_2 + \sum_{i=3}^{n} a_{1i}x_i \right) \left(a_{21}x_1 + \sum_{i=3}^{n} a_{2i}x_i \right) + \phi_1(x_3, \ldots, x_n),
$$

we have

$$
\phi(x_1, \ldots, x_n) \equiv \frac{2}{a_{12}} y_1 y_2 + \phi_1(y_3, \ldots, y_n).
$$

Put

$$
[z_1, \ldots, z_n]^T \equiv
\begin{bmatrix}
1 & 1 & 0 & \cdots & 0 \\
1 & -1 & 0 & \cdots & 0 \\
0 & 0 & 1 & \cdots & 0 \\
\vdots & \vdots & \vdots & \ddots & \vdots \\
0 & 0 & 0 & \cdots & 1
\end{bmatrix}
[y_1, \ldots, y_n]^T.
$$

Since

$$
\det
\begin{bmatrix}
1 & 1 & 0 & \cdots & 0 \\
1 & -1 & 0 & \cdots & 0 \\
0 & 0 & 1 & \cdots & 0 \\
\vdots & \vdots & \vdots & \ddots & \vdots \\
0 & 0 & 0 & \cdots & 1
\end{bmatrix}
= -2 \neq 0,
$$

$$R \equiv \begin{bmatrix} 1 & 1 & 0 & \cdots & 0 \\ 1 & -1 & 0 & \cdots & 0 \\ 0 & 0 & 1 & \cdots & 0 \\ \vdots & \vdots & \vdots & \ddots & \vdots \\ 0 & 0 & 0 & \cdots & 1 \end{bmatrix} \text{ is invertible, and hence}$$

$$[y_1, \ldots, y_n]^T \mapsto R[y_1, \ldots, y_n]^T$$

from $\mathbb{R}^3$ to $\mathbb{R}^3$ is a one-to-one linear transformation. Since

$$\phi(x_1, \ldots, x_n) \equiv \frac{2}{a_{12}} y_1 y_2 + \phi_1(y_3, \ldots, y_n),$$

we have

$$\begin{aligned} \phi(x_1, \ldots, x_n) &\equiv \frac{2}{a_{12}} \left(\frac{1}{2}(z_1 + z_2) \right) \left(\frac{1}{2}(z_1 - z_2) \right) + \phi_1(z_3, \ldots, z_n) \\ &\equiv \frac{1}{2a_{12}} (z_1)^2 + \frac{-1}{2a_{12}} (z_2)^2 + \phi_1(z_3, \ldots, z_n), \end{aligned}$$

and hence

$$\phi(x_1, \ldots, x_n) = \frac{1}{2a_{12}} (z_1)^2 + \frac{-1}{2a_{12}} (z_2)^2 + \phi_1(z_3, \ldots, z_n).$$

4.3.4 Conclusion (I) Let $A \equiv [a_{ij}]$ be a symmetric n-square real matrix. Let

$$\phi(x_1, \ldots, x_n) \equiv \sum a_{ij} x_i x_j = [x_1, \ldots, x_n] A [x_1, \ldots, x_n]^T$$

be a real quadratic form. Suppose that $a_{11} = 0$, $a_{22} = 0$, and $a_{12} \neq 0$. Then the one-to-one linear transformation

$$[z_1, \ldots, z_n]^T \equiv \left(\begin{bmatrix} 1 & 1 & 0 & \cdots & 0 \\ 1 & -1 & 0 & \cdots & 0 \\ 0 & 0 & 1 & \cdots & 0 \\ \vdots & \vdots & \vdots & \ddots & \vdots \\ 0 & 0 & 0 & \cdots & 1 \end{bmatrix} \begin{bmatrix} 0 & a_{12} & a_{13} & \cdots & a_{1n} \\ a_{21} & 0 & a_{23} & \cdots & a_{2n} \\ 0 & 0 & 1 & \cdots & 0 \\ \vdots & \vdots & \vdots & \ddots & \vdots \\ 0 & 0 & 0 & \cdots & 1 \end{bmatrix} \right) [x_1, \ldots, x_n]^T$$

reduces $\phi(x_1, \ldots, x_n)$ to $\frac{1}{2a_{12}}(z_1)^2 + \frac{-1}{2a_{12}}(z_2)^2 + \phi_1(z_3, \ldots, z_n)$, where $\phi_1(z_3, \ldots, z_n)$ is a quadratic form.

This result is also known as **Lagrangian reduction**.

By repeated application of Lagrangian reduction, we get the following result.

4.3.5 Conclusion (II) Let $A \equiv [a_{ij}]$ be a symmetric n-square real matrix. Let

$$\phi(x_1, \ldots, x_n) \equiv \sum a_{ij} x_i x_j = [x_1, \ldots, x_n] A [x_1, \ldots, x_n]^T$$

be a real quadratic form. Then there exists a one-to-one linear transformation

$$[y_1, \ldots, y_n]^T \equiv Q [x_1, \ldots, x_n]^T$$

such that $\phi(x_1, \ldots, x_n)$ reduces to a form $[y_1, \ldots, y_n] (\text{diag}(c_1, \ldots, c_n)) [x_1, \ldots, x_n]^T$.

4.3.6 Theorem Let $A \equiv [a_{ij}]$ be a symmetric n-square real matrix. Let

$$\phi(x_1, \ldots, x_n) \equiv \sum a_{ij} x_i x_j = [x_1, \ldots, x_n] A [x_1, \ldots, x_n]^T$$

be a real quadratic form. Let A be invertible. Let ϕ be a definite form. Then each $a_{ii} (i = 1, \ldots, n)$ is nonzero.

Proof Suppose to the contrary that there exists a diagonal entry of A that is 0. We seek a contradiction. For simplicity, suppose that $a_{11} = 0$.

Case I: ϕ is a positive definite form. Since A is a symmetric n-square real matrix and $\phi (= [x_1, \ldots, x_n] A [x_1, \ldots, x_n]^T)$ is a definite form, there exists, by 4.2.17, a real invertible matrix R such that

$$RAR^T = I_n.$$

It follows that

$$A = R^{-1} (R^T)^{-1} = R^{-1} (R^{-1})^T.$$

Since

$$\begin{aligned}
\phi(x_1, \ldots, x_n) &\equiv \sum a_{ij} x_i x_j = [x_1, \ldots, x_n] A [x_1, \ldots, x_n]^T \\
&= [x_1, \ldots, x_n] R^{-1} (R^{-1})^T [x_1, \ldots, x_n]^T \\
&= ([x_1, \ldots, x_n] R^{-1}) ([x_1, \ldots, x_n] R^{-1})^T \\
&= [y_1, \ldots, y_n] [y_1, \ldots, y_n]^T,
\end{aligned}$$

where

$$[y_1, \ldots, y_n] \equiv [x_1, \ldots, x_n] R^{-1}, .$$

it follows that

$$\phi(x_1,\ldots,x_n) = [y_1,\ldots,y_n][y_1,\ldots,y_n]^T = (y_1)^2 + \cdots + (y_n)^2.$$

Since R is invertible, we have that

$$[x_1,\ldots,x_n] \mapsto [x_1,\ldots,x_n]R^{-1}$$

from $\mathbb{R}^3$ to $\mathbb{R}^3$ is a one-to-one linear transformation, and hence

$$[1,0,\ldots,0]R^{-1} \neq [0,0,\ldots,0].$$

It follows that $([1,0,\ldots,0]R^{-1})([1,0,\ldots,0]R^{-1})^T > 0$. Since $\phi(x_1,\ldots,x_n) \equiv \sum a_{ij}x_ix_j$, we have

$$0 < ([1,0,\ldots,0]R^{-1})([1,0,\ldots,0]R^{-1})^T = \underbrace{\phi(1,0,\ldots,0) = a_{11} \cdot 1 \cdot 1 + 0 + \cdots + 0}$$

$$= a_{11} = 0.$$

Thus we have obtained a contradiction.

Case II: ϕ is a negative definite form. This case is similar to Case I.   ∎

4.3.7 Theorem Let $B \equiv [b_{ij}]$ be a symmetric n-square real matrix. Let B be invertible. Let

$$\psi(x_1,\ldots,x_n) \equiv \sum b_{ij}x_ix_j = [x_1,\ldots,x_n]B[x_1,\ldots,x_n]^T$$

be a real quadratic form. Let ψ be positive definite. Let P be a real orthogonal n-square real matrix. Then the $(1,1)$-entry in PBP^T is nonzero. Similarly, the $(2,2)$-entry in PBP^T is nonzero, etc.

Proof Suppose to the contrary that the $(1,1)$-entry in PBP^T is 0. We seek a contradiction.

Since $B \equiv [b_{ij}]$ is a symmetric n-square real matrix and

$$\psi(x_1,\ldots,x_n) \equiv \sum b_{ij}x_ix_j = [x_1,\ldots,x_n]B[x_1,\ldots,x_n]^T$$

is a positive definite form, by 4.2.17, there exists a real invertible matrix C such that $CBC^T = I_n$. It follows that $B = C^{-1}(C^T)^{-1}\left(= C^{-1}(C^{-1})^T\right)$. Now,

$$PBP^T = P\left(C^{-1}(C^{-1})^T\right)P^T = (PC^{-1})(PC^{-1})^T,$$

so $PBP^T = (PC^{-1})(PC^{-1})^T$. Since P and C are invertible, PC^{-1} is invertible.

Suppose that $PC^{-1} \equiv \begin{bmatrix} c_{11} & \cdots & c_{1n} \\ \vdots & \ddots & \vdots \\ c_{n1} & \cdots & c_{nn} \end{bmatrix}$, where each c_{ij} is a real number. Now,

$$PBP^T = \begin{bmatrix} c_{11} & \cdots & c_{1n} \\ \vdots & \ddots & \vdots \\ c_{n1} & \cdots & c_{nn} \end{bmatrix} \begin{bmatrix} c_{11} & \cdots & c_{1n} \\ \vdots & \ddots & \vdots \\ c_{n1} & \cdots & c_{nn} \end{bmatrix}^T$$

$$= \begin{bmatrix} c_{11} & \cdots & c_{1n} \\ \vdots & \ddots & \vdots \\ c_{n1} & \cdots & c_{nn} \end{bmatrix} \begin{bmatrix} c_{11} & \cdots & c_{n1} \\ \vdots & \ddots & \vdots \\ c_{1n} & \cdots & c_{nn} \end{bmatrix}.$$

Since the $(1,1)$-entry in

$$\begin{bmatrix} c_{11} & \cdots & c_{1n} \\ \vdots & \ddots & \vdots \\ c_{n1} & \cdots & c_{nn} \end{bmatrix} \begin{bmatrix} c_{11} & \cdots & c_{n1} \\ \vdots & \ddots & \vdots \\ c_{1n} & \cdots & c_{nn} \end{bmatrix} \left(= PBP^T \right)$$

is $(c_{11})^2 + \cdots + (c_{1n})^2$, the $(1,1)$-entry. in PBP^T is $(c_{11})^2 + \cdots + (c_{1n})^2$. By assumption, the $(1,1)$-entry. in PBP^T is 0, so $(c_{11})^2 + \cdots + (c_{1n})^2 = 0$. Since each c_{ij} is a real number, we have $c_{1i} = 0 (i = 1, \ldots, n)$. It follows that $\det(PC^{-1}) = 0$, and hence PC^{-1} is not invertible. This is a contradiction. ■

4.3.8 Note Let $A \equiv \begin{bmatrix} a_{ij} \end{bmatrix}$ be a symmetric n-square real matrix. Let $B \equiv \begin{bmatrix} b_{ij} \end{bmatrix}$ be a symmetric n-square real matrix. Let B be invertible. Let

$$\phi(x_1, \ldots, x_n) \equiv \sum a_{ij} x_i x_j = [x_1, \ldots, x_n] A [x_1, \ldots, x_n]^T$$

be a real quadratic form. Let

$$\psi(x_1, \ldots, x_n) \equiv \sum b_{ij} x_i x_j = [x_1, \ldots, x_n] B [x_1, \ldots, x_n]^T$$

be a real quadratic form. Let ψ be a positive definite form. Let λ_1 be a root of the λ-equation of ϕ and ψ, that is, $\det[A - \lambda_1 B] = 0$.

   Since $\det[A - \lambda_1 B] = 0$, the characteristic equation $\det[(A - \lambda_1 B) - \lambda I_n] = 0$ of $(A - \lambda_1 B)$ is satisfied by $\lambda = 0$, and hence 0 is an eigenvalue of $(A - \lambda_1 B)$. Since A, B are symmetric n-square real matrices, $(A - \lambda_1 B)$ is also a symmetric n-square real matrix. It follows, by 4.2.12, that there exists a real orthogonal matrix P such that

$$P(A - \lambda_1 B)P^T = \mathrm{diag}(\mu_1, \ldots, \mu_n),$$

where $\mu_1, \ldots, \mu_n$ are the eigenvalues of $(A - \lambda_1 B)$. Since 0 is an eigenvalue of $(A - \lambda_1 B)$, one of the μ_is is 0. For simplicity, suppose that $\mu_1 = 0$. Thus

$$(A - \lambda_1 B) = P^T(\operatorname{diag}(0, \mu_2, \ldots, \mu_n))P.$$

It follows that

$$[x_1, \ldots, x_n](A - \lambda_1 B)[x_1, \ldots, x_n]^T \equiv [x_1, \ldots, x_n](P^T(\operatorname{diag}(0, \mu_2, \ldots, \mu_n))P)[x_1, \ldots, x_n]^T$$

$$\equiv ([x_1, \ldots, x_n]P^T)(\operatorname{diag}(0, \mu_2, \ldots, \mu_n))([x_1, \ldots, x_n]P^T)^T$$

$$\equiv [y_1, \ldots, y_n](\operatorname{diag}(0, \mu_2, \ldots, \mu_n))[y_1, \ldots, y_n]^T,$$

where $[y_1, \ldots, y_n] \equiv [x_1, \ldots, x_n]P^T$. Since P is invertible,

$$[x_1, \ldots, x_n]^T \mapsto P[x_1, \ldots, x_n]^T$$

from $\mathbb{R}^3$ to $\mathbb{R}^3$ is a one-to-one linear transformation. Clearly, $[y_1, \ldots, y_n](\operatorname{diag}(0, \mu_2, \ldots, \mu_n))[y_1, \ldots, y_n]^T$ is a real quadratic form independent of y_1, so we can denote $[y_1, \ldots, y_n](\operatorname{diag}(0, \mu_2, \ldots, \mu_n))[y_1, \ldots, y_n]^T$ by $\phi_1(y_2, \ldots, y_n)$. Thus

$$[x_1, \ldots, x_n](A - \lambda_1 B)[x_1, \ldots, x_n]^T \equiv \phi_1(y_2, \ldots, y_n).$$

Hence in the reduced form $\phi_1(y_2, \ldots, y_n)$ of $\phi(x_1, \ldots, x_n) - \lambda_1 \psi(x_1, \ldots, x_n)$, the coefficient of $(y_1)^2$ is zero.

By 4.3.7, the $(1, 1)$-entry in PBP^T is nonzero. It follows that the coefficient of $(y_1)^2$ in the real quadratic form

$$(\psi(x_1, \ldots, x_n) = [x_1, \ldots, x_n]B[x_1, \ldots, x_n]^T$$
$$\equiv ([y_1, \ldots, y_n]P)B([y_1, \ldots, y_n]P)^T) \equiv [y_1, \ldots, y_n](PBP^T)[y_1, \ldots, y_n]^T$$

is nonzero. Thus in the reduced form, say $\psi_1(y_1, \ldots, y_n)$, of $\psi(x_1, \ldots, x_n)$, the coefficient of $(y_1)^2$ is nonzero.

Now we can suppose that

$$\psi_1(y_1, \ldots, y_n) = [y_1, \ldots, y_n][d_{ij}][y_1, \ldots, y_n]^T,$$

where $d_{11} \neq 0$. Next, by 4.3.2 the one-to-one linear transformation

$$[z_1, \ldots, z_n]^T = \begin{bmatrix} d_{11} & d_{12} & \cdots & d_{1n} \\ 0 & 1 & \cdots & 0 \\ \vdots & \vdots & \ddots & \vdots \\ 0 & 0 & \cdots & 1 \end{bmatrix} [y_1, \ldots, y_n]^T$$

reduces $\psi_1(y_1, \ldots, y_n)$ to $\frac{1}{d_{11}}(z_1)^2 + \phi_2(z_2, \ldots, z_n)$, where $\phi_2(z_2, \ldots, z_n)$ is a quadratic form. Here we can write

$$\left. \begin{array}{c} z_1 = d_{11}y_1 + d_{12}y_2 + \cdots + d_{1n}y_n \\ z_2 = y_2 \\ z_3 = y_3 \\ \vdots \\ z_n = y_n \end{array} \right\},$$

or

$$\left. \begin{array}{c} y_1 = \frac{1}{d_{11}}z_1 + \frac{-d_{12}}{d_{11}}z_2 + \cdots + \frac{-d_{1n}}{d_{11}}z_n \\ y_2 = z_2 \\ y_3 = z_3 \\ \vdots \\ y_n = z_n \end{array} \right\}.$$

Hence

$$\phi(x_1, \ldots, x_n) - \lambda_1 \psi(x_1, \ldots, x_n) \equiv [x_1, \ldots, x_n](A - \lambda_1 B)[x_1, \ldots, x_n]^T$$
$$\equiv \underbrace{\phi_1(y_2, \ldots, y_n) \equiv \phi_1(z_2, \ldots, z_n)}.$$

Thus

$$\phi(x_1, \ldots, x_n) - \lambda_1 \psi(x_1, \ldots, x_n) \equiv \phi_1(z_2, \ldots, z_n).$$

Since $\psi(x_1, \ldots, x_n)$ reduces to $\psi_1(y_1, \ldots, y_n)$, and $\psi_1(y_1, \ldots, y_n)$ reduces to $\frac{1}{d_{11}}(z_1)^2 + \phi_2(z_2, \ldots, z_n)$, it follows that $\psi(x_1, \ldots, x_n)$ reduces to $\frac{1}{d_{11}}(z_1)^2 + \phi_2(z_2, \ldots, z_n)$. Thus

$$\psi(x_1, \ldots, x_n) \equiv \frac{1}{d_{11}}(z_1)^2 + \phi_2(z_2, \ldots, z_n).$$

It follows that

$$\phi(x_1, \ldots, x_n) \equiv \lambda_1 \left(\frac{1}{d_{11}} (z_1)^2 + \phi_2(z_2, \ldots, z_n) \right) + \phi_1(z_2, \ldots, z_n)$$
$$\equiv \lambda_1 \frac{1}{d_{11}} (z_1)^2 + \phi_3(z_2, \ldots, z_n),$$

where $\phi_3(z_2, \ldots, z_n) \equiv \lambda_1 \phi_2(z_2, \ldots, z_n) + \phi_1(z_2, \ldots, z_n)$.

4.3.9 Conclusion Let $A \equiv [a_{ij}]$, and $B \equiv [b_{ij}]$ be symmetric n-square real matrices. Let B be invertible. Let

$$\left. \begin{array}{l} \phi(x_1, \ldots, x_n) \equiv \sum a_{ij} x_i x_j = [x_1, \ldots, x_n] A [x_1, \ldots, x_n]^T \\ \psi(x_1, \ldots, x_n) \equiv \sum b_{ij} x_i x_j = [x_1, \ldots, x_n] B [x_1, \ldots, x_n]^T \end{array} \right\}$$

be a pair of real quadratic forms. Let ψ be positive definite. Let λ_1 be a root of the λ-equation of ϕ and ψ. Then there exists a one-to-one linear transformation $[x_1, \ldots, x_n] \mapsto [z_1, \ldots, z_n]$ such that the pair's reduced forms are

$$\left. \begin{array}{l} \phi(x_1, \ldots, x_n) \equiv \lambda_1 c_1 (z_1)^2 + \phi_1(z_2, \ldots, z_n) \\ \psi(x_1, \ldots, x_n) \equiv c_1 (z_1)^2 + \psi_1(z_2, \ldots, z_n) \end{array} \right\},$$

where c_1 is a nonzero real number.

Definition Let $B \equiv [b_{ij}]$ be a symmetric n-square real matrix. Let

$$\psi(x_1, \ldots, x_n) \equiv \sum_{i=1}^{n} \left(\sum_{j=1}^{n} b_{ij} x_i x_j \right) \equiv \sum_{i=1}^{n} \left(x_i \left(\sum_{j=1}^{n} b_{ij} x_j \right) \right)$$
$$\equiv [x_1, \ldots, x_n] \left([b_{ij}] [x_1, \ldots, x_n]^T \right)$$

be a real quadratic form. Suppose that $(c_1, \ldots, c_n) \neq (0, \ldots, 0)$, where each c_i is real. If $[b_{ij}] [c_1, \ldots, c_n]^T = [0, \ldots, 0]^T$, then we say that $(c_1, \ldots, c_n)$ *is a vertex of* $\psi(x_1, \ldots, x_n)$.

4.3.10 Let $B \equiv [b_{ij}]$ be a symmetric n-square real matrix. Let B be invertible. Let

$$\psi(x_1, \ldots, x_n) \equiv \sum_{i=1}^{n} \left(\sum_{j=1}^{n} b_{ij} x_i x_j \right) \equiv \sum_{i=1}^{n} \left(x_i \left(\sum_{j=1}^{n} b_{ij} x_j \right) \right)$$
$$\equiv [x_1, \ldots, x_n] \left([b_{ij}] [x_1, \ldots, x_n]^T \right)$$

be a real quadratic form. Let $\psi(x_1, \ldots, x_n)$ be an indefinite form.

Since $\psi(x_1, \ldots, x_n)$ is an indefinite form, $\psi(x_1, \ldots, x_n)$ is neither positive definite nor negative definite. If follows, by 4.2.25, that there exists a real invertible matrix $C \equiv [c_{ij}]$ such that the one-to-one transformation

$$[x_1, \ldots, x_n]^T = C[y_1, \ldots, y_n]^T$$

reduces the form $\psi(x_1, \ldots, x_n)$ to the normal form

$$(y_1)^2 + \cdots + (y_r)^2 - (y_{r+1})^2 - \cdots - (y_n)^2,$$

where $1 \leq r < n$. Put

$$[a_1, \ldots, a_n]^T = C[1, 0, \ldots, 0]^T.$$

Now, since C is invertible, $(a_1, \ldots, a_n) \neq (0, \ldots, 0)$. Also,

$$\psi(a_1, \ldots, a_n) = 1^2 + 0^2 + \cdots + 0^2 - 0^2 - \cdots - 0^2 = 1 > 0,$$

so $\psi(a_1, \ldots, a_n)$ is positive. Similarly, there exist nonzero $(b_1, \ldots, b_n)$ such that $\psi(b_1, \ldots, b_n)$ is negative.

4.3.11 Conclusion Let $B \equiv [b_{ij}]$ be a symmetric n-square real matrix. Let B be invertible. Let

$$\psi(x_1, \ldots, x_n) \equiv \sum_{i=1}^{n} \left(\sum_{j=1}^{n} b_{ij} x_i x_j \right) \equiv \sum_{i=1}^{n} \left(x_i \left(\sum_{j=1}^{n} b_{ij} x_j \right) \right)$$
$$\equiv [x_1, \ldots, x_n] \left([b_{ij}] [x_1, \ldots, x_n]^T \right)$$

be a real quadratic form. Let $\psi(x_1, \ldots, x_n)$ be an indefinite form. Then there exist nonzero real points $(a_1, \ldots, a_n)$ and $(b_1, \ldots, b_n)$ such that $\psi(a_1, \ldots, a_n)$ is positive and $\psi(b_1, \ldots, b_n)$ is negative.

4.3.12 Let $B \equiv [b_{ij}]$ be a symmetric n-square real matrix. Let B be invertible. Let

$$\psi(x_1, \ldots, x_n) \equiv \sum_{i=1}^{n} \left(\sum_{j=1}^{n} b_{ij} x_i x_j \right) \equiv \sum_{i=1}^{n} \left(x_i \left(\sum_{j=1}^{n} b_{ij} x_j \right) \right)$$
$$\equiv [x_1, \ldots, x_n] \left([b_{ij}] [x_1, \ldots, x_n]^T \right)$$

be a real quadratic form. Let $\psi(x_1, \ldots, x_n)$ be an indefinite form. Let $(a_1, \ldots, a_n)$ and $(b_1, \ldots, b_n)$ be nonzero real points such that $\psi(a_1, \ldots, a_n)$ is positive and $\psi(b_1, \ldots, b_n)$ is negative.

Observe that

$$\psi(a_1 + \lambda b_1, \ldots, a_n + \lambda b_n) = [a_1 + \lambda b_1, \ldots, a_n + \lambda b_n][b_{ij}][a_1 + \lambda b_1, \ldots, a_n + \lambda b_n]^T$$
$$= ([a_1, \ldots, a_n] + \lambda[b_1, \ldots, b_n])[b_{ij}]([a_1, \ldots, a_n] + \lambda[b_1, \ldots, b_n])^T$$
$$= [a_1, \ldots, a_n][b_{ij}][a_1, \ldots, a_n]^T + ([a_1, \ldots, a_n][b_{ij}][b_1, \ldots, b_n]^T$$
$$+ [b_1, \ldots, b_n][b_{ij}][a_1, \ldots, a_n])\lambda + [b_1, \ldots, b_n][b_{ij}][b_1, \ldots, b_n]^T \lambda^2$$

$$= (\psi(b_1, \ldots, b_n))\lambda^2 + \left(\sum_{i=1}^{n} \left(\sum_{j=1}^{n} b_{ij} a_i b_j \right) + \sum_{i=1}^{n} \left(\sum_{j=1}^{n} b_{ij} b_i a_j \right) \right) \lambda + \psi(a_1, \ldots, a_n)$$

$$= (\psi(b_1, \ldots, b_n))\lambda^2 + \left(\sum_{i=1}^{n} \left(\sum_{j=1}^{n} b_{ij} a_i b_j \right) + \sum_{i=1}^{n} \left(\sum_{j=1}^{n} b_{ji} b_i a_j \right) \right) \lambda + \psi(a_1, \ldots, a_n)$$

$$= (\psi(b_1, \ldots, b_n))\lambda^2 + \left(\sum_{i=1}^{n} \left(\sum_{j=1}^{n} b_{ij} a_i b_j \right) + \sum_{i=1}^{n} \left(\sum_{j=1}^{n} b_{ji} a_j b_i \right) \right) \lambda + \psi(a_1, \ldots, a_n)$$

$$= (\psi(b_1, \ldots, b_n))\lambda^2 + \left(\sum_{i=1}^{n} \left(\sum_{j=1}^{n} b_{ij} a_i b_j \right) + \sum_{j=1}^{n} \left(\sum_{i=1}^{n} b_{ij} a_i b_j \right) \right) \lambda + \psi(a_1, \ldots, a_n)$$

$$= (\psi(b_1, \ldots, b_n))\lambda^2 + \left(\sum_{i=1}^{n} \left(\sum_{j=1}^{n} b_{ij} a_i b_j \right) + \sum_{i=1}^{n} \left(\sum_{j=1}^{n} b_{ij} a_i b_j \right) \right) \lambda + \psi(a_1, \ldots, a_n)$$

$$= (\psi(b_1, \ldots, b_n))\lambda^2 + 2 \sum_{i=1}^{n} \left(\sum_{j=1}^{n} b_{ij} a_i b_j \right) \lambda + \psi(a_1, \ldots, a_n),$$

so

$$\psi(a_1 + \lambda b_1, \ldots, a_n + \lambda b_n)$$
$$\equiv (\psi(b_1, \ldots, b_n))\lambda^2 + 2 \sum_{i=1}^{n} \left(\sum_{j=1}^{n} b_{ij} a_i b_j \right) \lambda + \psi(a_1, \ldots, a_n).$$

Since $\psi(a_1, \ldots, a_n)$ is positive and $\psi(b_1, \ldots, b_n)$ is negative, the discriminant of

$$(\psi(a_1 + \lambda b_1, \ldots, a_n + \lambda b_n) =)(\psi(b_1, \ldots, b_n))\lambda^2$$
$$+ 2 \sum_{i=1}^{n} \left(\sum_{j=1}^{n} b_{ij} a_i b_j \right) \lambda + \psi(a_1, \ldots, a_n)$$

is positive, and hence there exist two distinct real numbers λ_1 and λ_2 such that

$$\left. \begin{array}{l} \psi(a_1 + \lambda_1 b_1, \ldots, a_n + \lambda_1 b_n) = 0 \\ \psi(a_1 + \lambda_2 b_1, \ldots, a_n + \lambda_2 b_n) = 0 \end{array} \right\}.$$

Clearly $\lambda_1 \lambda_2$ is negative, and hence λ_1, λ_2 are of opposite signs.

Also, $\psi(x_1, \ldots, x_n)$ vanishes at the two real points $(a_1 + \lambda_1 b_1, \ldots, a_n + \lambda_1 b_n)$ $(= (a_1, \ldots, a_n) + \lambda_1 (b_1, \ldots, b_n))$ and $(a_1 + \lambda_2 b_1, \ldots, a_n + \lambda_2 b_n)(= (a_1, \ldots, a_n)$ $+ \lambda_2 (b_1, \ldots, b_n))$. Since λ_1 and λ_2 are distinct real numbers and $(b_1, \ldots, b_n)$ are nonzero, $(a_1, \ldots, a_n) + \lambda_1(b_1, \ldots, b_n)$ and $(a_1, \ldots, a_n) + \lambda_2(b_1, \ldots, b_n)$ are distinct points.

Clearly, $[b_{ij}][a_1 + \lambda_1 b_1, \ldots, a_n + \lambda_1 b_n]^T \neq [0, \ldots, 0]^T$.

Proof Suppose to the contrary that $[b_{ij}][a_1 + \lambda_1 b_1, \ldots, a_n + \lambda_1 b_n]^T = [0, \ldots, 0]^T$. We seek a contradiction.

Since $[b_{ij}][a_1 + \lambda_1 b_1, \ldots, a_n + \lambda_1 b_n]^T = [0, \ldots, 0]^T$, and $[b_{ij}]$ is invertible, we have

$[a_1 + \lambda_1 b_1, \ldots, a_n + \lambda_1 b_n] = [0, \ldots, 0]$, and hence $(a_1, \ldots, a_n) + \lambda_1$ $(b_1, \ldots, b_n) = (0, \ldots, 0)$. It follows that $(a_1, \ldots, a_n) = (-\lambda_1 b_1, \ldots, -\lambda_1 b_n)$. Since $\psi(a_1, \ldots, a_n)$ is positive,

$$\psi(-\lambda_1 b_1, \ldots, -\lambda_1 b_n)$$
$$\left(= [-\lambda_1 b_1, \ldots, -\lambda_1 b_n]([b_{ij}][-\lambda_1 b_1, \ldots, -\lambda_1 b_n]^T) = (\lambda_1)^2 \psi(b_1, \ldots, b_n) \right)$$

is positive, and hence $(\lambda_1)^2 \psi(b_1, \ldots, b_n)$ is positive. Since λ_1 is real and $\psi(b_1, \ldots, b_n)$ is negative,

$(\lambda_1)^2 \psi(b_1, \ldots, b_n) \leq 0$. This is a contradiction. ∎

Thus we have shown that $[b_{ij}][a_1 + \lambda_1 b_1, \ldots, a_n + \lambda_1 b_n]^T \neq [0, \ldots, 0]^T$. Hence $(a_1, \ldots, a_n) + \lambda_1(b_1, \ldots, b_n)$ is different from the origin and is not a vertex of $\psi(x_1, \ldots, x_n)$. Similarly, $(a_1, \ldots, a_n) + \lambda_2(b_1, \ldots, b_n)$ is different from the origin and is not a vertex of $\psi(x_1, \ldots, x_n)$.

4.3.13 Conclusion Let $B \equiv [b_{ij}]$ be a symmetric n-square real matrix. Let B be invertible. Let

$$\psi(x_1, \ldots, x_n) \equiv \sum_{i=1}^{n}\left(\sum_{j=1}^{n} b_{ij} x_i x_j\right) \equiv \sum_{i=1}^{n}\left(x_i\left(\sum_{j=1}^{n} b_{ij} x_j\right)\right)$$
$$\equiv [x_1, \ldots, x_n]([b_{ij}][x_1, \ldots, x_n]^T)$$

be a real quadratic form. Let $\psi(x_1, \ldots, x_n)$ be an indefinite form. Let $(a_1, \ldots, a_n)$ and $(b_1, \ldots, b_n)$ be nonzero real points such that $\psi(a_1, \ldots, a_n)$ is positive and $\psi(b_1, \ldots, b_n)$ is negative. Then there exist two distinct real numbers λ_1 and λ_2 such that

1. λ_1, λ_2 are of opposite signs,
2. $\psi(a_1 + \lambda_1 b_1, \ldots, a_n + \lambda_1 b_n) = 0$,
3. $\psi(a_1 + \lambda_2 b_1, \ldots, a_n + \lambda_2 b_n) = 0$,
4. $(a_1 + \lambda_1 b_1, \ldots, a_n + \lambda_1 b_n)$ and $(a_1 + \lambda_2 b_1, \ldots, a_n + \lambda_2 b_n)$ are points different from origin and the vertices of $\psi(x_1, \ldots, x_n)$.

4.3.14 Theorem Let $B \equiv [b_{ij}]$ be a symmetric n-square real matrix. Let B be invertible. Let

$$\psi(x_1, \ldots, x_n) \equiv \sum_{i=1}^{n} \left(\sum_{j=1}^{n} b_{ij} x_i x_j \right) = \sum_{i=1}^{n} \left(x_i \left(\sum_{j=1}^{n} b_{ij} x_j \right) \right)$$
$$\equiv [x_1, \ldots, x_n] \left([b_{ij}] [x_1, \ldots, x_n]^T \right)$$

be a real quadratic form. Suppose that for every real point $(c_1, \ldots, c_n)$ that is different from the origin and the vertices of $\psi(x_1, \ldots, x_n)$, $\psi(c_1, \ldots, c_n) \neq 0$. Then $\psi(x_1, \ldots, x_n)$ is definite.

Proof Suppose to the contrary that $\psi(x_1, \ldots, x_n)$ is indefinite. We seek a contradiction.

By 4.3.11, there exist nonzero real points $(a_1, \ldots, a_n)$ and $(b_1, \ldots, b_n)$ such that $\psi(a_1, \ldots, a_n)$ is positive and $\psi(b_1, \ldots, b_n)$ is negative. By 4.3.13, there exist two distinct real numbers λ_1 and λ_2 such that

1. λ_1, λ_2 are of opposite signs,
2. $\psi(a_1 + \lambda_1 b_1, \ldots, a_n + \lambda_1 b_n) = 0$,
3. $\psi(a_1 + \lambda_2 b_1, \ldots, a_n + \lambda_2 b_n) = 0$,
4. $(a_1 + \lambda_1 b_1, \ldots, a_n + \lambda_1 b_n)$ and $(a_1 + \lambda_2 b_1, \ldots, a_n + \lambda_2 b_n)$ are points different from origin and the vertices of $\psi(x_1, \ldots, x_n)$.

Since $(a_1 + \lambda_1 b_1, \ldots, a_n + \lambda_1 b_n)$ is a point different from origin and the vertices of $\psi(x_1, \ldots, x_n)$, by assumption, $\psi(a_1 + \lambda_1 b_1, \ldots, a_n + \lambda_1 b_n) \neq 0$. This is a contradiction. ∎

4.3.15 Theorem Let $B \equiv [b_{ij}]$ be a symmetric n-square real matrix. Let B be not invertible. Let

$$\psi(x_1, \ldots, x_n) \equiv \sum_{i=1}^{n} \left(\sum_{j=1}^{n} b_{ij} x_i x_j \right) = \sum_{i=1}^{n} \left(x_i \left(\sum_{j=1}^{n} b_{ij} x_j \right) \right)$$
$$\equiv [x_1, \ldots, x_n] \left([b_{ij}] [x_1, \ldots, x_n]^T \right)$$

be a real quadratic form. Then there exists a real point $(a_1, \ldots, a_n)$ different from the origin such that $\psi(a_1, \ldots, a_n) = 0$.

Proof Since B is not invertible, $\text{rank}(B) < n$. By 4.2.17, there exists a real invertible matrix R such that

$$RBR^T = \text{diag}\left(\underbrace{1, \ldots 1, -1, \ldots, -1}_{<n}, 0, \ldots, 0 \right).$$

It follows that

$$B = R^{-1} \left(\text{diag} \left(\underbrace{1, \ldots 1, -1, \ldots, -1}_{<n}, 0, \ldots, 0 \right) \right) (R^T)^{-1}$$

$$= R^{-1} \left(\text{diag} \left(\underbrace{1, \ldots 1, -1, \ldots, -1}_{<n}, 0, \ldots, 0 \right) \right) (R^{-1})^T,$$

and hence

$$\psi(x_1, \ldots, x_n)$$
$$\equiv [x_1, \ldots, x_n] B [x_1, \ldots, x_n]^T$$
$$\equiv [x_1, \ldots, x_n] \left(R^{-1} \left(\text{diag} \left(\underbrace{1, \ldots 1, -1, \ldots, -1}_{<n}, 0, \ldots, 0 \right) \right) (R^{-1})^T \right) \times [x_1, \ldots, x_n]^T$$
$$\equiv [y_1, \ldots, y_n] \text{diag} \left(\underbrace{1, \ldots 1, -1, \ldots, -1}_{<n}, 0, \ldots, 0 \right) [y_1, \ldots, y_n]^T,$$

where $[y_1, \ldots, y_n] \equiv [x_1, \ldots, x_n] R^{-1}$. Since R is invertible, $[0, \ldots, 0, 1]R$ is nonzero. Put $[a_1, \ldots, a_n] \equiv [0, \ldots, 0, 1]R$. Thus $(a_1, \ldots, a_n) \neq (0, \ldots, 0)$. Also,

$$\psi(a_1, \ldots, a_n)$$
$$= [a_1, \ldots, a_n] \left(R^{-1} \left(\text{diag} \left(\underbrace{1, \ldots 1, -1, \ldots, -1}_{<n}, 0, \ldots, 0 \right) \right) (R^{-1})^T \right) [a_1, \ldots, a_n]^T$$
$$= [0, \ldots, 0, 1] \left(\text{diag} \left(\underbrace{1, \ldots 1, -1, \ldots, -1}_{<n}, 0, \ldots, 0 \right) \right) [0, \ldots, 0, 1]^T = 0,$$

so $\psi(a_1, \ldots, a_n) = 0$.   ∎

4.3.16 Theorem Let $B \equiv \left[b_{ij} \right]$ be a symmetric n-square real matrix. Let B be invertible. Let

$$\psi(x_1, \ldots, x_n) \equiv \sum_{i=1}^n \left(\sum_{j=1}^n b_{ij} x_i x_j \right) \equiv \sum_{i=1}^n \left(x_i \left(\sum_{j=1}^n b_{ij} x_j \right) \right)$$
$$\equiv [x_1, \ldots, x_n] \left(\left[b_{ij} \right] [x_1, \ldots, x_n]^T \right)$$

be a real quadratic form. Let $\psi(x_1, \ldots, x_n)$ be an indefinite form. Then there exists a real point $(c_1, \ldots, c_n)$ different from the origin such that $\psi(c_1, \ldots, c_n) = 0$.

Proof By 4.3.11, there exist nonzero real points $(a_1, \ldots, a_n)$ and $(b_1, \ldots, b_n)$ such that $\psi(a_1, \ldots, a_n)$ is positive and $\psi(b_1, \ldots, b_n)$ is negative. By 4.3.13, there exist two distinct real numbers λ_1 and λ_2 such that

1. λ_1, λ_2 are of opposite signs,
2. $\psi(a_1 + \lambda_1 b_1, \ldots, a_n + \lambda_1 b_n) = 0$,
3. $\psi(a_1 + \lambda_2 b_1, \ldots, a_n + \lambda_2 b_n) = 0$,
4. $(a_1 + \lambda_1 b_1, \ldots, a_n + \lambda_1 b_n)$ and $(a_1 + \lambda_2 b_1, \ldots, a_n + \lambda_2 b_n)$ are points different from origin and the vertices of $\psi(x_1, \ldots, x_n)$.

Let us take $(c_1, \ldots, c_n) \equiv (a_1 + \lambda_1 b_1, \ldots, a_n + \lambda_1 b_n)$. Now, $\psi(c_1, \ldots, c_n) = 0$, and $(c_1, \ldots, c_n)$ is a real point different from the origin. ∎

4.3.17 Theorem Let $B \equiv [b_{ij}]$ be a symmetric n-square real matrix. Let B be invertible. Let

$$\psi(x_1, \ldots, x_n) \equiv \sum_{i=1}^{n}\left(\sum_{j=1}^{n} b_{ij}x_i x_j\right) = \sum_{i=1}^{n}\left(x_i\left(\sum_{j=1}^{n} b_{ij}x_j\right)\right)$$
$$\equiv [x_1, \ldots, x_n]\left([b_{ij}][x_1, \ldots, x_n]^T\right)$$

be a real quadratic form. Let $\psi(x_1, \ldots, x_n)$ be a definite form. Let $(a_1, \ldots, a_n)$ be a real point that is different from the origin. Then $\psi(a_1, \ldots, a_n) \neq 0$.

Proof Case I: $\psi(x_1, \ldots, x_n)$ is a positive definite form. Since $\psi(x_1, \ldots, x_n)$ is a definite form, $\text{rank}(B) = n$. By 4.2.17, there exists a real invertible matrix R such that

$$RBR^T = \text{diag}\left(\underbrace{1, \ldots, 1}_{n}\right).$$

It follows that

$$B = R^{-1}\left(\text{diag}\left(\underbrace{1, \ldots, 1}_{n}\right)\right)(R^T)^{-1} = \underbrace{R^{-1}\left(\text{diag}\left(\underbrace{1, \ldots, 1}_{n}\right)\right)(R^{-1})^T},$$

and hence

$$\psi(x_1, \ldots, x_n) \equiv [x_1, \ldots, x_n]B[x_1, \ldots, x_n]^T$$
$$\equiv [x_1, \ldots, x_n]\left(R^{-1}\left(\text{diag}\left(\underbrace{1, \ldots, 1}_{n}\right)\right)(R^{-1})^T\right) \times [x_1, \ldots, x_n]^T$$
$$\equiv [y_1, \ldots, y_n][y_1, \ldots, y_n]^T,$$

where $[y_1, \ldots, y_n] \equiv [x_1, \ldots, x_n]R^{-1}$. Since R is invertible and $(a_1, \ldots, a_n) \neq (0, \ldots, 0)$, $[a_1, \ldots, a_n]R^{-1}$ is nonzero. Put $[b_1, \ldots, b_n] \equiv [a_1, \ldots, a_n]R^{-1}$. Thus $(b_1, \ldots, b_n) \neq (0, \ldots, 0)$. It follows that $(b_1)^2 + \cdots + (b_n)^2 \neq 0$. Also,

$$\psi(a_1, \ldots, a_n) = [a_1, \ldots, a_n]\left(R^{-1}\left(\operatorname{diag}\left(\underbrace{1, \ldots, 1}_{n}\right)\right)\right)(R^{-1})^T[a_1, \ldots, a_n]^T$$

$$= [b_1, \ldots, b_n]\left(\operatorname{diag}\left(\underbrace{1, \ldots, 1}_{n}\right)\right)[b_1, \ldots, b_n]^T = (b_1)^2 + \cdots + (b_n)^2 \neq 0,$$

so $\psi(a_1, \ldots, a_n) \neq 0$.

Case II: $\psi(x_1, \ldots, x_n)$ is a negative definite form. This case is similar to Case I. ∎

4.3.18 Note Let $A \equiv [a_{ij}]$, and $B \equiv [b_{ij}]$ be symmetric n-square real matrices. Let B be invertible. Let

$$\left.\begin{array}{l}\phi(x_1, \ldots, x_n) \equiv \sum a_{ij}x_ix_j = [x_1, \ldots, x_n]A[x_1, \ldots, x_n]^T \\ \psi(x_1, \ldots, x_n) \equiv \sum b_{ij}x_ix_j = [x_1, \ldots, x_n]B[x_1, \ldots, x_n]^T\end{array}\right\}$$

be a pair of real quadratic forms. Let ψ be positive definite. Let λ_1 be a root of the λ-equation of ϕ and ψ.

Then by 4.3.9, there exists a one-to-one linear transformation $[x_1, \ldots, x_n] \mapsto [z_1, \ldots, z_n]$ such that the pair's reduced forms are

$$\left.\begin{array}{l}\phi(x_1, \ldots, x_n) \equiv \lambda_1 c_1(z_1)^2 + \phi_1(z_2, \ldots, z_n) \\ \psi(x_1, \ldots, x_n) \equiv c_1(z_1)^2 + \psi_1(z_2, \ldots, z_n)\end{array}\right\},$$

where c_1 is a nonzero real number.

Clearly, the matrix associated with ψ_1 is invertible.

Proof Suppose to the contrary that the matrix associated with ψ_1 is not invertible. We seek a contradiction.

By 4.3.15, there exists a real point $(d_2, \ldots, d_n) \neq (0, \ldots, 0)$ such that $\psi_1(d_2, \ldots, d_n) = 0$. Suppose that $([x_1, \ldots, x_n] =)[\alpha_1, \ldots, \alpha_n] \mapsto [0, d_2, \ldots, d_n]$ $(= [z_1, \ldots, z_n])$. Since $(0, d_2, \ldots, d_n) \neq (0, \ldots, 0)$, and the linear transformation $[x_1, \ldots, x_n] \mapsto [z_1, \ldots, z_n]$ is one-to-one, $(\alpha_1, \ldots, \alpha_n)$ is nonzero. Also,

$$\psi(\alpha_1, \ldots, \alpha_n) = c_1(0)^2 + \psi_1(d_2, \ldots, d_n) = 0,$$

so $\psi(\alpha_1, \ldots, \alpha_n) = 0$. Since $(\alpha_1, \ldots, \alpha_n)$ is nonzero and ψ is positive definite, by 4.3.17, $\psi(\alpha_1, \ldots, \alpha_n) \neq 0$. This is a contradiction.   ∎

Clearly, ψ_1 is definite.

Proof Suppose to the contrary that ψ_1 is indefinite. We seek a contradiction.

By 4.3.16, there exists a real point $(d_2, \ldots, d_n) \neq (0, \ldots, 0)$ such that $\psi_1(d_2, \ldots, d_n) = 0$. Suppose that $([x_1, \ldots, x_n] =)[\alpha_1, \ldots, \alpha_n] \mapsto [0, d_2, \ldots, d_n]$ $(= [z_1, \ldots, z_n])$. Since $(0, d_2, \ldots, d_n) \neq (0, \ldots, 0)$, and the linear transformation $[x_1, \ldots, x_n] \mapsto [z_1, \ldots, z_n]$ is one-to-one, $(\alpha_1, \ldots, \alpha_n)$ is nonzero. Also,

$$\psi(\alpha_1, \ldots, \alpha_n) = c_1(0)^2 + \psi_1(d_2, \ldots, d_n) = 0,$$

so $\psi(\alpha_1, \ldots, \alpha_n) = 0$. Since $(\alpha_1, \ldots, \alpha_n)$ is nonzero and ψ is positive definite, by 4.3.17, $\psi(\alpha_1, \ldots, \alpha_n) \neq 0$. This is a contradiction. ∎

4.3.19 Conclusion Let $A \equiv [a_{ij}]$, and $B \equiv [b_{ij}]$ be symmetric n-square real matrices. Let B be invertible. Let

$$\left. \begin{array}{l} \phi(x_1, \ldots, x_n) \equiv \sum a_{ij}x_ix_j = [x_1, \ldots, x_n]A[x_1, \ldots, x_n]^T \\ \psi(x_1, \ldots, x_n) \equiv \sum b_{ij}x_ix_j = [x_1, \ldots, x_n]B[x_1, \ldots, x_n]^T \end{array} \right\}$$

be a pair of real quadratic forms. Let ψ be positive definite. Let λ_1 be a root of the λ-equation of ϕ and ψ. Let $[x_1, \ldots, x_n] \mapsto [z_1, \ldots, z_n]$ be a one-to-one linear transformation such that the pair's reduced forms are

$$\left. \begin{array}{l} \phi(x_1, \ldots, x_n) \equiv \lambda_1 c_1(z_1)^2 + \phi_1(z_2, \ldots, z_n) \\ \psi(x_1, \ldots, x_n) \equiv c_1(z_1)^2 + \psi_1(z_2, \ldots, z_n) \end{array} \right\},$$

where c_1 is a nonzero real number. Then

1. the matrix associated with ψ_1 is invertible,
2. ψ_1 is definite.

4.3.20 Theorem Let $A \equiv [a_{ij}]$, and $B \equiv [b_{ij}]$ be symmetric n-square real matrices. Let B be invertible. Let

$$\left. \begin{array}{l} \phi(x_1, \ldots, x_n) \equiv \sum a_{ij}x_ix_j = [x_1, \ldots, x_n]A[x_1, \ldots, x_n]^T \\ \psi(x_1, \ldots, x_n) \equiv \sum b_{ij}x_ix_j = [x_1, \ldots, x_n]B[x_1, \ldots, x_n]^T \end{array} \right\}$$

be a pair of real quadratic forms. Let ψ be positive definite. Let $\lambda_1, \ldots, \lambda_n$ be the roots of the λ-equation of ϕ and ψ. Let $[z_1, \ldots, z_n] = [x_1, \ldots, x_n]Q$ be a one-to-one linear transformation such that the pair's reduced forms are

$$\left. \begin{array}{l} \phi(x_1, \ldots, x_n) \equiv \lambda_1 c_1(z_1)^2 + \phi_1(z_2, \ldots, z_n) \\ \psi(x_1, \ldots, x_n) \equiv c_1(z_1)^2 + \psi_1(z_2, \ldots, z_n) \end{array} \right\},$$

where c_1 is a nonzero real number and Q is an invertible n-square real matrix. Then $\lambda_2, \ldots, \lambda_n$ are the roots of the λ-equation of ϕ_1 and ψ_1.

Proof Let A_1 be the $(n-1)$-square real symmetric matrix associated with the quadratic form $\phi_1(z_2, \ldots, z_n)$. Let B_1 be the $(n-1)$-square real symmetric matrix associated with the quadratic form $\psi_1(z_2, \ldots, z_n)$. Clearly,

$$[x_1, \ldots, x_n]A[x_1, \ldots, x_n]^T$$

$$= [z_1, \ldots, z_n]\begin{bmatrix} \lambda_1 c_1 & 0 \\ 0 & A_1 \end{bmatrix}[z_1, \ldots, z_n]^T$$

$$= ([x_1, \ldots, x_n]Q)\begin{bmatrix} \lambda_1 c_1 & 0 \\ 0 & A_1 \end{bmatrix} \times ([x_1, \ldots, x_n]Q)^T$$

$$= [x_1, \ldots, x_n]\left(Q\begin{bmatrix} \lambda_1 c_1 & 0 \\ 0 & A_1 \end{bmatrix}Q^T\right)[x_1, \ldots, x_n]^T,$$

so

$$A = Q\begin{bmatrix} \lambda_1 c_1 & 0 \\ 0 & A_1 \end{bmatrix}Q^T.$$

Similarly,

$$B = Q\begin{bmatrix} c_1 & 0 \\ 0 & B_1 \end{bmatrix}Q^T.$$

Since Q is invertible, $\det(Q)$ is a nonzero real number. Since $\lambda_1, \ldots, \lambda_n$ are the roots of the λ-equation of ϕ and ψ, we have

$$\det(A - \lambda B) = (\lambda_1 - \lambda)(\lambda_2 - \lambda) \cdots (\lambda_n - \lambda).$$

It suffices to show that

$$\det(A_1 - \lambda B_1) = (\text{nonzero constant})(\lambda_2 - \lambda) \cdots (\lambda_n - \lambda).$$

Since

$$(\lambda_1 - \lambda)(\lambda_2 - \lambda) \cdots (\lambda_n - \lambda) = \det(A - \lambda B)$$

$$= \det\left(Q\begin{bmatrix} \lambda_1 c_1 & 0 \\ 0 & A_1 \end{bmatrix}Q^T - \lambda Q\begin{bmatrix} c_1 & 0 \\ 0 & B_1 \end{bmatrix}Q^T\right)$$

$$= \det\left(Q\begin{bmatrix} \lambda_1 c_1 - \lambda c_1 & 0 \\ 0 & A_1 - \lambda B_1 \end{bmatrix}Q^T\right)$$

$$= \det(Q) \cdot \det\begin{bmatrix} \lambda_1 c_1 - \lambda c_1 & 0 \\ 0 & A_1 - \lambda B_1 \end{bmatrix} \cdot \det(Q^T)$$

$$= \det(Q) \cdot \det \begin{bmatrix} \lambda_1 c_1 - \lambda c_1 & 0 \\ 0 & A_1 - \lambda B_1 \end{bmatrix} \cdot \det(Q)$$

$$= (\det(Q))^2 \cdot \det \begin{bmatrix} \lambda_1 c_1 - \lambda c_1 & 0 \\ 0 & A_1 - \lambda B_1 \end{bmatrix}$$

$$= (\det(Q))^2 \cdot (\lambda_1 c_1 - \lambda c_1) \cdot \det(A_1 - \lambda B_1)$$

$$= (\det(Q))^2 \cdot (\lambda_1 - \lambda) c_1 \cdot \det(A_1 - \lambda B_1),$$

we have

$$\det(A_1 - \lambda B_1) = \frac{1}{c_1 (\det(Q))^2} (\lambda_2 - \lambda) \cdots (\lambda_n - \lambda).$$

∎

4.3.21 Note Let $A \equiv [a_{ij}]$ and $B \equiv [b_{ij}]$ be symmetric n-square real matrices. Let B be invertible. Let

$$\left. \begin{aligned} \phi(x_1, \ldots, x_n) &\equiv \sum a_{ij} x_i x_j = [x_1, \ldots, x_n] A [x_1, \ldots, x_n]^T \\ \psi(x_1, \ldots, x_n) &\equiv \sum b_{ij} x_i x_j = [x_1, \ldots, x_n] B [x_1, \ldots, x_n]^T \end{aligned} \right\}$$

be a pair of real quadratic forms. Let ψ be positive definite. Let $\lambda_1, \ldots, \lambda_n$ be the roots of the λ-equation of ϕ and ψ. Let $[z_1, \ldots, z_n] = [x_1, \ldots, x_n] Q$ be a one-to-one linear transformation such that the pair's reduced forms are

$$\left. \begin{aligned} \phi(x_1, \ldots, x_n) &\equiv \lambda_1 c_1 (z_1)^2 + \phi_1(z_2, \ldots, z_n) \\ \psi(x_1, \ldots, x_n) &\equiv c_1 (z_1)^2 + \psi_1(z_2, \ldots, z_n) \end{aligned} \right\},$$

where c_1 is a nonzero real number and Q is an invertible n-square real matrix.

By 4.3.19,

1. the matrix associated with ψ_1 is invertible,
2. ψ_1 is definite.

Next, by 4.3.20,

3. $\lambda_2, \ldots, \lambda_n$ are the roots of the λ-equation of ϕ_1 and ψ_1.

Again, by repeating the same procedure, there exists a one-to-one linear transformation $[z_1, z_2, \ldots, z_n] \mapsto [w_1, w_2, \ldots, w_n]$ such that $z_1 = w_1$, and the pair's reduced forms are

$$\left. \begin{aligned} \phi_1(z_2, \ldots, z_n) &\equiv \lambda_2 c_2 (w_2)^2 + \phi_2(w_3, \ldots, w_n) \\ \psi_1(z_2, \ldots, z_n) &\equiv c_2 (w_2)^2 + \psi_2(w_3, \ldots, w_n) \end{aligned} \right\},$$

where c_2 is a nonzero real number. Also,

1. the matrix associated with ψ_2 is invertible,
2. ψ_2 is definite,
3. $\lambda_3, \ldots, \lambda_n$ are the roots of the λ-equation of ϕ_2 and ψ_2.

It follows that

$$\left.\begin{array}{l} \phi(x_1, \ldots, x_n) \equiv \lambda_1 c_1 (w_1)^2 + \lambda_2 c_2 (w_2)^2 + \phi_2 (w_3, \ldots, w_n) \\ \psi(x_1, \ldots, x_n) \equiv c_1 (w_1)^2 + c_2 (w_2)^2 + \psi_2 (w_3, \ldots, w_n) \end{array}\right\}.$$

On repeating the above procedure, we get a one-to-one linear transformation $[x_1, \ldots, x_n] \mapsto [v_1, \ldots, v_n]$ such that the pair's reduced forms are

$$\left.\begin{array}{l} \phi(x_1, \ldots, x_n) \equiv \lambda_1 c_1 (v_1)^2 + \cdots + \lambda_n c_n (v_n)^2 \\ \psi(x_1, \ldots, x_n) \equiv c_1 (v_1)^2 + \cdots + c_n (v_n)^2 \end{array}\right\},$$

where each c_i is a nonzero real number.

4.3.22 Conclusion Let $A \equiv [a_{ij}]$, and $B \equiv [b_{ij}]$ be symmetric n-square real matrices. Let B be invertible. Let

$$\left.\begin{array}{l} \phi(x_1, \ldots, x_n) \equiv \sum a_{ij} x_i x_j = [x_1, \ldots, x_n] A [x_1, \ldots, x_n]^T \\ \psi(x_1, \ldots, x_n) \equiv \sum b_{ij} x_i x_j = [x_1, \ldots, x_n] B [x_1, \ldots, x_n]^T \end{array}\right\}$$

be a pair of real quadratic forms. Let ψ be positive definite. Let $\lambda_1, \ldots, \lambda_n$ be the roots of the λ-equation of ϕ and ψ. Then there exists a one-to-one linear transformation $[x_1, \ldots, x_n] \mapsto [v_1, \ldots, v_n]$ such that the pair's reduced forms are

$$\left.\begin{array}{l} \phi(x_1, \ldots, x_n) \equiv \lambda_1 c_1 (v_1)^2 + \cdots + \lambda_n c_n (v_n)^2 \\ \psi(x_1, \ldots, x_n) \equiv c_1 (v_1)^2 + \cdots + c_n (v_n)^2 \end{array}\right\},$$

where each c_i is a nonzero real number.

4.3.23 Theorem Let $A \equiv [a_{ij}]$, and $B \equiv [b_{ij}]$ be symmetric n-square real matrices. Let B be invertible. Let

$$\left.\begin{array}{l} \phi(x_1, \ldots, x_n) \equiv \sum a_{ij} x_i x_j = [x_1, \ldots, x_n] A [x_1, \ldots, x_n]^T \\ \psi(x_1, \ldots, x_n) \equiv \sum b_{ij} x_i x_j = [x_1, \ldots, x_n] B [x_1, \ldots, x_n]^T \end{array}\right\}$$

be a pair of real quadratic forms. Let ψ be positive definite. Let $\lambda_1, \ldots, \lambda_n$ be the roots of the λ-equation of ϕ and ψ. Then there exists a one-to-one linear transformation $[x_1, \ldots, x_n] \mapsto [y_1, \ldots, y_n]$ such that the pair's reduced forms are

$$\left.\begin{array}{l} \phi(x_1, \ldots, x_n) \equiv \pm \left(\lambda_1 (y_1)^2 + \cdots + \lambda_n (y_n)^2 \right) \\ \psi(x_1, \ldots, x_n) \equiv \pm \left((y_1)^2 + \cdots + (y_n)^2 \right) \end{array}\right\}.$$

Proof By   4.3.22,   there   exists   a   one-to-one   linear   transformation
$[x_1, \ldots, x_n] \mapsto [v_1, \ldots, v_n]$ such that the pair's reduced forms are

$$\left. \begin{aligned} \phi(x_1, \ldots, x_n) &\equiv \lambda_1 c_1 (v_1)^2 + \cdots + \lambda_n c_n (v_n)^2 \\ \psi(x_1, \ldots, x_n) &\equiv c_1 (v_1)^2 + \cdots + c_n (v_n)^2 \end{aligned} \right\},$$

where each c_i is a nonzero real number. On applying the one-to-one linear
transformation

$$\left. \begin{aligned} v_1 &= \frac{1}{\sqrt{|c_1|}} y_1 \\ &\vdots \\ v_n &= \frac{1}{\sqrt{|c_n|}} y_n \end{aligned} \right\},$$

we get the following reduced forms:

$$\left. \begin{aligned} \phi(x_1, \ldots, x_n) &\equiv \lambda_1 \tfrac{c_1}{|c_1|} (y_1)^2 + \cdots + \lambda_n \tfrac{c_n}{|c_n|} (y_n)^2 \\ \psi(x_1, \ldots, x_n) &\equiv \tfrac{c_1}{|c_1|} (y_1)^2 + \cdots + \tfrac{c_n}{|c_n|} (y_n)^2 \end{aligned} \right\}.$$

Here each $\frac{c_i}{|c_i|}$ is equal to 1 or -1. So if $\psi(x_1, \ldots, x_n)$ is positive definite, then each
$\frac{c_i}{|c_i|}$ is equal to 1, and hence

$$\left. \begin{aligned} \phi(x_1, \ldots, x_n) &\equiv \lambda_1 (y_1)^2 + \cdots + \lambda_n (y_n)^2 \\ \psi(x_1, \ldots, x_n) &\equiv (y_1)^2 + \cdots + (y_n)^2 \end{aligned} \right\}.$$

Similarly, if $\psi(x_1, \ldots, x_n)$ is negative definite, then

$$\left. \begin{aligned} \phi(x_1, \ldots, x_n) &\equiv -\left(\lambda_1 (y_1)^2 + \cdots + \lambda_n (y_n)^2 \right) \\ \psi(x_1, \ldots, x_n) &\equiv -\left((y_1)^2 + \cdots + (y_n)^2 \right) \end{aligned} \right\}.$$

∎

Exercises

1. Let V be any n-dimensional vector space over the field F. Let $T : V \to V$ be a
   linear transformation. Show that there exists a positive integer k such that

$$i \geq k \Rightarrow \mathcal{N}(T^k) = \mathcal{N}(T^i),$$

   and $\mathcal{N}(T^{k-1})$ is a proper subset of $\mathcal{N}(T^k)$.
2. Let V be any n-dimensional vector space over the field F. Let $T : V \to V$ be a
   linear transformation. Show that $\mathrm{ran}(T^3)$ is invariant under T.

3. Let V be any n-dimensional inner product space over the field $\mathbb{C}$. Let $T : V \rightarrow V$ be a normal linear transformation. Suppose that all the eigenvalues of T are real. Show that T is Hermitian.

4. Let A be a 6-square complex matrix. Suppose that A is a nonnegative definite matrix. Show that there exists a unitary 6×6 matrix U such that

   a. $A = U(\text{diag}(\lambda_1, \ldots, \lambda_6))U^*$,
   b. $\lambda_1, \ldots, \lambda_6$ are the eigenvalues of the matrix A,
   c. each λ_i is a nonnegative real number,
   d. $\det(A)$ is a nonnegative real number.

5. Let A be a 6-square complex matrix. Let A be symmetric and unitary. Show that there exists a symmetric unitary complex matrix S such that $S^2 = A$.

6. Let $[a_{ij}]$ be an n-square real symmetric matrix. Let

$$\phi(x_1, \ldots, x_n) \equiv \sum_{i,j} a_{ij} x_i x_j$$

be a real quadratic form. Show that there exists a real invertible matrix $[c_{ij}]$ such that the transformation

$$\left.\begin{array}{c} x_1 = c_{11} y_1 + \cdots + c_{1n} y_n \\ \vdots \\ x_n = c_{n1} y_1 + \cdots + c_{nn} y_n \end{array}\right\}$$

reduces the form $\sum_{i,j} a_{ij} x_i x_j$ to the form

$$(y_1)^2 + \cdots + (y_r)^2 - (y_{r+1})^2 - \cdots - (y_{r+s})^2.$$

7. Let $A \equiv [a_{ij}]$, and $B \equiv [b_{ij}]$ be symmetric 5-square real matrices. Let B be invertible. Let

$$\left.\begin{array}{l} \phi(x_1, \ldots, x_5) \equiv [x_1, \ldots, x_5]A[x_1, \ldots, x_5]^T \\ \psi(x_1, \ldots, x_n) \equiv [x_1, \ldots, x_5]B[x_1, \ldots, x_5]^T \end{array}\right\}$$

be a pair of real quadratic forms. Let ψ be positive definite. Let $\lambda_1, \ldots, \lambda_5$ be the roots of the λ-equation of ϕ and ψ. Show that there exists a one-to-one linear transformation $[x_1, \ldots, x_5] \mapsto [v_1, \ldots, v_5]$ such that the pair's reduced forms are

$$\left.\begin{array}{l} \phi(x_1, \ldots, x_5) \equiv \lambda_1 c_1 (v_1)^2 + \cdots + \lambda_5 c_5 (v_5)^2 \\ \psi(x_1, \ldots, x_5) \equiv c_1 (v_1)^2 + \cdots + c_5 (v_5)^2 \end{array}\right\},$$

where each c_i is a nonzero real number.

8. Let A be an n-square complex matrix. Suppose that A is a nonnegative definite matrix. Show that the square root of A exists.
9. Let V be any n-dimensional vector space over the field F. Let $T : V \to V$ be a linear transformation. Show that there exists a positive integer k such that

$$V = \mathcal{N}(T^k) \oplus \operatorname{ran}(T^k).$$

10. Suppose that $A \equiv [a_{ij}]$ is a symmetric n-square real matrix. Let

$$\phi(x_1, \ldots, x_n) = [x_1, \ldots, x_n]A[x_1, \ldots, x_n]^T$$

be a real quadratic form. Suppose that $a_{11} \neq 0$. Show that the one-to-one linear transformation

$$[y_1, \ldots, y_n]^T = \begin{bmatrix} a_{11} & a_{12} & \cdots & a_{1n} \\ 0 & 1 & \cdots & 0 \\ \vdots & \vdots & \ddots & \vdots \\ 0 & 0 & \cdots & 1 \end{bmatrix} [x_1, \ldots, x_n]^T$$

reduces $\phi(x_1, \ldots, x_n)$ to

$$\frac{1}{a_{11}}(y_1)^2 + \phi_1(y_2, \ldots, y_n),$$

where $\phi_1(y_2, \ldots, y_n)$ is a quadratic form.

Bibliography

1. M. Artin, *Algebra* (Prentice Hall, 2008)
2. P.R. Halmos, *Finite-Dimensional Vector Spaces* (Springer, 2011)
3. I.N. Herstein, *Topics in Algebra*, 2nd edn. (Wiley-India, 2008)
4. N. Jacobson, *Lectures in Abstract Algebra* (D. Van Nostrand Company, Inc., 1965)
5. I.S. Luthar, I.B.S. Passi, *Field Theory* (Narosa, 2008)
6. F. Zhang, *Matrix Theory* (Springer, 1999)

© Springer Nature Singapore Pte Ltd. 2020

R. Sinha, *Galois Theory and Advanced Linear Algebra*,

https://doi.org/10.1007/978-981-13-9849-0

Printed in the United States
By Bookmasters